Atmospheric Chemistry: Theory, Models and Applications

Atmospheric Chemistry: Theory, Models and Applications

Editor: Michael Bradshaw

CALLISTO REFERENCE

www.callistoreference.com

Callisto Reference,
118-35 Queens Blvd., Suite 400,
Forest Hills, NY 11375, USA

Visit us on the World Wide Web at:
www.callistoreference.com

ISBN: 978-1-63239-931-1 (Hardback)

Cataloging-in-Publication Data

Atmospheric chemistry : theory, models and applications / edited by Michael Bradshaw.
    p. cm.
Includes bibliographical references and index.
ISBN 978-1-63239-931-1
1. Atmospheric chemistry. 2. Environmental chemistry. I. Bradshaw, Michael.
QC879.6 .A86 2018
551.511--dc23

Table of Contents

Preface

The branch of chemistry that studies the atmosphere is referred to as atmospheric chemistry. The main elements of atmospheric chemistry are observations, lab measurements and atmospheric modeling. Changes occur in the Earth's atmosphere because of natural processes such as volcano emissions and lightning and anthropologic reasons. Some of these changes can be harmful to the ecology. Such changes and their causes are studied under atmospheric chemistry. This book aims to equip students and experts with the advanced topics and upcoming concepts in this area. Its detailed analyses and data will prove immensely beneficial to professionals and students involved in this area at various levels.

This book has been the outcome of endless efforts put in by authors and researchers on various issues and topics within the field. The book is a comprehensive collection of significant researches that are addressed in a variety of chapters. It will surely enhance the knowledge of the field among readers across the globe.

It gives us an immense pleasure to thank our researchers and authors for their efforts to submit their piece of writing before the deadlines. Finally in the end, I would like to thank my family and colleagues who have been a great source of inspiration and support.

Editor

Traffic-Related Air Pollution and the Onset of Myocardial Infarction: Disclosing Benzene as a Trigger?

Denis Bard[1]*, Wahida Kihal[1], Charles Schillinger[2], Christophe Fermanian[1], Claire Ségala[3], Sophie Glorion[1], Dominique Arveiler[4], Christiane Weber[5]

1 Department of Epidemiology and Biostatistics, École des Hautes Études en Santé Publique, Rennes and Sorbonne Paris Cité, Paris, France, **2** Association pour la Surveillance de la Qualité de l'Air en Alsace-ASPA, Schiltigheim, France, **3** SEPIA-Santé, Baud, France, **4** Department of Epidemiology and Public Health (EA3430), University of Strasbourg, Strasbourg, France, **5** Laboratoire Image, Ville, Environnement (LIVE UMR7362 CNRS), Faculté de géographie et d'aménagement, University of Strasbourg, Strasbourg, France

Abstract

Background and Objectives: Exposure to traffic is an established risk factor for the triggering of myocardial infarction (MI). Particulate matter, mainly emitted by diesel vehicles, appears to be the most important stressor. However, the possible influence of benzene from gasoline-fueled cars has not been explored so far.

Methods and Results: We conducted a case-crossover study from 2,134 MI cases recorded by the local Coronary Heart Disease Registry (2000–2007) in the Strasbourg Metropolitan Area (France). Available individual data were age, gender, previous history of ischemic heart disease and address of residence at the time of the event. Nitrogen dioxide, particles of median aerodynamic diameter <10 μm (PM_{10}), ozone, carbon monoxide and benzene air concentrations were modeled on an hourly basis at the census block level over the study period using the deterministic ADMS-Urban air dispersion model. Model input data were emissions inventories, background pollution measurements, and meteorological data. We have found a positive, statistically significant association between concentrations of benzene and the onset of MI: per cent increase in risk for a 1 μg/m^3 increase in benzene concentration in the previous 0, 0–1 and 1 day was 10.4 (95% confidence interval 3–18.2), 10.7 (2.7–19.2) and 7.2 (0.3–14.5), respectively. The associations between the other pollutants and outcome were much lower and in accordance with the literature.

Conclusion: We have observed that benzene in ambient air is strongly associated with the triggering of MI. This novel finding needs confirmation. If so, this would mean that not only diesel vehicles, the main particulate matter emitters, but also gasoline-fueled cars –main benzene emitters–, should be taken into account for public health action.

Editor: Stephania Ann Cormier, University of Tennessee Health Science Center, United States of America

Funding: This work was supported by French Agency for Food, Environmental and Occupational Health & Safety (ANSES); Institute for Public Health Research (IRESP); Fondation Coeur et Artères; French Environment and Energy Management Agency (ADEME); and SITA Corporation. The funders had no role in study design, data collection and analysis, decision to publish, or preparation of the manuscript.

Competing Interests: The authors have the following interests. Wahida Kihal's Ph.D. was financially supported by both ADEME and the SITA Corporation. Charles Schillinger is employed by Association pour la Surveillance de la Qualité de l'Air en Alsace-ASPA, and Claire Ségala by SEPIA-Santé. There are no patents, products in development or marketed products to declare.

* Email: denis.bard@ehesp.fr

Introduction

The effects of traffic-related air pollution on cardiorespiratory mortality have been consistently established since the late 1980s [1]. Further studies specifically investigated the association between exposure to traffic and the onset of myocardial infarction (MI), one of the most frequent causes of death. Since the publication of the seminal paper by Peters *et al.* (2001) [2], which shows an association between exposure to traffic and the onset of myocardial infarction (MI), many studies have been published on the issue. The latest review and meta-analysis to date considered the association between short-term exposure to traffic-related air pollutants and subsequent MI risk [3]. The authors retained 34 studies, considering the effects of various air pollutants, either alone or in association, *e.g.* particles with aerodynamic diameter < 10 μm (PM_{10}), particles with diameter <2.5 μm ($PM_{2.5}$), black carbon/black smoke, ozone (O_3), carbon monoxide (CO), nitrogen oxides and sulphur dioxide. Using a random-effect model to estimate the meta-relative risk and 95% confidence interval, a significant, positive association appeared between all analyzed pollutants, with the exception of ozone, and MI risk, although published studies are quite inconsistent regarding both the direction of the association and statistical significance for all pollutants. Thus, the present state of knowledge strongly supports the role of exposure to traffic-related air pollution in triggering the

Table 1. Myocardial infarction events (ICD-9: 410) collected by the Bas-Rhin Coronary Heart Disease Register, Strasbourg Metropolitan Area, France, 2000–2007.

Age group	Females (n = 492) n (%)	Males (n = 1,642) n (%)	Total (N = 2,134) n (%)
35–54	136 (27.7)	637 (38.8)	773 (36.2)
55–74	356 (72.3)	1,005 (61.2)	1,361 (63.8)

onset of MI. In addition, PM_{10} is the air pollutant most consistently associated with myocardial infarction onset [4,5].

None of the above reviews mentions benzene, although gasoline- fueled engines emit this pollutant [6,7]. The literature addressing the acute cardiovascular effects of benzene in occupational settings is scarce, e.g. [8,9] but lends some support to an association between exposure to benzene and arrhythmias. However, in the Kotseva and Popov (1998) paper, benzene concentrations seem to have been very high (up to 65 mg/m^3). The authors provide few details on the study population. In addition, probable co-exposures to various stressors are poorly discussed. To our knowledge, a single group investigated the cardiovascular effects of traffic-related air pollution in a mortality study in Taiwan, addressing exposure to benzene [10]. The authors found a significant association between benzene concentration and cardiovascular mortality (ICD-9-CM 410–411, 414, 430–437, same-day association: lag 0). However, this study considered only fatal cases of cardiovascular diseases. Furthermore, exposure was loosely defined on an ecological basis (air concentrations were measured at a single monitoring station for a study population defined as those living within a 10 km radius). Thus, the potential for error in exposure assessment was high.

The aim of the present study is to investigate the possible association between traffic-related benzene emissions as well as 'classical' traffic-related pollutants (NO_2, PM_{10}, O_3, CO, SO_2) and the onset of myocardial infarction. Study design was time-stratified case-crossover, using a very small area as the statistical unit.

Methods

Setting

The Strasbourg Metropolitan Area (SMA), an urban area of 28 municipalities (316 km^2), is located in the Bas-Rhin district in northeastern France, with a population of about 450,000 inhabitants. It is subdivided into 190 census blocks of average population 2,000 (range 2–4,885) and a median area of 0.45 km^2 (range 0.05–19.60). These blocks are the smallest administrative geographic unit in France for which socioeconomic and demographic information from the national census is available. They are devised as to be homogeneous for population size, socioeconomic characteristics and land use. Sixteen blocks of population size <250 were excluded of the dataset (0.8% of the SMA population), for the sake of compliance with French confidentiality regulations.

Cases

Cases were all MI events (ICD-9: 410) either fatal or non-fatal, occurring in the age group 35–74 years between January 1, 2000 and December 31, 2007, ascertained by the local Bas-Rhin Coronary Heart Disease, a collaborating center of the WHO MONICA Project [11]. MIs were documented events, which have definitively been diagnosed as such whether clinically or at

necropsy. Individual data available were age, gender, previous history of ischemic heart disease, and address of residence at the time of the event. Cases were geocoded to their census block of residence using ArcGIS version 9.1 (ESRI, Redlands, CA).

Assessment of exposure to air pollution

Nitrogen dioxide (NO_2), particles of median aerodynamic diameter <10 μm (PM_{10}), ozone (O_3), carbon monoxide (CO), sulphur dioxide (SO_2) and benzene air concentrations were modeled on an hourly basis at the census block level over the whole study period using the deterministic ADMS-Urban air dispersion model (Atmospheric Dispersion Modeling System) [12]. Model input data comprised of emissions inventories, background pollution measurements and meteorological data. More details can be found elsewhere [13]. Model performance assessment took place on two occasions. First, we compared predictions to measures at monitoring stations on a yearly basis for NO_2, O_3, and SO_2. However, there was no routine station measurement for CO. Mean differences between measured and modeled values were −2% (range −10% to 9%) for NO_2, −4% (−10% to 10%) for O_3, −1% (−2% to 8%) for PM_{10}. Second, we used passive samplers to measure benzene, PM_{10}, and NO_2 concentrations at the census block level (61 measurements points for NO_2 and benzene, four occasions throughout the year, seven points for PM_{10} on eight occasions). Measurement points were selected as to cover the SMA and compared to the predicted value for the census block, both on a yearly and hourly basis. For both circumstances and whatever the pollutant, the mean error was small: −1% (range −39% to 42%) for NO_2, 0% (−26% to 11%) for PM_{10}, −10% (−33% to 30%) for benzene. One exception was SO_2, of which concentrations were low (maximum 11 $μg/m^3$), and modeling performance was poor. Accordingly, we did not consider SO_2 further in the analysis.

Proven or likely confounders

Daily meteorological variables (temperature, atmospheric pressure, and relative humidity) were obtained from the French meteorological service (Météo France); weekly influenza-like case counts came from the Sentinelles network [14] of the French National Institute of Health and Medical Research.

Statistical analysis

Associations between MI events and air pollution were assessed with a case-crossover model [15]. Control days were defined according to a monthly time-stratified design [16]. For a MI occurring on a given weekday (e.g., a Monday), control days were the same days of the week throughout the rest of the month (thus, three or four days; here, the other Mondays of the month). Associations between MI events and ambient air pollution concentrations modeled by census block were estimated, adjusting for holidays, meteorological variables (daily maximum temperature, maximum

Table 2. Air pollutants daily concentrations (µg/m³) and meteorological parameters in the Strasbourg (France) Metropolitan Area, 2000–2007.

Pollutant	Mean	SD	Minimum	Q1	Median	Q3	Maximum
NO_2	33.4	13.45	2.6	23.3	32.4	42.5	120.2
O_3	63.3	36.85	1.1	35.45	59.41	85.0	228.3
PM_{10}	21.1	9.94	1.65	14.1	19.3	26.1	107.5
CO	596.7	83.5	501.1	540.4	573.5	626.8	1800.5
Benzene	1.8	1.1	0.1	0.9	1.5	2.4	19.6
Minimum temperature (°C)	7.1	-	−15.3	2.0	7.5	12.5	21.8
Maximum atmospheric pressure (hPa)	1007.9	-	970.3	999.4	1005.9	1016.2	1043.1

atmospheric pressure, and mean relative humidity), and influenza epidemics. We tested the influence of the various lags reported in the literature between air pollution indicators, treated as continuous variables, and MI events: average of the day of the event: (lag 0), average of the day of the event and the 1st previous day (lag 0–1), and average of the previous day (lag 1).

The daily air pollution indicator considered for NO_2, PM_{10}, CO and benzene was the 24-hour average concentration, and for ozone, it was the maximum daily value of the 8-hour moving average. The analysis for ozone considered MI events occurring between April 1 and September 30 of each year, because of the very low concentrations of this pollutant in winter. Associations were assessed for cases of all ages and then for cases aged 35–54 and 55–74 years, respectively (categorizing was only for these two age groups, for ensuring confidentiality of the health data geocoded by census blocks).

We employed conditional logistic regression for analyses. Results are expressed as the percent increase in MI risk per 10 µg/m³ increase in pollutant concentrations for NO_2, PM_{10} and O_3, per 100 µg/m³ increase for CO, and per 1 µg/m³ increase for benzene. All statistical analyses were performed with SAS 9.1 software (SAS Institute, Cary, NC).

Ethics statement
The French data protection authority (CNIL) approved the study.

Results

Over the 8 years of the study period, the Bas-Rhin Coronary Heart Disease Register ascertained 2,141 MI events. Seven (0.3%) events could not be geocoded and were excluded from the analyses. Thus, 2,134 cases were analyzed (Table 1), among whom 21.9% of females and 24.0% of males had a previous history of ischemic disease (ICD-9: 410–414). The small number of subjects with a previous history of ischemic disease precluded a specific analysis of this group, because of a very limited statistical power.

Air pollution data appear in Table 2. The mean and range of air pollutants concentrations in the SMA for the study period were comparable to those observed in many metropolitan areas of Europe [17] or in the US [18]. All these pollutants were well correlated, as expected: Pearson's r for daily mean concentrations (Table 3) were in the range reported in the literature, the lowest being −0.16 (O_3 and PM_{10}) and the highest being 0.73 (NO_2 and PM_{10}), all statistically significant, in agreement with results reported in the literature [19,20].

We have found a positive, statistically significant association between incremental concentrations of benzene and the onset of MI for our study population (base model), for all lags tested (0, 0–1 and 1 day), slightly more marked for the first two lags studied (Table 4). The associations between the other individual pollutants and outcome were essentially inconclusive (although a negative, statistically significant association appeared for ozone, lag 1).

When examining the risk (excess odds ratio –eOR–, that is, per cent increase in risk for a 1 µg/m³ increase in benzene concentration) associated to incremental exposure to benzene in specific population segments, we observed that men aged 35–54 years were particularly at risk (lag 0, eOR = 15.3; 95% confidence interval [1.0–31.7]), as was the case for older women (age group 55–74 years) for all lags, all statistically significant but more marked for lag 0–1 (eOR = 29.6 [10.3–52.2]) (Table 5).

As regards to the effects of other ambient air pollutants, we have found a higher risk for younger men for NO_2 (lag 0, eOR = 9.3 [0–19.4]). In older women (55–74 years), NO_2 was strongly associated

Table 3. Ambient air pollution daily mean value correlation coefficients, Strasbourg (France) Metropolitan Area, 2000–2007.

	Pearson's r				
Pollutants	Benzene	NO_2	O_3	PM_{10}	CO
Benzene	1.00				
NO_2	0.64	1.00			
O_3	−0.51	−0.34	1.00		
PM_{10}	0.63	0.73	−0.16	1.00	
CO	0.60	0.72	−0.34	0.54	1.00

with MI risk for lag 0–1 (eOR = 15.0 [0.9–31.2]) and lag 1 (eOR = 15.4 [3–29.3]), as were PM_{10} (lag 0–1, eOR = 16.8 [2–33.7]; lag 1, eOR = 17.8 [4.2–33.1]). Models considering the May-September period for ozone were very similar to those covering the April-September period. Incorporating mean relative humidity and holidays as covariates led to results very similar to the above. In addition, influenza epidemics did not influence the results.

Discussion

Our finding of an association between ambient benzene and the triggering of MI has never been reported before. A possible explanation is that benzene is inconsistently measured in standard air monitoring systems, and so far essentially associated to long-term effects, such as cancer.

Among the strengths of our work is the accurate air pollution modeling at a very fine scale (census block) over the study period, diminishing as much as possible the potential for exposure misclassification. Another robust feature is case collection from a specialized registry, using internationally validated diagnosis ascertainment procedures. In addition, using a proven, robust case-crossover design, we observed associations (Table 4) between classically studied traffic-related pollutants (PM_{10}, NO_2, CO and ozone) and outcomes that were within the range reported in the literature [3], although non-significant for the study population as a whole, due to our limited sample size. Nonetheless, it appears a striking effect of benzene. However, this observation could be partially confounded by ultrafine particulate matter [21], unmeasured in this study.

Ambient air pollutants, in particular those produced by traffic, are systematically found being highly correlated. This is the case

for PM_{10} and $PM_{2.5}$, the latter (unmeasured in our study) being a sizeable constituent of the PM_{10} fraction [22]. In addition, unmeasured pollutants may confound associations [23]. Available methods aiming at disentangling the separate effects of individual pollutants do not provide so far a gold standard [24] and results remain difficult to interpret [25]. As most other authors, we assessed excess risks for each pollutant under study.

We observed the expected [2,26] baseline risk differences between genders, with a 3.3/1 male/female ratio in our study population. In subgroups analysis, younger males appear more at risk (benzene and NO_2, lag 0) than older ones, perhaps because the conditions or cumulative risk factors that contribute to a MI in this age group make them especially sensitive, to effects of benzene in particular. Older females appear also at higher risk with benzene, and with NO_2 and PM_{10} (Table 5). Such results have already been reported in the literature for the latter two pollutants [23,27] although not convincingly explained so far.

No data were available at the individual level on tobacco smoking and lifestyle, but these factors contribute concurrently to long term susceptibility, not to the very short-term circumstances triggering a MI.

Altogether, the overall consistency between our results and those published in the literature, associating exposure to usual traffic-related air pollutants and the triggering of MI, lends support to our finding of an association between this outcome and exposure to benzene.

As for limitations of this study, we acknowledge that there remains some room for exposure misclassification, since exposure was assimilated to the levels of air pollutants in the subjects' census block of residence. We have no data on the mobility of our study population. However, the lags showing a positive, statistically

Table 4. Exposure to air pollution and the onset of a myocardial infarction (MI) in the Strasbourg (France) Metropolitan Area, 2000–2007, base model[a].

	Lag 0		Lag 0–1		Lag 1	
Pollutant	eOR (95%CI)	p value	eOR (95%CI)	p value	eOR (95%CI)	p value
Benzene	10.4 (3.0, 18.2)	0.005	10.7 (2.7, 19.·2)	0.008	7.2 (0.3, 14.5)	0.04
PM_{10}	2.6 (−2.7, 8.2)		3.5 (−2.3, 9.7)		3.1 (−2.0, 8.5)	
NO_2	4.7 (−0.2, 9.9)	0.06	5.4 (−0.1, 11.2)	0.05	3.6 (−1.0, 8.5)	
CO	3.2 (−6.1, 13.3)		4.4 (−6.6, 16.7)		3.0 (−6.2, 13.1)	
O_3	−1.3 (−3.8, 1.3)		−2.7 (−5.5, 0.2)	0.07	**−3.1 (−5.7, −0.5)**	**0.02**

[a]Associations observed for different lag times; excess odds ratios (eOR) are expressed as per cent (95% confidence interval) increase for i) a 1 µg/m³ increase in benzene concentrations; ii) a 10 µg/m³ in NO_2, O_3 and PM_{10} concentrations and iii) a 100 µg/m3 increase in CO concentrations. Adjusted for the previous day maximum atmospheric pressure, same day minimum temperature and influenza epidemics.

Table 5. Exposure to air pollution and the onset of a myocardial infarction in the Strasbourg (France) Metropolitan Area, 2000–2007, by subgroups[a].

Gender (age group)	Pollutant	Lag 0	Lag 0–1	Lag 1
		eOR (95% CI)	eOR (95% CI)	eOR (95% CI)
Males (35–54)	Benzene	15.3 (1.0, 31.7)*	11.1 (−3.5, 27.9)	3.8 (−8.2, 17.4)
	PM_{10}	1.1 (−8.4, 11.6)	1.3 (−8.9, 12.6)	1.0 (−8.0, 10.9)
	NO_2	9.3 (0, 19.4)*	7.0 (−2.9, 17.8)	2.1 (−6.0, 10.9)
	CO	10.3 (−7.6, 31.7)	4.4 (−15.0, 28.1)	−3.0 (−18.1, 14.9)
	O_3	−0.4 (−4.9, 4·3)	−2.1 (−7.0, 3.2)	−3.0 (−7.5, 1.7)
Males (55–74)	Benzene	6.6 (−3.7, 18.0)	5.3 (−5.9, 17.8)	2.2 (−7.7, 13.1)
	PM_{10}	3.8 (−4.1, 12.2)	3.7 (−4.9, 13.0)	2.2 (−5.3, 10.2)
	NO_2	4.4 (−2.7, 12.0)	4.4 (−3.5, 13.1)	2.4 (−4.3, 9.7)
	CO	1.2 (−11.8, 16.0)	0.6 (−14.8, 18.7)	−0.3 (−13.3, 14.5)
	O_3	−0.5 (−4.2, 3.4)	−1.4 (−5.5, 3.0)	−1.7 (−5.5, 2.2)
Females (35–54)	Benzene	−13.9 (−36.8, 17.2)	−3.9 (−30.0, 31.8)	6.2 (−18.6, 38.7)
	PM_{10}	−15.8 (−32.8, 5.6)	−17.9 (−35.7, 4.7)	−13.7 (−30.1, 6.5)
	NO_2	−15.1 (−29.5, 2.3)	−12.9 (−29.0, 6.8)	−5.9 (−21.4, 12.6)
	CO	−30.5 (−54.2, 5.3)	−14.3 (−46.1, 36.3)	11.9 (−23.9, 64.5)
	O_3	−5.7 (−15.7, 5.5)	−8.5 (−19.4, 3.9)	−8.5 (−18.6, 2.9)
Females (55–74)	Benzene	21.5 (4.6, 41.0)*	29.6 (10.3, 52.2)**	27.1 (9.8, 47.1)**
	PM_{10}	9.5 (−3.1, 23.9)	16.8 (2.0, 33.7)*	17.8 (4.2, 33.1)*
	NO_2	7.2 (−4.6, 20.5)	15.0 (0.9, 31.2)*	15.4 (3.0, 29.3)*
	CO	9.1 (−11.6, 34.6)	22.1 (−5.5, 57.8)	23.3 (−1.7, 54.6)
	O_3	−3.8 (−10.1, 2.9)	−6.0 (−13.0, 1.5)	−5.9 (−12.2, 0.9)

[a]Associations observed for different lag times; excess odds ratios (eOR) are expressed as per cent (95% confidence interval) increase for *i*) a 1 $\mu g/m^3$ increase in benzene concentrations; *ii*) a 10 $\mu g/m^3$ in NO_2, O_3 and PM_{10} concentrations and *iii*) a 100 $\mu g/m^3$ increase in CO concentrations. Adjusted for the previous day maximum atmospheric pressure, same day minimum temperature and influenza epidemics.
*$p < 0.05$.
**$p < 0.001$.

significant association between benzene exposure and MI onset span from the same day to the previous day. Thus, exposure misclassification as regards time spent out of area of residence is limited, since people usually spend the major part of their time at home. We feel highly unlikely that such relative misclassification could account for the sizeable associations we have observed.

Mechanisms of action

The literature extensively addresses the underlying mechanisms of action of ambient air pollutants involved in the triggering of MI. Overall, it appears that changes in the synthesis or reactivity of nitric oxide that may be caused by environmental oxidants [28] or an increased endogenous production of reactive oxygen species are candidate mechanisms [29]. A recent review (although targeted to benzene-induced mutation mechanisms) indicates that oxidative stress is one mechanism-of-action of benzene as well [30], but the real contribution of such mechanism to the association we have observed remains to be assessed. In addition, short-term associations with ambient benzene have also been shown for asthma exacerbation [18], although providing no clues for a mechanism of action. The development of epigenetics may shed some light on intimate mechanisms [31].

Public health impact

In the above cited meta-analysis [3], the authors estimated the population attributable fractions (PAF) for those pollutants.

Assuming a 100% prevalence of exposure, the PAFs were of 4.5% for an incremental exposure of 1 mg/m^3 carbon monoxide, and ranging between 0.6% and 2.5% for a 10 $\mu g/m^3$ incremental exposure to the other pollutants. In an earlier analysis of the relative importance of triggers of MI [32], calculated from 14 studies an overall PAF for air pollution of 7.4%. That is, such PAF is of a magnitude similar to that of other well-documented triggers such as physical exertion, alcohol, and coffee. If the whole community were exposed, such a relatively limited PAF would have considerable public health impact. Provided our findings were replicated, this would be the case for benzene but with a much higher effect. However, we felt that calculating a PAF for benzene would be irrelevant in the absence of convergent studies.

Conclusion

We have observed a benzene-associated risk for the triggering of myocardial infarction, using a robust characterization of cases and of exposure. This association has not been documented previously. In addition, the strength of the association was greater for benzene as compared to traffic-related pollutants usually investigated, such as particulate matter. Of course, these results may be the product, of an unmeasured confounder (at least partially for ultrafine particles, which are strongly correlated to PM_{10}). If our findings were confirmed by others, this would mean that not only diesel vehicles, the main particulate matter emitters [21,33] but also gasoline-fueled cars –main benzene emitters-, should be taken into account for public health action.

Acknowledgments

We warmly thank Dr William Sherlaw and Ms Kristina Parkins for their editorial assistance.

References

1. Peters A, Pope CA 3rd (2002) Cardiopulmonary Mortality And Air Pollution. Lancet 360: 1184–1185.
2. Peters A, Dockery DW, Muller JE, Mittleman MA (2001) Increased Particulate Air Pollution And The Triggering Of Myocardial Infarction. Circulation 103: 2810–2815.
3. Mustafic H, Jabre P, Caussin C, Murad MH, Escolano S, et al. (2012) Main Air Pollutants And Myocardial Infarction: A Systematic Review And Meta-Analysis. Jama 307: 713–721.
4. Brook RD, Rajagopalan S, Pope CA 3rd, Brook JR, Bhatnagar A, et al. (2010) Particulate Matter Air Pollution And Cardiovascular Disease: An Update To The Scientific Statement From The American Heart Association. Circulation 121: 2331–2378.
5. Baccarelli A, Benjamin EJ (2011) Triggers Of Mi For The Individual And In The Community. Lancet 377: 694–696.
6. Keenan JJ, Gaffney SH, Galbraith DA, Beatty P, Paustenbach DJ (2010) Gasoline: A Complex Chemical Mixture, Or A Dangerous Vehicle For Benzene Exposure? Chem Biol Interact 184: 293–295.
7. Faroon O, Wilbur S (2008) Special Issue On The Toxicology And Epidemiology Of Benzene. Toxicol Ind Health 24: 261–262.
8. Kotseva K, Popov T (1998) Study Of The Cardiovascular Effects Of Occupational Exposure To Organic Solvents. Int Arch Occup Environ Health 71 Suppl: S87–91.
9. Kurppa K, Hietanen E, Klockars M, Partinen M, Rantanen J, et al. (1984) Chemical Exposures At Work And Cardiovascular Morbidity. Atherosclerosis, Ischemic Heart Disease, Hypertension, Cardiomyopathy And Arrhythmias. Scand J Work Environ Health 10: 381–388.
10. Tsai DH, Wang JL, Chuang Kj, Chan CC (2010) Traffic-Related Air Pollution And Cardiovascular Mortality In Central Taiwan. Sci Total Environ 408: 1818–1823.
11. Tunstall-Pedoe H, Kuulasmaa K, Amouyel P, Arveiler D, Rajakangas AM, et al. (1994) Myocardial Infarction And Coronary Deaths In The World Health Organization Monica Project. Registration Procedures, Event Rates, And Case-Fatality Rates In 38 Populations From 21 Countries In Four Continents. Circulation 90: 583–612.
12. Carruthers D, Edmunds H, Lester A, McHugh C, Singles R (1998) Use And Validation Of Adms-Urban In Contrasting Urban And Industrial Locations. Int J Environ Pollut 14: 2000.
13. Havard S, Deguen S, Zmirou-Navier D, Schillinger C, Bard D (2009) Traffic-Related Air Pollution And Socioeconomic Status: A Spatial Autocorrelation Study To Assess Environmental Equity On A Small-Area Scale. Epidemiology 20: 223–230.
14. Flahault A, Blanchon T, Dorléans Y, Toubiana L, Vibert JF, et al. (2006) Virtual Surveillance Of Communicable Diseases: A 20-Year Experience In France. Stat Methods Med Res 15: 413–421.
15. Maclure M, Mittleman MA (2000) Should We Use A Case-Crossover Design? Annu Rev Public Health 21: 193–221.
16. Janes H, Sheppard L, Lumley T (2005) Case-Crossover Analyses Of Air Pollution Exposure Data: Referent Selection Strategies And Their Implications For Bias. Epidemiology 16: 717–726.
17. European Environment Agency (2012) Air Quality In Europe – 2012 Report. Copenhagen: European Environment Agency. 108 P.
18. Delfino RJ, Gong H Jr, Linn WS, Pellizzari ED, Hu Y (2003) Asthma Symptoms In Hispanic Children And Daily Ambient Exposures To Toxic And Criteria Air Pollutants. Environ Health Perspect 111: 647–656.
19. Fusco D, Forastiere F, Michelozzi P, Spadea T, Ostro B, et al. (2001) Air Pollution And Hospital Admissions For Respiratory Conditions In Rome, Italy. Eur Respir J 17: 1143–1150.
20. Schwartz J (1999) Air Pollution And Hospital Admissions For Heart Disease In Eight U.S. Counties. Epidemiology 10: 17–22.
21. Huang C, Lou D, Hu Z, Feng Q, Chen Y, et al. (2013) A Pems Study Of The Emissions Of Gaseous Pollutants And Ultrafine Particles From Gasoline- And Diesel-Fueled Vehicles. Atmospheric Environment 77: 703–710.
22. Marcazzan GM, Vaccaro S, Valli G, Vecchi R (2001) Characterisation Of Pm_{10} And $Pm_{2.5}$ Particulate Matter In The Ambient Air Of Milan (Italy). Atmospheric Environment 35: 4639–4650.
23. Bhaskaran K, Hajat S, Haines A, Herrett E, Wilkinson P, et al. (2009) Effects Of Air Pollution On The Incidence Of Myocardial Infarction. Heart 95: 1746–1759.
24. Billionnet C, Sherrill D, Annesi-Maesano I (2012) Estimating The Health Effects Of Exposure To Multi-Pollutant Mixture. Ann Epidemiol 22: 126–141.
25. Sacks JD, Ito K, Wilson We, Neas LM (2012) Impact Of Covariate Models On The Assessment Of The Air Pollution-Mortality Association In A Single- And Multipollutant Context. Am J Epidemiol 176: 622–634.
26. Bhaskaran K, Armstrong B, Hajat S, Haines A, Wilkinson P, et al. (2012) Heat And Risk Of Myocardial Infarction: Hourly Level Case-Crossover Analysis Of Minap Database. Bmj 345: E8050.
27. Peters A, Von Klot S, Heier M, Trentinaglia I, Hormann A, et al. (2004) Exposure To Traffic And The Onset Of Myocardial Infarction. N Engl J Med 351: 1721–1730.
28. Murugesan K, Baumann S, Wissenbach DK, Kliemt S, Kalkhof S, et al. (2013) Subtoxic And Toxic Concentrations Of Benzene And Toluene Induce Nrf2-Mediated Antioxidative Stress Response And Affect The Central Carbon Metabolism In Lung Epithelial Cells A549. Proteomics 13: 3211–3221.
29. Bhatnagar A (2006) Environmental Cardiology: Studying Mechanistic Links Between Pollution And Heart Disease. Circ Res 99: 692–705.
30. Mc Hale CM, Zhang L, Smith MT (2012) Current Understanding Of The Mechanism Of Benzene-Induced Leukemia In Humans: Implications For Risk Assessment. Carcinogenesis 33: 240–252.
31. Bollati V, Baccarelli A, Hou L, Bonzini M, Fustinoni S, et al. (2007) Changes In Dna Methylation Patterns In Subjects Exposed To Low-Dose Benzene. Cancer Research 67: 876–880.
32. Nawrot TS, Perez L, Kunzli N, Munters E, Nemery B (2011) Public Health Importance Of Triggers Of Myocardial Infarction: A Comparative Risk Assessment. Lancet 377: 732–740.
33. Pant P, Harrison RM (2013) Estimation Of The Contribution Of Road Traffic Emissions To Particulate Matter Concentrations From Field Measurements: A Review. Atmospheric Environment 77: 78–97.

Author Contributions

Conceived and designed the experiments: DB WK C. Ségala DA CW. Performed the experiments: WK. Analyzed the data: WK C. Schillinger CF SG C. Ségala. Contributed reagents/materials/analysis tools: C. Schillinger. Wrote the paper: DB WK CW.

2

Dual p38/JNK Mitogen Activated Protein Kinase Inhibitors Prevent Ozone-Induced Airway Hyperreactivity in Guinea Pigs

Kirsten C. Verhein[1*¤a], Francesco G. Salituro[3¤b], Mark W. Ledeboer[3], Allison D. Fryer[2], David B. Jacoby[2]

1 Department of Physiology and Pharmacology, Oregon Health and Science University, Portland, Oregon, United States of America, 2 Division of Pulmonary and Critical Care Medicine, Oregon Health and Science University, Portland, Oregon, United States of America, 3 Vertex Pharmaceuticals, Inc., Cambridge, Massachusetts, United States of America

Abstract

Ozone exposure causes airway hyperreactivity and increases hospitalizations resulting from pulmonary complications. Ozone reacts with the epithelial lining fluid and airway epithelium to produce reactive oxygen species and lipid peroxidation products, which then activate cell signaling pathways, including the mitogen activated protein kinase (MAPK) pathway. Both p38 and c-Jun NH_2 terminal kinase (JNK) are MAPK family members that are activated by cellular stress and inflammation. To test the contribution of both p38 and JNK MAPK to ozone-induced airway hyperreactivity, guinea pigs were pretreated with dual p38 and JNK MAPK inhibitors (30 mg/kg, ip) 60 minutes before exposure to 2 ppm ozone or filtered air for 4 hours. One day later airway reactivity was measured in anesthetized animals. Ozone caused airway hyperreactivity one day post-exposure, and blocking p38 and JNK MAPK completely prevented ozone-induced airway hyperreactivity. Blocking p38 and JNK MAPK also suppressed parasympathetic nerve activity in air exposed animals, suggesting p38 and JNK MAPK contribute to acetylcholine release by airway parasympathetic nerves. Ozone inhibited neuronal M_2 muscarinic receptors and blocking both p38 and JNK prevented M_2 receptor dysfunction. Neutrophil influx into bronchoalveolar lavage was not affected by MAPK inhibitors. Thus p38 and JNK MAPK mediate ozone-induced airway hyperreactivity through multiple mechanisms including prevention of neuronal M_2 receptor dysfunction.

Editor: Heinz Fehrenbach, Research Center Borstel, Germany

Funding: This work was funded by the American Heart Association 0810148Z (to KCV) and the National Institutes of Health HL55543 (to ADF), ES014601 (to ADF), HL54659 (to DBJ), HL071795 (to DBJ), and RR023424 (to DBJ). The funders had no role in study design, data collection and analysis, decision to publish, or preparation of the manuscript. No additional external funding was received for this study.

Competing Interests: FGS and MWL were employed by Vertex Pharmaceuticals at the time of these studies. FGS is currently employed by Sage Therapeutics.

* E-mail: verheinkc@niehs.nih.gov

¤a Current address: National Institute of Environmental Health Sciences, Research Triangle Park, North Carolina, United States of America
¤b Current address: Sage Therapeutics, Cambridge, Massachusetts, United States of America

Introduction

Over half the United States population lives in counties with unhealthy levels of ozone, a major component of smog [1]. Epidemiological studies demonstrate a significant link between exposure to ground level ozone and pulmonary hospitalizations. Exposure to ozone in excess of 0.16 ppm is associated with increased airway reactivity, lung inflammation and exacerbation of asthma in both adults and children [2,3,4].

Ozone induced hyperreactivity is demonstrated by increased reactivity to inhaled methacholine and other agonists, including those causing reflex bronchoconstriction in man [5,6,7]. In animals, ozone induced airway hyperreactivity is demonstrated by increased bronchoconstriction to intravenous methacholine, but this effect is mediated largely via increased acetylcholine release from parasympathetic nerves, since it is blocked by vagal section [8,9]. Direct stimulation of the vagus nerves results in bronchoconstriction that is potentiated in ozone exposed animals and that is associated with loss of function of neural M_2 muscarinic

receptors that normally inhibit acetylcholine release [10,11]. Inflammatory cells, especially eosinophils through release of the M_2 inhibitor major basic protein, mediate loss of neuronal M_2 function and airway hyperreactivity in ozone exposed guinea pigs [11].

However, ozone is unlikely to contact inflammatory cells [12]. At the airway epithelial layer, ozone forms reactive oxygen species and lipid peroxides in lungs of humans and animals [13,14]. These end products activate cell signaling pathways, including mitogen activated protein kinase pathways (MAPK) [15]. Activation of the MAPK pathway results in inflammation [16], mucus hypersecretion [17] and airway hyperreactivity [18].

MAPK signaling pathways are important in many cell processes including differentiation, proliferation, activation, degranulation, and migration. Three MAPK subfamilies have been well characterized: ERK, JNK, and p38. The extracellular signal-regulated kinase (ERK) pathway is usually activated by mitogens and growth factors while p38 and c-Jun NH_2 terminal kinase (JNK) pathways are associated with chronic inflammation and are typically activated by inflammatory cytokines, heat shock, and cellular stress [19,20]. Activation of MAPK signaling induces inflammatory cytokine and chemokine production in airway epithelial cells, inflammatory cells, and airway smooth muscle cells [16,21,22]. Humans with severe asthma have increased activated p38 in airway epithelium compared to mild asthmatics or healthy controls, as demonstrated by increased immunostaining of phosphorylated p38 in airway biopsies [23].

Inhibition of MAPKs is protective in allergen challenge models of asthma. Inhibition of p38, either pharmacologically or with antisense oligonucleotides, partially prevents airway hyperreactivity after sensitization and challenge in mice [18,24]. Eosinophil influx into bronchoalveolar lavage is the dominant event in antigen challenged animals, and is prevented by a p38 inhibitor in guinea pigs and mice [25]. Blocking p38 also prevents IL-13 induced mucus metaplasia in human and mouse airway epithelial cells [17,26].

Less is known about the role of the MAP kinases in ozone-induced hyperreactivity. Inhibiting p38 prevents ozone-induced airway hyperreactivity in mice while inhibiting JNK is partially protective [27,28]. Ozone-induced increases in inflammatory cells in bronchoalveolar lavage are significantly inhibited in *Jnk1* knockout mice [29].

The experiments described here use three different MAPK inhibitors to test whether dual inhibition of both p38 and JNK MAPK pathways prevents ozone-induced inflammation and subsequent airway hyperreactivity in guinea pigs.

Methods

Ethics Statement

Guinea pigs were handled in accordance with the standards established by the United States Animal Welfare Act set forth in National Institutes of Health guidelines. All protocols were approved by Oregon Health and Science University Animal Care and Use Committee (protocol A984).

Animals

Specific pathogen-free female Hartley guinea pigs (300–470 g; Elm Hill Breeding Labs, Chelmsford, MA) were shipped in filtered crates, housed in high efficiency particulate filtered air, and fed a normal diet.

Ozone Exposure

Guinea pigs were exposed to 2 ppm ozone or filtered air for 4 hours as described previously [11]. Physiological measurements,

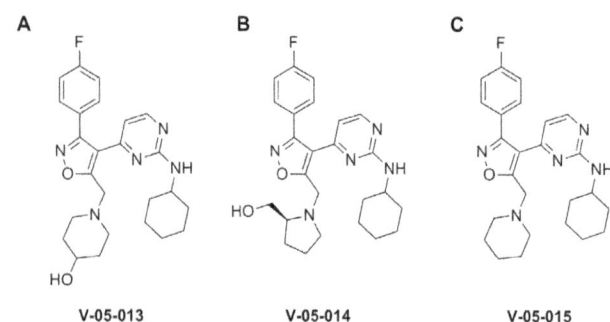

Table 1. Ki values for dual p38 and JNK MAPK inhibitors.

Ki (nM)	V-05-013	V-05-014	V-05-015
p38	13	15	8*
JNK1	170	300	120
JNK2	10	15	4
JNK3	5	10	8

All compounds have a Ki greater than 1 µM for all other kinases tested.
*This one value is an IC50, not a Ki.

airway inflammation, and histological measurements were made one day after a single ozone exposure.

Treatment of Guinea Pigs with p38 and JNK MAPK Inhibitors

Animals were given 30 mg/kg intraperitoneally of the dual p38 and JNK MAPK inhibitors V-05-013, V-05-014, or V-05-015 (Vertex Pharmaceuticals, Cambridge, MA) one hour before ozone exposure (Figure 1). These compounds were chosen because of their overall kinase selectivity profile. They are potent and selective dual inhibitors of p38 and JNK (see below and table 1) and they do not show activity against a panel of other kinases at concentrations <1 µM (see characterization data below). Inhibitors were dissolved in 25% DMSO in phosphate buffered saline (PBS). Air exposed control animals were given 25% DMSO in PBS one hour before ozone exposure.

All three drugs have similar kinase inhibition profiles and exhibit potent affinity for both p38 and JNK. Affinity was measured using a kinase inhibition assay. Compounds were assayed for the inhibition of various kinases using a modification of a spectrophotometric coupled-enzyme assay [30]. In this assay, a fixed concentration of activated kinase (10–40 nM) was incubated with various concentrations of a potential inhibitor dissolved in DMSO for 10 minutes at 30°C in a buffer containing 0.1 M HEPES, pH 7.5, containing 10 mM $MgCl_2$, 2.5 mM phospho-enolpyruvate, 200 µM NADH, 2 mM DTT, 30 µg/mL pyruvate kinase, 10 µg/mL lactate dehydrogenase, and 200 µM–500 µM EGF receptor peptide. The EGF receptor peptide has the sequence KRELVEPLTPSGEAPNQALLR. The reaction was initiated by the addition of ATP equal to the ATP Km of the kinase, and the assay plate is inserted into the spectrophotometer's assay plate compartment that was maintained at 30°C. The decrease of absorbance at 340 nm was monitored as a function of time for 10 minutes. The rate data as a function of inhibitor concentration was either fit as an IC50 or to a competitive inhibition kinetic model to determine the compound K_i (see table 1).

Proton NMR spectra for the compounds was recorded on a Bruker Advance instrument with a QNP probe using TMS as the internal standard in the indicated deuterated solvent. LC–MS analyses were performed on a Waters ZQ or ZMD or QuatroII mass spectrometer using the electrospray (ESI) ionization technique. Samples were introduced into the mass spectrometer using chromatography. Methods (LC-MS) consisted of the following: 5–95% Water/Acetontrile (0.1% TFA) over 0.6 min on a Waters Acquity CSH C18, 1.7 µm, 2.1×50 mm column, with a flow rate of 0.6 mL/min. NMR spectra for each compound are below.

V-05-013:1H NMR (300 MHz, MeOD) 8.20 (d, J = 5.6 Hz, 1H), 7.62 (dd, J = 5.3, 8.6 Hz, 1H), 7.29 (t, J = 8.6 Hz, 1H), 6.64-

Figure 1. Chemical structures of dual p38 and JNK MAPK inhibitors.

Table 2. Baseline cardiovascular and pulmonary parameters.

Group	n	Heart Rate (beats/min)	Blood Pressure (mmHg) Systolic	Diastolic	Pulmonary Inflation Pressure (mmH$_2$O)
Air	7	306±6	40±4	20±3	110±4
Ozone	5	309±10	48±4	23±4	248±21 *
Ozone+V-05-013	5	301±7	40±3	20±2	190±15 * ‡
Ozone+V-05-014	4	286±16	38±4	21±1	225±5 *
Ozone+V-05-015	3	353±7	40±1	18±1	213±7 *
Air+V-05-013	4	291±7	32±2	17±1	102±3
Air+V-05-014	5	306±9	38±2	19±1	102±2
Air+V-05-015	5	327±14	40±3	22±3	94±5

Values are means ± SEM. Baseline pulmonary inflation pressure significantly increased after ozone exposure. Treatment with dual p38/JNK MAPK inhibitor V-05-013 significantly reduced the ozone-induced increase in baseline pulmonary inflation pressure. *$p < 0.05$ Significantly different from air exposure. ‡$p < 0.05$ Significantly different from ozone exposure.

6.47 (m, 1H), 5.03 (s, H), 4.18-4.00 (m, 1H), 4.18-4.00 (m, 1H), 3.80-3.35 (m, 5H), 1.71 (m, 1H), 1.52-1.23 (m, 5H); LC-MS (method X) $t_R = 0.60$ min., (M+H$^+$) 452.37. Ki or IC50 was determined to be >1 μM for the following kinases: AKT3, AurA, CDK2, ERK-2, EphA, FLT3, IGF1R, IRAK, ITK, JAK2, JAK3, KDR, MAPKAP2, cMET, MKK4, MKK6, MKK7, PIM1, PKA, PLK1, PRAK, ROCK1, SRC, SYK, TIE2, ZAP70; GSK3β Ki = 0.73 μM.

V-05-014:1H NMR (400 MHz, DMSO-d6) 11.04 (s, 0.3 H), 10.22 (s, 0.2 H), 8.33 (s, 1 H), 8.08 (s, 1 H), 7.38–7.79 (m, 4 H), 6.68 (s, 1 H), 6.20 (s, 1 H), 5.15–5.23 (m, 0.7 H), 4.86–4.91 (m, 1H), 3.78 (s, 3 H), 3.53 (s, 1 H), 3.33 (s, 1 H), 1.56–2.13 (m, 9 H), 1.09–1.28 (m, 6 H); LC-MS (method Y) ^{t}R = 2.47 min., (M+H$^+$) 452.25; LC-MS (method X) $t_R = 0.63$ min., (M+H$^+$) 452.37. Ki or IC50 was determined to be >1 μM for the following kinases: AKT3, AurA, COT, CDK2, ERK2, EphA, FAK, GSK3β, IGF1R, IRAK, JAK3, KDR, LCK, MAPKAP2, cMET, MKK4, MKK6, MKK7, NIK, PDK1, PIM1, PKA, PLK1, PRAK, ROCK1, SRC, SYK, TIE2, ZAP70.

V-05-015:1H NMR (300 MHz, MeOD) 8.20 (d, J = 5.1 Hz, 1H), 7.62 (dd, J = 5.3, 8.6 Hz, 2H), 7.29 (t, J = 8.7 Hz, 2H), 6.68-6.42 (m, 1H), 5.00 (s, 2H), 3.75-3.53 (m, 3H), 3.34-3.15 (m, 2H), 2.10-1.77 (m, 9H), 1.88 (m, 1H), 1.65-1.20 (m, 7H); LC-MS (method X) $t_R = 0.64$ min., (M+H$^+$) 436.38. Ki or IC50 was determined to be >1 μM for the following kinases: AKT3, AurA, CDK2, ERK2, FLT3, GSK3b, IGF1R, IRAK, JAK2, JAK3, KDR, LCK, MAPKAP2, cMET, MKK4, MKK6, MKK7, PDK1, PIM1, PKA, PLK1, PRAK, ROCK1, SRC, SYK, ZAP70.

Measurement of Pulmonary Inflation Pressure

One day after exposure to ozone, guinea pigs were anesthetized with 1.9 g/kg urethane i.p. (Sigma-Aldrich, St. Louis, MO). This dose produces a deep anesthesia lasting 8–10 hours [31] though no experiments lasted longer than 4 hours.

Physiological measurements were made as previously described [32]. The jugular veins were cannulated for intravenous administration of drugs and the right carotid artery was cannulated to measure heart rate and blood pressure. Both vagus nerves were cut and distal ends placed on platinum electrodes submerged in liquid paraffin. Animals were tracheostomized, ventilated (1 ml/100 g body weight, 100 breaths per minute) and paralyzed with a constant infusion of succinylcholine (10 μg/kg/min iv, Sigma-

Aldrich). Pulmonary inflation pressure was measured at the trachea and bronchoconstriction was measured as the increase in pressure over basal inflation pressure produced by the ventilator.

Measurement of Vagally Induced Bronchoconstriction

Electrical stimulation of both vagus nerves (10V, 0.2 ms pulse width, 1–25 Hz, 5 sec duration at 1 minute intervals) produced frequency dependent bronchoconstriction and bradycardia due to release of acetylcholine onto muscarinic receptors. To confirm vagally induced bronchoconstriction was cholinergic, atropine (1 mg/kg iv, Sigma-Aldrich) was given at the end of each experiment.

Measurement of Smooth Muscle M$_3$ Muscarinic Receptor Function

Recovery from vagal stimulation was confirmed by pulmonary inflation pressure and heart rate returning to baseline before measuring smooth muscle M$_3$ muscarinic receptor function (5–10 minutes after cessation of vagal stimulation). In vagotomized guinea pigs, M$_3$ muscarinic receptor function on airway smooth muscle was tested by measuring bronchoconstriction after administration of acetylcholine (1–10 μg/kg iv, Sigma-Aldrich).

Measurement of Neuronal M$_2$ Muscarinic Receptor Function

Recovery from administration of intravenous acetylcholine was confirmed by pulmonary inflation pressure and heart rate returning to baseline before measuring M$_2$ muscarinic receptor function (5–10 minutes after the last dose of acetylcholine). To test the function of neuronal M$_2$ muscarinic receptors, vagally induced bronchoconstriction was measured before and after administration of gallamine (0.1–10 mg/kg iv, Sigma-Aldrich) an M$_2$ receptor antagonist. Electrical stimulation of both vagus nerves (3–30V, 0.2 ms pulse width, 15 Hz, 5 sec duration at 1 minute intervals) produced reproducible, frequency dependent, bronchoconstrictions. In the presence of normally functioning M$_2$ receptors, gallamine (0.1–10 mg/kg iv) will block them, resulting in increased vagally induced bronchoconstriction [33]; an effect that is suppressed if M$_2$ receptors are not responding to endogenous acetylcholine [10].

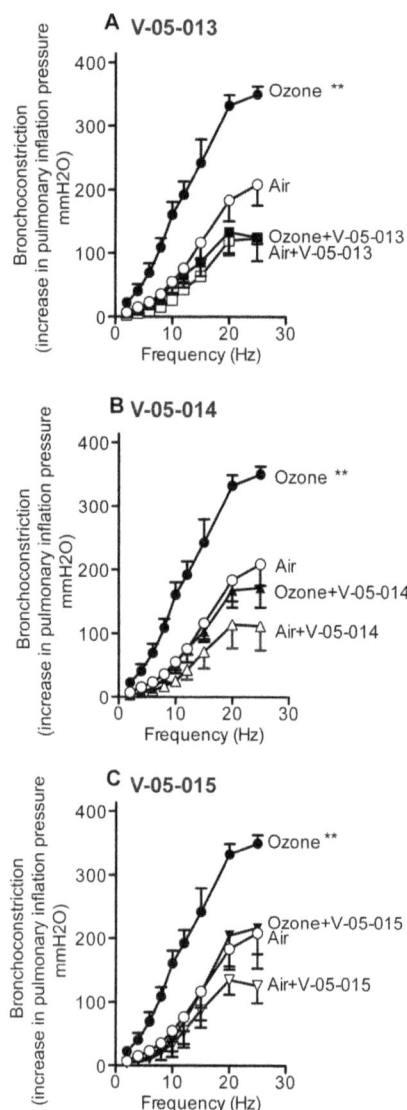

Figure 2. Blocking p38 and JNK MAPK completely prevented ozone-induced airway hyperreactivity mediated by the vagus nerves. In anesthetized and vagotomized guinea pigs, stimulation of the vagus nerves (10V, 0.2 ms pulse width, 1–25 Hz, 5 sec duration at 1 minute intervals) caused frequency dependent bronchoconstriction (A open circles; measured as an increase in inflation pressure in mmH$_2$O) that is significantly potentiated one day post-ozone exposure (A closed circles). Pretreatment with dual MAPK inhibitors V-05-013 (A closed squares), V-05-014 (B closed triangles), or V-05-015 (C closed inverted triangles) completely prevented ozone-induced airway hyperreactivity. All three dual MAPK inhibitors suppressed parasympathetic nerve activity (A open squares, B open triangles, C open inverted triangles). Ozone and air exposed control data are the same in A-C. *p<0.05, **p<0.01 Significantly different from air exposed controls. Data are mean ± SEM. n = 4–7.

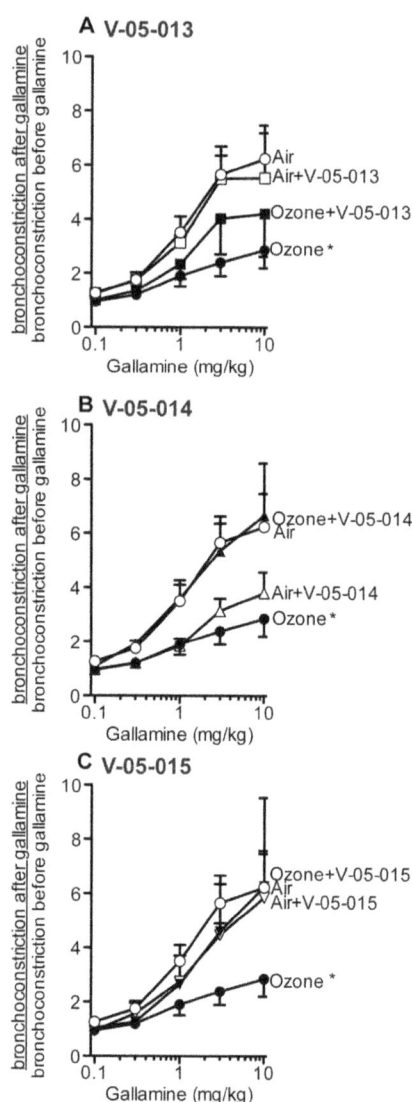

Figure 3. In control (air exposed) guinea pigs electrical stimulation of the vagus nerves (3–30V, 0.2 ms pulse width, 15 Hz, 5 sec duration at 1 minute intervals) resulted in vagally induced bronchoconstriction (measured as an increase in pulmonary inflation pressure; 16±1 mmH$_2$O). An M$_2$ receptor antagonist, gallamine, potentiated vagally induced bronchoconstriction up to 6-fold in air exposed animals (open circles) demonstrating that functional M$_2$ receptors were limiting acetylcholine release. The potentiation by gallamine was decreased in ozone-exposed animals, demonstrating M$_2$ receptors were dysfunctional after ozone exposure (closed circles). V-05-013 partially prevented M$_2$ receptor dysfunction (C closed squares), while V-05-014 (B closed triangles) and V-05-015 (C closed inverted triangles) completely protected M$_2$ receptor function. Vagally induced bronchoconstriction in the absence of gallamine was not different from control among all groups. Ozone and air exposed controls are the same in A–C. *p<0.05, **p<0.01 Significantly different from air exposed controls. Data are mean ± SEM. n = 4–7.

Bronchoalveolar Lavage (BAL)

At the end of each experiment, the lungs were lavaged with five 10 ml aliquots of phosphate buffered saline (PBS) that contained 100 μg isoproterenol (Sigma-Aldrich). Lavage fluid was centrifuged (400 g, 10 min) and the pellets were resuspended in PBS. Cells were counted using a hemocytometer and slides made from centrifuged lavaged cells were stained with Hemacolor (EMD Chemicals, Gibbstown, NJ) and used to determine cell differentials.

Drugs

Acetylcholine, succinylcholine, and urethane were purchased from Sigma (St. Louis, MO) and were dissolved and diluted in PBS.

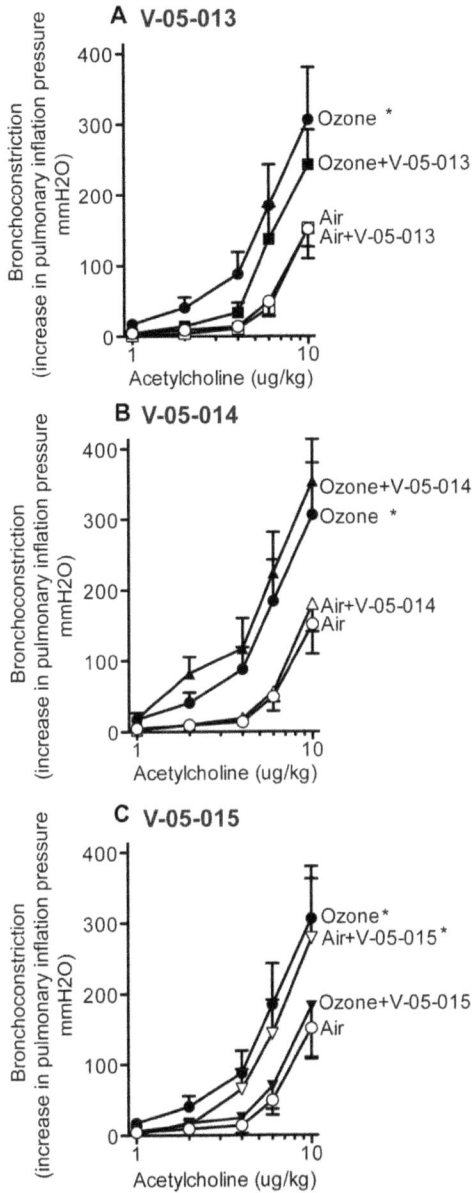

Figure 4. Bronchoconstriction (measured as an increase in inflation pressure in mmH₂O) in response to intravenous acetylcholine was significantly potentiated one day post-ozone (closed circles) compared to air exposed controls (open circles), and was not blocked by V-05-013 (A closed squares), or V-05-014 (B closed triangles). V-05-015 (C closed inverted triangles) attenuated ozone-induced smooth muscle hyperreactivity while air-exposed animals pretreated with V-05-015 were hyperreactive to intravenous acetylcholine (C open inverted triangles). Ozone and air exposed controls are the same in A-C. *p<0.05, **p<0.01 Significantly different from air exposed controls. Data are mean ± SEM. n = 3–7.

Data Analysis and Statistics

All data are expressed as means ± SE. *In vivo* frequency response and dose response curves were compared using two-way ANOVA for repeated measures. Baseline data were analyzed by one-way ANOVA with Bonferroni's correction. A P value of less than 0.05 was considered significant. Analyses were made with GraphPad Prism (version 5.0; GraphPad Software, La Jolla, CA).

Results

Baselines

One day after ozone exposure, baseline pulmonary inflation pressure was significantly increased compared to air-exposed controls (Table 2). All the dual p38 and JNK inhibitors partially attenuated the ozone induced increase in baseline airway inflation pressure, although the attenuation only reached statistical significance in the group treated with V-05-013. None of the MAPK inhibitors affected baseline inflation pressure in air-exposed controls. Neither ozone nor the MAPK inhibitors affected baseline heart rate or blood pressure.

Airway Physiology

Ozone significantly potentiated bronchoconstriction in response to electrical stimulation of the vagus nerves compared to air-exposed controls as previously reported (Figure 2). Treatment with any of the dual MAPK inhibitors prevented ozone induced airway hyperreactivity (Figures 2A–C). Vehicle treatment had no effect on vagally mediated bronchoconstriction in either air or ozone exposed animals (data not shown). M_2 muscarinic receptors were dysfunctional in ozone treated animals as gallamine, an M_2 selective inhibitor, potentiated bronchoconstriction in response to vagal stimulation in air-exposed animals but not in ozone-exposed animals (Figure 3); an effect that is consistent with decreased function of neuronal M_2 muscarinic receptors [34]. Ozone induced M_2 receptor dysfunction was prevented by treatment with V-05-014 and V-05-015 (Figure 3B–C), and attenuated by treatment with V-05-013 (Figure 3A). Airway smooth muscle responses to intravenous acetylcholine were potentiated by ozone (Figure 4). This was not prevented by any of the MAPK inhibitors, but was partially attenuated by V-05-015 (Figure 4C). V-05-015 also produced a paradoxical increase in airway response to intravenous acetylcholine in air-exposed animals.

Ozone exposure potentiated falls in heart rate in response to vagal stimulation compared to air-exposed controls (Figure 5A–C). Ozone and air-exposed controls are the same in figure 5A–C. Separate pretreatment with all three dual MAPK inhibitors prevented the ozone-induced potentiation in falls in heart rate and had no effect in air-exposed animals (Figure 5A–C). Falls in heart rate in response to intravenous acetylcholine were not affected by either ozone or the MAPK inhibitor (Figure 5D–F). Ozone and air exposed controls are the same in figure 5D–F.

Bronchoalveolar Lavage and Peripheral Blood

One day after ozone exposure neutrophils were increased in bronchoalveolar lavage (Figure 6D). All the MAPK inhibitors slightly, though not significantly, attenuated the ozone induced increase in neutrophils (Figure 6D). None of the other inflammatory cell types were affected by either ozone or the MAPK inhibitors (Figure 6).

There were no significant differences between inflammatory cells in peripheral blood after either ozone exposure, or treatment with the dual MAPK inhibitors (Figure 7).

Discussion

Ozone induces airway hyperreactivity, measured as potentiation of vagally induced bronchoconstriction, in guinea pigs one day after exposure confirming previous studies [35,36]. Ozone also significantly potentiated bronchoconstriction in response to intravenous acetylcholine; an effect that has also been previously reported [10]. Blocking both p38 and JNK MAPK with three different, but related, inhibitors prevented vagally mediated

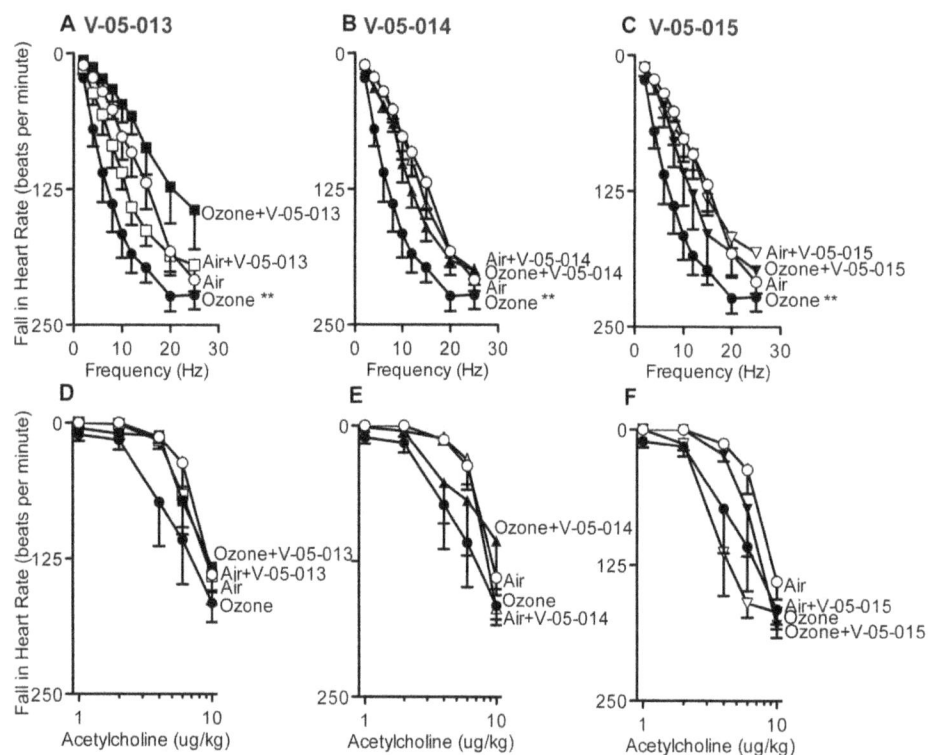

Figure 5. Ozone potentiated vagally mediated falls in heart rate (measured as beats/minute; A–C closed circles) compared to air exposed animals (A–C open circles). Separate pretreatment with all three dual MAPK inhibitors (V-05-013: A closed squares; V-05-014: B closed squares; V-05-015: C closed squares) prevented the ozone-induced potentiation of frequency induced falls in heart rate. Fall in heart rate following intravenous acetylcholine administration was not changed by either ozone, or MAPK inhibitors (D–E). Ozone and air exposed controls are the same for A–C, and are the same for D–F. **p<0.01 Significantly different from air exposed controls. Data are mean ± SEM. n=3–7.

hyperreactivity in ozone-exposed animals but had no effect on inflammatory cell numbers in bronchoalveolar lavage. The prevention of vagally mediated hyperreactivity was associated with prevention of ozone induced M_2 receptor dysfunction that was complete in animals treated with V-05-014 and V-05-015, and partial in animals treated with V-05-013. Ozone induced hyperreactivity to intravenous acetylcholine was partially attenuated by treatment with the MAPK inhibitors.

All three MAPK inhibitors were administered at a dose of 30 mg/kg i.p. one hour before ozone. While the compounds are active with submicromolar potencies, preliminary studies suggested there is a significant shift in *in vivo* potency from the tens of nanomolar to hundreds of nanomolar IC50 s presumably the result of plasma protein binding (unpublished data). Nonetheless, the compounds were chosen because they exhibit adequate pharmacokinetic profiles (table 1) to test our hypothesis *in vivo*. The relatively high *in vivo* clearances and half lives are somewhat limiting, leading to the need for a sufficient dose to demonstrate a role for MAPKs in ozone induced hyperreactivity. However, as with most compounds, different effects could occur at lower doses.

Treatment of air-exposed guinea pigs with any of the three MAPK inhibitors decreased the airway response to vagal stimulation slightly. This effect was most pronounced at high frequency stimulation, but could not be explained by changes in M_2 receptor function, as the effects of gallamine were not potentiated by the MAPK inhibitors in air exposed animals. This effect was also not due to decreased smooth muscle responsiveness, as the effects of intravenous acetylcholine were

not decreased by the MAPK inhibitors. Although response to the MAPK inhibitors was variable in air-exposed animals the overall effect with ozone exposure was prevention of ozone-induced airway hyperreactivity. These minor differences may be due to off target effects of the inhibitors, or to the dose of inhibitors used in this study. Thus, in air exposed guinea pigs, p38 and JNK MAPK inhibitors inhibit vagally induced bronchoconstriction by suppressing release of acetylcholine from airway parasympathetic nerves.

The mechanism for this decreased acetylcholine release is unknown. p38 and JNK are involved in nerve regeneration and development [37,38] but whether they inhibit ganglionic transmission, action potentials or transmitter release (by a mechanism separate from M_2 receptors, since there was no change in the response to gallamine) is not well studied. In *Aplysia*, activation of p38 by the peptide neurotransmitter FMRFa leads to long-term depression in sensory neurons in the pleural ganglia [39], although the mechanism is not known. In *Drosophila* motor neurons, expression of constitutively active JNK decreases neurotransmitter release [40] while in primary cultures of rat cortical neurons, IL-1β signaling activates p38, decreasing synaptophysin, a protein involved in synaptic transmission [41]. These varied and sometimes contradictory effects of MAPKs on neural function and transmitter release may be involved in the effects we observed. In neutrophils, activation of p38 MAPK is required for granule exocytosis after stimulation by CXCR1/2 ligands [42]; if neurotransmitter exocytosis were similarly mediated by MAPK, kinase inhibitors would block secretion. Thus, the role of MAPK is cell type dependent and additionally may differ between central

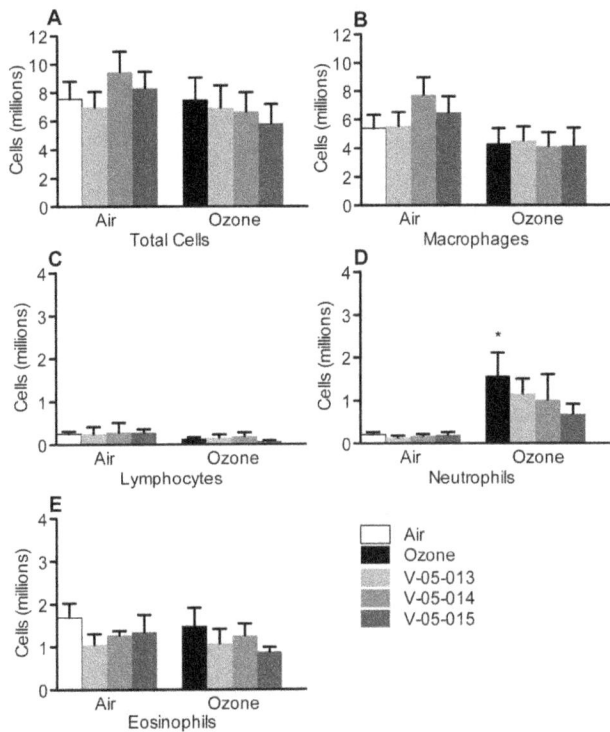

Figure 6. Ozone exposure increased neutrophils in bronchoalveolar lavage (D closed bar). No other inflammatory cell type number was affected by either ozone or the dual p38/JNK MAPK inhibitors. *$p<0.05$ Significantly different from air exposed controls. Data are mean $\pm$ SEM. n = 3–6.

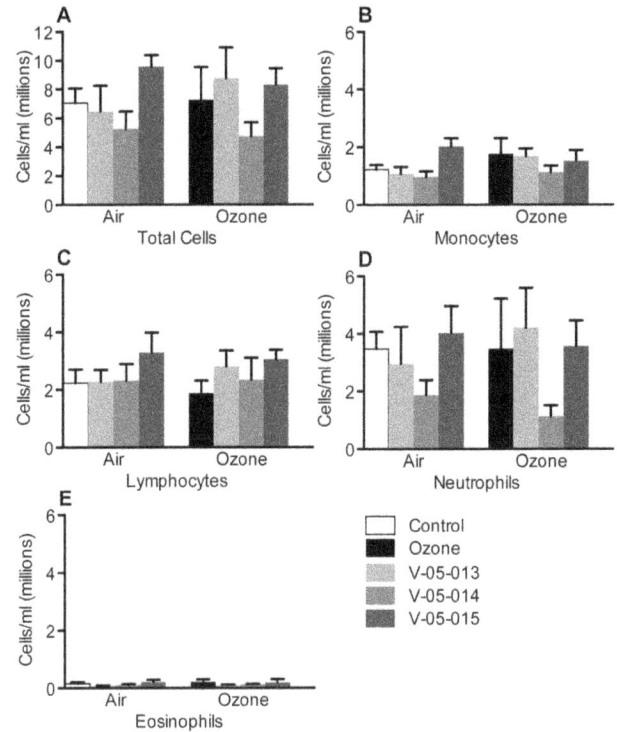

Figure 7. Neither ozone nor the dual p38/JNK MAPK inhibitors affected inflammatory cell numbers in peripheral blood. Data are mean $\pm$ SEM. n = 3–6.

neurons where kinases inhibit neurotransmission and peripheral neurons, where they have not been well studied. The data in this paper suggest p38 or JNK MAPK may additionally play a previously unrecognized role in release of acetylcholine from lung parasympathetic nerves.

None of the MAPK inhibitors completely reversed the ozone-induced increase in baseline pulmonary inflation pressure, which is commonly due to airway edema and not increased vagal tone. Ozone also significantly increased the numbers of neutrophils in bronchoalveolar lavage compared to air exposed controls confirming previously published data [35,36]. However, blocking both p38 and JNK MAPK did not prevent the neutrophil influx. No other inflammatory cell population in the lavage was effected by ozone or by the p38 and JNK MAPK inhibitors. Thus, prevention of ozone-induced airway hyperreactivity did not occur via a decrease in airway inflammatory cells.

Previously we have shown major basic protein, released from eosinophils, inhibits neuronal M_2 muscarinic receptor function, thereby increasing acetylcholine release and subsequently leading to increased bronchoconstriction and airway hyperreactivity after ozone exposure [43,44]. Depletion of eosinophils with an antibody to IL-5, or blocking major basic protein with heparin, prevents M_2 receptor dysfunction and ozone-induced airway hyperreactivity one day post-ozone exposure [11]. Thus, although neutrophils are the cells that increase in the bronchoalveolar lavage after ozone, it is tissue eosinophils around airway nerves that mediate ozone-induced hyperreactivity. In eosinophils, eotaxin and IL-5 signal through both

ERK and p38 MAPK activation [45,46]. Inhibition of p38 reduces eosinophil degranulation as measured by decreased eosinophil cationic protein release [45]. Major basic protein has also been shown to alter smooth muscle contractility [47]. Thus, while not tested directly in this study, blocking eosinophil degranulation with MAPK inhibitors could also contribute to preventing smooth muscle hyperreactivity.

Thus, p38 and JNK MAPK inhibitors inhibit ozone-induced hyperreactivity by multiple mechanisms. Exposure to high levels of environmental ozone increases hospitalizations from asthma exacerbations. Over 4 million children and 10 million adults with asthma live in counties with unhealthy levels of ozone, and those with asthma are an especially susceptible population to the adverse health effects of ozone [1]. Our data show that treatment with p38 and JNK inhibitors, immediately prior to ozone exposure prevented subsequent development of airway hyperreactivity. Currently there is no specific therapy for ozone related asthma complications and our data suggest both p38 and JNK are potential targets for additional therapeutic candidates; and that inhibitors could be tested as prophylactic treatment for asthma exacerbations on days with anticipated high ozone.

Author Contributions

Conceived and designed the experiments: KCV FGS MWL ADF DBJ. Performed the experiments: KCV. Analyzed the data: KCV. Contributed reagents/materials/analysis tools: FGS MWL. Wrote the paper: KCV ADF DBJ MWL. Designed and characterized the molecules used in this study: FGS MWL.

References

1. Association AL (2009) State of the Air Report.
2. Hiltermann JT, Lapperre TS, van Bree L, Steerenberg PA, Brahim JJ, et al. (1999) Ozone-induced inflammation assessed in sputum and bronchial lavage fluid from asthmatics: a new noninvasive tool in epidemiologic studies on air pollution and asthma. Free Radic Biol Med 27: 1448–1454.
3. Bell ML, McDermott A, Zeger SL, Samet JM, Dominici F (2004) Ozone and short-term mortality in 95 US urban communities, 1987–2000. Jama 292: 2372–2378.
4. Lewis TC, Robins TG, Dvonch JT, Keeler GJ, Yip FY, et al. (2005) Air pollution-associated changes in lung function among asthmatic children in Detroit. Environ Health Perspect 113: 1068–1075.
5. Kreit JW, Gross KB, Moore TB, Lorenzen TJ, D'Arcy J, et al. (1989) Ozone-induced changes in pulmonary function and bronchial responsiveness in asthmatics. J Appl Physiol 66: 217–222.
6. Wagner EM, Jacoby DB (1999) Methacholine causes reflex bronchoconstriction. J Appl Physiol 86: 294–297.
7. Foster WM, Brown RH, Macri K, Mitchell CS (2000) Bronchial reactivity of healthy subjects: 18–20 h postexposure to ozone. J Appl Physiol 89: 1804–1810.
8. Lee LY, Bleecker ER, Nadel JA (1977) Effect of ozone on bronchomotor response to inhaled histamine aerosol in dogs. J Appl Physiol 43: 626–631.
9. Mitchell HW, Adcock J (1988) Vagal mechanisms and the effect of indomethacin on bronchoconstrictor stimuli in the guinea-pig. Br J Pharmacol 94: 522–527.
10. Schultheis AH, Bassett DJ, Fryer AD (1994) Ozone-induced airway hyperresponsiveness and loss of neuronal M2 muscarinic receptor function. J Appl Physiol 76: 1088–1097.
11. Yost BL, Gleich GJ, Fryer AD (1999) Ozone-induced hyperresponsiveness and blockade of M2 muscarinic receptors by eosinophil major basic protein. J Appl Physiol 87: 1272–1278.
12. Pryor WA (1992) How far does ozone penetrate into the pulmonary air/tissue boundary before it reacts? Free Radic Biol Med 12: 83–88.
13. Hamilton RF Jr, Hazbun ME, Jumper CA, Eschenbacher WL, Holian A (1996) 4-Hydroxynonenal mimics ozone-induced modulation of macrophage function ex vivo. Am J Respir Cell Mol Biol 15: 275–282.
14. Kirichenko A, Li L, Morandi MT, Holian A (1996) 4-hydroxy-2-nonenal-protein adducts and apoptosis in murine lung cells after acute ozone exposure. Toxicol Appl Pharmacol 141: 416–424.
15. Kumagai T, Nakamura Y, Osawa T, Uchida K (2002) Role of p38 mitogen-activated protein kinase in the 4-hydroxy-2-nonenal-induced cyclooxygenase-2 expression. Arch Biochem Biophys 397: 240–245.
16. Cui CH, Adachi T, Oyamada H, Kamada Y, Kuwasaki T, et al. (2002) The role of mitogen-activated protein kinases in eotaxin-induced cytokine production from bronchial epithelial cells. Am J Respir Cell Mol Biol 27: 329–335.
17. Atherton HC, Jones G, Danahay H (2003) IL-13-induced changes in the goblet cell density of human bronchial epithelial cell cultures: MAP kinase and phosphatidylinositol 3-kinase regulation. Am J Physiol Lung Cell Mol Physiol 285: L730–739.
18. Nath P, Leung SY, Williams A, Noble A, Chakravarty SD, et al. (2006) Importance of p38 mitogen-activated protein kinase pathway in allergic airway remodelling and bronchial hyperresponsiveness. Eur J Pharmacol 544: 160–167.
19. Denhardt DT (1996) Signal-transducing protein phosphorylation cascades mediated by Ras/Rho proteins in the mammalian cell: the potential for multiplex signalling. Biochem J 318 (Pt 3): 729–747.
20. Kyriakis JM, Avruch J (1996) Protein kinase cascades activated by stress and inflammatory cytokines. Bioessays 18: 567–577.
21. Kalesnikoff J, Huber M, Lam V, Damen JE, Zhang J, et al. (2001) Monomeric IgE stimulates signaling pathways in mast cells that lead to cytokine production and cell survival. Immunity 14: 801–811.
22. Peng Q, Matsuda T, Hirst SJ (2004) Signaling pathways regulating interleukin-13-stimulated chemokine release from airway smooth muscle. Am J Respir Crit Care Med 169: 596–603.
23. Liu W, Liang Q, Balzar S, Wenzel S, Gorska M, et al. (2008) Cell-specific activation profile of extracellular signal-regulated kinase 1/2, Jun N-terminal kinase, and p38 mitogen-activated protein kinases in asthmatic airways. J Allergy Clin Immunol 121: 893–902 e892.
24. Duan W, Chan JH, McKay K, Crosby JR, Choo HH, et al. (2005) Inhaled p38alpha mitogen-activated protein kinase antisense oligonucleotide attenuates asthma in mice. Am J Respir Crit Care Med 171: 571–578.
25. Underwood DC, Osborn RR, Kotzer CJ, Adams JL, Lee JC, et al. (2000) SB 239063, a potent p38 MAP kinase inhibitor, reduces inflammatory cytokine production, airways eosinophil infiltration, and persistence. J Pharmacol Exp Ther 293: 281–288.
26. Fujisawa T, Ide K, Holtzman MJ, Suda T, Suzuki K, et al. (2008) Involvement of the p38 MAPK pathway in IL-13-induced mucous cell metaplasia in mouse tracheal epithelial cells. Respirology 13: 191–202.
27. Williams AS, Issa R, Leung SY, Nath P, Ferguson GD, et al. (2007) Attenuation of ozone-induced airway inflammation and hyper-responsiveness by c-Jun NH2 terminal kinase inhibitor SP600125. J Pharmacol Exp Ther 322: 351–359.
28. Williams AS, Issa R, Durham A, Leung SY, Kapoun A, et al. (2008) Role of p38 mitogen-activated protein kinase in ozone-induced airway hyperresponsiveness and inflammation. Eur J Pharmacol 600: 117–122.
29. Cho HY, Morgan DL, Bauer AK, Kleeberger SR (2007) Signal transduction pathways of tumor necrosis factor–mediated lung injury induced by ozone in mice. Am J Respir Crit Care Med 175: 829–839.
30. Fox T, Coll JT, Xie X, Ford PJ, Germann UA, et al. (1998) A single amino acid substitution makes ERK2 susceptible to pyridinyl imidazole inhibitors of p38 MAP kinase. Protein Sci 7: 2249–2255.
31. Green C (1982) Animal Anesthesia. London: Elsevier.
32. Fryer AD, Stein LH, Nie Z, Curtis DE, Evans CM, et al. (2006) Neuronal eotaxin and the effects of CCR3 antagonist on airway hyperreactivity and M2 receptor dysfunction. J Clin Invest 116: 228–236.
33. Fryer AD, Maclagan J (1984) Muscarinic inhibitory receptors in pulmonary parasympathetic nerves in the guinea-pig. Br J Pharmacol 83: 973–978.
34. Fryer AD, Wills-Karp M (1991) Dysfunction of M2-muscarinic receptors in pulmonary parasympathetic nerves after antigen challenge. J Appl Physiol 71: 2255–2261.
35. Yost BL, Gleich GJ, Jacoby DB, Fryer AD (2005) The changing role of eosinophils in long-term hyperreactivity following a single ozone exposure. Am J Physiol Lung Cell Mol Physiol 289: L627–635.
36. Verhein KC, Jacoby DB, Fryer AD (2008) IL-1 receptors mediate persistent, but not acute, airway hyperreactivity to ozone in guinea pigs. Am J Respir Cell Mol Biol 39: 730–738.
37. Hirai S, Kawaguchi A, Suenaga J, Ono M, Cui DF, et al. (2005) Expression of MUK/DLK/ZPK, an activator of the JNK pathway, in the nervous systems of the developing mouse embryo. Gene Expr Patterns 5: 517–523.
38. Agthong S, Koonam J, Kaewsema A, Chentanez V (2009) Inhibition of MAPK ERK impairs axonal regeneration without an effect on neuronal loss after nerve injury. Neurol Res.
39. Guan Z, Kim JH, Lomvardas S, Holick K, Xu S, et al. (2003) p38 MAP kinase mediates both short-term and long-term synaptic depression in aplysia. J Neurosci 23: 7317–7325.
40. Etter PD, Narayanan R, Navratilova Z, Patel C, Bohmann D, et al. (2005) Synaptic and genomic responses to JNK and AP-1 signaling in Drosophila neurons. BMC Neurosci 6: 39.
41. Li Y, Liu L, Barger SW, Griffin WS (2003) Interleukin-1 mediates pathological effects of microglia on tau phosphorylation and on synaptophysin synthesis in cortical neurons through a p38-MAPK pathway. J Neurosci 23: 1605–1611.
42. Rittner HL, Labuz D, Richter JF, Brack A, Schafer M, et al. (2007) CXCR1/2 ligands induce p38 MAPK-dependent translocation and release of opioid peptides from primary granules in vitro and in vivo. Brain Behav Immun 21: 1021–1032.
43. Fryer AD, Jacoby DB (1992) Function of pulmonary M2 muscarinic receptors in antigen-challenged guinea pigs is restored by heparin and poly-L-glutamate. J Clin Invest 90: 2292–2298.
44. Evans CM, Fryer AD, Jacoby DB, Gleich GJ, Costello RW (1997) Pretreatment with antibody to eosinophil major basic protein prevents hyperresponsiveness by protecting neuronal M2 muscarinic receptors in antigen-challenged guinea pigs. J Clin Invest 100: 2254–2262.
45. Kampen GT, Stafford S, Adachi T, Jinquan T, Quan S, et al. (2000) Eotaxin induces degranulation and chemotaxis of eosinophils through the activation of ERK2 and p38 mitogen-activated protein kinases. Blood 95: 1911–1917.
46. Adachi T, Choudhury BK, Stafford S, Sur S, Alam R (2000) The differential role of extracellular signal-regulated kinases and p38 mitogen-activated protein kinase in eosinophil functions. J Immunol 165: 2198–2204.
47. White SR, Ohno S, Munoz NM, Gleich GJ, Abrahams C, et al. (1990) Epithelium-dependent contraction of airway smooth muscle caused by eosinophil MBP. Am J Physiol 259: L294–303.

3

The Association of Ambient Air Pollution and Physical Inactivity in the United States

Jennifer D. Roberts[1]*, Jameson D. Voss[1,2], Brandon Knight[1]

1 Department of Preventive Medicine and Biometrics, F. Edward Hebert School of Medicine, Uniformed Services University, Bethesda, Maryland, United States of America, **2** Epidemiology Consult Service, United States Air Force School of Aerospace Medicine, Wright-Patterson Air Force Base, Ohio, United States of America

Abstract

Background: Physical inactivity, ambient air pollution and obesity are modifiable risk factors for non-communicable diseases, with the first accounting for 10% of premature deaths worldwide. Although community level interventions may target each simultaneously, research on the relationship between these risk factors is lacking.

Objectives: After comparing spatial interpolation methods to determine the best predictor for particulate matter ($PM_{2.5}$; PM_{10}) and ozone (O_3) exposures throughout the U.S., we evaluated the cross-sectional association of ambient air pollution with leisure-time physical inactivity among adults.

Methods: In this cross-sectional study, we assessed leisure-time physical inactivity using individual self-reported survey data from the Centers for Disease Control and Prevention's 2011 Behavioral Risk Factor Surveillance System. These data were combined with county-level U.S. Environmental Protection Agency air pollution exposure estimates using two interpolation methods (Inverse Distance Weighting and Empirical Bayesian Kriging). Finally, we evaluated whether those exposed to higher levels of air pollution were less active by performing logistic regression, adjusting for demographic and behavioral risk factors, and after stratifying by body weight category.

Results: With Empirical Bayesian Kriging air pollution values, we estimated a statistically significant 16–35% relative increase in the odds of leisure-time physical inactivity per exposure class increase of $PM_{2.5}$ in the fully adjusted model across the normal weight respondents (*p*-value<0.0001). Evidence suggested a relationship between the increasing dose of $PM_{2.5}$ exposure and the increasing odds of physical inactivity.

Conclusions: In a nationally representative, cross-sectional sample, increased community level air pollution is associated with reduced leisure-time physical activity particularly among the normal weight. Although our design precludes a causal inference, these results provide additional evidence that air pollution should be investigated as an environmental determinant of inactivity.

Editor: Jonatan R. Ruiz, University of Granada, Spain

Funding: Funding for this study was provided by a Uniformed Services University of the Health Sciences intramural start-up grant for newly appointed faculty. The funders had no role in study design, data collection and analysis, decision to publish, or preparation of the manuscript.

Competing Interests: The authors have declared that no competing interests exist.

* E-mail: jennifer.roberts@usuhs.edu

Introduction

Worldwide, physical inactivity accounts for more than three million annual deaths and 6–10% of major non-communicable diseases, such as coronary heart disease, type-II diabetes and breast and colorectal cancers [1–5]. Similarly, physical inactivity is strongly associated with obesity and a portion of physical inactivity related mortality is attributed to obesity [6–9]. In the U.S., two-thirds of adults are overweight or obese and approximately six percent are extremely obese, which is a body mass index greater than or equal to 40.0 kg/m² [10,11]. While a majority of Americans are overweight or obese, sub-populations are disproportionately impacted. For instance, there are racial, ethnic, geographic and economic disparities in the obesity prevalence throughout the U.S. [12,13]. Research into how the built environment may impact these disparities has shown conflicting

results. [14,15]. One explanation is that individual determinants interact with one another in a dynamic system, which suggests future research needs to account for the way factors interrelate with one another in the real world by using an ecological perspective.

Granted, modifiable lifestyle factors such as the increased consumption of unhealthy foods and physical inactivity are important independent contributors to the increasing burden of non-communicable disease. Other insidious factors, however, such as poor air quality, may influence physical inactivity, but current research has not adequately established this role. While not yet considered an environmental determinant of inactivity, there is little confusion about the unfavorable effects of acute and chronic air pollution exposure, particularly from particulate matter (PM) and ozone (O_3), on both the respiratory and cardiovascular systems [16–19]. While some harms likely remain uncharacterized, research has shown that exposure to $PM_{2.5}$ (particulate matter <2.5 μm in aerodynamic diameter), PM_{10} (particulate matter <10 μm in aerodynamic diameter) and O_3 is associated with reduced exercise capacity, higher resting blood pressure, lower ventilator function and decrements in exercise performance [20–23]. Although there is abundant research illustrating these effects in a resting, inactive state, among athletes or normal weight subjects, the data examining the effects of poor air quality in real world settings are meager. Thus, the generalizability of these findings is in question particularly when over 60% of the U.S. population is overweight.

Another important gap is the difficulty in determining the geographic pattern of air pollution exposure. Although the U.S. Environmental Protection Agency (U.S. EPA) monitors and reports air pollution levels throughout the U.S., it is challenging to appreciate how these readings translate to air pollution exposures across standard geographic units, such as U.S. counties.

Thus, the overall aim of this study was to assess the association between ambient air pollution and leisure-time physical activity. Additionally, this association was examined after stratifying by body weight category.

Materials and Methods

In this cross-sectional study, our two sources of data consisted of (1) annual summary measurements of 2011 ambient air quality monitoring data from the U.S. EPA Air Quality System (AQS) Data Mart and (2) Behavioral Risk Factor Surveillance System (BRFSS) survey data collected in 2011 throughout the U.S. that provided self-reported levels of leisure-time physical activity, demographic information and residential location. Using these data sources, we compared spatial interpolation methods to determine the best predictor for county-level $PM_{2.5}$, PM_{10} and O_3 exposures throughout the U.S and we evaluated the possible cross-sectional relationship of ambient air pollution with physical inactivity in the full study population and after stratifying by body weight.

Physical Inactivity

BRFSS, a state-based telephone health survey system, collects data on behavioral and other health risk factors. As a whole, BRFSS, uses a methodology to collect a representative sample of the U.S. non-institutionalized adult population. As guided by the Centers for Disease Control and Prevention (CDC), data are collected from all 50 States, District of Columbia, Puerto Rico, U.S. Virgin Islands, and Guam. Although the CDC and other researchers have described the complex survey design in great detail, it should be noted that for many states, BRFSS is implemented through the use of disproportionate stratified random sampling (DSS) [24–26]. In order to account for the relatively recent rise in the proportion of U.S. households without landline telephones, adjustments were actualized during the fielding of the recent 2011 BRFSS to include households that rely on cellular telephones. Additionally, a more sophisticated weighting methodology known as "raking" was implemented. Raking, in contrast to the previously used poststratification method, forms individual variable adjustments in a series of data processing iterations, and thus reduces the risk of potential bias [27,28].

Using BRFSS 2011 data, we assessed leisure-time physical inactivity through responses to the question, "During the past month, other than your regular job, did you participate in any physical activities or exercises such as running, calisthenics, golf, or walking for exercise?" The responses to this question were either

Table 1. Annual means of $PM_{2.5}$, PM_{10} and O_3 Empirical Bayesian Kriging (EBK) interpolated ambient air pollution concentrations by natural breaks classes.

Air Pollution Classes	$PM_{2.5}$ (ug/m³)	PM_{10} (ug/m³)	O_3 (ppb)
Class 1	3.49–6.52	5.00–13.40	26.93–37.83
Class 2	6.53–8.45	13.41–17.59	37.84–42.40
Class 3	8.46–9.85	17.60–21.27	42.41–45.83
Class 4	9.86–10.89	21.28–26.31	45.84–49.89
Class 5	10.90–15.38	26.32–52.88	49.90–56.94

"yes", "no", "don't know/not sure", and "refused". A response of "no" was defined as leisure-time physical inactivity.

In this study, inclusion criteria for the BRFSS data were as follows: (1) geographically located within the contiguous U.S including the District of Columbia; (2) responses from respondents who either were categorized as normal weight, overweight or obese; and (3) respondents from counties with both county and state Federal Information Processing Standard (FIPS) codes. Missing, "refused" and "don't know" responses were also excluded from the analysis.

Air Pollution Exposure

The air quality data collected by the U.S. EPA AQS contains air monitoring measurements for criteria air pollutants from 1957 to present [29]. The database contains several million observations from thousands of monitors throughout the U.S. [29]. In addition to descriptive and geographic information about the monitoring sites, quality assurance information is also available.

For $PM_{2.5}$ and PM_{10}, annual summaries for 2011 were obtained using the standard 1-hour or 24-hour collection periods. Due to the strong seasonal and diurnal patterns that exist for ground level O_3, U.S. EPA requires that monitoring locations collect only during specified months of the year as determined by their geographic location. Thus, the National Ambient Air Quality Standard (NAAQS) for ozone is based on an 8-hour averaging time. For inclusion in this study, the 2011 annual O_3 summaries calculated using daily maximum 8-hour averages over the effective monitoring season were selected.

To confirm geographical accuracy, the monitoring data were mapped along with 2011 U.S. Census counties. County FIPS codes from the monitoring data were compared to county FIPS from the enclosing Census county. Of the 3945 records, four monitoring locations were found not to have a FIPS match. The spatial locations of these four were examined visually and the discrepant cases were located less than 500 meters from the county border, suggesting the discrepancy was due to error introduced during the import and processing of the air pollution data. Thus, no records were excluded from further analysis based on locational accuracy.

In order to ensure the accuracy and reliability of the air pollution concentrations, U.S. EPA inclusion criteria were applied. The inclusion criteria for the AQS data were as follows: (1) for $PM_{2.5}$ and PM_{10}: availability of greater than 75% of observations was required; (2) for O_3: availability of greater than 75% of valid days in the effective monitoring season was required [29].

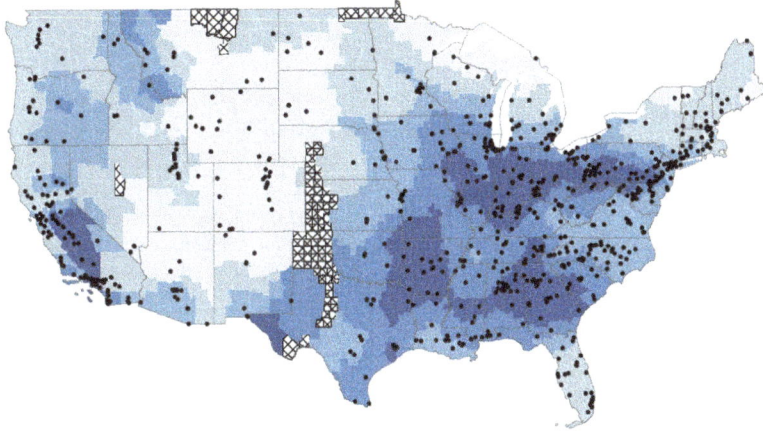

PM_{2.5}

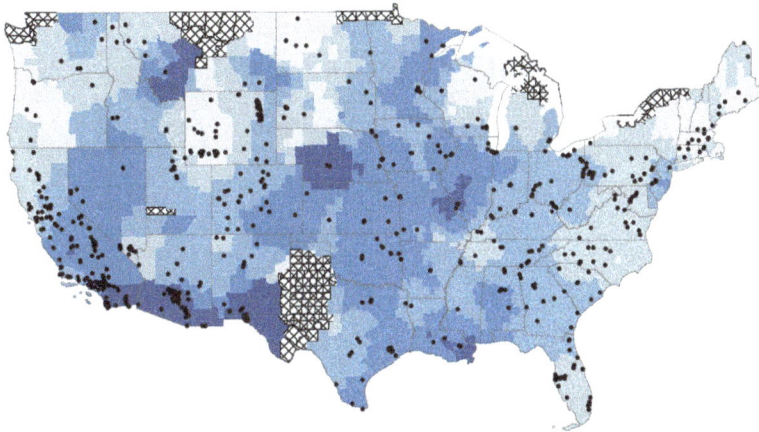

PM₁₀

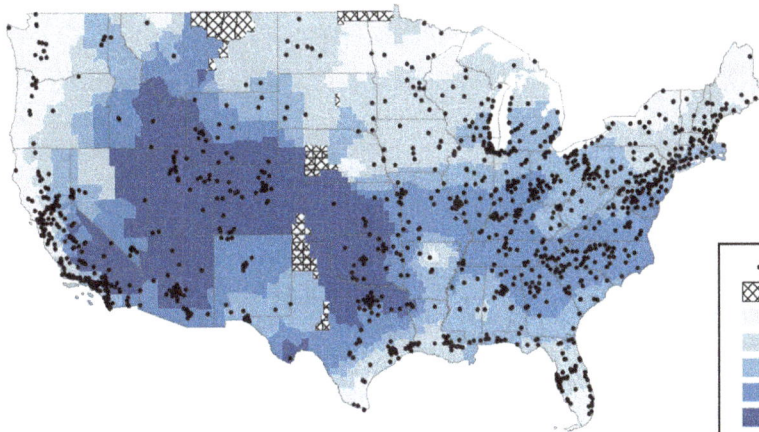

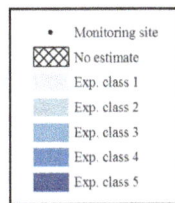

O₃

Figure 1. U.S. map of annual mean Empirical Bayesian Kriging (EBK) interpolated ambient air pollution concentrations by natural breaks classes.

Air Pollution Modeling

Since BRFSS data are available at the U.S. county level, the air pollution data provided from the discrete monitoring stations were modeled to estimate county-level average exposures. Studies examining the relationship between air pollution and health outcomes have implemented a variety of techniques to estimate pollution from U.S. EPA monitoring data, including various interpolation and spatiotemporal regression models [30–32]. With the use of ArcGIS 10.1, Inverse Distance Weighting (IDW) and Empirical Bayesian Kriging (EBK) were employed to perform spatial interpolation, creating continuous surfaces for the three air pollution parameters, which were then compared in order to select the best method for inclusion in subsequent analysis. A search window of 250 km was selected in order to ensure that air pollution estimates were generated for a substantial percentage of U.S. counties in the study area, while also maximizing the prediction precision.

The first method, IDW, is a deterministic method that imposes a model of spatial autocorrelation and calculates interpolation weights for each known point (in this case, each monitoring

Table 2. Prevalence of demographic factors by air pollution exposure class.

Demographic Factors	Prevalence (%)[a]					
	$PM_{2.5}$[b]		PM_{10}[b]		O_3[b]	
	Class 1 (n = 32,438)	Class 5 (n = 57,600)	Class 1 (n = 42,539)	Class 5 (n = 16,081)	Class 1 (n = 31,333)	Class 5 (n = 45,901)
Age*						
18–24 years	9.07	10.71	10.07	11.07	9.77	11.29
25–34 years	17.04	17.55	15.06	18.46	16.62	18.92
35–44 years	19.05	19.13	17.13	19.31	19.13	19.54
45–54 years	20.91	20.36	21.80	19.20	20.47	19.92
55–64 years	17.48	15.87	17.79	15.20	16.52	15.04
≥65 years	16.45	16.38	18.16	16.76	17.49	15.29
Sex						
Male	51.71	51.15	51.84	50.12	51.28	51.32
Female	48.29	48.85	48.16	49.88	48.72	48.68
Race/Ethnicity*						
White/non-Hispanic	75.81	64.28	85.48	59.33	62.54	65.18
Black/non-Hispanic	2.93	13.15	3.68	9.56	5.41	7.24
Hispanic	13.62	15.90	5.38	24.29	19.92	20.64
Asian/Pacific Islander	2.52	4.29	2.44	4.25	8.80	3.20
American Indian	3.56	0.87	1.12	0.98	1.19	2.18
Multi/Other	1.56	1.52	1.90	1.60	2.14	1.56
Educational Level*						
Not graduated from high school	10.49	15.19	9.19	15.50	12.02	14.58
High school graduate	27.25	29.59	28.06	25.45	23.85	25.75
Attended college	33.57	29.64	31.66	33.72	32.81	34.26
College graduate or higher	28.70	25.58	31.09	25.33	31.32	25.42
Annual Income Level*						
<$25,000	28.27	32.44	25.21	31.80	28.63	29.91
≥$25,000 to <$50,000	26.84	25.14	25.24	25.79	24.47	26.28
≥$50,000	44.89	42.42	49.55	42.41	46.90	43.81
Marital Status*						
Married/partnered	61.09	57.44	59.35	56.30	58.53	60.55
Divorced/widowed/separated	20.15	19.62	19.24	20.86	19.52	19.27
Never married	18.76	22.93	21.41	22.84	21.95	20.18

[a]Proportions based on frequency-weighted final weight variable rounded to nearest integer.
[b]Air pollution based on five natural breaks air pollution classes (Classes 2-4 not shown).
*p-value <0.0001 Chi-square test of homogeneity – All three pollutants.

Table 3. Prevalence of risk or geographic factors by air pollution exposure class.

| Risk/Geographic Factors | Prevalence (%)[a] | | | | | |
| | $PM_{2.5}$[b] | | PM_{10}[b] | | O_3[b] | |
	Class 1 (n = 32,438)	Class 5 (n = 57,600)	Class 1 (n = 42,539)	Class 5 (n = 16,081)	Class 1 (n = 31,333)	Class 5 (n = 45,901)
Smoking*						
Current and former smoker	48.24	44.80	48.27	44.08	43.39	42.92
Never smoked	51.76	55.20	51.73	55.92	56.61	57.08
Body Mass Index[§]						
Normal weight	37.13	33.41	35.63	34.65	36.51	35.37
Overweight	37.58	36.93	37.72	37.27	37.41	36.83
Obese	25.29	29.65	26.65	28.08	26.08	27.81
Disability*						
No	76.32	77.11	75.24	76.57	74.98	77.00
Yes	23.68	22.89	24.76	23.43	25.02	23.00
General Health Status*						
Excellent/good	84.69	81.50	85.38	81.94	82.93	83.52
Fair/poor	15.31	18.50	14.62	18.06	17.07	16.48
Asthma Currently[¶]						
No	91.09	91.72	89.88	91.80	91.03	91.51
Yes	8.91	8.28	10.12	8.20	8.97	8.49
Seasonality*						
Quarter 1 (January to March)	17.13	23.89	27.70	25.40	27.27	20.62
Quarter 2 (April to June)	27.36	23.40	26.88	27.16	25.48	29.27
Quarter 3 (July to September)	30.13	26.86	23.71	21.94	23.80	24.91
Quarter 4 (October to December)	25.39	25.86	21.71	25.49	23.45	25.20
U.S. Geographic Region*						
Northeast	6.79	10.89	68.89	0.00	7.82	0.00
Southeast	0.00	20.73	0.58	1.26	19.51	0.00
Midwest	11.04	33.58	5.59	19.47	4.02	4.04
Southwest	56.50	12.90	1.61	16.26	5.58	59.12
West	25.67	21.89	23.53	62.74	63.07	36.84
Metropolitan County Classification*						
Rural counties	29.84	14.03	21.43	5.37	14.17	12.67
Counties with <250,0000	21.87	9.14	10.12	7.61	8.04	8.49
Counties with 250,000–1 Million	14.86	21.19	23.40	12.59	20.00	20.94
Counties with ≥1 Million	33.43	55.63	45.06	74.44	57.79	57.90

[a]Proportions based on frequency-weighted final weight variable rounded to nearest integer.
[b]Air pollution based on five natural breaks air pollution classes (Classes 2–4 not shown).
*p-value <0.0001 Chi-square test of homogeneity – All three pollutants.
[§]p-value <0.0001 Chi-square test of homogeneity – Only $PM_{2.5}$ and O_3.
[¶]p-value <0.0001 Chi-square test of homogeneity – Only PM_{10} and O_3.

station) as a function of distance between the known points and the predicted points within a specific search window. There is an inverse relationship between the interpolation weights and the distance from the interpolated points to each known point. Hence, the values that are closer to the prediction location have more weight or influence on the predicted values than those farther away. We chose to calculate weights that change linearly in order to create a smooth surface. The second method, EBK, is an implementation of the kriging class of geostatistical methods that allow the development of a statistical autocorrelation model using a sample data set. Common kriging methods such as ordinary,

simple, and universal kriging, require selection of model parameters from an empirical variogram. The empirical variogram is used to calculate interpolating weights such that the mean square error is minimized. EBK automates the parameter selection process through simulation and subsetting and generates accurate results from moderately non-stationary data, indicating that the mean and variance do not differ with geographical position [33,34].

A one square kilometer grid was overlaid on a map of the counties that comprise the contiguous U.S. Each interpolation method produced air pollution estimates at the center of each

Table 4. Association between Empirical Bayesian Kriging (EBK) interpolated ambient $PM_{2.5}$ exposure class and physical inactivity by body weight subset, logistic regression model.

$PM_{2.5}$ Exposure Class	Obese Weight (n=99,699)			Overweight (n=127,720)			Normal Weight (n=116,927)		
	A[a]	B[b]	C[c]	A	B	C	A	B	C
	OR (95% CI)			OR (95% CI)			OR (95% CI)		
Class 1	Referent	Referent	Referent	Referent	Referent	Referent	Referent	Referent	Referent
Class 2	1.19 (1.06, 1.33)	1.11 (0.98, 1.25)	1.14 (0.99, 1.31)	1.14 (1.02, 1.28)	1.10 (0.97, 1.29)	1.12 (0.98, 1.29)	1.26 (1.11, 1.42)	1.16 (1.02, 1.32)	1.16 (1.01, 1.34)
Class 3	1.21 (1.09, 1.35)	1.14 (1.02, 1.29)	1.17 (1.02, 1.34)	1.14 (1.02, 1.27)	1.13 (0.99, 1.28)	1.13 (0.98, 1.30)	1.32 (1.18, 1.47)	1.24 (1.09, 1.40)	1.29 (1.11, 1.49)
Class 4	1.33 (1.20, 1.49)	1.15 (1.02, 1.30)	1.17 (1.02, 1.35)	1.40 (1.25, 1.56)	1.24 (1.09, 1.40)	1.24 (1.07, 1.43)	1.59 (1.42, 1.78)	1.30 (1.14, 1.48)	1.35 (1.16, 1.58)
Class 5	1.35 (1.21, 1.50)	1.21 (1.07, 1.36)	1.23 (1.07, 1.42)	1.22 (1.09, 1.37)	1.13 (0.99, 1.31)	1.13 (0.98, 1.31)	1.49 (1.33, 1.67)	1.27 (1.12, 1.44)	1.35 (1.16, 1.58)

Logistic Regression Models*

[a]Model A unadjusted.
[b]Model B adjusted for age, sex, race/ethnicity, education, annual income, marital status, seasonality and geographic region.
[c]Model C adjusted for age, sex, race/ethnicity, education, annual income, marital status, seasonality, geographic region, general health status, smoking, disability, asthma, urbanization, and the other two air pollutants.
*Wald Chi-square p-value <0.0001 for all three models.

square of the grid. County exposure estimates were calculated by averaging the interpolated values spatially located within the county borders. U.S. counties that did not have a single interpolated estimate within its borders were excluded from further analysis.

Cross-validation was completed for each interpolation method and for each pollutant to test the generalization performance and provide a quantitative comparison of the IDW and EBK methods. This was performed by omitting values for a single monitor and then using the remaining monitors to interpolate the concentration at the removed monitor's location. To identify and select the most appropriate interpolation method to use for further analysis, the cross-validations were compared in terms of prediction root mean square error (RMSE) and prediction mean absolute error (MAE).

Upon this selection, the annual mean of $PM_{2.5}$, PM_{10} and O_3 concentrations were transformed from continuous variables to categorical variables using natural breaks classification method with Jenks optimization. Natural breaks are data-specific classes, which are based on natural groupings inherent in the data and where boundaries are set based on relatively large differences in the data values by reducing the within and maximizing the between class variance. Finally, we linked annual average concentrations of $PM_{2.5}$, PM_{10} and O_3 with the 2011 BRFSS data using FIPS codes as the linking unit.

Statistical Analysis

BRFSS uses a complex survey design with stratification, multistage clustering and sampling weights. Therefore, statistical analysis was performed in STATA MP/12.1 using the -svyset- commands. The weighted prevalence of leisure-time physical inactivity was calculated by physical demographic and risk factor categories. Additionally, the weighted prevalence of demographic and other risk factor variables were calculated by air pollution exposure class. Chi-square tests for homogeneity were performed to investigate the association of the demographic and behavioral risk variables with physical inactivity and air pollution exposure.

We considered whether adults who were exposed to higher levels of $PM_{2.5}$, PM_{10} and O_3 concentrations exhibited higher levels of physical inactivity by performing logistic regression. Additionally, we examined this association after stratifying the data into three subgroups [e.g. (1) normal weight (body mass index (BMI) 18.5 to 24.9); (2) overweight (BMI 25.0 to 29.9), and (3) obese (BMI 30 or higher)] as defined and categorized by the BRFSS data. The air pollution variables were analyzed both as continuous and natural breaks categorical variables using three models for each pollutant. Model A examined the effect on physical inactivity with ambient exposure to $PM_{2.5}$, PM_{10} and O_3 without the adjustment of any confounders. Model B adjusted for age, sex, race/ethnicity, education, annual income, marital status, seasonality and geographic region. Along with the aforementioned confounders, Model C, also adjusted for general health status, smoking, disability, asthma, urbanization and the other two air pollutants. Additionally, we calculated Pearson's correlation coefficients to examine relations between the three pollutant measures.

Ethics Statement

The Uniformed Services University of the Health Sciences, Human Research Protections Program Office, determined that this research was non-human subjects research consistent with 32 CFR 219.102.

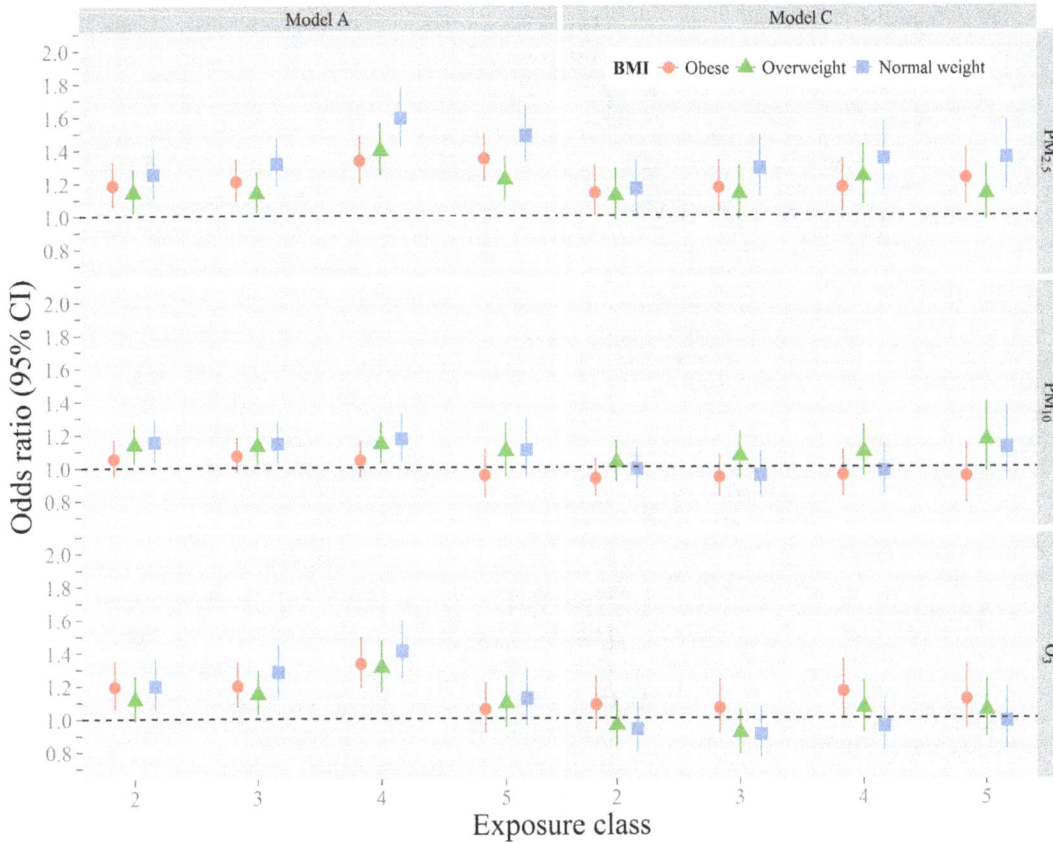

Figure 2. Association between air pollution exposure classes and the odds ratios of physical inactivity [†, ††, †††]. † Model A unadjusted. †† Model C adjusted for age, sex, race/ethnicity, education, annual income, marital status, seasonality, geographic region, general health status, smoking, disability, asthma, urbanization, and the other two air pollutants. ††† Exposure class 1 is referent.

Results

The 2011 BRFSS dataset encompassed 504,408 observations. Based on the inclusion and exclusion criteria, data from 48 states and the District of Columbia were included in the analysis (N = 329,628 subjects from 2249 U.S. counties). A total of 24.5% (n = 80,825) responded "no" to participating in any physical activity during the previous month. The prevalence of leisure-time physical inactivity was higher among older respondents and among females; however with respect to race and ethnicity, Black/non-Hispanic respondents demonstrated the highest prevalence of physical inactivity, which was closely followed by Hispanic respondents. The highest levels of reported physical inactivity were also observed among the respondents with lower levels of education, those who reported being divorced, widowed or separated, and, those receiving less than $50,000 in annual income. Respondents who reported obesity, disability, asthma, or prior history of smoking were also more likely to report physical inactivity. Physical inactivity was highest during the months of January through March and October through December. In addition, physical inactivity was highest in the Southeastern part of the U.S. while also being lowest in the West. There was also a much higher level of physical inactivity among respondents who resided in rural counties or non-metropolitan counties (Data not shown, p-value <0.001).

The two interpolation methods, IDW and EBK, were evaluated based on accuracy and precision for predicting annual ambient $PM_{2.5}$, PM_{10} and O_3 concentrations. Results showed that the IDW and EBK models were similar in regards to RMSE and MAE with EBK trending toward increased precision for all three air quality parameters. Thus, EBK was identified as the most appropriate method. Furthermore, kriging interpolation methods have been recognized by U.S. EPA as possessing the greatest merit in predicting air pollution concentrations in unknown locations [35].

The mean of $PM_{2.5}$, PM_{10} and O_3 annual mean concentrations were 9.50 ug/m^3 [SD: 1.80], 19.52 ug/m^3 [SD: 4.16], and 45 ppb [SD: 4.48] for the BRFSS counties, respectively. All of the pollutants were positively correlated with $PM_{2.5}$ and PM_{10} having the strongest correlation (r: 0.29) followed by O_3 and PM_{10} (r: 0.23). Furthermore, the air pollution variables were transformed into categorical variables using the Jenks' natural breaks methodology (Table 1). While the highest concentrations of $PM_{2.5}$ were found in the upper Atlantic, Midwest, and the South, along with a small cluster in Southern California, the higher concentration PM_{10} counties were clustered throughout the U.S., particularly in the Southwest (Figure 1). The highest two classes of O_3 were clustered in the middle and western part of the country (Figure 1). When comparing air pollution by natural breaks class, the most evident differences were observed within the race/ethnicity, U.S geographic region and metropolitan county classification categories among all of the air pollutants (Tables 2–3). Generally, White/non-Hispanic respondents and those living in rural counties or with a population of less than 250,000 were exposed to lower levels of $PM_{2.5}$ and PM_{10} (Tables 2–3).

Association of Physical Inactivity and Air Pollution Exposure by Body Weight

When considering $PM_{2.5}$ as a continuous variable, the odds of leisure-time physical inactivity significantly increased with increasing concentration of $PM_{2.5}$ across all models and strata. For the fully adjusted Model C, we estimated a 2.4% relative increase in the odds of physical inactivity per $\mu g/m^3$ increase of $PM_{2.5}$ exposure among the obese respondents [OR = 1.02 (95% CI: 1.00, 1.05)]. Similarly, increasing concentration of PM_{10} among the normal weight respondents was also associated with higher odds of inactivity [OR = 1.01 (95% CI: 1.00, 1.02)]. We also estimated increases in the odds of physical inactivity per unit increase of each air pollutant for the entire combined dataset and those results were also found to be statistically significant for $PM_{2.5}$ in all models (Data not shown, p-value <0.01).

Alternatively, the associations with inactivity were also modeled using natural breaks exposure classes for each of the three pollutants ($PM_{2.5}$, PM_{10} and O_3). There was a statistically significant 16–35% relative increase in the odds of physical inactivity per increase from the lowest $PM_{2.5}$ exposure class in the fully adjusted Model C among the normal weight respondents (Table 4), which was a stronger association than the other body weight strata. Figure 2 illustrates a relationship between the odds of physical inactivity and increasing dose as represented by exposure class. For O_3, statistical significance was observed only in the unadjusted Model A for all three weight groups (Figure 2). Results found using the full dataset were similar to that of the normal weight stratum (Data not shown).

Lastly, the relationship between air pollution and several other covariates was notable. For instance, odds ratios for physical inactivity increased strongly with increasing age and BMI classes (Table 5). By contrast, higher levels of education and income classes decreased the odds of physical inactivity. Furthermore, respondents in the Western part of the U.S had 35% higher odds of being active than those in the Northeast (Table 6). There was also a 32–33% higher odds of activity found among respondents during the warmer months of the year (Table 6). Finally, the odds of physical inactivity decreased with increasing urbanization (Table 6).

Discussion

In this nationally representative, cross-sectional sample, increased ambient levels of $PM_{2.5}$, PM_{10} and O_3 were associated with reduced physical activity. This association was significant in all models of adjustment for $PM_{2.5}$. Our research demonstrated an association between increasing ordinal air pollution classes and increasing odds of inactivity among adults. Remarkably, the most compelling relationships were evident among normal weight as opposed to overweight or obese respondents. These findings suggest that the presence of air pollution may discourage normal weight individuals from engaging in leisure-time physical activity.

Our research exhibited an inverse relationship between air pollution exposure and leisure-time physical activity among Americans. This is consistent with findings from similar studies that also examined the association of physical inactivity and air pollution and found a direct relationship with increasing $PM_{2.5}$ and the increasing prevalence of physical inactivity (p-value <0.01) [36]. This prior work, however, only examined crude levels of $PM_{2.5}$ and generalized findings across all weight categories. Another study modeled other air pollutants, including O_3 and nitrogen oxides (NO_x), and physical inactivity limited to Southern California, but did not consider air pollution as a determinant of inactivity [37].

Table 5. Association of demographic factors with physical inactivity.

Demographic Factors	OR (95% CI)
Age*	
18–24 years	Referent
25–34 years	1.51 (1.36, 1.67)
35–44 years	1.72 (1.55,1.91)
45–54 years	1.74 (1.57, 1.92)
55–64 years	1.82 (1.64, 2.01)
≥65 years	2.00 (1.80, 2.21)
Sex*	
Male	Referent
Female	1.15 (1.10, 1.19)
Race/Ethnicity*	
White/non-Hispanic	Referent
Black/non-Hispanic	1.19 (1.13, 1.27)
Hispanic	1.22 (1.14, 1.30)
Asian/Pacific Islander	1.80 (1.58.2.05)
American Indian	0.92 (0.79, 1.07)
Multi/Other	0.91 (0.80, 1.03)
Educational Level*	
Not graduated from high school	Referent
High school graduate	0.84 (0.79, 0.90)
Attended college	0.62 (0.58, 0.66)
College graduate or higher	0.42 (0.39, 0.45)
Annual Income Level*	
<$25,000	Referent
≥$25,000 to <$50,000	0.96 (0.91, 1.00)
≥$50,000	0.72 (0.69, 0.76)
Marital Status	
Married/partnered	Referent
Divorced/widowed/separated	1.02 (0.98, 1.06)
Never married	0.97 (0.92, 1.03)

Adjusted for all above variables in the table and all three air pollution variables.
*Joint adjusted Wald test p-value <0.0001.

We add to this literature by using more recent data, sophisticated interpolation methods increasing the coverage and modeling reliability, and clarifying the association across weight strata. Our research revealed a higher prevalence of physical inactivity among obese and overweight, as compared to normal weight, respondents. When stratifying by body weight category, the association between air pollution and leisure-time physical activity varied minimally by body weight category with the exception of $PM_{2.5}$ where the magnitude of association trended higher among the normal weight respondents. One reason for this finding may be due to the fact that normal or healthy weight adults already are more physically active than obese or overweight adults and therefore the reduction in activity is greater. For instance, obesity related disability may create a disparity in discretionary activity - a lean population is active when the conditions are favorable while those who are disabled are inactive regardless.

Although a causal mechanism cannot be elucidated by our study design, the association of physical inactivity and ambient air

Table 6. Association of risk or geographic factors with physical inactivity.

Risk/Geographic Factors	OR (95% CI)
Smoking	
Never smoked*	Referent
Current and former smoker	1.18 (1.14, 1.22)
Body Mass Index*	
Normal weight	Referent
Overweight	1.04 (1.00,1.09)
Obese	1.44 (1.38, 1.50)
Disability*	
No	Referent
Yes	1.42 (1.37, 1.48)
General Health Status*	
Excellent/good	Referent
Fair/poor	1.74 (1.66, 1.82)
Asthma Currently	
No	Referent
Yes	0.98 (0.92, 1.03)
Seasonality*	
Quarter 1 (January to March)	Referent
Quarter 2 (April to June)	0.68 (0.65, 0.72)
Quarter 3 (July to September)	0.67 (0.63, 0.70)
Quarter 4 (October to December)	0.84 (0.80, 0.88)
U.S. Geographic Region*	
Northeast	Referent
Southeast	1.01 (0.96, 1.06)
Midwest	0.90 (0.86, 0.95)
Southwest	0.89 (0.84, 0.96)
West	0.65 (0.61, 0.69)
Metropolitan County Classification*	
Rural counties	Referent
Counties with <250,0000	0.93 (0.89, 0.99)
Counties with 250,000–1 Million	0.90 (0.85, 0.94)
Counties with ≥1 Million	0.88 (0.84, 0.92)

Adjusted for all above variables in the table and all three air pollution variables.
*Joint adjusted Wald test p-value <0.0001.

pollution as mediated by body weight category is plausible. Positive associations have been found with exposure to air pollutants and direct health effects on the respiratory and cardiovascular systems, such as increased blood pressure, asthma exacerbations, cardiac arrhythmia, and decreased lung function [17,38–41]. Therefore, adverse health effects from increasing levels of air pollution could reduce one's capacity for physical activity. In addition to the physiological effects (e.g. difficulty breathing), the possibility of a psychosocial effect (e.g. smog appearance disincentivizing physical activity) could be contributing factor to this association. With readily available information, the U.S. population is likely more aware of the health risk associated with exposure to high levels of air pollution. Hence, this awareness may ultimately discourage individuals from engaging in outdoor physical activity.

While these are important findings, one study limitation was potential misclassification of exposure based on the air pollution modeling, interpolated estimates and annual means. Because we applied the same methods uniformly throughout the entire U.S. we suspect this biases to the null as non-differential measurement error, but we cannot rule out differential misclassification given the geographic variation in our exposure and outcome. Furthermore, the air pollution data used were at the county-level and did not provide information at an individual daily exposure level. Yet, since $PM_{2.5}$ and O_3 are often more homogeneous air pollutants in their distribution over large regions, we believe this misclassification was minimized.

Our study used self-reported leisure-time physical activity and self-reported data are often subject to certain biases. Since physical activity was assessed as a dichotomous variable over a month timeframe, the accuracy of responses may not have been compromised. However, the BRFSS physical activity question provided examples of exercise such as running and calisthenics. Physical activity includes not only the participation of sports and exercise, but also walking or biking on a daily basis by means of an active commute or transport. With the BRFSS examples, respondents may not have considered their less obvious physical activities, such as walking to work because the survey question asked respondents about participation in physical activities outside of their "regular job". Additionally, other built environmental factors, such as neighborhood walkability or safety, may have influenced one's level of physical activity.

The major strength of this study is the use of our novel EBK estimations for air pollution exposures. Unlike other kriging methods, EBK allows for automated model fitting and more accurate and robust predictions. Another strength of this study is the use of a nationally representative sample with extensive demographic, behavioral and risk factor data. The combination of BRFSS data with U.S. EPA air pollution data brought to light novel findings. Lastly, we were able to examine the relationships of three air pollutants and their influence with each other on physical inactivity.

The findings of this research emphasized the phenomenon that there is a complex interplay among many risk factors, behavioral and demographic variables, which are associated with physical activity. Thus, the complexity limits the applications of observational research as it can raise questions of causality and directionality. Because $PM_{2.5}$ is a modifiable exposure with cost effective mitigation strategies, future research could evaluate causality by cluster randomizing the timing of $PM_{2.5}$ reduction interventions and assessing the short-term impact on leisure-time physical inactivity [42,43].

Conclusions

We present evidence that as air pollution concentrations increase, American adults, especially those who are lean, are less likely to be physically active. Given the public health emphasis on community level determinants of inactivity, additional research should determine if environmental air pollution is a modifiable risk factor for inactivity. We postulate those interventions which improve physical activity and reduce air pollution such as transportation interventions will have both primary and secondary benefits.

Acknowledgments

The authors gratefully acknowledge the technical assistance of the U.S. Environmental Protection Agency, Air Quality System (AQS) Team and the biostatistical support provided by Dr. Cara Olsen of Uniformed Services University.

References

1. Pratt M, Sarmiento OL, Montes F, Ogilvie D, Marcus BH, et al. (2012) The implications of megatrends in information and communication technology and transportation for changes in global physical activity. Lancet 380: 282–293.
2. Lee IM, Shiroma EJ, Lobelo F, Puska P, Blair SN, et al. (2012) Effect of physical inactivity on major non-communicable diseases worldwide: an analysis of burden of disease and life expectancy. Lancet 380: 219–229.
3. Awatef M, Olfa G, Rim C, Asma K, Kacem M, et al. (2011) Physical activity reduces breast cancer risk: a case-control study in Tunisia. Cancer Epidemiol 35: 540–544.
4. Haggar FA, Boushey RP (2009) Colorectal cancer epidemiology: incidence, mortality, survival, and risk factors. Clin Colon Rectal Surg 22: 191–197.
5. Arsenault BJ, Rana JS, Lemieux I, Despres JP, Kastelein JJ, et al. (2010) Physical inactivity, abdominal obesity and risk of coronary heart disease in apparently healthy men and women. Int J Obes (Lond) 34: 340–347.
6. Pietilainen KH, Kaprio J, Borg P, Plasqui G, Yki-Jarvinen H, et al. (2008) Physical inactivity and obesity: a vicious circle. Obesity (Silver Spring) 16: 409–414.
7. Lee IM, Djousse L, Sesso HD, Wang L, Buring JE (2010) Physical activity and weight gain prevention. JAMA 303: 1173–1179.
8. Herring MP, Puetz TW, O'Connor PJ, Dishman RK (2012) Effect of exercise training on depressive symptoms among patients with a chronic illness: a systematic review and meta-analysis of randomized controlled trials. Arch Intern Med 172: 101–111.
9. Motl RW, Birnbaum AS, Kubik MY, Dishman RK (2004) Naturally occurring changes in physical activity are inversely related to depressive symptoms during early adolescence. Psychosom Med 66: 336–342.
10. Ogden C, Carroll M, Kit BK, Flegal KM (2012) Prevalence of obesity in the United States, 2009-2010 - NCHS Data Brief. Centers for Disease Control and Prevention. Available: http://www.cdc.gov/nchs/data/databriefs/db82.pdf. Accessed 2013 February 1.
11. Fryar CD, Carroll MD, Ogden C (2012) Prevalence of overweight, obesity, and extreme obesity among adults: United States, Trends 1960–1962 through 2009–2010 - NCHS Health E-Stat. Centers for Disease Control and Prevention. Available: http://www.cdc.gov/nchs/data/hestat/obesity_adult_09_10/obesity_adult_09_10.htm. Accessed 2013 January 15.
12. Singh GK, Kogan MD, van Dyck PC (2010) Changes in state-specific childhood obesity and overweight prevalence in the United States from 2003 to 2007. Arch Pediatr Adolesc Med 164: 598–607.
13. CDC (2011) Obesity rates among low-income preschool children. Centers for Disease Control and Prevention. Available: http://www.cdc.gov/obesity/childhood/data.html. Accessed 2012 March 12.
14. Casazza K, Fontaine KR, Astrup A, Birch LL, Brown AW, et al. (2013) Myths, presumptions, and facts about obesity. N Engl J Med 368: 446–454.
15. Voss JD, Masuoka P, Webber BJ, Scher AI, Atkinson RL (2013) Association of elevation, urbanization and ambient temperature with obesity prevalence in the United States. Int J Obes (Lond) doi:10.1038/ijo.2013.5 [Online 30 January 2013].
16. Brunekreef B, Holgate ST (2002) Air pollution and health. Lancet 360: 1233–1242.
17. Gent JF, Triche EW, Holford TR, Belanger K, Bracken MB, et al. (2003) Association of low-level ozone and fine particles with respiratory symptoms in children with asthma. JAMA 290: 1859–1867.
18. Laden F, Neas LM, Dockery DW, Schwartz J (2000) Association of fine particulate matter from different sources with daily mortality in six U.S. cities. Environ Health Perspect 108: 941–947.
19. Pope CA, 3rd, Burnett RT, Thurston GD, Thun MJ, Calle EE, et al. (2004) Cardiovascular mortality and long-term exposure to particulate air pollution: epidemiological evidence of general pathophysiological pathways of disease. Circulation 109: 71–77.
20. Cakmak S, Dales R, Leech J, Liu L (2011) The influence of air pollution on cardiovascular and pulmonary function and exercise capacity: Canadian Health Measures Survey (CHMS). Environ Res 111: 1309–1312.
21. Marr LC, Ely MR (2010) Effect of air pollution on marathon running performance. Med Sci Sports Exerc 42: 585–591.
22. Cutrufello PT, Rundell KW, Smoliga JM, Stylianides GA (2011) Inhaled whole exhaust and its effect on exercise performance and vascular function. Inhal Toxicol 23: 658–667.
23. Rundell KW, Caviston R (2008) Ultrafine and fine particulate matter inhalation decreases exercise performance in healthy subjects. J Strength Cond Res 22: 2–5.
24. Mokdad AH, Stroup DF, Giles WH (2003) Public health surveillance for behavioral risk factors in a changing environment. Recommendations from the Behavioral Risk Factor Surveillance Team. Centers for Disease Control and Prevention Morbidity and Mortality Weekly Report (MMWR) 52: 1–12.
25. Nelson DE, Holtzman D, Waller M, Leutzinger CL, Condon K (1998) Objectives and design of the Behavioral Risk Factor Surveillance System. American Statistical Association 1998 Proceedings of the Section on Survey Methods. Alexandria, VAAmerican Statistical Association. pp. 214–218.
26. VDH (2012) BRFSS Methodology. Virginia Department of Health. Available: http://www.vahealth.org/brfss/methodology.htm. Accessed 2012 December 1.
27. CDC (2012) Methodologic changes in the Behavioral Risk Factor Surveillance System in 2011 and potential effects on prevalence estimates. Morbidity and Mortality Weekly Report MMWRCenters for Disease Control and Prevention. pp. 410–413.
28. Battaglia MP, Frankel MR, Link MW (2008) Improving standard poststratification techniques for random-digit-dialing telephone surveys. Survey Research Methods 2: 11–19.
29. EPA (2011) AQS Data Dictionary. U.S. Environmental Protection Agency. Available: http://www.epa.gov/ttn/airs/airsaqs/manuals/AQS Data Dictionary.pdf. Accessed 2012 December 20.
30. Marshall JD, Nethery E, Brauer M (2008) Within-urban variability in ambient air pollution: Comparison of estimation methods. Atmospheric Environment 42: 1359–1369.
31. Son JY, Bell ML, Lee JT (2010) Individual exposure to air pollution and lung function in Korea: spatial analysis using multiple exposure approaches. Environ Res 110: 739–749.
32. Hystad P, Demers PA, Johnson KC, Brook J, van Donkelaar A, et al. (2012) Spatiotemporal air pollution exposure assessment for a Canadian population-based lung cancer case-control study. Environ Health 11,: 22.
33. Krivoruchko K (2012) Empirical Bayesian Kriging: Implemented in ArcGIS Geostatistical Analyst. ESRI. Available: http://www.esri.com/news/arcuser/1012/files/ebk.pdf. Accessed 2012 December 15.
34. Pilz J, Spöck G (2008) Why do we need and how should we implement Bayesian kriging methods. Stoch Environ Res Risk Assess 22: 621–632.
35. EPA (2004) Developing spatially interpolated surfaces and estimating uncertainty. U.S. Environmental Protection Agency. Available: http://www.epa.gov/airtrends/specialstudies/dsisurfaces.pdf. Accessed 2012 February 21.
36. Wen XJ, Balluz LS, Shire JD, Mokdad AH, Kohl HW (2009) Association of self-reported leisure-time physical inactivity with particulate matter 2.5 air pollution. J Environ Health 72: 40–44; quiz 45.
37. Hankey S, Marshall JD, Brauer M (2012) Health impacts of the built environment: within-urban variability in physical inactivity, air pollution, and ischemic heart disease mortality. Environ Health Perspect 120: 247–253.
38. Alexeeff SE, Litonjua AA, Suh H, Sparrow D, Vokonas PS, et al. (2007) Ozone exposure and lung function: effect modified by obesity and airways hyperresponsiveness in the VA normative aging study. Chest 132: 1890–1897.
39. Delfino RJ, Tjoa T, Gillen DL, Staimer N, Polidori A, et al. (2010) Traffic-related air pollution and blood pressure in elderly subjects with coronary artery disease. Epidemiology 21: 396–404.
40. Peters A, Liu E, Verrier RL, Schwartz J, Gold DR, et al. (2000) Air pollution and incidence of cardiac arrhythmia. Epidemiology 11: 11–17.
41. Pope CA, 3rd, Burnett RT, Thun MJ, Calle EE, Krewski D, et al. (2002) Lung cancer, cardiopulmonary mortality, and long-term exposure to fine particulate air pollution. JAMA 287: 1132–1141.
42. Brook RD, Rajagopalan S, Pope CA 3rd, Brook JR, Bhatnagar A, et al. (2010) Particulate matter air pollution and cardiovascular disease: An update to the scientific statement from the American Heart Association. Circulation 121: 2331–2378.
43. EPA (2008) Appendix G: Health-based cost-effectiveness of reductions in ambient PM 2.5 associated with illustrative PM NAAQS attainment strategies. U.S. Environmental Protection Agency. Available: http://www.epa.gov/ttnecas1/regdata/RIAs/Appendix G-Health Based Cost Effectiveness Analysis.pdf. Accessed 2012 March 7.

Author Contributions

Conceived and designed the experiments: JDR JDV BK. Performed the experiments: JDR BK. Analyzed the data: JDR BK. Contributed reagents/materials/analysis tools: JDR BK. Wrote the paper: JDR JDV BK.

4

Aerosol Chemistry over a High Altitude Station at Northeastern Himalayas, India

Abhijit Chatterjee[1], Anandamay Adak[1], Ajay K. Singh[2], Manoj K. Srivastava[3], Sanjay K. Ghosh[2,4], Suresh Tiwari[5], Panuganti C. S. Devara[6], Sibaji Raha[1,2,4]*

1 Environmental Sciences Section, Bose Institute, Kolkata, India, **2** Center for Astroparticle Physics and Space Science, Bose Institute, Kolkata and Darjeeling, India, **3** Department of Geophysics, Banaras Hindu University, Varanasi, India, **4** Department of Physics, Bose Institute, Kolkata, India, **5** Indian Institute of Tropical Meteorology, New Delhi, India, **6** Indian Institute of Tropical Meteorology, Pune, India

Abstract

Background: There is an urgent need for an improved understanding of the sources, distributions and properties of atmospheric aerosol in order to control the atmospheric pollution over northeastern Himalayas where rising anthropogenic interferences from rapid urbanization and development is becoming an increasing concern.

Methodology/Principal Findings: An extensive aerosol sampling program was conducted in Darjeeling (altitude ~2200 meter above sea level (masl), latitude 27°01′N and longitude 88°15′E), a high altitude station in northeastern Himalayas, during January–December 2005. Samples were collected using a respirable dust sampler and a fine dust sampler simultaneously. Ion chromatograph was used to analyze the water soluble ionic species of aerosol. The average concentrations of fine and coarse mode aerosol were found to be 29.5 ± 20.8 µg m^{-3} and 19.6 ± 11.1 µg m^{-3} respectively. Fine mode aerosol dominated during dry seasons and coarse mode aerosol dominated during monsoon. Nitrate existed as NH_4NO_3 in fine mode aerosol during winter and as $NaNO_3$ in coarse mode aerosol during monsoon. Gas phase photochemical oxidation of SO_2 during premonsoon and aqueous phase oxidation during winter and postmonsoon were the major pathways for the formation of SO_4^{2-} in the atmosphere. Long range transport of dust aerosol from arid regions of western India was observed during premonsoon. The acidity of fine mode aerosol was higher in dry seasons compared to monsoon whereas the coarse mode acidity was higher in monsoon compared to dry seasons. Biomass burning, vehicular emissions and dust particles were the major types of aerosol from local and continental regions whereas sea salt particles were the major types of aerosol from marine source regions.

Conclusions/Significance: The year-long data presented in this paper provide substantial improvements to the heretofore poor knowledge regarding aerosol chemistry over northeastern Himalayas, and should be useful to policy makers in making control strategies.

Editor: Juan A. Añel, Universidade de Vigo, Spain

Funding: This study was supported through a grant (No. IR/S2/PF-01/2003) of the Science & Engineering Research Council, Department of Science & Technology, Government of India under the IRHPA (Intensification of Research in High Priority Areas) scheme. The funders had no role in study design, data collection and analysis, decision to publish, or preparation of the manuscript.

Competing Interests: The authors have declared that no competing interests exist.

* E-mail: sibajiraha@bic.boseinst.ernet.in

Introduction

Atmospheric aerosol is linked to visibility reduction, adverse health effects and heat balance of the Earth, directly by reflecting and absorbing solar radiation and indirectly by influencing the properties and cloud processes and, possibly, by changing the heterogeneous chemistry of reactive greenhouse gases [1]. The combined global radiative forcing due to increases in major greenhouse gases (CO_2, CH_4 and N_2O) is $+2.3$ Wm^{-2}. Anthropogenic contributions to aerosols (primarily sulphate, organic carbon, nitrate and dust) together produce a cooling effect, with a total direct radiative forcing of -0.5 Wm^{-2} and an indirect cloud albedo forcing of -0.7 Wm^{-2} [1]. Thus aerosols compensate by ~50% for the mean global radiative forcing due to greenhouse gases warming. The large range of uncertainty in estimating the aerosol forcing reflects the poor state of knowledge regarding the sources, distributions and properties of atmospheric aerosol.

Increasing pollutant emissions associated with the fast-growing economies of southeastern Asian countries have led to the progressive increase of aerosol concentrations above the natural background [2]. Satellite observations have shown that the light-absorbing aerosol hazes (which is about 3–5 mm thick) over India intensify over the Thar desert and the polluted Indo-Gangetic plain (IGP). The IGP has a sharp boundary to the north, where the Himalayas act as a barrier, extending thousands of miles southward and over the north Indian ocean [3,4]. Aerosol rich boundary layer air can be transported to higher altitudes by valley breezes on the Himalayan slopes [2].

The transport of optically-active aerosol to the higher Himalayas is a matter of concern, since most of the glaciers in the region have been retreating since 1850 [5] with increasing melting rates, and are in danger of completely disappearing in the next decades [6]. If the retreat of the Himalayan glaciers continues unabated, it will exacerbate the water stress in northern India, especially during the dry season [7]. The rising anthropogenic interferences from rapid urbanization and development in the Himalayas affect both the landscape and the atmospheric environments and are the causes of increasing concern [8,9].

A short-term sampling program in the Nguzompa glacier basin near Mt. Everest [10], a two-week sampling project in Hidden Valley in the Himalayas of western Nepal [11], a year-long sampling of atmospheric aerosol at a remote Himalayan site and a rural Middle-Mountain site in Nepal [12], a study on the effect of mineral dust and carbonaceous species on the aerosol composition in Nepal Himalayas [13] and a study on the seasonal variation of total suspended particulates (TSP) in Manali, northwestern Himalayan range [14] are the major research studies carried out in the Nepal Himalayan and in other northwestern Himalayan sites. But as far as the northeastern Himalayas are concerned, the region still lacks systematic studies focused on chemical characterization of aerosols.

A strong seasonal variation in aerosol chemistry is expected in this northeastern Himalayan region. During premonsoon and summer months, due to enhanced convection, aerosols are lofted to elevated altitudes in the troposphere. Together with the westerly premonsoon winds, enhanced convection and the steep pressure gradient across the Himalayan-Gangetic region steer aerosols aloft. With the onset of rainy season (the Arabian Sea and Bay of Bengal branches of the South West summer monsoon), the heavy dust loading significantly diminishes due to aerosol washout from the atmosphere and enhances the loading of sea salt aerosols to a significant level. During winter and postmonsoon, northeasterly winds from the subcontinent bring anthropogenic aerosols over the Himalayan region. In addition to that, massive biomass burning during winter also plays a role in the loading of anthropogenic aerosols over northeastern Himalayas. These distinctly different seasonal behaviors of aerosol in northeastern Himalayas prompted us to make a year-long study on the formation and distribution of atmospheric aerosols over Darjeeling, a high altitude station in northeastern Himalayas.

In order to understand the seasonal nature of the predominant water soluble ionic species in fine (aerodynamic diameter less than 2.5 μm) and coarse mode (aerodynamic diameter more than 2.5 μm) aerosol a year-long aerosol sampling was done during January–December 2005. This study presents the possible formation mechanisms of secondary ionic species in different seasons, distribution of primary ionic species, long range transport of dust aerosol and the interaction between transported marine aerosol and locally generated anthropogenic aerosol. Finally, an attempt was made to find out the possible types of aerosols from different source regions.

The primary focus of this study was to determine the relative contribution of natural and anthropogenic components on the total aerosol loading and their distribution between fine and coarse mode aerosols at a high altitude hill station in northeastern Himalayas which could provide the scientific basis for controlling atmospheric pollution over this geo-politically and environmentally important region.

1. Site description

Darjeeling is one of the most popular tourist hill-stations in eastern India with a population of ∼100,000. The overall areas of the Darjeeling district and Darjeeling Township are about 1200 and 11.44 squared kilometers, respectively. Darjeeling Township is located at an average altitude of ∼2000 meter above sea level (masl) and surrounded by different types of topography of the lower-eastern-Himalayas. The southern region comprises the marshy low-lying area at an average height of ∼100–300 masl. The apex is formed by the Phalut ridge (altitude of 3800 masl) at the border between Nepal and India. The eastern frontier lies along two rivers, locally called Tista and Rangeet.

The sampling site (shown in Figure 1) is located on the terrace of a three-storied building on our institute premises. This site (latitude: 27°01′N, longitude: 88°15′E with an altitude of 2194 masl) is at an altitude of about 200 m above the main township and is a remote area compared to the main township with a limited number of residential houses and forested areas dominated by juniper and varieties of pine in the immediate vicinity of the observatory. The closest street with significant road traffic is about 200 m away from the study site. The area, within a radius of ∼10 km, is occupied by several major and minor tea processing units operated by furnace oil and coal and several tea gardens where several ammoniated fertilizers are used. Wood and biomass burning in the nearby villages, automobile exhaust (mainly tourist vehicles) throughout the year and the exhaust from the "Toy Train" (Darjeeling Himalayan Railway), which is enlisted as an UN (United Nations) world-heritage and still runs on coal as its fuel, are the major sources of air pollution at this hill station.

2. Prevailing meteorology

The monthly variations of surface meteorological parameters like temperature (°C), relative humidity (%), wind speed (m s^{-1}) along with total rainfall (mm) over the entire study period (Jan–Dec) are shown in Figure 2 and the surface wind directions presented seasonally in different seasons, namely, winter (Dec–Feb), premonsoon (Mar–May), monsoon (Jun–Sep) and postmonsoon (Oct–Nov) in Figure 3. The average temperature was found to be 15±4°C with minimum of 7°C during December and maximum of 20°C during June. In general, the relative humidity was high across the whole study period with an average of 81%. The dry season (Jan–May, Oct–Dec) remained moderately dry with an average relative humidity of 76% compared to the wet season with an average relative humidity of 91%. The total rainfall during the entire study period (Jan–Dec) was found to be 2220 mm, 80% of which was during southwest monsoon (1783 mm) with scanty or no rainfall during winter (20 mm) and premonsoon (304 mm). The surface wind pattern during winter was mainly easterly and northeasterly with average speed of 0.84 m s^{-1} and during monsoon it was mainly southwesterly with an average speed of 1.18 m s^{-1}. In order to know the wind pattern variations in different seasons, the monthly mean wind vectors (at 850 hPa level) for four different seasons; winter (Dec–Feb), premonsoon (Mar–May), monsoon (Jun–Sep) and postmonsoon (Oct–Nov) are shown in Figure 4, for the region covering equator to 40°N and 40–130°E. The NCEP/NCAR reanalysis data clearly show the contrasting wind patterns between winter and monsoon whereas premonsoon and postmonsoon represents the transition phase in the circulation patterns. Winter shows the weak northeasterly wind from the continental area covering densely populated cities including semi arid regions whereas monsoon shows strong southwesterly wind originating from Arabian Sea. These distinctly different wind fields impart extreme temporal variability in aerosol characteristics.

Methods

1. Collection of aerosol samples and determination of mass concentration

For the collection of total respirable suspended particulate matter (aerodynamic diameter less than 10 μm) from ambient air,

Figure 1. Map of the sampling station.

a respirable dust sampler was used. The sampler (model APM 460BL) was manufactured by Envirotech Instrument Pvt Ltd, which pioneered the development of indigenous air monitoring instruments all over India. The sampler collected the samples with a flow rate of 1.4 m³ min⁻¹. The sampler was fitted with a cyclone, which was used for fractioning the dust into two fractions.

Total respirable particulate matter was collected on the filter paper (EPM 2000 filter paper from Whatman of 8″×10″ dimension) while particulate matter of aerodynamic diameter more than 10 μm was collected in a cup placed under the cyclone.

For collection of fine particulate matter or fine mode aerosol, a fine dust sampler was used from the same company (model APM

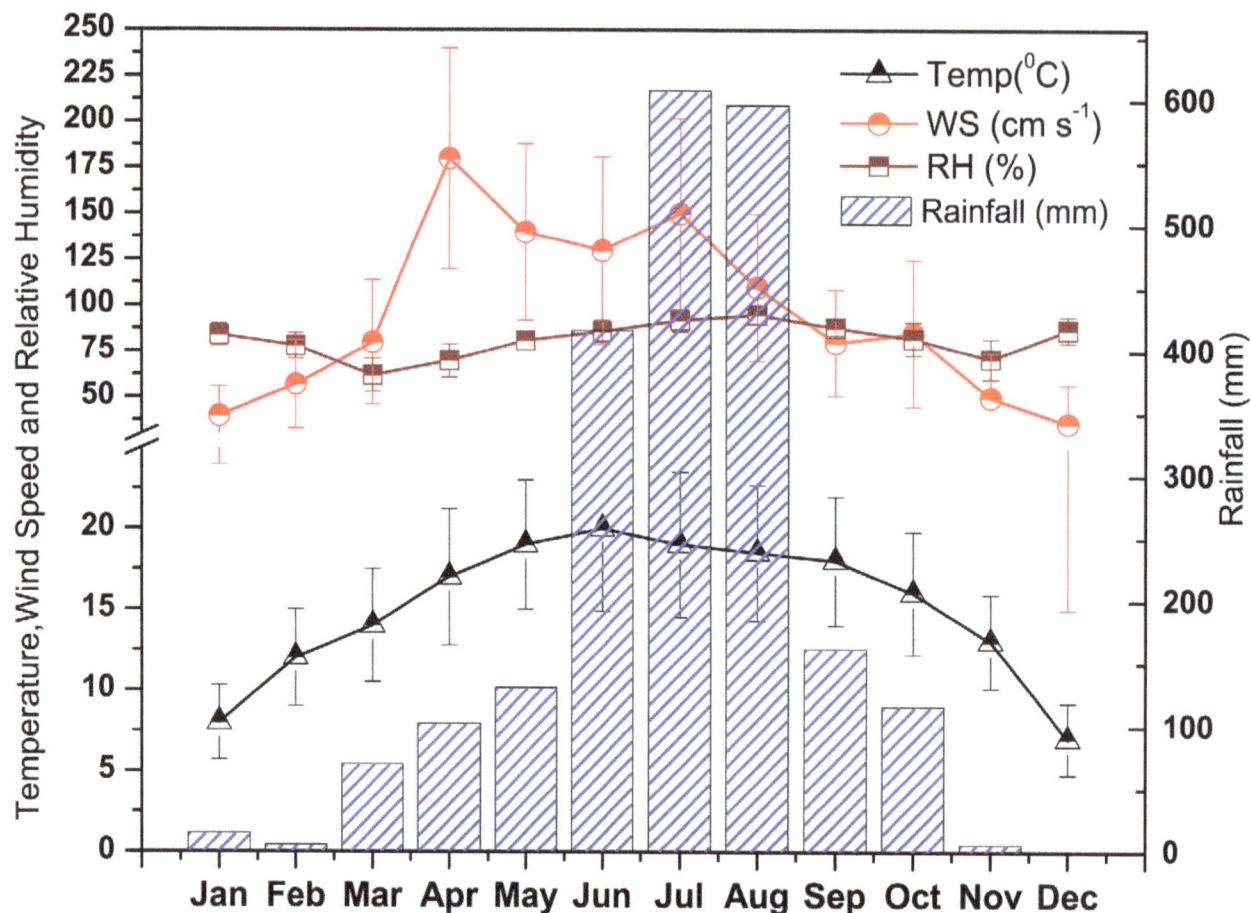

Figure 2. Seasonal variations of temperature (Temp), wind speed (WS), relative humidity (RH) and rainfall over Darjeeling.

550). The flow rate was $1 \text{ m}^3 \text{ hr}^{-1}$. After entering the air particles through the inlet of the sampler the coarse particulate matter (aerodynamic diameter more than 2.5 μm) was removed using a GF/A (Glass Fiber) filter paper of 37 mm diameter immersed in silicone oil, used as an impaction surface. The impaction surface was placed above the main aerosol collection filter. The fine particles were collected on a PTFE filter paper of 47 mm diameter.

On average, an aerosol sample (both for fine and total respirable particulate matter) was collected on every 3rd day during dry seasons (Jan–May, Oct–Dec) and every 4th day during monsoon (Jun–Sep). Thus a total of 111 samples were collected (81 during dry season and 30 during monsoon). Although collection of samples on daily basis would be better in carrying out this kind of aerosol study, collection of samples of more than 111 were beyond our scope. Each sampling was started at 0900 hrs (local time) and run for ~24 hrs. Both the samplers were placed on the terrace of a three-storied building (~15 m above ground level) on our institute premises.

The mass concentration of aerosol was determined by the gravimetric measurement. The filters were placed in desiccators for ~24 hrs before and after the sampling to remove the absorbed water and weighed in a controlled environment chamber after taking the filters out of the desiccators before and after the sampling using a semi-micro balance (Sartorius, Model ME 235 P). The aerosol mass (μg) was determined by the differences between initial and final weight of the filter and the concentration

($\mu\text{g m}^{-3}$) was determined dividing the aerosol mass by total volume of air (m^3).

2. Meteorological parameters

The meteorological parameters were recorded with the help of an automatic weather station of Lawrence & Mayo (Model: AWS-PC) and all the data were run by LYNX-software of version V0007. The weather station was run continuously and the data were recorded at the interval of half an hour throughout the year covering all the sampling events. The weather station was equipped with a tower and all the sensors of wind speed and its direction, relative humidity (RH) and temperature were fitted with that tower at a height of 15 m from the ground level. The rainfall data was obtained from Indian Meteorological Department, India.

3. Chemical analysis

For the analysis of water soluble ions, chromatographic separation method was used [15]. One-half of the filters were soaked in 20 ml Milli-Q water (18.2 MΩ resistivity) for ~30 min. and ultrasonicated for 20 min. The solutions were made up to known volume (100 ml) using Milli-Q water. The solutions were then kept in polypropelene bottles and kept at ~4°C until analysis. Prior to their use, the bottles were cleaned repeatedly using distilled water and soaked for ~72 hrs. The major ions, namely anions (Cl^-, NO_3^- and SO_4^{2-}) and cations (Na^+, NH_4^+, K^+, Ca^{2+} and Mg^{2+}) were quantitatively determined by Ion Chromatograph (DIONEX-2000, USA) using analytical column IonPac ® AS15

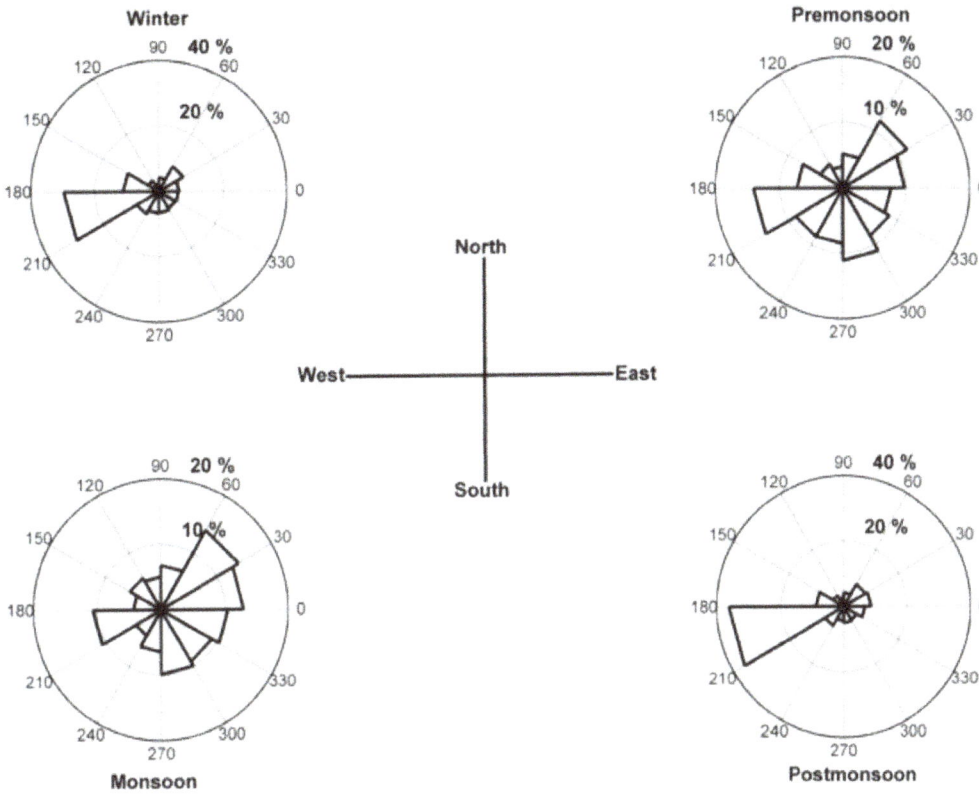

Figure 3. Rose diagram of surface wind speed and direction over the entire study period in Darjeeling.

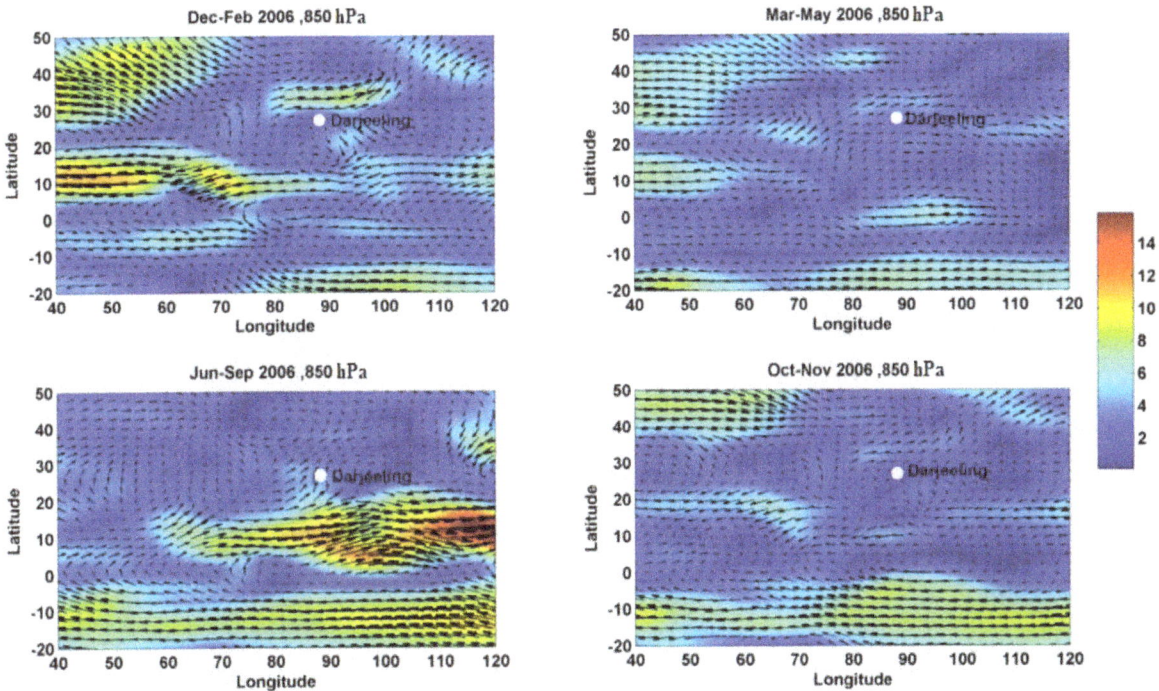

Figure 4. Mean wind at 850 hPa level obtained from NCEP/NCAR reanalysis during different seasons over Darjeeling.

with micro-membrane suppressor ASRS ultra II 2mm, 38 mM. KOH as eluent and triple distilled water as regenerator for anions. Similarly, the IonPac ® CS17 column with micro-membrane suppressor CSRS ultra II 2mm, 6 mM methansulfonic acid as eluent and triple distilled water as regenerator were used for cations. For calibration purpose, the standards were procured from Dionex for cations and anions. Detection limits of the ionic species, concentrations corresponding to three times the standard deviation of five replicate blank level measurements for Na^+, K^+, Ca^{2+}, Mg^{2+}, NH_4^+, Cl^-, NO_3^- and SO_4^{2-} were 0.009, 0.0013, 0.003, 0.0015, 0.0024, 0.009, 0.005 and 0.008 $\mu g\ m^{-3}$ respectively. The precision estimated from the standard deviation of repeat measurements of standard and samples was 2% for Na^+, K^+ and Ca^{2+}; 3% for Mg^{2+}; 5% for NH_4^+; 2% for Cl^-, SO_4^{2-} and 4% for NO_3^-. Trace gas SO_2 was measured using an on-line SO_2 analyzer (Horiba, APSA-360A) throughout all the sampling events at five minute interval.

Results and Discussion

1. Seasonal variation of particulate matter

The average concentration of fine mode aerosol in Darjeeling was found to be $29.5\pm20.8\ \mu g\ m^{-3}$ varying between 3.6 $\mu g\ m^{-3}$ and 61 $\mu g\ m^{-3}$, whereas coarse mode aerosol ranged between 5.4 $\mu g\ m^{-3}$ and 32 $\mu g\ m^{-3}$ with an average of $19.6\pm11.1\ \mu g\ m^{-3}$. The large variation in concentrations of both fine

and coarse mode aerosol (Figure 5) during the entire study period could be due to the thermodynamic conditions in the planetary boundary layer (PBL), which either favor or adversely affect pollutants dispersion. Ambient weather conditions, such as air temperature, relative humidity and short wave radiation, could also influence the chemical reactions leading to secondary aerosol formation. Stable atmospheric conditions with a low mixing layer height may result in significantly enhanced particulate concentrations [16]. Aerosol shows higher concentrations during winter months and minimum concentrations during monsoon. In winter, very frequent and persistent thermal inversion and fog situations at ground level caused a considerable amount of aerosol to accumulate in the lower layers of the atmosphere [17]. Aerosol concentrations during winter were largely affected due to massive biomass burning over Darjeeling. The higher emission of K^+, SO_4^{2-} and carbonaceous species (not analyzed for the present study) could enhance the aerosol loading in the atmosphere during winter. The sharp fall in fine mode aerosol concentrations and high precipitation amount (1783 mm) during monsoon indicates the wash out effect of aerosol and its components. On the other hand, the coarse mode aerosol did not show a sharp decrease in concentrations during monsoon due to the contribution of sea salt aerosol (Na^+, Cl^-, Mg^{2+}). Non-sea-sulphate and nitrate were also found to enhance the coarse mode aerosol concentrations during monsoon (discussed later in detail). The ratios of fine to coarse mode aerosol concentrations were found to widely vary between

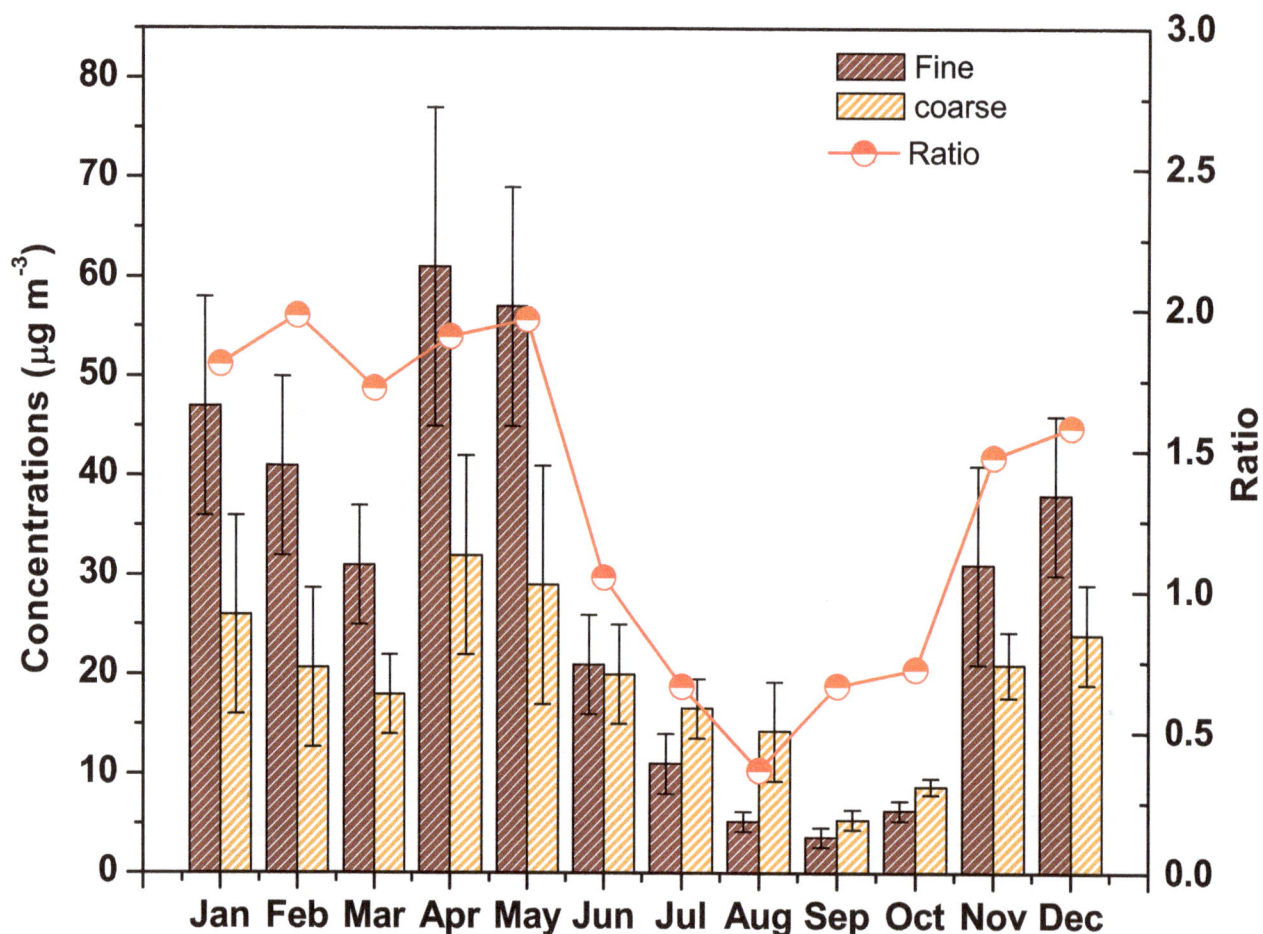

Figure 5. Seasonal variations of aerosol over Darjeeling.

Table 1. Comparison of this study with other high altitude stations in India and Nepal.

	Darjeeling Northeastern Himalayas, India		Nagarkot, Nepal Himalayas	Langtang, Nepal Himalayas	Phortse, Nepal Himalayas	Jiri, Nepal Himalayas	Manora Peak, India	Khumbu valley, Nepal	Mt Abu, Western India
References	This study		[13]	[13]	[12]	[12]	[42]	[7]	[19]
Altitude(masl)	2194		2150	3920	4100	1900	1950	5079	1680
Aerosol type	Fine	Coarse	Fine	Fine	PM_{10}	PM_{10}	TSP	PM_{10}	TSP
Samples	111	111	NA	NA	17	13	NA	13	25
Na^+	0.66±0.43	2.20±2.00	0.20±0.16	0.12±0.18	0.03	0.08	NA	NA	0.28
NH_4^+	0.88±0.76	0.05±0.04	1.50±1.00	0.54±0.56	0.28	0.35	0.52	0.14	0.37
K^+	1.20±0.80	0.31±0.21	0.62±0.58	0.20±0.26	0.15	0.25	0.23	0.02	0.20
nsK^+	1.15±0.82	0.23±0.20	0.61	NA	NA	NA	NA	NA	NA
Ca^{2+}	0.13±0.01	0.40±0.20	0.31±0.29	0.66±0.89	0.12	0.56	0.75	0.10	1.70
$nsCa^{2+}$	0.11±0.10	0.38±0.21	NA	NA	NA	NA	NA	NA	NA
Mg^{2+}	0.12±0.06	0.31±0.17	0.04±0.03	0.05±0.05	0.03	0.08	NA	NA	0.09
$nsMg^{2+}$	0.04±0.03	0.14±0.18	NA	NA	NA	NA	NA	NA	NA
Cl^-	1.21±1.00	2.35±1.50	0.05±0.05	0.06±0.10	0.01	0.01	NA	NA	0.31
$nsCl^-$	0.44±0.41	−0.18±1.10	NA	NA	NA	NA	NA	NA	NA
NO^{3-}	3.31±2.30	0.95±0.20	1.20±1.80	0.78±0.10	0.70	0.48	0.50	0.08	0.74
SO_4^{2-}	3.80±2.90	2.50±2.10	3.80±1.00	1.41±1.30	0.98	0.41	2.60	0.22	2.71
$nsSO_4^{2-}$	3.64±2.80	1.92±1.84	NA	NA	NA	NA	NA	NA	NA

NA: Not Available; Concentrations of ionic species are given in µg m^{-3}.

0.36 (during August) and 4.4 (during January) with an average of 1.4±1.1. On an average, the average ratio of 1.72 was observed during dry seasons and 0.76 during monsoon. Thus, fine mode aerosol dominated over coarse mode aerosol during dry seasons whereas coarse mode aerosol dominated over fine mode aerosol during monsoon. A peak was observed in aerosol concentrations both in fine and coarse mode during two premonsoon months, April and May. This could be attributed to the long range transport of dust aerosol from arid regions of western India and even from Arabian deserts (discussed later in detail) during

Table 2. Comparison of this study with other urban sites in India.

	Darjeeling, Northeastern Himalayas, India		Mumbai, India	Pune, India	Agra, India	Ahmedabad, India
References	This study		[43]	[9]	[44]	[19]
Altitude(masl)	2194					
Aerosol type	Fine	Coarse	PM_{10}	PM_{10}	TSP	TSP
Samples	111	111	NA	NA	NA	25
Na^+	0.66±0.43	2.20±2.00	2.20	0.48	2.97	0.81
NH_4^+	0.88±0.76	0.05±0.04	NA	2.14	6.52	0.48
K^+	1.20±0.80	0.31±0.20	8.90	0.43	2.50	0.76
nsK^+	1.15±0.82	0.23±0.20	NA	NA	NA	0.73
Ca^{2+}	0.13±0.01	0.40±0.20	6.20	2.50	3.02	2.96
$nsCa^{2+}$	0.11±0.10	0.38±0.20	NA	NA	NA	2.93
Mg^{2+}	0.12±0.06	0.31±0.17	2.20	0.23	1.24	0.25
$nsMg^{2+}$	0.04±0.03	0.14±0.18	NA	NA	NA	0.15
Cl^-	1.21±1.00	2.35±1.50	2.60	1.80	6.78	0.99
$nsCl^-$	0.44±0.41	−0.18±1.10	NA	NA	NA	NA
NO^{3-}	3.31±2.30	0.95±0.20	6.00	2.91	8.37	2.07
SO_4^{2-}	3.80±2.90	2.50±2.10	6.20	2.98	14.7	4.57
$nsSO_4^{2-}$	3.64±2.80	1.92±1.84	NA	NA	NA	4.36

NA: Not available; Concentrations of ionic species are given in µg m^{-3}.

Table 3. Inter-ionic correlations in fine and coarse mode aerosols over Darjeeling during dry seasons (n = 81).

	Na$^+$	NH$_4^+$	K$^+$	nsK$^+$	Ca^{2+}	nsCa^{2+}	Mg^{2+}	nsMg^{2+}	Cl$^-$	NO$_3^-$	SO$_4^{2-}$	nsSO$_4^{2-}$
Na$^+$		F=0.35 C=0.14	F=0.24 C=0.21	F=0.16 C=0.13	F=0.26 C=0.23	F=0.18 C=0.39	F=0.13 C=0.19	F=0.11 C=0.29	F=0.38 C=0.44	F=0.17 C=0.11	F=0.19 C=0.37	F=0.14 C=0.26
NH$_4^+$			F=0.12 C=0.16	F=0.19 C=0.31	F=0.11 C=0.10	F=0.17 C=0.21	F=0.22 C=0.20	F=0.19 C=0.14	F=0.23 C=0.10	F=0.74 C=0.51	F=0.44 C=0.42	F=0.65 C=0.44
K$^+$					F=0.18 C=0.38	F=0.21 C=0.34	F=0.16 C=0.18	F=0.14 C=0.34	F=0.26 C=0.37	F=0.11 C=0.15	F=0.19 C=0.27	F=0.29 C=0.27
nsK$^+$					F=0.29 C=0.31	F=0.31 C=0.21	F=0.11 C=0.20	F=0.13 C=0.42	F=0.82 C=0.57	F=0.19 C=0.27	F=0.39 C=0.17	F=0.78 C=0.37
Ca^{+2}							F=0.55 C=0.62	F=0.48 C=0.42	F=0.19 C=0.22	F=0.19 C=0.27	F=0.22 C=0.31	F=0.11 C=0.25
nsCa^{2+}							F=0.39 C=0.54	F=0.64 C=0.74	F=0.10 C=0.17	F=0.10 C=0.22	F=0.31 C=0.37	F=0.20 C=0.12
Mg^{+2}									F=0.23 C=0.31	F=0.10 C=0.17	F=0.32 C=0.31	F=0.25 C=0.27
nsMg^{2+}									F=0.30 C=0.12	F=0.16 C=0.17	F=0.15 C=0.17	F=0.11 C=0.26
Cl$^-$										F=0.19 C=0.15	F=0.29 C=0.48	F=0.78 C=0.37
NO$_3^-$											F=0.58 C=0.33	F=0.55 C=0.41
SO$_4^{-2}$												
nsSO$_4^{2-}$												

premonsoon. According to the elevated heat pump (EHP) mechanism proposed in [18] the dust aerosols mixed with carbonaceous aerosols primarily from Indo-Gangetic Plain (IGP) reaches the foothills of the Himalayas and are vertically advected to elevated altitudes. This causes significant loading of aerosol over the Himalayan region during premonsoon. The carbonaceous aerosol components and the trace elements (mainly the mineral dust component) data, though not analyzed for the present study, would provide better information on the high aerosol loading over Darjeeling during premonsoon.

2. Water-soluble ionic species in aerosol

2.1. Comparison with other studies over Himalayan region and Indian subcontinent. Water-soluble inorganic ions (Na$^+$, NH$_4^+$, K$^+$, Ca^{2+}, Mg^{2+}, Cl$^-$, NO$_3^-$, SO$_4^{2-}$) in fine and coarse mode aerosol over Darjeeling have been compared with the data reported in other high altitude stations in India and Nepal and other sites in Indian subcontinent.

For comparison with other high altitude stations, seven such stations (five in Nepal and two in India) have been chosen (Table 1). It is observed that the concentrations of sodium and chloride are several times higher than all the other hill stations indicating the strong influence of sea salt aerosol over Darjeeling. The most interesting feature is that in addition to the strong influence of sea salt aerosol, massive coal and biomass burning (domestic, industrial and from Darjeeling Himalayan Railways) throughout the year enhanced the concentration of non-sea-Cl$^-$ in fine mode aerosol over Darjeeling. The concentrations of ammonium, nitrate and sulphate, the secondary anthropogenic ionic species in aerosol over Darjeeling, are 3–15 times higher than all the other hill stations except Nagarkot. The concentration of NH$_4^+$ in Nagarkot is higher than all the hill stations due to the close proximity of the sampling station to agricultural land and animal husbandry. The higher vehicular activities due to the high influx of tourists in Darjeeling

could be the reason behind the higher loading of nitrate. Strong influence of massive biomass burning in Darjeeling was observed in the very high concentrations of non-sea-SO$_4^{2-}$ and non-sea-K$^+$. When we compared the Ca^{2+} concentration, we noticed that our data do not differ significantly from the other high altitude stations. The concentration of Ca^{2+} varied between 0.12 µg m^{-3} and 0.75 µg m^{-3} in various stations in Himalayas (including Darjeeling), located between 1900 (Jiri, Nepal) and 5079 (NCO-P, Nepal) masl. Therefore, Ca^{2+} seems to be more homogeneously distributed vertically in Himalayan regions compared to other ionic species. The loading of dust aerosol enriched with Ca^{2+} in Himalayan region is mostly due to the long-range transport of dust aerosol originated in the western part of India (discussed later). However, the Ca^{2+} concentration over Mt. Abu in western India shows the highest concentration which is due to the widespread marble queries and stone-crushing mills in plain lands along the Aravali range in western India [19]. Mg^{2+} in the present study shows the highest concentration compared to the other sites. But if we look at the fine mode Mg^{2+} only, we find that our data do not significantly differ from the other sites (varying between 0.03 and 0.12 µg m^{-3}) indicating homogeneous vertical distribution of Mg^{2+}, like Ca^{2+}, in Himalayan regions.

The comparison of the data of aerosol ionic species of the present study with four urban sites (including a mega city, Mumbai) in India is shown in Table 2. It is observed from the table that, Na$^+$ and Cl$^-$ are higher in concentrations than Pune and Ahmedabad and comparable to Agra, but the most interesting feature is that the concentrations of Na$^+$ and Cl$^-$ in Darjeeling are found to be higher than at Mumbai, a coastal city in western India. At this stage, we could not find a fool-proof explanation for this most interesting observation; further, long-term study is required to understand this phenomenon. NH$_4^+$ and NO$_3^-$ in Darjeeling are found to show lower concentrations than at Mumbai, Pune or Agra. K$^+$ and SO$_4^{2-}$ show higher concentrations in Darjeeling

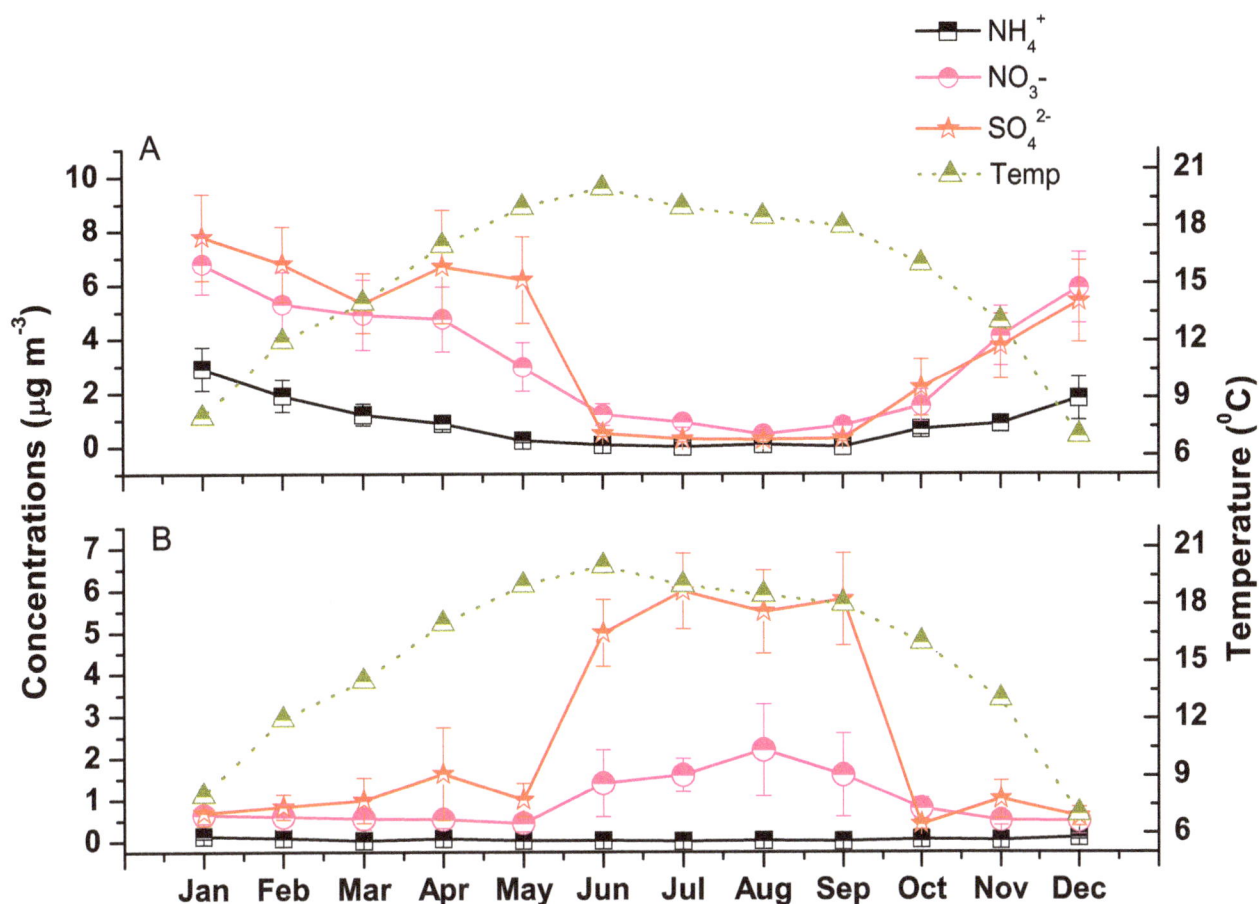

Figure 6. Seasonal variations of secondary ionic species and temperature over Darjeeling.

than at Pune and Ahmedabad whereas we observed a similarity in SO_4^{2-} concentrations between Darjeeling and Mumbai. The concentrations of Ca^{2+} over all the urban sites are found to be higher than that over Darjeeling. The very high Ca^{2+} concentration over Ahmedabad is due to its close proximity to Thar deserts. On the other hand, the concentration of Mg^{2+} in Darjeeling is found to be higher than Pune and Ahmedabad and lower than Mumbai and Agra.

Conclusively, we observed that Darjeeling in northeastern Himalayas is strongly influenced by the sea salt and anthropogenic aerosol which are higher than all the high altitude stations in Indian and Nepal Himalayan region and even higher than most of the urban regions in India, compared with the present study. On the other hand, dust aerosol is found to be homogeneously distributed vertically between all the Himalayan regions but is significantly lower than urban regions in India.

2.2. Secondary ionic species in fine and coarse mode aerosol. NH_4^+, NO_3^- and SO_4^{2-}, the secondary components of aerosol, constituted $67.8 \pm 5.9\%$ fine mode and $36.4 \pm 10.07\%$ coarse mode aerosol in Darjeeling. The average concentrations of NH_4^+, NO_3^- and SO_4^{2-} in fine and coarse mode aerosol were 0.88 ± 0.76 μg m^{-3}, 3.31 ± 2.25 μg m^{-3}, 3.8 ± 2.9 μg m^{-3} and 0.05 ± 0.04 μg m^{-3}, 0.95 ± 0.17 μgm^{-3}, 2.5 ± 2.1 μg m^{-3}, respectively.

NH_4^+ and NO_3^- in fine mode aerosol showed higher enrichment (~15 times and 4 times, respectively) compared to coarse mode aerosol. This is due to the fact that gas to particle conversion of gas phase HNO_3 and NH_3 to particulate NO_3^- and

NH_4^+ are more feasible in nucleation (<0.1μm) and accumulation mode (>0.1 μm <2.5 μm) particles [20]. NH_4^+ and NO_3^- show good correlation (with correlation coefficient, $R^2 = 0.74$ and number of samples, n = 81) (Table 3) between each other in fine mode aerosol during dry seasons. Figure 6 shows the month wise variations in secondary species concentrations along with ambient temperature both in fine and coarse mode aerosol. The variation of fine mode NH_4^+ and NO_3^- are similar in nature showing a gradual decrease in concentrations from the month of January with a minimum during monsoon and a gradual increase till December which is exactly opposite in nature with respect to the ambient temperature variations. The higher NH_4^+ concentrations during winter is under the influence of NE wind transporting the large-scale pollutants whereas the lower concentration of NH_4^+ in monsoon suggests the dominant occurrence of NH_4^+ in gas phase [19].

The higher concentration of particulate ammonium and nitrate in winter was due to the shifting from the gas phase of nitric acid to the particulate phase of nitrate at lower temperature. The formation of particulate NH_4NO_3 is given by the following equilibrium

$$NH_3(g) + HNO_3(g) \leftrightarrow NH_4NO_3(s)$$

The equilibrium shifts towards the left side at higher temperature as NH_4NO_3 volatilizes when temperature increases [21]. We observed a gradual decrease in fine mode ammonium and nitrate

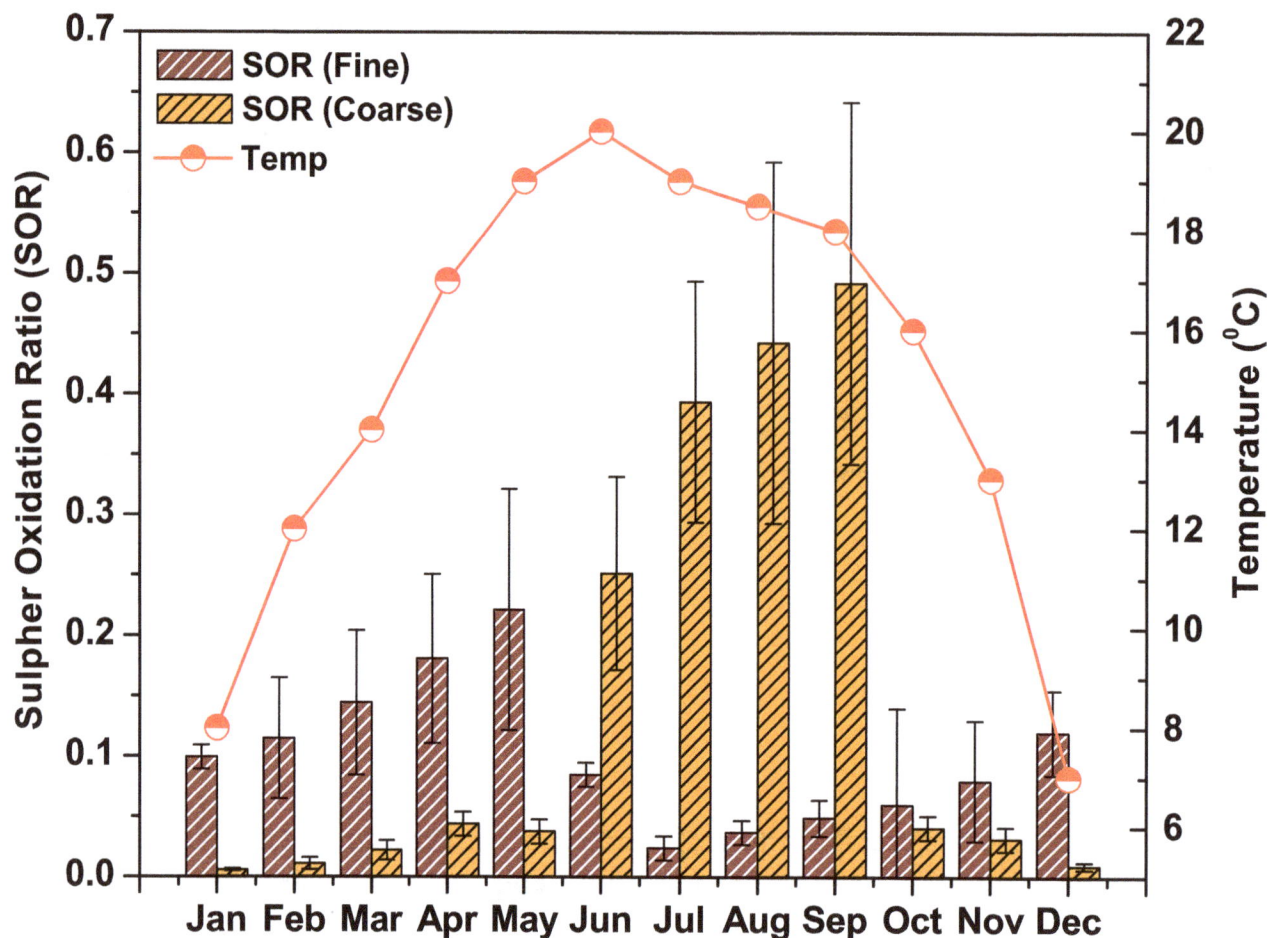

Figure 7. Different mechanisms of sulphate production over Darjeeling in different seasons.

concentrations from January as temperature increases. Also, when there is significant production of sulphate by reaction between OH radical and SO_2, then sulphate acts as a sink for ammonia. Once ammonia becomes ammonium bisulphate or ammonium sulphate, ammonium becomes unavailable for nitrate [22]. We observed the high sulphate production in fine mode aerosol through SO_2 oxidation by OH radical (discussed in section 4.2) during premonsoon. While fine mode nitrate shows minimum concentrations during monsoon due to the wash out effect or below cloud scavenging, the coarse mode NO_3^- shows higher concentrations during monsoon compared to the dry seasons. The presence of sea salt aerosol may favor the formation of coarse mode NO_3^- as $NaNO_3$ by up taking of nitrogen oxides and HNO_3 on the surface of sea salt aerosol [23]. Many investigators have observed that the depletion of Cl^- from sea salt aerosol and the simultaneous occurrence is particularly pronounced in coarse particle, typical of sea salt aerosol [23]. The role of nitrate in chloride depletion from sea salt aerosol has been discussed in detail in section 4.2.

The sulfur oxidation ratio defined as $SOR = \dfrac{non-sea-SO_4^{2-}}{(non-sea-SO_4^{2-}+SO_2)}$ was used as an indicator of the secondary transformation process and the formation route and source of sulphate in the atmosphere. Figure 7 shows the month wise variation of SOR with temperature for both fine and coarse mode aerosol. The ratio varied between 0.03 and 0.22 with an average of 0.11 ± 0.07 in fine mode aerosol and between 0.006 and

0.49 with an average of 0.15 ± 0.14 in coarse mode aerosol. SORs in fine mode aerosol were found to increase gradually with the temperature and higher values were observed during premonsoon. When the ratio value is greater than 0.10, then there would be the occurrence of the photochemical oxidation of SO_2 in the atmosphere [24]. This indicates that gas phase photochemical oxidation of SO_2 followed by the condensation and absorption into the particle phase was the most important pathway for the production of fine mode SO_4^{2-} in the atmosphere during premonsoon (SOR>0.1). The gas phase oxidation of SO_2 to SO_4^{2-} by OH radical is a strong function of temperature [25]. The comparatively lower SOR during the winter months for both the fine and coarse mode aerosols indicate some aqueous phase transformation processes such as metal catalyzed oxidation of SO_2, aqueous phase H_2O_2/O_3 oxidation of SO_2 etc. Similar observation was also made in [26]. On the other hand, the very high SOR in coarse mode aerosol during monsoon could be due to the absorption of SO_2 by the soil dust particles at higher relative humidity during monsoon to form coarse mode sulphate of crustal origin [19,27]. Table 4 shows strong correlations (with correlation coefficient, $R^2 = 0.68$ and number of samples, n = 30) non-sea-SO_4^{2-} and non-sea-Ca^{2+} and between non-sea-SO_4^{2-} and non-sea-Mg^{2+} ($R^2 = 0.68$, n = 30) in coarse mode aerosol during monsoon.

2.3. Primary ionic species in fine and coarse mode aerosol. Na^+ and Cl^- constituted $16.6\pm5.46\%$ fine mode and

Table 4. Inter-ionic correlations in fine and coarse mode aerosols over Darjeeling during monsoon (n = 30).

	Na^+	NH_4^+	K^+	nsK^+	Ca^{2+}	$nsCa^{2+}$	Mg^{2+}	$nsMg^{2+}$	Cl^-	NO_3^-	SO_4^{2-}	$nsSO_4^{2-}$
Na^+		F=0.15 C=0.14	F=0.14 C=0.11	F=0.17 C=0.22	F=0.10 C=0.11	F=0.19 C=0.23	F=0.15 C=0.23	F=0.17 C=0.31	F=0.68 C=0.89	F=0.12 C=0.83	F=0.29 C=0.39	F=0.78 C=0.21
NH_4^+			F=0.16 C=0.19	F=0.21 C=0.41	F=0.23 C=0.31	F=0.11 C=0.22	F=0.09 C=0.13	F=0.08 C=0.03	F=0.15 C=0.10	F=0.14 C=0.31	F=0.11 C=0.12	F=0.10 C=0.14
K^+				F=0.11 C=0.18	F=0.16 C=0.11	F=0.05 C=0.11	F=0.03 C=0.02	F=0.13 C=0.09	F=0.11 C=0.19	F=0.23 C=0.16	F=0.09 C=0.07	
nsK^+					F=0.39 C=0.13	F=0.10 C=0.21	F=0.11 C=0.08	F=0.10 C=0.06	F=0.08 C=0.02	F=0.19 C=0.31	F=0.29 C=0.11	F=0.33 C=0.41
Ca^{+2}						F=0.20 C=0.32	F=0.11 C=0.12	F=0.09 C=0.06	F=0.10 C=0.25	F=0.21 C=0.30	F=0.14 C=0.15	
$nsCa^{2+}$							F=0.19 C=0.04	F=0.16 C=0.13	F=0.15 C=0.12	F=0.13 C=0.21	F=0.08 C=0.06	F=0.13 C=0.72
Mg^{+2}									F=0.33 C=0.68	F=0.13 C=0.03	F=0.12 C=0.06	F=0.07 C=0.06
$nsMg^{2+}$									F=0.11 C=0.03	F=0.16 C=0.02	F=0.04 C=0.10	F=0.09 C=0.68
Cl^-										F=0.21 C=0.15	F=0.31 C=0.58	F=0.09 C=0.07
NO_3^-											F=0.15 C=0.15	F=0.16 C=0.13
SO_4^{-2}												
$nsSO_4^{2-}$												

46.2±11.14% coarse mode aerosol. The average concentrations of Na^+ and Cl^- in fine and coarse mode aerosol were found to be 0.66±0.43 μg m^{-3}, 1.21±1.00 μg m^{-3} and 2.16±2.01 μg m^{-3}, 2.35±1.5 μg m^{-3} respectively. The seasonal distribution of Na^+ and Cl^- in coarse mode aerosol is similar in nature with the higher concentrations during monsoon. But the concentration of fine mode Cl^- was found to be higher during the winter months and minimum during monsoon (Figure 8).

During monsoon, the correlation between coarse mode Na^+ and Cl^- was found to be very strong ($R^2 = 0.89$, n = 30) (Table 4). The month wise variation of Na^+ and Cl^- (Figure 8) for both fine and coarse mode aerosol show that during monsoon, coarse mode Na^+ and Cl^- show higher concentrations whereas fine mode Na^+ and Cl^- show minimum concentrations. This indicates that both the coarse mode Na^+ and Cl^- have a common source *i.e.* sea salt particle, which could be transported by the southwest monsoon. Monsoon air masses reaching Darjeeling originate in the Bay of Bengal and incorporate a large amount of Na^+ and Cl^-. The inter-tropical convergence zone (ITCZ) is aligned at a rather northerly position during the monsoon season. The air masses are vertically raised by convective motion and transported horizontally by the upper air southerly monsoon flow to the Himalayas [12]. The higher concentrations of fine and coarse mode Cl^- compared to Na^+ during the dry seasons (Jan–May, Oct–Dec) is believed to be associated with biomass and coal burning (domestic and railway). We also observed strong correlations between non-sea-K^+ and Cl^- ($R^2 = 0.82$, n = 81) and between non-sea-SO_4^{2-} and Cl^- ($R^2 = 0.78$, n = 81) during dry seasons (Table 3) indicating the common biomass-burning source.

The average concentrations of K^+ in fine and coarse mode aerosol were 1.17±0.83 μg m^{-3} and 0.31±0.17 μg m^{-3}, respectively. Biomass burning could have the highest abundance of fine mode K^+ of all source emissions. The fine mode K^+ could be released in the atmosphere by the burning of vegetative scrap [28].

The average concentration of fine mode non-sea-K^+ during winter was much higher (2.12±0.4 μg m^{-3}) than premonsoon (1.37±0.8 μg m^{-3}), monsoon (1.15±0.7 μg m^{-3}) and post-monsoon (1.08±0.62 μg m^{-3}) which was due to the massive biomass burning around Darjeeling especially during night times in winter. Figure 9(A) shows the month wise variations of fine mode non-sea-K^+ which is similar to fine mode non-sea-SO_4^{2-} variations. A strong correlation ($R^2 = 0.78$, n = 81) between non-sea-K^+ and non-sea-SO_4^{2-} was also observed in fine mode aerosol during dry seasons (Table 3) indicating their common source of biomass burning.

The average concentrations of Ca^{2+} and Mg^{2+} in fine mode aerosol were 0.13±0.1 μg m^{-3} and 0.12±0.09 μg m^{-3} respectively and in coarse mode aerosol were 0.39±0.19 μg m^{-3} and 0.31±0.17 μg m^{-3} respectively. Figure 9(B) shows the monthly variations of non-sea-Ca^{2+} and non-sea-Mg^{2+} both in fine and coarse mode aerosol. Strong correlations were observed between non-sea-Ca^{2+} and non-sea-Mg^{2+} (Table 3) during dry seasons indicating their enrichment in aerosol mainly as dust aerosol. Non-sea-Ca^{2+} and non-sea-Mg^{2+} both show peaks during pre-monsoon both in fine and coarse mode aerosol. The higher concentrations of those mineral components during pre-monsoon could be related to the long-range transport of dust aerosol. The dust aerosols driven by the premonsoon westerlies are vertically advected to elevated altitudes (~5 km) against the foothills of the Himalayas due to the enhanced convection and steep pressure gradient across the Himalayan-Gangetic region [29]. According to [29], the dust-rich aerosols from the Indo-Gangetic plain can "climb" the slope of the Himalayas in the premonsoon season. Similar concentrations of Ca^{2+} at two different altitudes, 800 and 3920 masl at two stations in Nepal Himalayan region during premonsoon was also observed showing the long-range transport from southwestern Asia and northern Africa [13]. Very high concentrations of coarse mode Ca^{2+} over a station near Thar deserts was also observed [30,31].

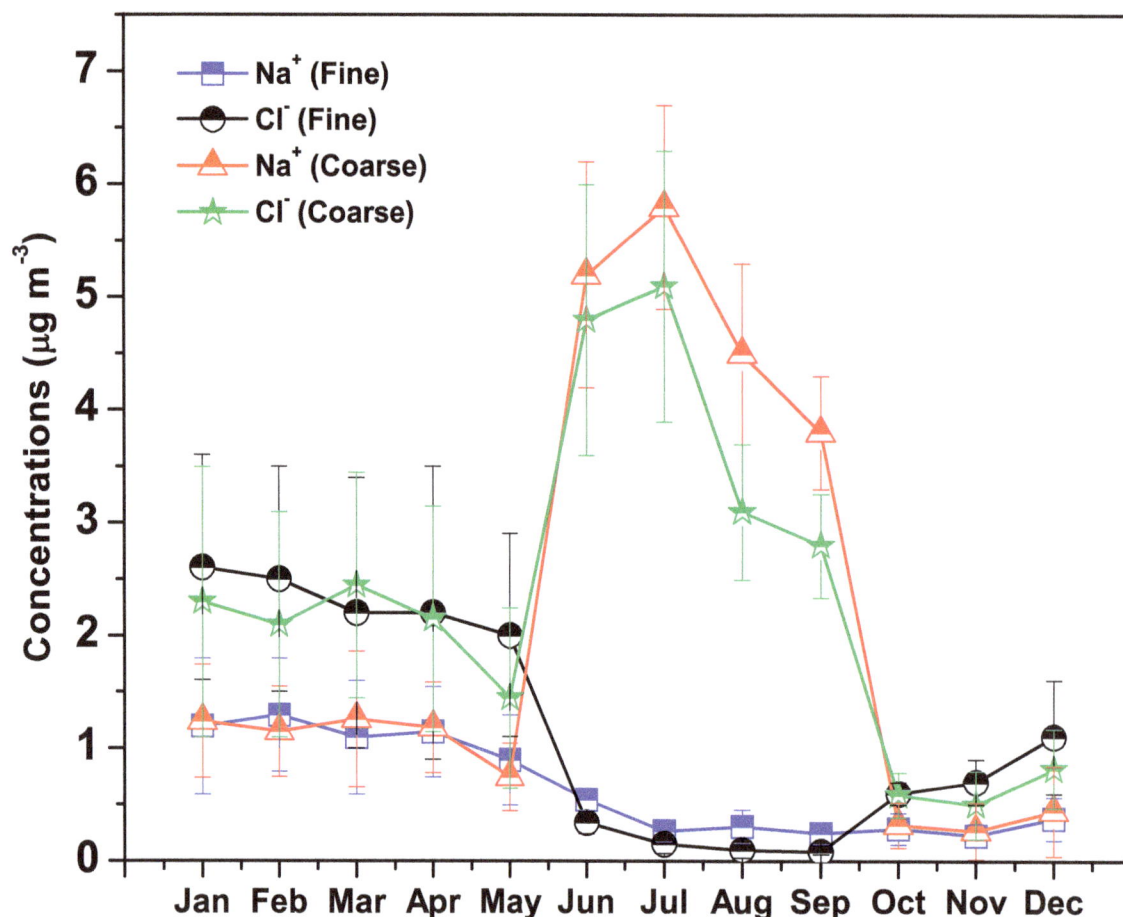

Figure 8. Seasonal variations of primary ionic species over Darjeeling.

Using HYSPLIT_4 (Hybrid Single Particle Lagrangian Integrated Trajectory, www.arl.noaa.gov/ready/hysplit4.html) model developed by NOAA/ARL, back trajectories were computed for all sampling events at 0500 UTC. Two distinct source regions are identified, arid and semi-arid regions of western India including Thar deserts (45%) and upwind regions of Arabian deserts (32%), which are shown in figure 10A and 10B as representative figures. The percentage distributions of source regions (45% from Thar deserts and 32% from Arabian deserts) were determined based on the ratio of the number of events of the respective regions (using HYSPLIT) to the total number of sampling events. The loading of Ca^{2+} and Mg^{2+} over Darjeeling from Thar deserts was found to be higher than the Arabian deserts. The correlation of Ca^{2+} and Mg^{2+} with trace metals like Fe, Al, Si during premonsoon would be useful to put the evidence towards the transport of dust aerosol from distant regions. However the studies on the effect of dust aerosol on the total aerosol loading over Himalayan region and its effect on the optical properties of aerosol during premonsoon are under development.

The higher concentrations of Ca^{2+} over Ahmedabad (located near Thar desert in western India) compared to Darjeeling and other Himalayan regions (Table 1 and 2) indicate that the dust aerosol reach Himalayas after significant dilution. With the onset of rainy season (the Arabian Sea and Bay of Bengal branches of the southwest summer monsoon), the heavy dust loading significantly diminishes due to aerosol washout from the atmosphere. Non-sea-Ca^{2+} and non-sea-Mg^{2+} both show mini-

mum concentrations indicating the below cloud scavenging of dust aerosol due to heavy rain during monsoon (1783 mm). However, the negative concentration of coarse mode non-sea-Mg^{2+} during monsoon indicates that the entire coarse mode Mg^{2+} was from the marine source during monsoon.

2.4. Chloride depletion: interaction between marine and urban aerosol. Chloride depletion results when the acidic species, mainly nitrate, sulphate and some organic acids react with sea-salt particles and replace Cl^- in the form of HCl gas. NOx transforms into gaseous nitrous and nitric acid, which react with NaCl in sea-salt aerosol to form $NaNO_3$ and HCl. SO_2 oxidation and to a lesser extent H_2SO_4 vapor condensation on sea salt aerosols can also lead to chloride depletion. The reaction pathways are given as follows [32]:

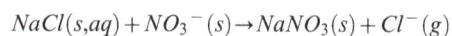

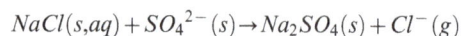

$$NaCl(s,aq) + NO_3^-(s) \rightarrow NaNO_3(s) + Cl^-(g)$$

$$NaCl(s,aq) + SO_4^{2-}(s) \rightarrow Na_2SO_4(s) + Cl^-(g)$$

Those reactions given above occur especially when polluted urban aerosols and maritime aerosols are mixed with each other. Thus the extent of chloride depletion is an indication of the interaction between marine and urban aerosol as well as important in estimating the amount of nitrate and sulphate formed on sea-salt particles. Figure 11 shows the monthly variations of chloride depletion and it is clearly observed from the figure that only monsoon shows chloride depletion both in fine and coarse mode

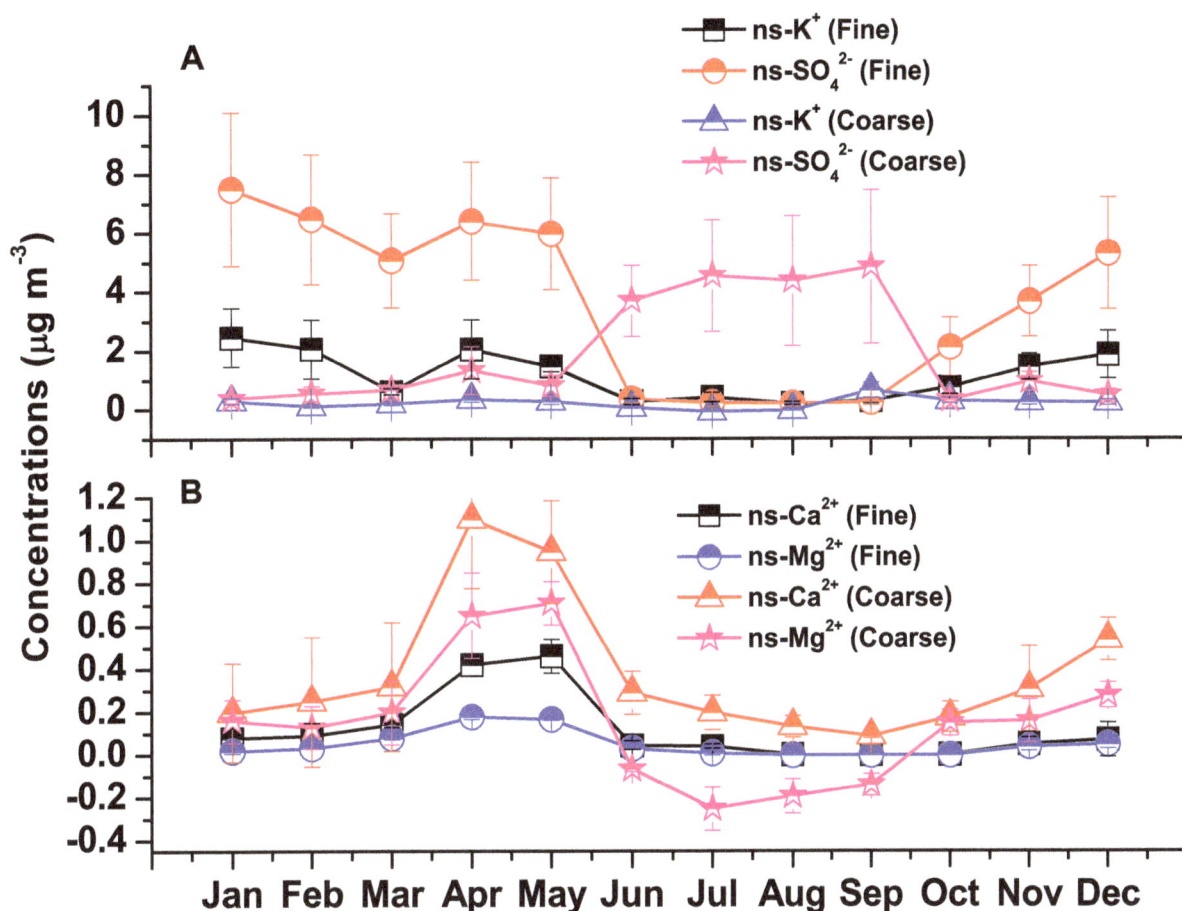

Figure 9. Seasonal variations of non-sea (ns) aerosol ionic species. A) Anthropogenic aerosols (biomass burning) and B) Natural aerosols (dust).

aerosol. The percentage of Cl^- depletion (Cl^-_{dep}) can be calculated as follows [32]:

$$Cl^-_{dep}(\%) = \frac{\left([Cl^-]_{seawater,estimated} - [Cl^-]_{aerosol,measured}\right)}{[Cl^-]_{seawater,estimated}} \times 100$$

$$= \frac{\left(1.17[Na^+]_{aerosol,measured} - [Cl^-]_{aerosol,measured}\right)}{1.17[Na^+]_{aerosol,measured}} \times 100$$

Here, the source of measured Na^+ concentration in aerosol is assumed to be sea water only. In this study, the average depletion of Cl^- in fine mode aerosol (61.2±13.3%) during monsoon was higher than coarse mode aerosol (31.02±9.6%). It was suggested that the surface reaction mechanism is the principal explanation for higher depletion of smaller particles [33,34]. The dynamics of the chloride depletion reactions favor smaller particles because of their larger surface area distribution and longer atmospheric residence time. [32] observed that the Cl^- depletion decreases from 98% to 10% as particle size increases from 1.8 μm to 18 μm. [33] reported a reduction of chloride depletion of 90–100% at particle size of 1–2 μm to less than 40% at particle size of 8–15 μm for sea-salt particles in Finland.

An attempt was made to estimate the contribution of nitrate and non-sea-sulphate to the chloride depletion in fine and coarse mode aerosol. A ratio of the measured Na^+ concentrations in aerosol to the estimated original Cl^- concentration in sea salt was used to determine the contribution of nitrate and non-sea-sulphate to chloride depletion. According to [33] the original chloride concentrations in sea salt can be determined as $[Cl^-]_{original} = [Cl^-]_{measured} + [NO_3^-]_{measured}$, assuming that chloride depletion results from nitrate formation process and the amount of measured nitrate is equal to the amount of lost chloride or $[Cl^-]_{original} = [Cl^-]_{measured} + [non-sea-SO_4^2-]_{measured}$, assuming that chloride depletion results from non-sea-sulphate formation process and the amount of measured non-sea-sulphate is equal to the amount of lost chloride or $[Cl^-]_{original} = [Cl^-]_{measured} + [NO_3^-]_{measured} + [non-sea-SO_4^{2-}]_{measured}$, assuming both nitrate and non-sea-sulphate are responsible for the depletion of chloride. Figure 12 shows the scatter plot of the ratios during monsoon for fine and coarse mode aerosol along with the ratio of measured $[Na^+]$ to measured $[Cl^-]$ in aerosol. The ratio, $[Na^+]/[Cl^-]$ in unreacted original sea water is 0.85.

Naturally the ratio, $[Na^+]/[Cl^-]$, in fine and coarse mode aerosol were higher than 0.85 because of chloride depletion. Interestingly, the ratios $[Na^+]/([Cl^-]+[NO_3^-])$ for coarse mode aerosol were found to be close to 0.85. This indicates that the measured coarse mode nitrate during monsoon was involved in chloride depletion. A strong correlation between coarse mode Na^+ and coarse mode NO_3^- during monsoon ($R^2 = 0.83$, n = 30) (Table 4) added further evidence to the above fact. On the other hand, the ratios $[Na^+]/([Cl^-]+[NO_3^-])$ for fine mode aerosol were quite lower than 0.85 indicating the overestimation of nitrate, *i.e.* the measured nitrate was not totally the nitrate from sea-salt but also from some other sources.

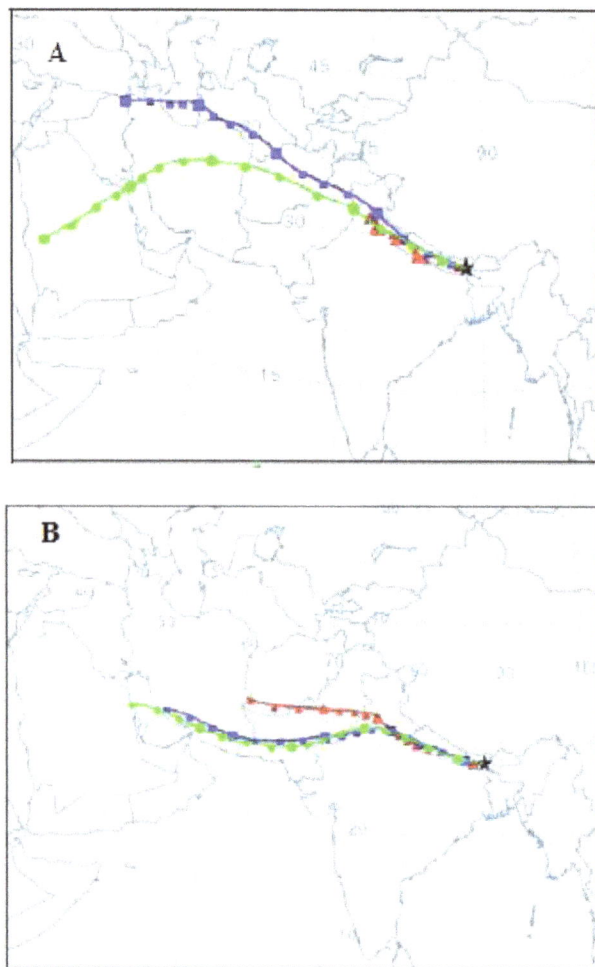

Figure 10. Long range transport of dust aerosol. A) From Thar deserts and B) From Arabian deserts.

The ratios, $[Na^+]/([Cl^-]+[non\text{-}sea\text{-}SO_4^{2-}])$ in fine mode aerosol were close to 0.85. This indicates that the measured fine mode non-sea-sulphate during monsoon was totally involved in chloride depletion. We observed a strong correlation between Na^+ and fine mode non-sea-SO_4^{2-} ($R^2 = 0.78$, n = 30) (Table 4) during monsoon. Thus, we observed that non-sea-sulphate depleted chloride from fine mode sea salt aerosol whereas nitrate depleted chloride from coarse mode sea salt aerosol.

The chloride depletion from sea salt aerosol *i.e.* the interaction between marine (natural) and urban (anthropogenic) aerosol has major implications not only in changing the chemical composition of aerosol in different size modes but also it governs the rate of dry atmospheric removal of the species involved in chloride depletion. For example, here in this study the chloride depletion from coarse mode sea salt aerosol was found to govern the dry atmospheric removal of nitrate from the atmosphere. The dry depositional flux of nitrate in fine and coarse mode aerosol was calculated as follows:

$$\text{Dry depositional flux } \left(mg\,m^{-2}\,day^{-1}\right) =$$

$$\text{concentration}\left(mg\,m^{-3}\right) \times \text{deposition velocity}\left(m\,day^{-1}\right)$$

Here, deposition velocity of nitrate in fine mode aerosol was considered to be 0.1 cm s^{-1} and that in coarse mode was

considered to be 1 cm s^{-1} [35]. It is relevant to state that these deposition velocities can be in error up to a factor of 3 [36]. The nitrate was found to be enriched mainly in fine mode aerosol as NH_4NO_3 during winter with the average deposition flux of 518 ± 65 mg m^{-2} day^{-1}. On the other hand, coarse mode nitrate was found to be enriched as $NaNO_3$ after depleting chloride from sea salt aerosol during monsoon with a higher deposition flux of 1468 ± 300 mg m^{-2} day^{-1}. Thus we observed that the interaction between natural marine aerosol and anthropogenic urban aerosol enhanced the rate of dry atmospheric removal of nitrate.

3. Acidity of aerosol: ionic balance and ionic ratios

Strong acidity in terms of nmol H$^+$ per m^3 of air is a parameter that characterizes the absolute acidity of atmospheric aerosols [37]. This is obtained from the total H$^+$ derived from the aqueous extract of atmospheric aerosols and is estimated using an ionic balance of the inorganic ionic species. On the other hand, free acidity (pH) characterizes the relative acidity of atmospheric aerosols and depends on the water content and composition of aerosol and relative humidity [36]. Here in this study we have given more emphasis on ionic balance in explaining the aerosol acidity. Although strong acidity is due to the strong acidic components in aerosols like nitrate and sulphate, we have estimated the acidity (total H$^+$ in nmol m^{-3}) using all the ionic species according to [38].

$$[H^+]_{total} = \left(2 \times [SO_4^{2-}] + [NO_3^-] + [Cl^-]\right) -$$
$$\left([Na^+] + [K^+] + [NH_4^+] + 2 \times [Ca^{2+}] + 2 \times [Mg^{2+}]\right)$$

Figure 13 shows the month wise variations of $[H^+]_{total}$ and the equivalent ratio of the sum of cations to the sum of anions (Σ^+/Σ^-) in fine and coarse mode aerosol. It was observed that fine mode aerosol acidity ($[H^+]_{total}$) was much higher during dry seasons (73 nmol m^{-3}) compared to monsoon (-4.1 nmol m^{-3}) whereas coarse mode aerosol acidity was higher during monsoon (19 nmol m^{-3}) compared to dry seasons (-34.6 nmol m^{-3}). The ratio, Σ^+/Σ^-, ranged between 0.46 to 1.43 (Average: 0.84, Standard deviation: 0.29) in fine mode aerosol and between 0.9 to 1.95 (Average: 1.31, Standard deviation: 0.34) in coarse mode aerosol. The cation deficiency in fine mode aerosol could be attributed to H$^+$ whereas anion deficiency in coarse mode aerosol could be attributed to unanalyzed organic acid ions, carbonate, bicarbonate etc. The higher concentrations of nitrate, non-sea-sulphate and chloride during dry seasons increased $[H^+]_{total}$ in fine mode aerosol whereas the lower concentrations of non-sea-magnesium and non-sea-calcium along with higher concentration of nitrate during monsoon increased $[H^+]_{total}$ in coarse mode aerosol. Thus, the acidity of fine mode aerosol was mainly controlled by the species originated from fossil fuel and biomass burning whereas the acidity of coarse mode aerosol was mainly controlled by dust particles.

4. Sources of aerosol

4.1. Source apportionment by Principal Component Analysis. It is very important, and difficult as well, to identify the exact sources of the aerosol components in Darjeeling, the area under concern at the northeastern Himalayas, where composite anthropogenic activities including biomass burning and vehicular emission play a central role in loading of air pollutants in the atmosphere. Principal component analysis (PCA), a multivariate analysis technique [39] was used to identify possible sources of aerosols in Darjeeling. Each principal component (PC) shows correlation of each variable as loadings (loading greater than 0.5 was

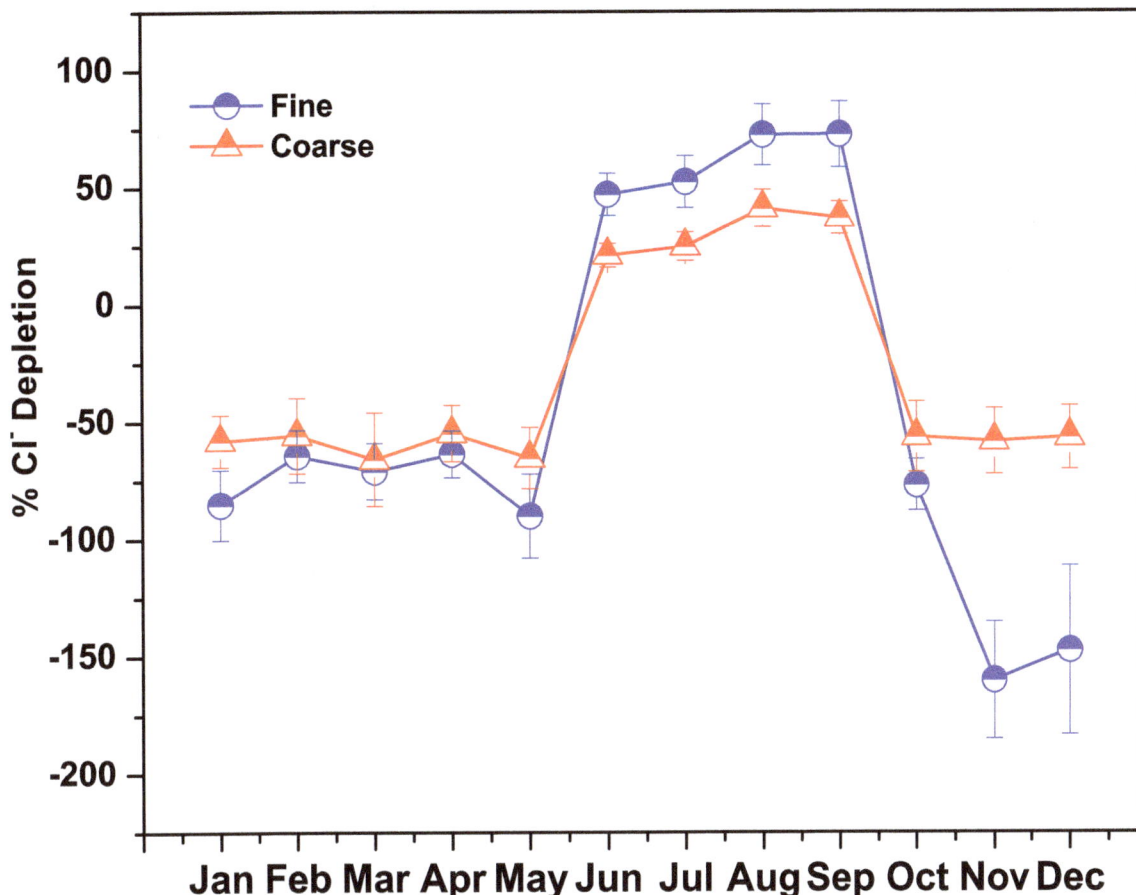

Figure 11. Seasonal variation of chloride depletion from sea-salt aerosol.

considered to be statistically significant in this study). Since higher loading of particular variable in a PC can help in identifying the possible sources [40], the number of PCs selected (sources identified) should represent the sources that are relevant in the receptor domain. PCA was performed using the statistical software, SPSS (Statistical Package for the Social Sciences) [41] of version 16.0.2 using the data sets over the entire study period (n = 66).

For fine mode aerosol, three PCs were extracted (Figure 14). The first PC (PC1) shows the heavy loading of K^+, Cl^- and SO_4^{2-} with wind speed having 41.5% variance of the data set. These species in PC1 correspond to the massive biomass burning throughout the year mainly during winter in Darjeeling. The coal engine which is used for the Darjeeling Himalayan Railways is also a major source of these species in the atmosphere. Thus, PC1 indicates the loading of K^+, Cl^- and SO_4^{2-} as non-sea-K^+, non-sea-Cl^- and non-sea-SO_4^{2-} respectively. The negative loading of wind speed in this PC indicates the dispersion of these fine mode species favored by higher wind speed. The second PC (PC2) shows the heavy loading of NH_4^+, NO_3^-, SO_4^{2-} with 33.3% of the data variance. This PC is associated with the formation of secondary anthropogenic particles in the atmosphere. Different agricultural activities and usage of different ammoniated fertilizers in several tea gardens and also in tea processing plants, animal manure and human activities are the major sources for the emission of NH_3 and/or NH_4^+ in the atmosphere. The third PC (PC3) is moderately loaded with NH_4^+ and highly loaded with NO_3^- along with temperature with 16.4% of the data variance. This

indicates the emission of NOx from vehicular exhaust and its subsequent transformation to particulate nitrate mostly as ammonium nitrate. The negative loading of temperature in PC3 clearly indicates the inverse relation of formation of particulate nitrate with temperature as discussed earlier.

For coarse mode aerosol, four PCs were extracted (Figure 14). The first PC (PC1) is highly loaded with the Na^+, Cl^- and Mg^{2+} having 40.5% data variance. This indicates the loading of sea salt aerosol. The second PC (PC2) shows the heavy loading of Ca^{2+}, Mg^{2+} along with wind speed with the moderate loading of K^+. This explains 34.2% of the data variance and indicates the enrichment of calcium, magnesium and a fraction of potassium as non-sea-calcium, non-sea-magnesium and non-sea-potassium from soil dust particles. The loading of wind speed in this PC indicates the resuspension of soil dust in the atmosphere favored by the higher wind speed. The third PC (PC3) shows the loading of Na^+ and NO_3^- with the data variance of 14.6%. This indicates the interaction of sea salt aerosol (NaCl) with nitrate (particulate or gaseous) and the production of coarse mode nitrate through the chloride displacement reactions discussed earlier. The fourth PC (PC4) shows the higher loading and association of SO_4^{2-} only with 6.4% of data variance. The loading of sulphate singly in PC4 indicates the enrichment of non-sea-sulphate in coarse mode which is due to the aqueous phase transformation of SO_2 to sulphate in high relative humidity during monsoon. The higher concentration of non-sea-sulphate during monsoon in coarse mode aerosol has already been discussed.

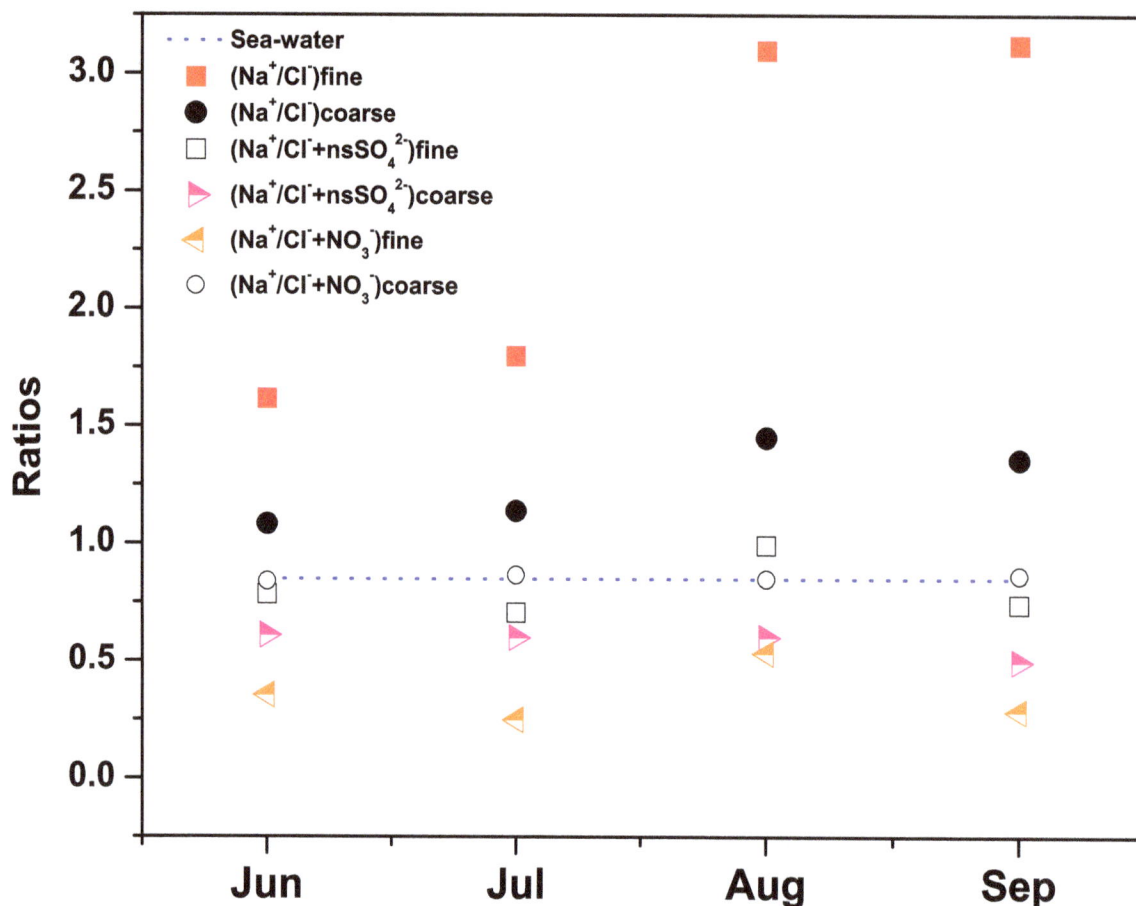

Figure 12. Modified ratios of sodium to chloride indicating contribution of nitrate and non-sea (ns) sulphate to chloride depletion.

4.2. Source identification by air mass parcels using HYSPLIT back trajectory. In order to find out the possibility of atmospheric transport of different species from outside of Darjeeling, backward trajectories for air parcels arriving at the sampling site during the sampling period were calculated with the HYSPLIT_4 model. Back trajectories were computed for all sampling events at 0500 UTC. The Final Run (FNL) meteorological data was used for the trajectory calculation.

Different air masses travel through different regions and bring different chemical components with the aerosol, thus the distribution of chemical components among different air masses could shed some light on their possible sources. Based on the transport pathways of air masses, three typical air mass trajectories, representing local, continental (other than local), and marine were found in Darjeeling and have been shown in Figure 14. Marine air parcels are mainly from the South East (Bay of Bengal) and South West (Arabian Sea) and could bring a large amount of marine aerosol. Continental air masses are mainly from North West and could bring several crustal species from inland emissions. Table 5 presents the distributions of mass concentrations of major ionic species along with $[H^+]_{total}$ (nmol m^{-3}) and Σ^+/Σ^- ratio both in fine and coarse mode aerosol for three types of air masses.

The higher contribution of local and continental sources to the fine mode aerosol was observed with a little influence of marine sources while the concentration of coarse mode aerosol was found to be higher from the marine source regions than the local and

other continental sources. The concentrations of Na$^+$ and Cl$^-$ follow the same sequence and their coarse mode concentrations were found to be the highest for marine air parcel, indicating the strong influence from the sea. The higher concentrations of fine mode Cl$^-$ for continental and local sources indicate the influence of biomass and coal burning. The higher concentrations of coarse mode Ca^{2+} and Mg^{2+} from local and continental sectors indicate the resuspension of local and wind blown soil dust particles. The concentration of coarse mode Mg^{2+} from marine sectors shows the influence of sea salt Mg^{2+}. Local and continental sectors show the higher concentrations of fine and coarse mode K$^+$ indicating the strong influence of biomass burning and soil dust aerosol respectively. The higher concentrations of fine mode NH$_4^+$, NO$_3^-$ and SO$_4^{2-}$ from local and continental sectors show a very strong influence of anthropogenic activities whereas coarse mode NO$_3^-$ shows higher concentration from marine sectors. This coarse mode nitrate from marine sectors is not sea salt nitrate rather it indicates the interaction between sea salt aerosol and gas phase HNO$_3$ and its association with sodium through chloride displacement reaction discussed earlier. Fine mode aerosol shows higher $[H^+]_{total}$ and lower Σ^+/Σ^- (104.8 nmol m^{-3} and 0.59, respectively) from the local compared to continental (39.2 nmol m^{-3} and 0.79, respectively) and marine (-11.3 nmol m^{-3} and 1.2, respectively) source regions whereas coarse mode aerosol shows maximum $[H^+]_{total}$ and minimum Σ^+/Σ^- ratio (8.03 nmol m^{-3} and 0.96, respectively) from marine source compared to local (-57.5 nmol m^{-3} and 1.51, respectively) and continental sources

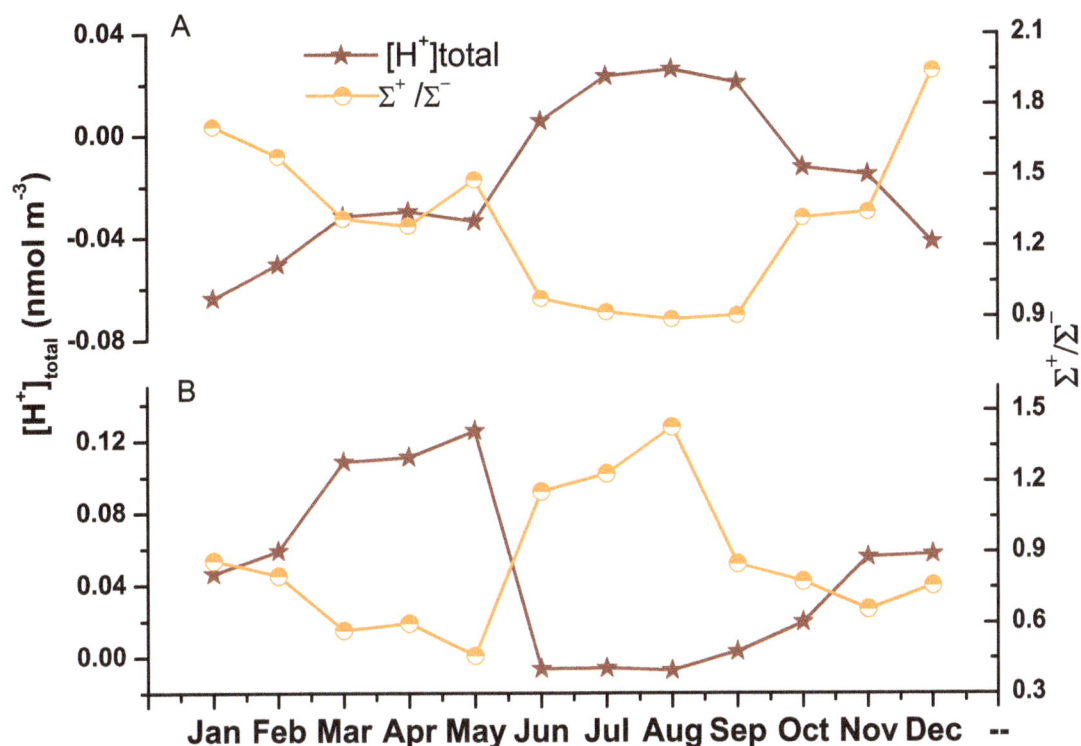

Figure 13. Seasonal variation of acidity of aerosol and ionic ratios.

(-2.6 nmol m^{-3} and 1.03, respectively). Thus, local and continental fine mode aerosols were more acidic than coarse mode aerosol whereas marine fine mode aerosols were less acidic compared to coarse mode aerosol.

Based on the Principal Component Analysis (for source types) and HYSPLIT trajectory model (for source regions) the percentage distribution of several source types of water soluble ions from different source regions (local, continental and marine) in fine and coarse mode aerosol were determined and depicted in Figure 15. In this figure, the percentage distribution of source regions was determined based on the ratio of the number of events of the respective regions (using HYSPLIT) to the total number of events. The figure also describes the percentage distribution of ionic species between primary and secondary species. The percentage of the primary species was determined based on the ratio of sum of the concentrations of chloride, sodium, potassium, calcium and magnesium to the total ionic concentration of aerosol whereas ammonium, nitrate and sulphate were used for the determination of percentage of secondary species in a similar way. Principal component analysis was done using data set from each sector (local, continental and marine) separately for fine and coarse mode aerosol. That means the percentage distribution of various source types (shown in Figure 15) is actually the percentage of variation obtained from factor analysis by PCA.

Figure 15 shows that, 80% of the ionic species originated from the local and other continental sources and 20% was from the marine source. Biomass burning and vehicular emissions (non-sea-potassium, non-sea-sulphate, non-sea-chloride, nitrate) were the major sources for fine mode local and continental aerosols whereas dust particles (non-sea-calcium and magnesium) were the major source for coarse mode local and continental aerosols. On the other hand, the major source for fine and coarse mode marine aerosol was sea salt aerosols enriched mainly with sodium and chloride.

Conclusions

The major findings of this study can be summarized as:

(1) The average concentration of fine mode aerosol is found to be higher than coarse mode aerosol during dry seasons whereas monsoon shows higher loading of coarse mode aerosol compared to fine mode.

(2) This study discusses the formation pathways of major secondary aerosol components. Nitrate is found to exist as ammonium nitrate in fine mode aerosol during winter whereas during monsoon nitrate exists as sodium nitrate in coarse mode aerosol. Photochemical oxidation of SO_2 is the major pathway for the formation of sulphate during premonsoon whereas some other aqueous phase transformation processes are important pathways during winter in sulphate formation.

(3) There is a major contribution of non-sea-sulphate and nitrate in the chloride depletion from fine and coarse mode sea salt aerosols, respectively. The chloride depletion, i.e. the interaction between marine and urban aerosol, is found to govern the dry removal of nitrate from the atmosphere. The nitrate which is found to be enriched mostly in fine mode during dry seasons gets enriched in coarse mode during monsoon because of the chloride displacement from coarse mode sea salt aerosol. Thus nitrate is found to be deposited (dry or free fall deposition) at a faster rate during monsoon compared to dry seasons.

(4) The acidity of fine mode aerosol is found to be higher in dry seasons compared to monsoon whereas the acidity of coarse mode acidity is found to be higher in monsoon compared to dry seasons. Non-sea-sulphate and nitrate are found to govern the acidity of fine mode aerosol whereas non-sea-calcium and non-sea-magnesium govern the coarse mode acidity.

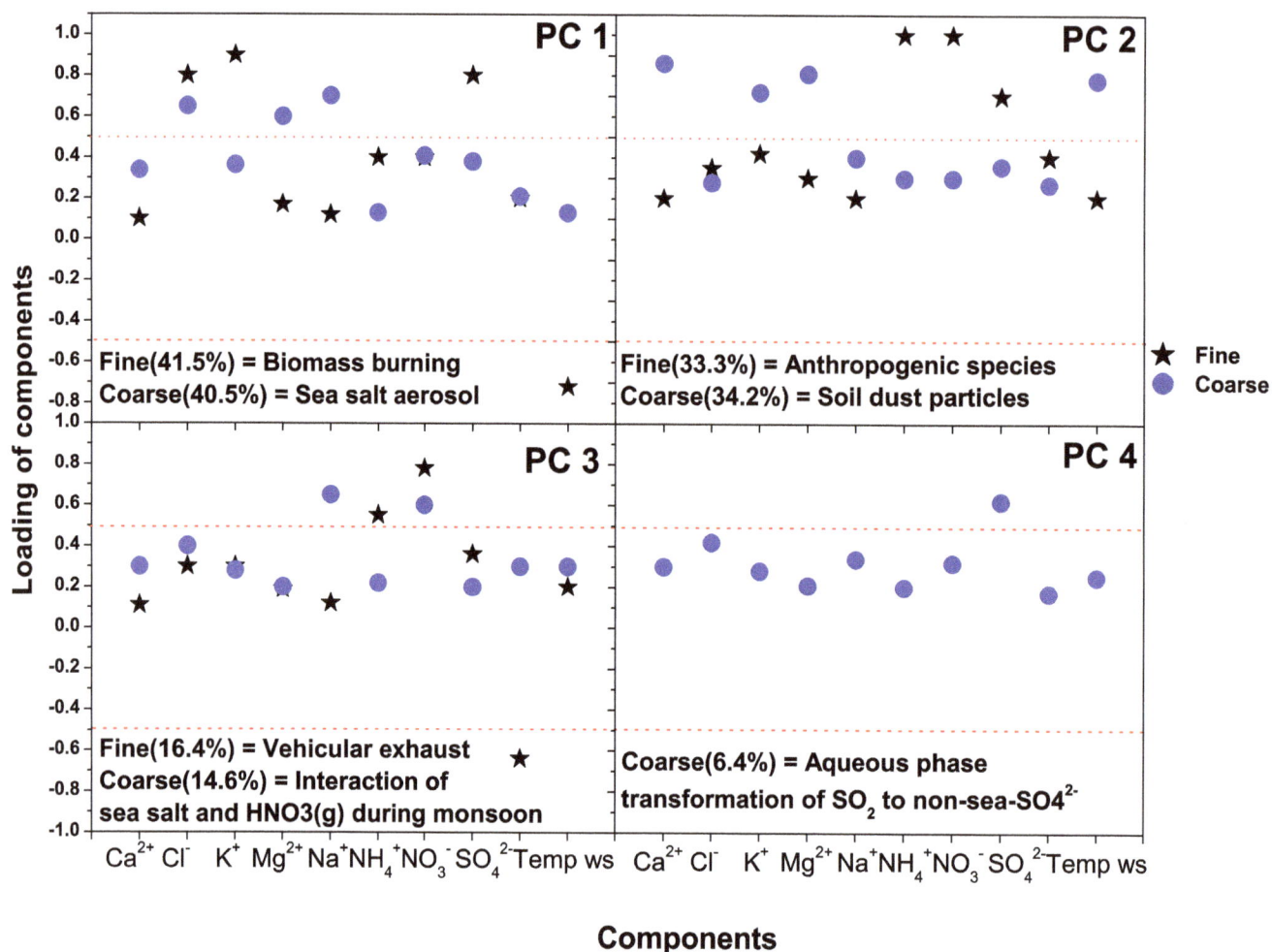

Figure 14. Aerosol source apportionment by Principal Component Analysis.

(5) Three major source regions like local, continental (mainly from north-western part of India) and marine (Bay of Bengal and Arabian Sea) sources are identified based on HYSPLIT backward trajectory model. Local and other continental source regions contribute 80% with high loading of secondary anthropogenic species whereas the contribution from marine source regions is 20% enriched with primary sea salt particles.

(6) Biomass burning and vehicular emissions are the major sources for fine mode local and continental aerosols with higher acidity whereas soil dust particles are the major source for coarse mode local and continental aerosols with lower acidity. On the other hand, the major source for fine and coarse mode marine aerosol is sea salt aerosols with higher acidity in coarse mode.

Table 5. Concentrations of water soluble ionic species in fine and coarse mode aerosols from different source regions.

		Aerosol ($\mu g\ m^{-3}$)	Na^+ ($\mu g\ m^{-3}$)	NH_4^+ ($\mu g\ m^{-3}$)	K^+ ($\mu g\ m^{-3}$)	Ca^{2+} ($\mu g\ m^{-3}$)	Mg^{2+} ($\mu g\ m^{-3}$)	Cl^- ($\mu g\ m^{-3}$)	NO_3^- ($\mu g\ m^{-3}$)	SO_4^{2-} ($\mu g\ m^{-3}$)	$[H^+]_{total}$ (nmol m^{-3})	$\sum^+/\sum^-$
Local	Fine	34.1	0.4	1.3	1.7	0.10	0.13	2.4	5.1	5.8	104.8	0.59
	Coarse	18.2	1.9	0.08	0.6	0.61	0.45	1.6	1.6	2.1	−57.5	1.51
Continental	Fine	44.2	0.9	1.1	1.1	0.26	0.16	1.2	4.1	4.5	39.2	0.79
	Coarse	12.2	0.7	0.05	0.22	0.40	0.35	1.3	1.1	1.8	−2.6	1.03
Marine	Fine	9.2	0.7	0.2	0.7	0.03	0.07	0.7	0.8	2.1	−11.3	1.20
	Coarse	26.4	3.9	0.03	0.1	0.2	0.12	3.3	2.1	1.6	8.03	0.96

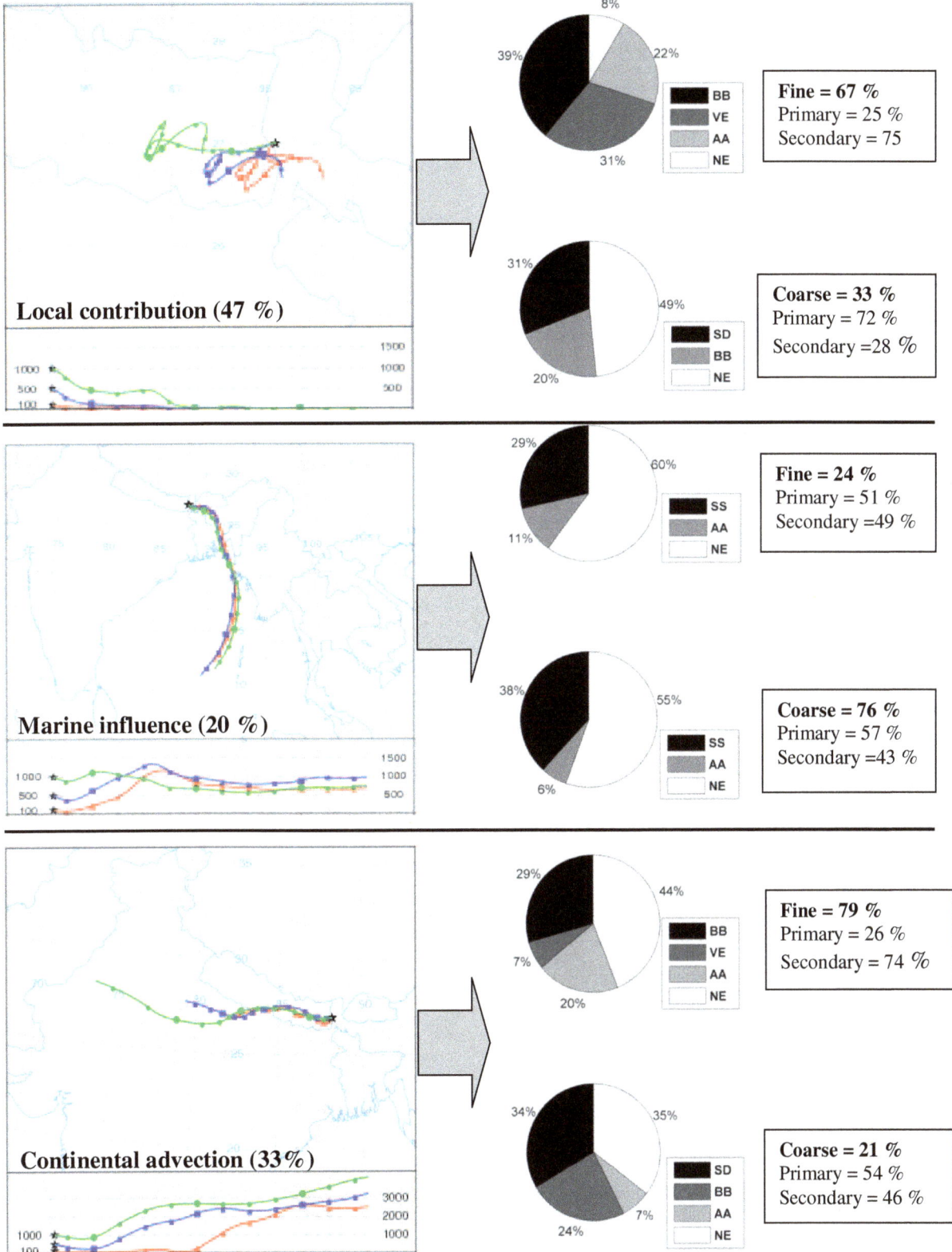

Figure 15. Several source types of aerosol from different source regions. Contributions from local and continental sources (with higher enrichment of fine aerosols) were found to be higher comparative to marine sources (with higher enrichment of coarse aerosols).

Acknowledgments

Sincere thanks are due to Mrs. Y. Yadav for her consistent efforts during the sampling and Mr. Bhaskar Roy for his support in the analytical work. Thanks are also due to Mr. D. K. Roy in Darjeeling for his overall logistic support.

Author Contributions

Conceived and designed the experiments: AC SR. Performed the experiments: AC AA AKS. Analyzed the data: AC MKS SKG. Contributed reagents/materials/analysis tools: ST PCSD. Wrote the paper: AC SKG SR.

References

1. IPCC (2007) Intergovernmental Panel on Climate Change. Fourth Assessment Report. Cambridge University Press, Cambridge, Section 2.2. 37 p.
2. Gautam R, Hsu NC, Lau KM, Tsay SC, Kafatos M (2009) Enhanced pre-monsoon warming over the Himalayan-Gangetic region from 1979 to 2007. Geophys Res Lett 36: L07704. doi: 10.1029/2009GL037641.
3. Ramanathan V, Li F, Ramana MV, Siva PS, Kim D, et al. (2007) Atmospheric Brown Clouds: Hemispherical and regional variations in long range transport, absorption and radiative forcing. J Geophys. Res 112: D24S91. doi:10.1029/2006JD008124.
4. Gautam R, Liu Z, Singh RP, Hsu NC (2009) Two contrasting dust-dominant periods over India observed from MODIS and CALIPSO. Geophys Res Lett 36: L06813. doi: 10.1029/2008GL036967.
5. Mayewski PA, Jeschke PA (1979) Himalayan and Trans-Himalayan glacier fluctuations since AD 1812. Arct Alp Res 11: 267–287.
6. Ashish A, Joshi V, Sharma A, Anthwal S (2006) Retreat of Himalayan glaciers indicator of climate change. Nature and Science 4: 53–60.
7. Decesari S, Facchini MC, Carbone CL, Giulianelli M, Rinaldi E, et al. (2009) Chemical composition of PM10 and PM1 at the high-altitude Himalayan station Nepal Climate Observatory-Pyramid (NCO-P) (5079 m a.s.l.). Atmos Chem Phys Discuss 9: 25487–25522.
8. Safai PD, Momin GA, Rao PSP, Pillai AG, Tiwari S, et al. (1995) Chemical nature of the aerosols at a rural site in India. IASTA Bulletin. S1–S5.
9. Momin GA, Rao PSP, Safai PD, Ali K, Naik MS, et al. (1999) Atmospheric aerosol characteristic Studies at Pune and Thiruvananthapuram during INDOEX programme-1998. Curr Sci 76: 985–989.
10. Wake CP, Dibb JE, Mayewski PA, Zhongqin L, Zichu X (1994) The chemical composition of aerosols over the Eastern Himalayas and Tibetan Plateau during low dust periods. Atmos Environ 28: 695–704.
11. Shrestha AB, Wake CP, Dibb JE (1997) Chemical composition of aerosol and snow in the High Himalayas during the summer monsoon season. Atmos Environ 31: 2815–2826.
12. Shrestha AB, Cameron PW, Dibb JE, Mayewski PA, Whitlow SI, et al. (2000) Seasonal variations in aerosol concentrations and compositions in the Nepal Himalayas. Atmos Environ 34: 3349–3363.
13. Carrico CM, Bergin HM, Shrestha AB, Dibb JE, Gomes L, et al. (2003) The importance of carbon and mineral dust to seasonal aerosol properties in the Nepal Himalayas. Atmos Environ 37: 2811–2824.
14. Gajanana K, Kuniyal JC, Momin GA, Rao PSP, Safai PD, et al. (2005) Trend of atmospheric aerosols over the north western Himalayan region, India. Atmos Environ 39: 4817–4825.
15. Chatterjee A, Dutta C, Sen S, Ghosh K, Biswas N, et al. (2006) Formation, transformation, and removal of aerosol over a tropical mangrove forest. J Geophys Res 111: D24302. doi: 10.1029/2006JD007144.
16. Kaupp H, McLachlan MS (1999) Atmospheric particle size distributions of polychlorinated dibenzo-p-dioxins and dibenzofurans (PCDD/Fs) and polycyclic aromatic hydrocarbons (PAHs) and their implications for wet and dry deposition. Atmos Environ 33: 85–95.
17. Giuliacci M (1988) Physical and dynamical climatology of the Po Valley, Regional Meteo. Service, Bologna (in Italian).
18. Lau KM, Kim MK, Kim KM (2006a) Asian monsoon anomalies induced by aerosol direct effects. Clim Dyn 26: 855–864.
19. Rastogi N, Sarin MM (2005) Long-term characterization of ionic species in aerosols from urban and high-altitude sites in western India: Role of mineral dust and anthropogenic sources. Atmos Environ 39: 5541–5554.
20. Mozurkewich M (1993) The dissociation constant of ammonium nitrate and its dependence on temperature, relative humidity and particle size. Atmos Environ 27: 261–270.
21. Wang Y, Zhuang G, Zhang X, Huang K, Xu C, et al. (2005) The ion chemistry, seasonal cycle, and sources of PM2.5 and TSP aerosol in Shanghai. Atmos Environ 39: 3771–3784.
22. Kim HS, Huh JB, Hopke PK, Holsen TM, Yi SM (2007) Characteristics of the major chemical constituents of PM2.5 and smog events in Seoul, Korea in 2003 and 2004. Atmos Environ 41: 6762–6770.
23. Dasgupta PK, Campbell SW, Al-Horr RS, Rahmat Ullah SM, Li J, et al. (2007) Conversion of sea salt aerosol to NaNO3 and the production of HCl: Analysis of temporal behavior of aerosol chloride/nitrate and gaseous HCl/HNO3 concentrations with AIM. Atmos Environ 41: 4242–4257.
24. Ohta S, Okita T (1990) A chemical characterization of atmospheric aerosol in Sapporo. Atmos Environ 24A: 815–822.
25. Seinfeld JH (1986) Atmospheric Chemistry and Physics of Air Pollution. Wiley, New York. 348 p.
26. Xiu G, Zhang D, Chen J, Huang X, Chen Z, et al. (2004) Characterization of major water-soluble inorganic ions in size-fractionated particulate matters in Shanghai campus ambient air. Atmos Environ 38: 227–236.
27. Kulshrestha MJ, Kulshrestha UC, Parashar DC, Vairamani M (2003) Estimation of SO_4^{2-} contribution by dry deposition of SO_2 onto the dust particles in India. Atmos Environ 37: 3057–3063.
28. Zelenka MP, Wilson WE, Chow JC, Lioy PJ (1994) A combined TTFA/CMB receptor modeling approach and its application to air pollution sources in China. Atmos Environ 28: 1425–1435.
29. Gautam R, Hsu NC, Lau KM, Tsay SC, Kafatos M (2009) Enhanced pre-monsoon warming over the Himalayan-Gangetic region from 1979 to 2007. Geophys Res Lett 36: L07704. doi:10.1029/2009GL037641.
30. Rastogi N, Sarin MM (2009) Quantitative chemical composition and characteristics of aerosols over western India: One-year record of temporal variability. Atmos Environ 43: 3481–3488.
31. Kumar A, Sudheer AK (2008) Mineral and anthropogenic aerosols in Arabian Sea-atmospheric boundary layer: Sources and spatial variability. Atmos Environ 42: 5169–5182.
32. Zhuang H, Chan CK, Fang M, Wexler AS (1999) Size distributions of particulate sulphate, nitrate and ammonium at a coastal site in Hong Kong. Atmos Environ 33: 843–853.
33. Pakkanen TA (1996) Study of formation of coarse particle nitrate aerosol. Atmos Environ 30: 2475–2482.
34. McInnes LM, Covert DS, Quinn PK, Germani MS (1994) Measurements of chloride depletion and sulfur enrichment in individual sea-salt particles collected from the remote marine boundary layer. J Geophys Res 99: 8257–8268.
35. Rastogi N, Sarin MM (2007) Chemistry of Precipitation Events and Inter-Relationship with Ambient Aerosols over a Semi-Arid Region in Western India. J Atmos Chem 56: 149–163.
36. Duce RA, Liss PS, Merril JT, Atlas EL, Buat-Menard P, et al. (1991) The atmospheric input of trace species to the world ocean. Global Biogeochem Cycles 5: 193–259.
37. Pathak RK, Peter KK, Chan LCK (2004) Characteristics of aerosol acidity in Hong Kong. Atmos Environ 38: 2965–2974.
38. Lippmann M, Xiong JQ, Li W (2000) Development of a continuous monitoring system for PM10 and components of PM2.5. Appl Occup Environ Hyg 15: 57–67.
39. Storch Von, Zwiers FW (1999) Statistical analysis in climate research. Cambridge University Press. 298 p.
40. Maenhaut W, Cafmeyer J (1987) Particle induced X-ray emission analysis and multivariate techniques: an application to the study of the sources of respirable atmospheric particles in Gent, Belgium. J Trace Microprobe Tech 5: 135–158.
41. SPSS (2003) Advanced Statistics Manual. SPSS/PC+V 3.0. SPSS Inc., 444 North Michigan Avenue, Chicago IL 60611.
42. Rengarajan R, Sarin MM, Sudheer AK (2007) Carbonaceous and inorganic species in atmospheric aerosols during wintertime over urban and high-altitude sites in North India. J Geophys Res 112: D21307. doi: 10.1029/2006JD008150.
43. Venkataraman C, Reddy CK, Josson S, Reddy MS (2002) Aerosol size and chemical characteristics at Mumbai, India, during the INDOEX-IFP (1999). Atmos Environ 36: 1979–1991.
44. Kulshrestha UC, Saxena A, Kumar N, Kumari KM, Srivastava SS (1998) Chemical composition and association of size differentiated aerosols at a suburban site in a semi-arid tract of India. J Atmos Chem 29: 109–118.

5

Calculation of the Relative Chemical Stabilities of Proteins as a Function of Temperature and Redox Chemistry in a Hot Spring

Jeffrey M. Dick[1]*, Everett L. Shock[1,2]

1 School of Earth and Space Exploration, Arizona State University, Tempe, Arizona, United States of America, 2 Department of Chemistry and Biochemistry, Arizona State University, Tempe, Arizona, United States of America

Abstract

Uncovering the chemical and physical links between natural environments and microbial communities is becoming increasingly amenable owing to geochemical observations and metagenomic sequencing. At the hot spring known as Bison Pool in Yellowstone National Park, the cooling of the water in the outflow channel is associated with an increase in oxidation potential estimated from multiple field-based measurements. Representative groups of proteins whose sequences were derived from metagenomic data also exhibit an increase in average oxidation state of carbon in the protein molecules with distance from the hot-spring source. The energetic requirements of reactions to form selected proteins used in the model were computed using amino-acid group additivity for the standard molal thermodynamic properties of the proteins, and the relative chemical stabilities of the proteins were investigated by varying temperature, pH and oxidation state, expressed as activity of dissolved hydrogen. The relative stabilities of the proteins were found to track the locations of the sampling sites when the calculations included a function for hydrogen activity that increases with temperature and is higher, or more reducing, than values consistent with measurements of dissolved oxygen, sulfide and oxidation-reduction potential in the field. These findings imply that spatial patterns in the amino acid compositions of proteins can be linked, through energetics of overall chemical reactions representing the formation of the proteins, to the environmental conditions at this hot spring, even if microbial cells maintain considerably different internal conditions. Further applications of the thermodynamic calculations are possible for other natural microbial ecosystems.

Editor: Jonathan H. Badger, J. Craig Venter Institute, United States of America

Funding: This material is based upon work supported by the National Science Foundation under grant EAR-0847616. The funders had no role in study design, data collection and analysis, decision to publish, or preparation of the manuscript.

Competing Interests: The authors have declared that no competing interests exist.

* E-mail: jmdick@asu.edu

Introduction

The imprints of distinct geochemical environments can be found in the molecular compositions of microbial genomes and their protein products. For example, transmembrane proteins of ancestral organisms were likely to be depleted in oxygen, paralleling the low oxygen content of Earth's atmosphere in the past [1]. Environmental imprints on proteins can also be found for spatially separated organisms living contemporaneously; the amino acid composition of proteins differ systematically between organisms living at different temperatures [2,3]. Together with temperature, the chemical properties of the environment are linked to the compositions of gene sequences in hot-spring microbial communities [4,5].

Although the sequences of proteins must satisfy a complex array of biological requirements, the different biosynthetic costs of amino acids are viewed as one contributing factor to actual patterns of amino acid usage [6,7]. In some studies, the biosynthetic costs of amino acids have been estimated from metabolic constraints including numbers of phosphate bonds and hydrogen atoms transferred during synthesis from precursors [6,8]. Those estimates depend on the growth medium and specific metabolic pathways but otherwise do not involve environmental

variables such as temperature and oxidation-reduction conditions. Nevertheless, it can be shown that the Gibbs energy change in overall chemical reactions to synthesize amino acids from from inorganic species depends on environmental conditions [9]. The calculations of energetics of overall synthesis reactions can now be done for proteins, where group additivity methods permit assessing standard Gibbs energies of proteins of any amino acid composition [10,11].

The goal of this study is to use thermodynamic tools to characterize simultaneously the chemical environment and metagenomically derived protein sequences in a hot spring exhibiting large gradients of temperature and oxidation-reduction, or redox, chemistry. The geochemical and biomolecular data are combined using a single model framework based on chemical reactions and their energy changes. The use of a metagenomic dataset in a location where extensive geochemical data are available permits calibration and testing of the model.

One setting where in-depth metagenomic and geochemical information are available is the hot spring known as "Bison Pool", a flowing, moderately alkaline hot spring in Yellowstone National Park [12,13]. The water at the source is boiling, and rapidly cools along the outflow channel as a result of exposure to the ambient conditions. Extensive chemical analysis of the water also reveals large gradients of chemical composition such as increase in pH, decrease in sulfide concentration, increase in dissolved oxygen and in oxidation-reduction potential of the water. Prior metagenomic sampling of the microbial communities at five sites from the source to approximately 22 meters down the outflow channel offers a window into the biomolecular composition of these communities. Although the metagenomic sequencing is of the DNA molecules, genes present in the metagenome provide a picture of the proteins that are likely to be used by the organisms.

The first major theme of this study concerns the changes in chemical composition of proteins along the outflow channel. The stoichiometric quantity we investigate is the average oxidation state of carbon in the proteins, which can be calculated directly from the chemical formulas of the proteins. In general, the average oxidation state of carbon in proteins increases down the outflow channel. This effect is present at the level of the whole metagenome and also within different functional classes of proteins. There is a positive correlation between the average oxidation state of carbon in proteins and the oxidation-reduction potential of the surrounding water.

The results of the stoichiometric calculations support a hypothesis that chemical compositions of the proteins reflect processes that tend to minimize the free energy of the system. We applied thermodynamic models to integrate molecular composition with temperature and multiple environmental chemical variables. The second major theme of the paper addresses the relative stabilities of the different classes of model proteins from each sampling site in terms of temperature, pH and oxidation-reduction potential. The major finding of this part of the study is that a redox gradient as a function of temperature traverses the stability fields of the proteins in a way that largely parallels the proteins' spatial distribution. This redox gradient, expressed as activity of dissolved hydrogen, generally parallels estimates derived from measurements of sulfide/sulfate concentrations, oxidation-reduction potential electrodes, and dissolved oxygen, but is more reducing than any of those.

These results help to outline the interrelationships between biomolecular composition and geochemistry in the Bison Pool ecosystem. One use of these models is to quantify gradients in oxidation potential between the water and the interiors of cells and/or biofilms at the temperatures found in the hot spring. Another is to establish the extent to which organisms minimize the energy expenditure involved in formation of biomolecules in specific chemical environments. Generalizing the methods and calculations described below can aid in resolving the effects of chemical gradients, energy minimization and other features of this hot spring and other geobiochemical systems.

Methods

Average oxidation state of carbon

The average nominal oxidation state of carbon, $\overline{Z}_C$, is a quantity related to the different electronegativities of elements involved in the covalent structure of an organic molecule. $\overline{Z}_C$ is equal to the sum of the nominal oxidation states of all the carbon atoms in a molecule divided by the number of carbon atoms. The concept of average oxidation state of carbon has found application in various contexts, ranging from balancing organic oxidation-reduction reactions [14] to characterization of organic matter in aerosols [15] and in terrestrial ecosystems [16]. Moreover, a correlation can be observed between the standard molal Gibbs energies of oxidation half-reactions and the average oxidation state of carbon of the organic molecules involved [17]. As with smaller

molecules, it is possible to interpret the chemical composition of proteins using the average oxidation state of carbon.

The rules for calculating the formal oxidation states on any carbon atom can be summarized as follows [18]. Each single bond to a more electronegative element (e.g., oxygen, nitrogen, sulfur) contributes $+1$ to the oxidation state of a particular carbon atom, while each single bond to a less electronegative element (e.g., hydrogen) contributes -1 to the oxidation state of a particular carbon atom, and a carbon-carbon bond counts (formally) as zero. Double bonds count doubly. Familiar, though extreme, examples are found with CO_2 (two double bonds to oxygen; $\overline{Z}_C = +4$) and CH_4 (four single bonds to hydrogen; $\overline{Z}_C = -4$). The concept of the average oxidation state can be extended to more complex molecules, for example acetic acid, CH_3COOH, which has $\overline{Z}_C = 0$. That value is consistent with an oxidation state of $+3$ on the first carbon (having three bonds to hydrogen) and -3 on the second carbon (having one double bond and one single bond to oxygen).

The definition of oxidation state cited in the IUPAC Gold Book [19,20] states that "... in ions the algebraic sum of the oxidation states of the constituent atoms must be equal to the charge on the ion" (i.e., positive or negative values for cations or anions, or zero for neutral species). We can adopt values for the formal charges of atoms other than carbon, -2 for oxygen, $+1$ for hydrogen, -3 for nitrogen, -2 for sulfur, that are consistent with this requirement for amino acids and proteins. The values for nitrogen and sulfur are those that would be assigned to the atoms if they were found in amine groups and sulfide groups, respectively [16]. Writing the formula of glycine as H_2NCH_2COOH, the oxidation state of the first carbon is -1 (one bond to nitrogen, two bonds to hydrogen) and that of the second carbon is $+3$ (one double bond and one single bond to oxygen), so the average oxidation state of carbon in the molecule is $+1$. The sum of the formal charges of the atoms, in the order indicated by the formula, is $2(+1) - 3 - 1 + 2(+1) + 3 - 2 - 2 + 1 = 0$, which is equal to the net charge of the molecule.

In many cases, the value of the average oxidation state of carbon is amenable to calculation using only the chemical formula of a molecule, instead of the more tedious accounting for each carbon. Let us use Z to stand for the total charge on an ion (which becomes zero for a neutral molecule) and let the average oxidation state of carbon be represented by $\overline{Z}_C$. Using formal oxidation states mentioned above for the elements other than carbon, the requirement for algebraic sums of oxidation states of the atoms can be expressed symbolically as

$$n_H - 2n_O - 2n_S - 3n_N + \overline{Z}_C n_C = Z, \qquad (1)$$

where n_C, n_H, n_N, n_O and n_S are the numbers of the respective subscripted elements in the chemical formula. Rearranging Eq. (1) gives

$$\overline{Z}_C = \frac{Z - n_H + 2(n_O + n_S) + 3n_N}{n_C}. \qquad (2)$$

This equation shows that the average oxidation state of carbon in proteins is effectively a linear combination of the elemental ratios H/C, N/C, O/C and S/C.

Note that ionization of the amino acid sidechains in proteins, and other ionization reactions involving only protons, have equal contributions to Z and n_H and produce no net effect on the value of $\overline{Z}_C$. Similarly, polymerization of amino acids, or other reactions involving only the gain or loss of a water molecule, produce no net effect on the value of $\overline{Z}_C$ [18]. On the other hand, Eq. (2), and the electronegativity rules outlined above, show that oxidation-reduction reactions in organic compounds are not limited to gain or loss of either hydrogen or oxygen, but that the addition of other heteroatoms (sulfur, nitrogen) to a compound also causes an increase in the overall oxidation state of the molecule [18].

The average oxidation states of carbon in the twenty common amino acids range from -1 (leucine, isoleucine) to $+1$ (glycine, aspartic acid, asparagine) and are summarized in Table 1. The values listed for the amino acids in Table 1 span a considerable range but other types of organic molecules are even more or less oxidized [16,17]. Proteins made up of these amino acids have an average oxidation state of carbon that can be computed as a weighted average of the $\overline{Z}_C$ values of the amino acids, or equivalently, using Eq. (2) and the chemical formulas of the proteins. It may be noted that other physical-chemical properties of the amino acids can be correlated with differences in average oxidation state of carbon. For example, four highly hydrophobic amino acids (isoleucine, valine, leucine and phenylalanine) [21] have negative average oxidation states of carbon, which is associated with the high H/C ratios of their sidechain groups.

Relative stabilities of proteins

Chemical thermodynamic methods, borrowed from geochemical modeling applications, can also be used to study the relative stabilities of model proteins from different sampling sites in the hot spring outflow channel. The methods are described conceptually below, followed by description of a specific example.

Four informal definitions help introduce the modeling strategy. 1) *Basis species* are a minimum set of chemical constituents that represent all of the chemical elements and the ionization state of proteins. 2) A *formation reaction* is a chemical reaction to form one mole of a protein from the basis species. The formation reactions of different proteins have different coefficients on basis species because the proteins themselves have different chemical formulas. 3) *Chemical affinity* is energy change during a reaction; positive values mean energy is released, and negative values mean that energy is consumed. A reaction with a higher chemical affinity is more favored to proceed to the product side. 4) *Chemical activity* is related fundamentally to chemical potential and can be thought of as the effective concentration of a basis species or protein.

Table 1. Average oxidation states of carbon and number of carbon atoms of the twenty amino acids commonly occurring in proteins.

Amino Acid	n_C	$\overline{Z}_C$	Amino Acid	n_C	$\overline{Z}_C$
Alanine	3	0.00	Methionine	5	-0.40
Cysteine	3	$0.6\overline{6}$	Asparagine	4	1.00
Aspartic Acid	4	1.00	Proline	5	-0.40
Glutamic Acid	5	0.40	Glutamine	5	0.40
Phenylalanine	9	$-0.4\overline{4}$	Arginine	6	$0.3\overline{3}$
Glycine	2	1.00	Serine	3	$0.6\overline{6}$
Histidine	6	$0.6\overline{6}$	Threonine	4	0.00
Isoleucine	6	-1.00	Valine	5	-0.80
Lysine	6	$-0.6\overline{6}$	Tryptophan	11	$-0.\overline{18}$
Leucine	6	-1.00	Tyrosine	9	$-0.2\overline{2}$

Let us define one system of interest as a collection of proteins with equal chemical activities interacting with a physical-chemical environment defined by constant values of temperature, pressure, and chemical activities of the basis species. What is the relative stability of one protein compared to another? If the activities of the proteins are equal, the affinities of the formation reactions of the proteins are generally unequal to each other, and the system is not in equilibrium. The most stable protein is identified as the one with the highest chemical affinity of its formation reaction. That is the protein whose formation, at a given chemical activity, releases the most energy, or requires the least energy input.

Now consider the outcome of hypothetical chemical reactions among the proteins, so that different proteins (chemical species) are formed and destroyed at each others' expense, and as a consequence the chemical activities of the proteins change. The temperature and pressure are maintained, and the system is open so that the activities of the basis species are buffered and therefore remain unchanged. One or more specific outcomes of the hypothetical progression of reactions is an assemblage of proteins in a (possibly metastable) equilibrium distribution. In this equilibrium, or minimum-energy state, the chemical affinities of the formation reactions of the proteins are all equal (but might be non-zero), so the hypothetical transformation of one protein to another involves no overall energy change. If the affinities of the formation reactions of the proteins are all equal but less than zero, then the proteins are less stable than the basis species; that system represents a type of metastable equilibrium and a local, not global, energy minimum. Since the system is at equilibrium, the most stable protein is identified as the one with the highest chemical activity – in terms of concentration it has a higher degree of formation compared to the other proteins.

The hypothetical systems described above consist of populations of proteins with either equal activities or equal affinities of formation. To a first approximation (under conditions of ideal mixing) the stabilities of the proteins relative to each other are the same in both cases, since the definition of the chemical environment – temperature, pressure and activities of the basis species – is unchanged. Therefore, it is helpful to conceptualize the systems with equal activities of proteins and equal affinities of protein-formation reactions as being different states of a more generic system, defined only by the chemical environment and the identities of the proteins, but not their chemical activities. The relationship between the equal-activity and equal-affinity reference states is shown schematically in Fig. 1. The relative stabilities of species A, B, and C are the same in both panels of the Figure. If an equal-activity reference state is adopted, greater stability goes with higher affinity (Fig. 1a). If the equal-affinity reference state is adopted, greater stability goes with higher activity (Fig. 1b).

In quantifying the relative stabilities of proteins, a choice can be made between the two reference states; either one is valid, but the relative stabilities of the proteins are revealed through different variables. To generate the figures in this paper relative stabilities were quantified using an equal-affinity, or metastable equilibrium reference state. The primary advantage of doing so is the production of equilibrium activity diagrams [22] that are interpreted as depicting the relative stabilities of the proteins in terms of temperature and activities of the basis species. The specific methods used to calculate relative stabilities of proteins starting with an equal-activity reference state, then using a reaction matrix or equilibrium distribution equation to quantify the activities of the proteins at metastable equilibrium are described below.

Thermodynamic definitions. An equation for the differential of Gibbs energy (dG) that takes account of reaction

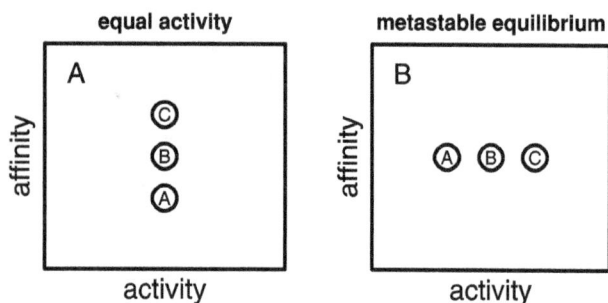

Figure 1. Relative stabilities portrayed in different reference states. In these qualitative diagrams, the same relative stabilities are shown in two different reference states. Species "C" is more stable than "B" is more stable than "A". Chemical activity is shown on the x-axis, and chemical affinity of formation reaction is shown on the y-axis. In the equal-activity, or non-equilibrium, reference state (left), the species with the most positive chemical affinity of formation is the most stable. In the equal-affinity, or metastable equilibrium, reference state (right), the species with the most positive chemical activity is the most stable.

progress in a system, formulated by de Donder [23,24] can be written as

$$dG = -SdT + VdP - Ad\xi, \qquad (3)$$

where S, T, V and P are entropy, temperature, volume and pressure, A is chemical affinity, and ξ is a reaction progress variable. The chemical affinity of the ith reaction can be expressed as

$$A_i = 2.303RT\log(K_i/Q_i) = -\Delta G_{r,i}, \qquad (4)$$

where R is the gas constant, 2.303 represents the natural logarithm of 10, and K_i and Q_i are the equilibrium constant and activity product of the ith reaction. The chemical affinity is equal to the negative of the overall Gibbs energy change of the reaction ($\Delta G_{r,i}$). K and Q can be calculated from

$$\log K_i = -\Delta G^\circ_{r,i}/2.303RT, \qquad (5)$$

where $\Delta G^\circ_{r,i}$ is the standard Gibbs energy of the ith reaction, and

$$\log Q_i = \sum_j v_{j,i}\log a_{j,i}, \qquad (6)$$

where $v_{j,i}$ and $a_{j,i}$ are the reaction coefficient (negative for reactants, positive for products) and activity of the jth species in the ith reaction. In the equations, all operations involving logarithms have a base of 10.

The standard state convention adopted for liquids, including H_2O, corresponds to unit activity of the pure substance at any temperature and pressure. The standard state convention adopted for aqueous species other than H_2O corresponds to unit activity of a hypothetical one molal solution referenced to infinite dilution at any temperature and pressure [25]. The conventional standard molal thermodynamic properties of both the aqueous electron and proton are taken to be zero at all temperatures and pressures [26].

Calculating relative stabilities of proteins

The case study described below is based on an example described previously [27] for calculating the equilibrium activities

of cell-surface glycoproteins (CSG) from *Methanococcus voltae* and *Methanocaldococcus jannaschii*. These methanogenic organisms are not likely to be present in detectable quantities at Bison Pool in Yellowstone National Park, but they nevertheless are common model organisms for studying microbial adaptations to differences in temperature and other environmental characteristics [28,29]. Clues about the organisms' environments may emerge from comparing the sequence and chemical properties of the two functionally homologous proteins.

The methods described here for calculating the amino acid and chemical composition as well as the standard molal properties of the proteins [11] are currently restricted to only the unfolded peptide molecules and not the carbohydrate constituent of the glycoproteins. In their uncharged states, the peptide chains of these two molecules have formulas of $C_{2575}H_{4097}N_{645}O_{884}S_{11}$ and $C_{2555}H_{4032}N_{640}O_{865}S_{14}$, respectively, with sequence lengths of 553 and 530 amino acid residues (UniProtKB accessions Q50833 for CSG_METVO and Q58232 for CSG_METJA; signal peptides were removed). At 25°C and pH = 7, the charges of the ionized proteins calculated using standard Gibbs energies of ionization of the amino acid sidechains and protein terminal groups [11] are -56.06 and -55.87. Although the calculation of protein charge, which is based on group additivity, does not take into consideration the effects of interactions between the ionizable groups, it does have the advantage of being sensitive to changes in temperature.

Writing formation reactions for residue equivalents. In general, the relative stabilities of proteins of different lengths are of interest. Because of the differences in size, the molal reaction energies can not be directly compared in calculations of relative stability. In many environments, the synthesis of larger molecules, per mole, demands more energy, so for proteins of otherwise equal chemical composition and thermodynamic properties (such as two proteins of different size but with the same relative frequencies of amino acids), the smaller one would generally be thought of as more stable. In more reduced settings, the overall synthesis of organic molecules can actually release energy [30], so the synthesis of larger molecules would be favored. Taking the polymeric nature of the proteins into account, the relative stabilities of proteins of different size can be assessed by first writing formation reactions that are normalized by numbers of amino acid residues. The reactions involve the residue equivalents of the proteins, which have chemical formulas and standard molal properties that are those of the protein divided by the sequence length of the protein. The following two formation reactions are written for the residue equivalents of the two protein homologs:

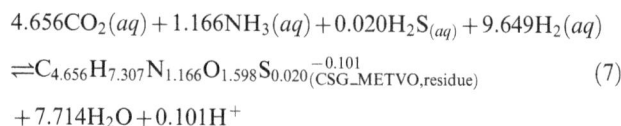

$$4.656CO_2(aq) + 1.166NH_3(aq) + 0.020H_2S_{(aq)} + 9.649H_2(aq)$$
$$\rightleftharpoons C_{4.656}H_{7.307}N_{1.166}O_{1.598}S_{0.020}^{-0.101}(CSG\_METVO,residue) \qquad (7)$$
$$+ 7.714H_2O + 0.101H^+$$

for the protein from *M. voltae* and

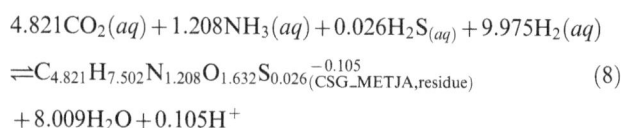

$$4.821CO_2(aq) + 1.208NH_3(aq) + 0.026H_2S_{(aq)} + 9.975H_2(aq)$$
$$\rightleftharpoons C_{4.821}H_{7.502}N_{1.208}O_{1.632}S_{0.026}^{-0.105}(CSG\_METJA,residue) \qquad (8)$$
$$+ 8.009H_2O + 0.105H^+$$

for that from *M. jannaschii*.

The reactions above involve the basis species $CO_{2(aq)}$, H_2O, $NH_{3(aq)}$, $H_2S_{(aq)}$, $H_{2(aq)}$ and H^+, which are the same as used in Ref. [27] except that $H_{2(aq)}$ is used here instead of $O_{2(g)}$. The choice of basis species determines the expression for chemical activities, i.e. the way in which the environmental chemical potentials are quantified. Similarly to components, the set of basis species is valid only if they represent a number of independent variables equal to the dimension of chemical variability in the system. There are unlimited combinations of basis species that would qualify, but the actual choice is usually made to facilitate comparisons with the natural system. For the calculations described in this paper $H_{2(aq)}$ is used instead of $O_{2(g)}$ because of the actual formation and metabolic significance of molecular hydrogen in the hot-spring ecosystem [31].

Writing the formation reactions normalized per residue offers insight into the consequences of changing environmental variables on the relative stabilities of the proteins. Because Reactions 7 and 8 are written per residue of the proteins, comparing the reaction coefficients to infer the effect of changing chemical variables on the relative stabilities of the proteins is consistent with an overall reaction between the proteins that is balanced on the protein backbone group. For example, more moles of $H_{2(aq)}$ are consumed in Reaction (8) compared to Reaction (7). From specific statements of Eqs. (4) and (6) for both reactions it follows that increasing the activity of $H_{2(aq)}$ would tend to favor formation of – that is, decrease the energy change of the reaction for – the homolog from *M. jannaschii* more strongly than that from *M. voltae*. The effect of changing activity of hydrogen on the relative stabilities of proteins from *M. jannaschii* and *M. voltae* parallels differences in oxidation state of the natural environments of these two organisms. A likely range of activities of dissolved hydrogen in the mixing zones of submarine hydrothermal vents and ocean water, representative of the environments inhabited by *M. jannaschii*, is $\sim 10^{-2}$ to $\sim 10^{-5}$ at $\sim 100°C$ [30]. In lower-temperature estuarine sediments, typical of the growth setting of *M. voltae*, lower hydrogen concentrations of $< 10^{-9}$ to $\sim 10^{-8}$ have been observed [32].

Inspection of Reactions 7 and 8 implies that increasing activity of $H_{2(aq)}$ tends to favor formation of the protein from *M. jannaschii* more strongly than that from *M. voltae*. This finding is the opposite of what is implied by the difference between the average oxidation state of carbon in CSG_METJA $(\overline{Z}_C = -0.138)$ and CSG_METVO $(\overline{Z}_C = -0.144)$; the protein from *M. jannaschii* actually has a higher average oxidation state of carbon. While the average oxidation state of carbon can be derived solely from the chemical composition of the protein, the formation reactions set the stage for understanding relative stabilities of the proteins in terms of reaction stoichiometry, energy, and their relationships to multiple chemical variables represented by the basis species.

Calculation of equilibrium constants. The equilibrium constants of each of the reactions can be calculated using the standard Gibbs energies of formation from the elements of the species in the reactions. In this study, the standard molal thermodynamic properties of aqueous species as a function of temperature and pressure were evaluated using the revised Helgeson-Kirkham-Flowers (HKF) equations of state [25,33,34,35]. The equations of state used for liquid H_2O were taken from Refs. [36,37,38] as implemented in a Fortran subroutine in the SUPCRT92 software package [39]. Values of the standard molal thermodynamic properties and of the equations of state parameters for the basis species other than H^+ and e^- were taken from Refs. [40,34,41].

The standard molal thermodynamic properties and equations of state parameters of the proteins can be calculated from amino acid group additivity [11]. In the present study, the CHNOSZ package [27] for the R software environment [42], which includes the group additivity equations for the proteins and the equations of

state for calculating standard molal thermodynamic properties as a function of temperature, was used for the calculations. Sample code for performing the calculations for this example is included in Supporting Dataset S1. Combining the sources of data outlined above, values of $\log K_7 = 63.1945$ and $\log K_8 = 65.7929$ at 25°C and 1 bar can be obtained for these two reactions.

Calculation of chemical affinities. The next step is to calculate the chemical affinities of the formation reactions in an equal-activity reference state. Let us use Eq. (4) to write

$$A_7/2.303RT = \log K_7 - \log\left(a_{\text{CSG\_METVO,residue}} a_{H_2O}^{7.714} a_{H+}^{0.101}\right) + \log\left(a_{CO_{2(aq)}}^{4.656} a_{NH_{3(aq)}}^{1.166} a_{H_2S_{(aq)}}^{0.020} a_{H_{2(aq)}}^{9.649}\right) \quad (9)$$

and

$$A_8/2.303RT = \log K_8 - \log\left(a_{\text{CSG\_METJA,residue}} a_{H_2O}^{8.009} a_{H+}^{0.105}\right) + \log\left(a_{CO_{2(aq)}}^{4.821} a_{NH_{3(aq)}}^{1.208} a_{H_2S_{(aq)}}^{0.026} a_{H_{2(aq)}}^{9.975}\right). \quad (10)$$

The activities of the basis species are set to reference values nominally representative of environmental conditions. The activities of the basis species used in this example are taken from Ref. [27]: $\log a_{H_2O} = 0$, $\log a_{CO_{2(aq)}} = -3$, $\log a_{NH_{3(aq)}} = -4$, $\log a_{H_2S} = -7$, $\log a_{H^+} = -7$ (pH = 7) and $\log a_{H_{2(aq)}} = -4.657486$. The value for $\log a_{H_{2(aq)}}$ was chosen so that the results would be numerically equivalent to those described in Ref. [27], where $\log f_{O_{2(g)}} = -80$ was specified instead. Substituting these values into Eqs. (9) and (10) allows us to write

$$A_7/2.303RT = 0.1893 - \log a_{\text{CSG\_METVO,residue}} \quad (11)$$

and

$$A_8/2.303RT = 0.5928 - \log a_{\text{CSG\_METJA,residue}}. \quad (12)$$

The activities of the residue equivalents are related to the activities of the proteins as follows. Activity of the ith protein $(a_{\text{protein},i})$ is related to concentration $(m_{\text{protein},i},$ for molality) by

$$a_{\text{protein},i} = \gamma_{\text{protein},i} m_{\text{protein},i} \quad (13)$$

where $\gamma_{\text{protein},i}$ stands for the activity coefficient of the ith protein. The total activity of residues in the ith protein $(a_{\text{residue},i})$ is given by

$$a_{\text{residue},i} = \gamma_{\text{residue},i} m_{\text{residue},i}, \quad (14)$$

where $\gamma_{\text{residue},i}$ stands for an activity coefficient. The total molality of residues associated with the ith protein is

$$m_{\text{residue},i} = n_{\text{residue},i} m_{\text{protein},i}, \quad (15)$$

where $n_{\text{residue},i}$ stands for the number of amino acid residues, or sequence length of the ith protein.

Because of the high concentration of metabolites and biomacromolecules in cells, the activity coefficients of proteins in their natural subcellular environments are probably significantly different from unity [43], but available methods for calculating non-ideal behavior of protein solutions are referenced to electrolyte solutions [44] and depend on structural parameters of proteins that would be difficult to deduce from metagenomic sequence fragments. Under these circumstances the activity coefficients of both residues and proteins can be approximated as unity, and Eqs. (13)–(15) can be combined to write

$$a_{\text{residue},i} = n_{\text{residue},i} a_{\text{protein},i}. \tag{16}$$

To characterize the affinities in an equal-activity reference state, activities of the proteins nominally given by $a_{\text{CSG\_METVO}} = a_{\text{CSG\_METJA}} = 10^{-3}$ are used for this example. Using Eq. (16), one then obtains reference activities of the residues given by $a_{\text{CSG\_METVO,residue}} = 0.553$ and $a_{\text{CSG\_METJA,residue}} = 0.530$ at 25°C and 1 bar.

The reference activities of the residues (computed from equal activities of proteins) can be substituted into Eqs. (11) and (12) to write $A_7/2.303RT = 0.446$ and $A_8/2.303RT = 0.868$. Therefore, on a per-residue basis, the homolog from *M. jannaschii* is more stable under the conditions (temperature, pressure, chemical activities of basis species) stated above. Decreasing the activity of hydrogen below a certain value, or changing the values of one or more variables in a specific manner determined by the reaction stoichiometry and Gibbs energy, would change the outcome so the homolog from *M. voltae* would be the more stable protein.

Calculation of the metastable equilibrium activities of proteins: Reaction-matrix approach. Casting the relative stabilities of the proteins into a metastable equilibrium reference state facilitates comparisons on equilibrium activity diagrams. One approach involves a reaction matrix, where a system of equations is constructed based on the formation reactions of the proteins. In metastable equilibrium, the affinities of the formation reactions are all equal. Let us denote this value by $A_{\text{met-equil}}$. Combining $A_7 = A_8 \equiv A_{\text{met-equil}}$ with Eqs. (11) and (12) permits writing

$$A_{\text{met-equil}}/2.303RT = 0.1893 - \log a_{\text{CSG\_METVO,residue}} \tag{17}$$

and

$$A_{\text{met-equil}}/2.303RT = 0.5928 - \log a_{\text{CSG\_METJA,residue}}. \tag{18}$$

So far this is a system of two equations with three unknowns. A third equation arises from the conservation of activity of residues in the system; recall that activities are additive only if the activity coefficients are unity. Assigning both proteins reference activities of 10^{-3}, the total activity of residues follows from Eq. (16):

$$a_{\text{CSG\_METVO,residue}} + a_{\text{CSG\_METJA,residue}} = 1.083. \tag{19}$$

The solution to the system of equations (17)–(19) is $a_{\text{CSG\_METVO,residue}} = 0.3066$, $a_{\text{CSG\_METJA,residue}} = 0.7764$ and $A_{\text{met-equil}}/2.303RT = 0.703$. It follows that the metastable equilibrium activities of the proteins (not the residues) are $\log a_{\text{CSG\_METVO}} = \log(0.3066/553) = -3.256$ and $\log a_{\text{CSG\_METJA}} = \log(0.7764/530) = -2.834$. As with the outcome of the equal-activity calculations described previously, CSG_METJA is found to be the more stable protein at the conditions of this example. Changes in temperature, pressure or activities of the basis species would alter these results; in some conditions, for example at more oxidizing conditions specified by lowering the activity of hydrogen, CSG_METVO would instead be the more stable protein.

Each additional protein that is added to the system represents another unknown and another equation like Eq. (17) or (18), so this method is applicable to systems with any number of proteins. Note however that Eqs. (17)–(19), or others that would be written for different systems of proteins, do not constitute a linear system of equations; the unknown activities are summed in the last equation, but the logarithms of activities appear in the former equations. In software, a root finder can be used to solve these equations, leading to slow performance when the relative stabilities of many proteins (hundreds or thousands) are being considered. This performance penalty would not hinder the calculations described in this paper because at most five model proteins for each of the sampling sites are being considered. However, it is useful to consider a different approach, described in the next section, that is computationally more direct and yields identical results.

Calculation of the metastable equilibrium activities of proteins: Boltzmann distribution. Let us define, for the per-residue formation reaction of the ith protein,

$$A_i^* \equiv A_i + 2.303RT \log a_{i,\text{residue}}. \tag{20}$$

It can be seen by comparison with Eqs. (4) and (6) that $_1A_i^*$ includes all contributions to the chemical affinity of the ith reaction except for the term associated with the activity of the residue equivalent of the protein of interest. For the per-residue formation reaction of the ith protein, it follows that

$$A_i^* = 2.303RT \log(K_i/Q_i^*), \tag{21}$$

where

$$\log Q_i^* = \sum_j v_{j,i} \log a_{j,i}, \tag{22}$$

where $\hat{j}$ enumerates all of the basis species, but not the protein, in the ith reaction.

In physical applications, the Boltzmann distribution gives the probabilities of occupation of specific energy levels for systems in thermal equilibrium [45]; analogously for chemical systems it can be used to derive the equilibrium distributions of species [46]. An expression for the Boltzmann distribution, written using the current notation, is

$$\frac{a_{i,\text{residue}}}{\sum a_{i,\text{residue}}} = \frac{e^{A_i^*/RT}}{\sum e^{A_i^*/RT}}. \tag{23}$$

Since the chemical affinity is the negative of the Gibbs energy of reaction, the exponents in Eq. (23) do not carry negative signs, unlike the energy terms in most common representations of the equation [45].

Following the case study above, it can be deduced from Eqs. (11), (12) and (20) that $A^*/RT = 0.4359$, and $A^*/RT = 1.3650$. From Eq. (19) it follows that $\sum a_i = 1.083$. Substituting these values into Eq. (23), one can directly calculate $a_{\text{CSG\_METVO,residue}} = 0.3066$ and $a_{\text{CSG\_METJA,residue}} = 0.7764$. These are the same as the values calculated above using the reaction-matrix approach, and can be combined with Eq. (16) to calculate the metastable equilibrium activities of the proteins. Application of Eq. (23) works as well for systems of three or more proteins and, compared to the reaction-matrix approach, leads to a more efficient implementation in software and faster calculations.

Equilibrium activity diagrams. In the example described above, the calculations were carried out at only a single point in temperature-pressure-chemical activity space. The stability calculations can also be performed when one considers the effects of changing temperature, pressure, and/or chemical activities of the basis species, singly or in combination. Interpreting the results of this type of calculation is facilitated by visualizing the relative stabilities of the proteins on equilibrium activity diagrams. In the Results described below, the lines on chemical speciation diagrams show the metastable equilibrium activities of the proteins as a function of a single or composite variable on the x-axis. Where two variables are being considered, the fields on predominance diagrams show the protein with the highest metastable equilibrium activity, as a function of the two variables on the x- and y-axes.

The thermodynamic calculations and stability diagrams reported below were made using the CHNOSZ software package [27]. The package encodes the equations of state, thermodynamic data, and the group additivity algorithms for proteins cited above. In recent versions of the software, the Boltzmann distribution was implemented for calculating the relative stabilities of proteins, and the calculations reported below use this method. The source code for the calculations reported in this paper, written in the R language [42] and utilizing the functions available in CHNOSZ, is available in the Supporting Information of this paper; Dataset S1 contains code for the example described above, and Dataset S2 contains the code used to produce the figures in the Results.

Contribution by energy of protein folding to uncertainty in chemical stability calculations

The group additivity algorithm adopted here for calculating the standard molal Gibbs energies of proteins is referenced to unfolded aqueous proteins [11]. Most proteins in their active forms adopt a folded conformation. The energy change for the folding reaction, or change of conformation, is commonly referred to as "protein stability" [47]. The latter nevertheless is distinct from the chemical stabilities being considered in this study, which are based on energies of protein formation. However, the energy change in the folding process contributes some uncertainty, assessed below, to the values adopted here for the standard Gibbs energies of the proteins.

It was estimated that the uncertainty in standard Gibbs energies of proteins inherent in the group additivity algorithm is of the order of five percent [11]. The values of ΔG_f° of the non-ionized forms CSG_METVO and CSG_METJA calculated using group additivity are -104102 and -101403 kJ mol^{-1} (-24881 and -24236 kcal mol^{-1}), respectively [27]; a nominal 5% uncertainty corresponds to ± 5205 and ± 5070 kJ mol^{-1} (± 1244 and ± 1211.8 kcal mol^{-1}). For proteins of comparable size, Gibbs energies of folding of ~ 40–80 kJ mol^{-1} (~ 10–20 kcal mol^{-1}) are not uncommon, depending on the temperature [48,47]. These values are approximately one-one hundredth the magnitude of the estimated uncertainties in the additive standard Gibbs energies of the unfolded proteins. Moreover, the effect of any systematic uncertainty that affects the standard Gibbs energies of the proteins in the same direction (as would the folding process) would tend to cancel in the relative stability calculations. Therefore, not accounting for the energy of protein folding contributes little to the overall uncertainty of the relative stability calculations.

Results

Description of field site and metagenomic sampling

Chemical and biological sampling was performed in July 2005 at the hot spring known as "Bison Pool" in the Sentinel Meadows in the Lower Geyser Basin of Yellowstone National Park [12]. "Bison Pool" is the unofficial name of a hot spring whose source pool is located at approximately 44.56961°N, 110.86513°W (WGS 84 datum), the closest officially named feature being called Rosette Geyser [49]. A map identifying the sampling sites referred to in this study, based on one found in Ref. [12], is shown in Fig. 2. The spring emits a continuous flow of boiling (93°C), moderately alkaline (pH $\sim$ 7.5) water, and emerges from within a base of sinter made of silica that has precipitated from the water. The winding outflow channel is occupied by a plethora of biofilms in a striking array of colors. At this and similar springs, the white and pink filaments found at higher temperatures harbor chemotrophic organisms such as *Aquificae* and some Archaea [50,49]. Yellow, orange and green biofilms (thick mats) found at lower temperatures are predominantly made up of photosynthetic communities of Cyanobacteria and relatives of *Chloroflexi* [51,13], although archaeal organisms can also be found at the lower temperatures [13].

The available metagenomic and geochemical data were obtained from five sampling sites from the source pool of the hot spring to 22 meters down the outflow channel. Site 3 is notable because it is within the "photosynthetic fringe", or the transition zone (ecotone) where bright colors indicate the onset of photosynthetic potential [52,12]. A summary of some of the field- and laboratory-based chemical analyses of the water relevant to this study is given in Table 2. Together with a decrease in temperature down the outflow channel, there is an increase in pH. An increase in the oxidation potential of the water is also apparent from the higher dissolved oxygen and lower sulfide concentrations observed in water sampled away from the source.

The biofilm samples used for metagenomic sequencing were collected at the same time as the water samples used for chemical analysis [12], except for field measurements of oxidation-reduction potential (ORP), which were obtained in 2009. Environmental DNA in the biofilm samples was shotgun-sequenced by the Joint Genome Institute using the Sanger method. The assembly and annotation of protein coding sequences was carried out through an automated pipeline in the Integrated Microbial Genomes with Microbiome Samples (IMG/M) system [53], and sequences used in this study were downloaded from the IMG/M website (http://img.jgi.doe.gov/m).

Amino acid compositions of model proteins

For this study, FASTA data files containing predicted protein sequences were downloaded from IMG/M using the taxonomic IDs BISONN, BISONS, etc. The letter codes for all the sampling sites are listed in Table 2, together with the total numbers of metagenomic reads and protein-coding sequences for each site.

Because they are derived from shotgun sequencing, most of the inferred protein-coding sequences are actually fragments of whole genes. It would be possible to select specific types of homologs, align the sequence fragments, and use the aligned positions in the stoichiometric and thermodynamic calculations described below. However, the calculation of the average oxidation states of carbon requires only the chemical formula of the proteins, and the calculation of the standard Gibbs energies of the proteins as described in the Methods only requires the amino acid compositions of the proteins. Therefore, in this study, model amino acid compositions were used to represent averages of groups of protein sequences in the metagenome.

The five "overall model proteins" have average amino acid compositions that were calculated as the average of all inferred protein sequences, including fragments, identified in the metagenome at each site. The amino acid compositions of the overall

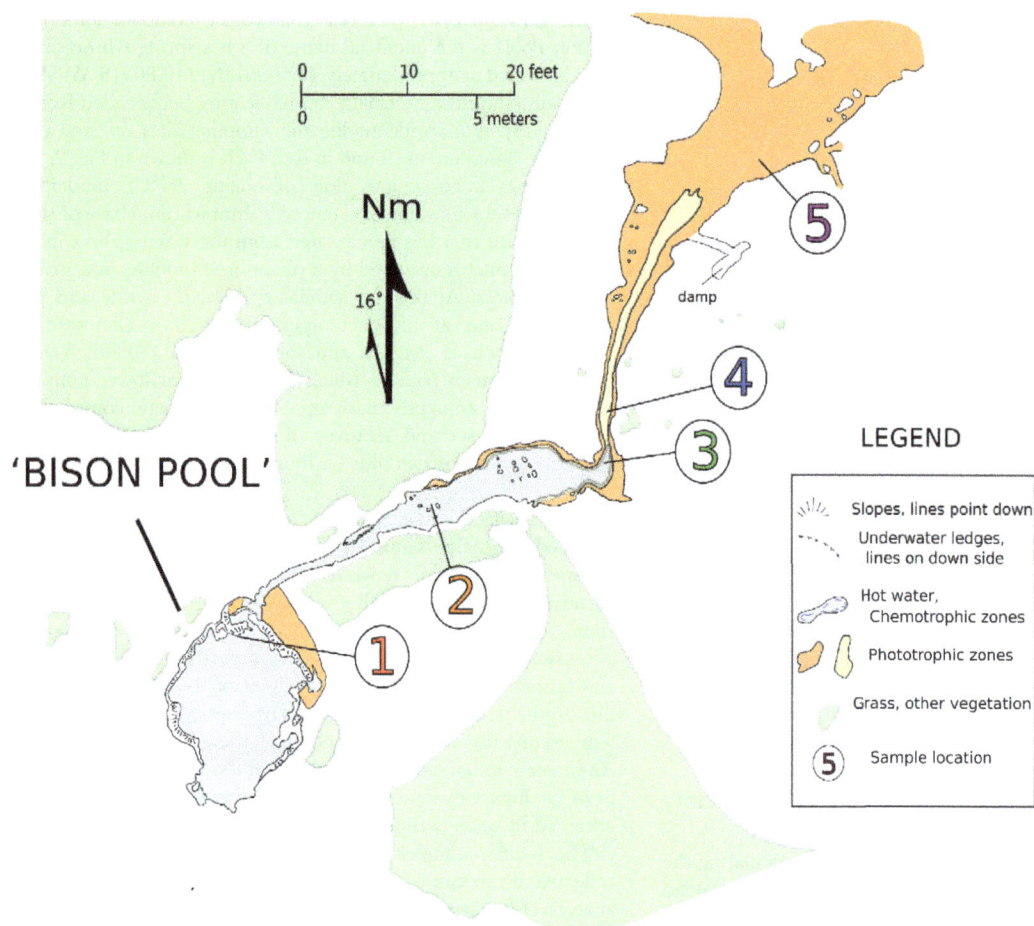

Figure 2. Map of the Bison Pool hot spring system. The map includes locations of the sites where biofilm and geochemical sampling was performed in the summer of 2005.

model proteins were calculated by summing the amino acid counts of all sequences at each site and dividing by the total number of sequences at each site. Accordingly, the model proteins are not whole proteins, but instead have fractional amino acid frequencies. The amino acid compositions are listed in Supporting Dataset S3.

The average amino acid compositions were also calculated for "classified model proteins" in twenty functional classes each corresponding to a keyword in the sequence annotations reported in IMG/M. The keywords were selected based on their frequencies in the annotations and represent a variety of functions

Table 2. Sampling site identification, distance from source pool, and summary of chemical and molecular sequence data at five sampling sites at Bison Pool[a].

Site	Code	Distance (m)	T (°C)	pH	Reads[b]	Protein sequences[b]	DO (mg/L)	DIC (ppm)	ΣS^{-2} (M)	SO_4^{-2} (M)
1	N	0	93.3	7.350	68350	40360	0.173	81.97	4.77E-06	2.10E-04
2	S	6	79.4	7.678	76642	50497	0.776	80.67	2.03E-06	2.03E-04
3	R	11	67.5	7.933	66798	43250	0.9	80.06	3.12E-07	1.98E-04
4	Q	14	65.3	7.995	123327	83790	1.6	78.79	4.68E-07	2.01E-04
5	P	22	57.1	8.257	90921	74082	2.8	78.75	2.18E-07	1.89E-04

[a]Temperature and pH were measured in the field with hand held temperature/conductivity (YSI, Yellow Springs, Ohio) and pH (WTW, Weilheim, Germany 300i pH meter with SenTix 41 pH electrode) meters. Dissolved inorganic carbon (DIC) was calculated from field titration of alkalinity. Dissolved oxygen (DO) and total sulfide (ΣS^{-2}) were measured in the field using a portable spectrophotometer (Hach, Loveland, Colorado). Sulfate was measured by ion chromatography in the lab (Dionex, Sunnyvale, California). For additional details see Ref. [55].
[b]Number of metagenomic reads and number of protein-coding genes available in files downloaded from IMG/M.

and cellular structures, but are neither comprehensive nor mutually exclusive. The keywords and number of identified sequences are listed in Table 3. The classification with the highest number of inferred protein sequences is "transferase", with a total of 15768 sequences across all five sampling sites, or ca. 5.4% of all of the protein sequences in the metagenome. The classification with the fewest number of sequences is "phosphatase", with a total of 2260 sequences, or about 0.8% of the metagenome. All the keyword searches were case-insensitive and any match was accepted (e.g., an annotation including the word "transporter" was matched by the "transport" keyword), except for "reductase", which was only matched to the beginning of a word in the annotation (e.g., annotations including the word "oxidoreductase" were not matched by the "reductase" keyword).

Average oxidation state of carbon of model proteins

To characterize the changes in the compositions of proteins across the sampling sites, we first calculated the elemental ratios and average oxidation number of carbon ($\overline{Z}_C$) of all the protein sequences available in the metagenome for each sampling site. The 95% confidence intervals around the mean values were calculated from a bootstrap analysis (nonparametric, ordinary bootstrap, 1000 replicates) performed using the "boot" package for the R software environment [42]. The results are plotted in Fig. 3 and the numerical values given in Supporting Dataset S4.

The S/C, O/C and N/C ratios shown in Fig. 3 exhibit an overall increase with distance from the hot-spring source, but there is a decrease in S/C and O/C of the proteins from site 3. The H/C ratio rises sharply between sites 1 and 2 and then decreases, with the proteins at site 3 again having a relatively lower value. The combined effect of the elemental ratios accounts for the trend in $\overline{Z}_C$ appearing in Fig. 3, which can be described as increasing with distance from the hot-spring source. The chemical compositions of the proteins at sites 4 and 5 are more similar to each other than to the other sites, and the overall trends for elemental ratios and $\overline{Z}_C$ show a slight reversal at these two sites.

The average oxidation state of carbon of overall and classified model proteins is shown in Fig. 4a. Because of the number of lines plotted in this figure only selected ones are labeled. The others can be identified by referring to the values of $\overline{Z}_C$ for the model proteins for site 1 listed in Table 3. Whatever the classification of the model protein, there in an increase in $\overline{Z}_C$ going from site 1 to site 5, and in most cases an increase between each of sites 1

through 4. Generally, there is a slight decrease in the value of $\overline{Z}_C$ between sites 4 and 5. The differences between the different classes of model proteins are profound: the oxidoreductases, transport and membrane proteins, and especially permeases, all have lower oxidation states of carbon than the others. This result is not surprising, given the greater abundance of hydrophobic sidechains in these predominantly membrane-associated [54] proteins. The model proteins that have the highest oxidation states of carbon are hydrolase at sites 1–3 and transposase at sites 4 and 5. That the transposase model proteins at sites 4 and 5 are more oxidized than other model proteins is noteworthy because at site 1 the transposase model protein has a value of $\overline{Z}_C$ that is only a little greater than that of the overall model protein.

Some relationships can be observed between the chemical composition of proteins and the chemical characteristics of the water. On the whole, there is a positive correlation between the values of $\overline{Z}_C$ of the model proteins and the field measurements of dissolved oxygen, and a negative correlation with total sulfide concentrations listed in Table 2. The correlations suggest that analysis of the energetics of the protein formation reactions could be used to combine protein composition and hot-spring chemistry in a thermodynamic model. In the following sections results are presented from a thermodynamic analysis that describes the relative stabilities of proteins in terms of temperature, pH, oxidation potential and other environmental variables.

Relative stabilities of model proteins: Metastable equilibrium

Residue-normalized formation reactions for the overall model proteins are listed in Table 4. These reactions are written in terms of the basis species HCO_3^-, $NH_{3(aq)}$, H_2O, $H_{2(aq)}$, HS^- and H^+, which correspond to inorganic sources of the major elements in the proteins. Because the number of amino acid residues in each of the model proteins is the average of the lengths of metagenomically derived protein sequences, including many fragments, the number of amino acids in each of the model proteins in Table 4 does not necessarily reflect the actual lengths of protein sequences in the microbial organisms in the hot spring.

Because the reactions in Table 4 are written for the formation of the residue equivalents of the model proteins, the reactions are effectively balanced with respect to the protein backbone group. Consequently, the stoichiometric reaction coefficients are independent of the sizes of the model proteins and can be compared

Table 3. Annotation terms, total number of sequences used to construct the classified model proteins, and $\overline{Z}_C$ of the classified model proteins (for site 1 only; entries are ordered by decreasing $\overline{Z}_C$).

Classification	Sequences	$\overline{Z}_{C,1}$	Classification	Sequences	$\overline{Z}_{C,1}$
hydrolase	4326	−0.188	kinase	6123	−0.206
transcription	3747	−0.189	signal	2377	−0.208
reductase	2905	−0.189	ATPase	7983	−0.214
dehydrogenase	9567	−0.190	transferase	15768	−0.218
synthetase	5979	−0.191	ribosomal	3598	−0.218
phosphatase	2260	−0.193	protease	2929	−0.219
peptidase	3635	−0.198	oxidoreductase	3803	−0.237
synthase	8561	−0.198	transport	11029	−0.258
periplasmic	2784	−0.203	membrane	5194	−0.262
transposase	2845	−0.206	permease	4886	−0.326

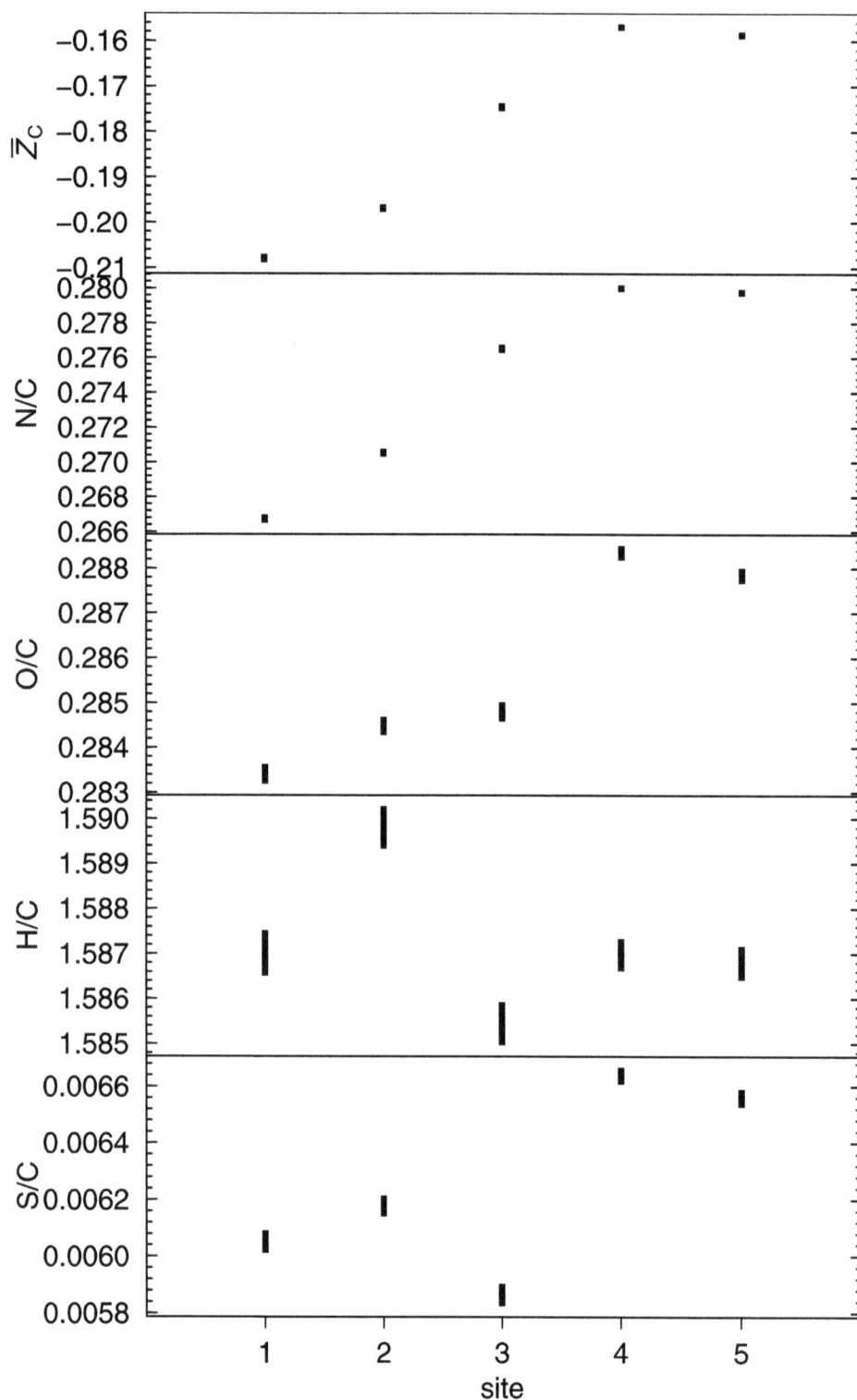

Figure 3. Elemental ratios and average oxidation state of carbon in protein sequences. Elemental ratios were calculated from the chemical formulas of all protein sequences available in the Bison Pool environmental genome, and average oxidation state of carbon ($\overline{Z}_C$) was calculated using Eq. (2). The bars are centered on the means, and the heights of the bars represent the 95% confidence intervals derived from a bootstrap analysis.

with each other in a first approximation to assess the effects of changing chemical conditions on the relative stabilities of the model proteins. For example, the coefficients on H_2, appearing on the reactant side of the reactions, decrease in order of increasing distance from the hot spring source, except for the last two sites, where the pattern is reversed. Therefore, increasing the chemical

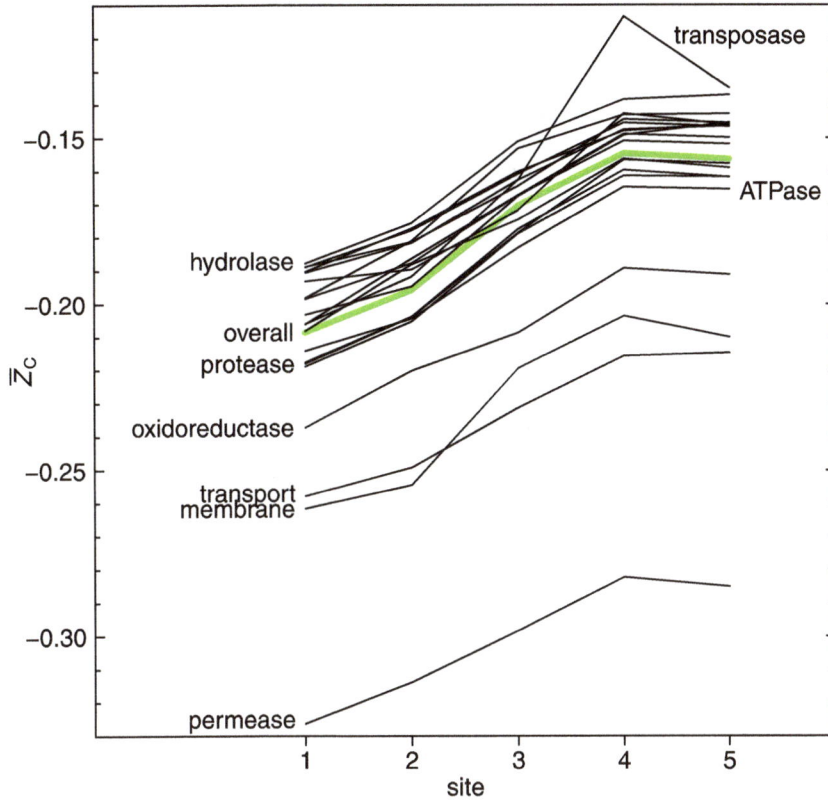

Figure 4. Average oxidation state of carbon in model proteins along the outflow channel. The values of $\overline{Z}_C$ calculated using Eq. (2) for the overall model proteins for each sampling site are shown by the bold green line, and those for each of the 20 classes of model proteins listed in Table 3 are shown by the thin lines. For clarity, only selected classes are labeled; hydrolase and protease are the ones with highest and lowest values of $\overline{Z}_C$ among the main group of lines at the top.

activity of H_2 tends to decrease the energy demand of forming overall model proteins at the high-temperature sites more strongly than the others.

The intensive variables used in the equilibrium calculations are temperature (T), pressure (P), and the chemical activities of the basis species and proteins. Activity coefficients of all species can be set to unity, so, for aqueous species, chemical activities were taken to be equivalent to concentrations in molal units. As noted above,

the activity coefficients of proteins in subcellular conditions are currently not amenable to general calculation. In addition, the concentrations of major ions in the water [55] are low enough that setting the activity coefficients of the basis species to unity is a tolerable first approximation. In the calculations reported below, the variables held constant were $P=1$ bar, $\log a_{H_2O}=0$, $\log a_{HCO_3^-}=-3$, $\log a_{NH_{3(aq)}}=-4$, $\log a_{HS^-}=-7$. The activities chosen for HCO_3^- and HS^- are based on the measurements

Table 4. Numbers of amino acid residues, average oxidation state of carbon, and per-residue formation reactions normalized per residue for overall model proteins for different sampling sites.[a]

Site	n_{AA}	$\overline{Z}_C$	Reaction
1	199.13	−0.208	5.125 HCO_3^-+1.357 NH_3+0.029 HS^-+10.784 H_2+5.154 H^+⇌13.924 H_2O+$C_{5.125}H_{8.099}N_{1.357}O_{1.451}S_{0.029}$
2	184.35	−0.196	5.076 HCO_3^-+1.366 NH_3+0.029 HS^-+10.649 H_2+5.105 H^+⇌13.788 H_2O+$C_{5.076}H_{8.033}N_{1.366}O_{1.441}S_{0.029}$
3	195.48	−0.171	5.023 HCO_3^-+1.388 NH_3+0.027 HS^-+10.473 H_2+5.050 H^+⇌13.639 H_2O+$C_{5.023}H_{7.933}N_{1.388}O_{1.429}S_{0.027}$
4	191.80	−0.154	4.966 HCO_3^-+1.389 NH_3+0.030 HS^-+10.315 H_2+4.996 H^+⇌13.467 H_2O+$C_{4.966}H_{7.854}N_{1.389}O_{1.430}S_{0.030}$
5	189.40	−0.156	4.972 HCO_3^-+1.389 NH_3+0.030 HS^-+10.333 H_2+5.002 H^+⇌13.487 H_2O+$C_{4.972}H_{7.863}N_{1.389}O_{1.429}S_{0.030}$

[a]The amino acid compositions of the model proteins are the bulk averages of all metagenomically derived protein sequences at each sampling site. The values of $\overline{Z}_C$ (average oxidation number of carbon) were calculated using Eq. (2).

of dissolved inorganic carbon and sulfide listed in Table 2, while that of NH_3 is a nominal value. The calculations of metastable equilibrium were referenced to unit total activity of the amino acid residues in the system. The other variables (T, pH and $\log a_{H_{2(aq)}}$) were used as exploratory variables as described below.

The results of computations of relative stabilities of the overall model proteins are depicted in Fig. 5. Values of temperature measured at each site in the hot spring (Table 2) are shown in Fig. 5a. Stability calculations were performed for a system composed of the five overall model proteins, using the pH measured at site 3. The most stable overall model proteins as a function of temperature and $\log a_{H_{2(aq)}}$ are shown in the equilibrium predominance diagram in Fig. 5b. The temperature range in this diagram is somewhat larger than the measured range of temperature in the hot spring, and the range of $\log a_{H_{2(aq)}}$ was set to encompass the stability fields of the proteins.

Values of pH measured at each site are shown in Fig. 5c. Stability calculations using the temperature measured at site 3 lead to the equilibrium predominance diagram shown in Fig. 5d. The pH range in this diagram is slightly larger than the measured range of pH in the hot spring.

Note that in Figs. 5b and d, only the overall model proteins from sites 1, 2 and 4 appear, going from high to low values of $\log a_{H_{2(aq)}}$ in that order. Increasing pH at constant $\log a_{H_{2(aq)}}$ and

temperature (Fig. 5d) moves toward the relative stability fields for the overall model proteins that are more distal from the hot-spring source; this pattern is congruent with the pH differences between sampling sites in the hot spring. On the other hand, decreasing temperature at constant $\log a_{H_{2(aq)}}$ and pH (Fig. 5b) moves toward the relative stability fields for the overall model proteins that are more proximal to the hot-spring source; this pattern is incongruent with the temperature gradient in the outflow channel of the hot spring.

If it were representative of the chemical gradients in the hot spring, the thermodynamic model used here would generate a pattern of relative stabilities of the model proteins that reflects their geographical distribution. It is apparent from the above findings that this is not possible if $\log a_{H_{2(aq)}}$ is constant along the outflow channel. Instead, the dashed lines in Figs. 5b and d and the equilibrium chemical activities of the proteins shown in Figs. 5e and f are consistent with changing $\log a_{H_{2(aq)}}$ in the model together with the measured changes in temperature and pH and $\log a_{H_{2(aq)}}$, and were derived by considering the relative stabilities of many classes of model proteins as described below.

Operational equation for activity of hydrogen

Figure 6 contains $\log a_{H_{2(aq)}}$-temperature equilibrium predominance diagrams for the 20 classes of model proteins listed in

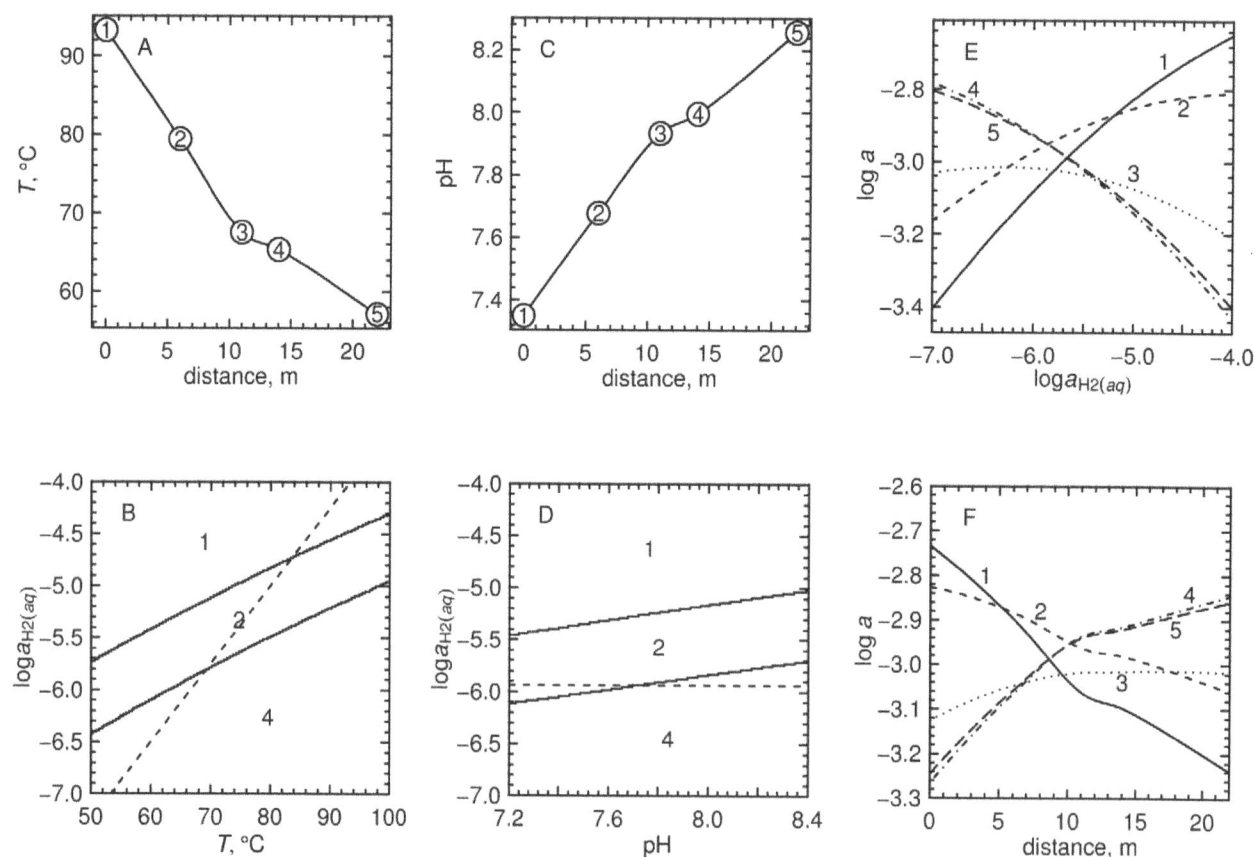

Figure 5. Analysis of relative stabilities of overall model proteins for sampling sites. Panel (a) is a plot of the measured temperatures as a function of distance with a smooth curve connecting the points. Panel (b) is a plot of the relative stabilities of the overall model proteins as a function of $\log a_{H_{2(aq)}}$ and temperature at the pH of site 3 (7.933). Panel (c) is a plot of the measured pHs as a function of distance with a smooth curve connecting the points. Panel (d) is a plot of the relative stabilities of the overall model proteins as a function of $\log a_{H_{2(aq)}}$ and pH at the temperature of site 3 (67.5°C). The dashed lines in (b) and (d) depict Eq. (24). Panel (e) is a plot of the relative stabilities of the overall model proteins as a function of $\log a_{H_{2(aq)}}$ at the temperature and pH of site 3. Panel (f) is a plot of the relative stabilities of the overall proteins as a function of the combination of changes with distance of measured temperature and pH and modeled $\log a_{H_{2(aq)}}$ (using Eq. 24).

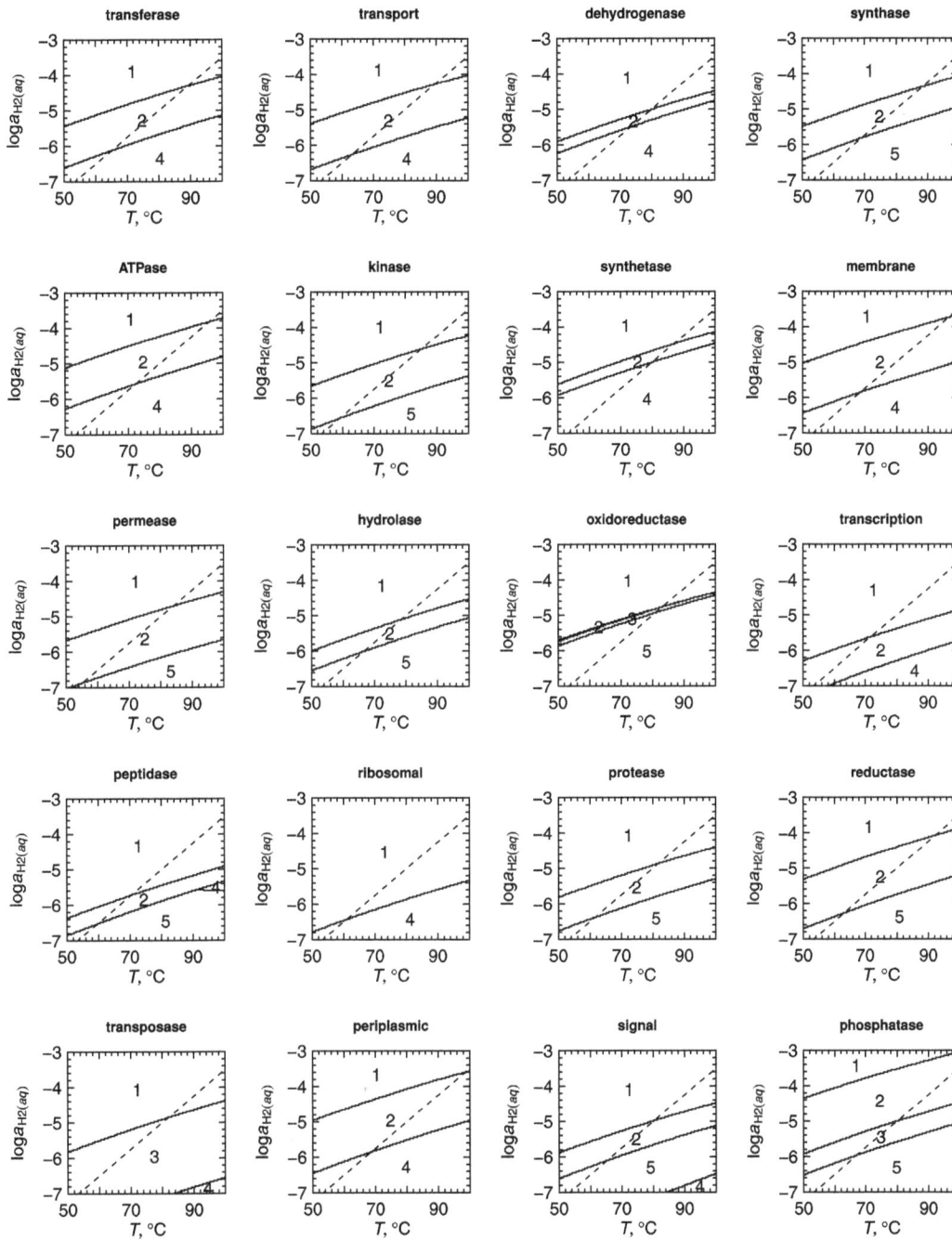

Figure 6. Equilibrium predominance diagrams for classified model proteins. For each class of model proteins, the fields in these predominance diagram represent the model protein with the most positive equilibrium activity as a function of $\log a_{H_{2(aq)}}$ and temperature. pH was set to the value listed in Table 2 for site 3 and the chemical activities of the other basis species were set to reference values specified in the Results section. The dashed line in each diagram indicates values of $\log a_{H_{2(aq)}}$ calculated using Eq. (24).

Table 3. These figures were constructed in a manner analogous to Fig. 5b. For each class of model proteins shown in Fig. 6 the protein from site 1 is relatively stable at higher values of $\log a_{H_{2(aq)}}$, and the protein from site 4 or 5 occupies the low-$\log a_{H_{2(aq)}}$ portion of the diagram. Other details differ between the various classes of model proteins; for example, the model protein for phosphatase at site 1 is stable relative to the model proteins for phosphatase at other sites only at highly reduced conditions. Compared to other

classes of proteins, the predominance field for the model protein for oxidoreductase at site 2 is much smaller, and that for the ribosomal proteins at site 2 does not even appear. In spite of these differences, the overall resemblance of the plots in Fig. 6 to each other and to Fig. 5b indicates that different subsets of the metagenome have similar relative stability relationships.

The dashed lines in Fig. 5b and d and Fig. 6 denote values of $\log a_{H_{2(aq)}}$ that are given by the function

$$\log a_{H_{2(aq)}} = -11 + 3/(40 \times T(^\circ C)). \qquad (24)$$

This equation is the result of a graphical regression of the plots in Fig. 5b and Fig. 6 such that the values of $\log a_{H_{2(aq)}}$ as a function of temperature give a progression in the stability fields of a majority of the model proteins that is similar to the geographical distribution of the proteins. For example, in Fig. 5b, the line given by Eq. (24) encounters the stability fields for the overall model proteins for sites 1, 2 and 4, in that order, with decreasing temperature.

Note that the plot in Fig. 5d is drawn for a single temperature (that for site 3 listed in Table 2), so the line for $\log a_{H_{2(aq)}}$ is horizontal in this figure. The line crosses the predominance field boundary between sites 2 and 4 at pH ~ 7.7, which is close to the measured pH of site 3 (Table 2). The absence of a predominance field for the model protein for site 3 in this figure and in most plots shown in Fig. 6 suggests that generally the proteins at site 3 are less stable than the model proteins for other sites, even under the specific conditions of site 3.

Effects of other variables

The model developed here incorporates spatial gradients of pH and activity of hydrogen, but in principle other chemical variables could also contribute significantly to the relative stabilities of the proteins. For example, the concentration of dissolved sulfide decreases by more than an order of magnitude between the source of the hot spring and the most distal sampling site (Table 2). However, the consequences of that gradient on the relative stabilities of any two proteins is proportional to the difference in the reaction coefficient of sulfide between the two reactions. In Table 4, the largest difference between reaction coefficients on HS^- is only 0.003, which is much smaller than the differences in the relative stabilities of the proteins ($A/2.303RT$ for any of the sites listed the Table). Accordingly, the sulfide concentration in the model was set to a constant since the changes seen in the hot spring have a smaller effect on the relative stabilities of the proteins compared to temperature, pH and H_2. For comparison, the difference between reaction coefficients on H_2 in the formation reactions of overall model proteins 1 and 2 is 0.134, while the difference in relative stabilities of the proteins is 0.126, so a one-order-of-magnitude decrease in the activity of H_2, by itself, is enough to increase the relative stability of the model protein for site 2 over site 1.

Conversely, although the coefficients on HCO_3^- in the reactions shown in Table 4 differ from each other more than those on HS^-, the measured concentrations of dissolved inorganic carbon differ by not more than 4 ppm between the sites (Table 2). Therefore, the small changes in the corresponding activities of HCO_3^- would not be expected to significantly alter the chemical affinities of the reactions relative to each other, so, as with sulfide, the activity of HCO_3^- was set to a constant value.

The calculations described above used a nominal value of 10^{-4} for the activity of NH_3. Total ammonia was not detected above 0.01 mg/L (0.6 μmol) in spectrophotometric analysis of water at the hot spring [55], so it is reasonable to ask whether lowering the activity NH_3 to $\sim 10^{-7}$ has a major effect on the calculations of relative stabilities. The interdependence of equilibrium activities of the basis species can be assessed by writing a reaction representing the transformation between two proteins, for example the overall model proteins for sites 2 and 4. Adding reaction 4 in Table 4 to the opposite of reaction 2 yields the following reaction:

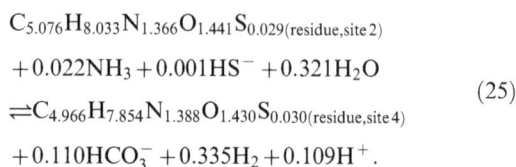

$$C_{5.076}H_{8.033}N_{1.366}O_{1.441}S_{0.029(residue,site 2)}$$
$$+ 0.022NH_3 + 0.001HS^- + 0.321H_2O$$
$$\rightleftharpoons C_{4.966}H_{7.854}N_{1.388}O_{1.430}S_{0.030(residue,site 4)} \qquad (25)$$
$$+ 0.110HCO_3^- + 0.335H_2 + 0.109H^+ .$$

It follows from Eqs. (4)–(6) that, if the chemical affinity of Reaction 25 and the activities of all other reactants and products remain unchanged, decreasing $\log a_{NH_{3(aq)}}$ by 3 results in a decrease in the calculated $\log a_{H_{2(aq)}}$ of 0.197 (i.e., $\frac{0.022}{0.335} \times 3$). Likewise, recon-naissance calculations indicate that decreasing the activity of NH_3 to 10^{-7} generally results in a lowering of the equal-activity lines shown in Fig. 6, with a more pronounced effect for the lower equal-activity lines, which in some cases shift downwards by at about half a $\log a_{H_{2(aq)}}$ unit. Therefore, refinement of the calculations described here may yield results that support modifying Eq. (24) to have a somewhat steeper slope and lower intercept. Nevertheless, without any direct measurements of the total ammonia concentration, such refinements remain speculative.

Relative stabilities of model proteins: Chemical affinities

Chemical affinities for the per-residue formation reactions of the overall model proteins in an equal-activity reference state calculated for sites 1 to 5 are listed in Table 5. The calculations used values of temperature and pH listed in Table 2, $\log a_{H_{2(g)}}$ for each of these sites taken from Eq. 24, and activities of the other basis species given above. The activities of the residue equivalents of each of the model proteins were set to unity. All of the reactions listed in Table 4 are endergonic reactions, which is apparent from the negative values of chemical affinity that are shown in Table 5. However, the chemical affinities are less negative at the higher-temperature, more reduced conditions, which is consistent with a previous comparison of the energetics of biomass synthesis under oxic and anoxic conditions [56].

Examination of Table 5 shows that the reaction with the greatest chemical affinity at site 1 is that for the overall model protein for that site. In contrast, at the conditions of sites 3–5 the formation, per residue, of the model protein for site 4 is the least energetically demanding. The affinities listed in Table 5 are calculated for the formation reactions of the proteins normalized per residue, but

Table 5. Chemical affinities for the reactions in Table 4 at the model conditions for each site.[a]

Reaction	Site 1	Site 2	Site 3	Site 4	Site 5
1	**−18.720**	**−27.894**	−35.386	−36.822	−42.265
2	−18.846	−27.914	−35.319	−36.740	−42.120
3	−19.120	−28.053	−35.349	−36.749	−42.051
4	−19.270	−28.080	**−35.276**	**−36.657**	**−41.888**
5	−19.254	−28.078	−35.285	−36.668	−41.907

[a]The chemical affinities of the reactions are in dimensionless values (i.e., $A/2.303RT$) calculated using the temperature and pH of the sampling sites listed in Table 2, $\log a_{H_{2(aq)}}$ from Eq. (24), chemical activities of the other basis species described in the Results, and chemical activities of the proteins equal to unity. The charges and Gibbs energies of ionized proteins (not shown in Table 4) calculated using group additivity were considered in the calculations of chemical affinity. Bold entries in each columns indicate the reaction with the highest calculated chemical affinity.

comparison of the relative stabilities of the proteins (as is done on the predominance diagrams) requires further accounting for the relative lengths of the model proteins. Equation (16) was used to account for different lengths when constructing equilibrium activity diagrams for proteins. Analogously, by subtracting the logarithm (base 10) of the number of amino acids present in each of the model proteins from the values listed in Table 4, one obtains length-corrected affinities per residue that can be compared with the equilibrium activity diagrams. For example, performing this operation on the second column of Table 5 shows that the model protein from site 2 is the most stable, even though the per-residue reaction for site 1 has a greater chemical affinity, before applying the length correction. The outcome is in accord with the progression, from sites 1 to 2 to 4, in the relative stabilities of the overall model proteins apparent in Figs. 5b and d.

Combined analysis of temperature, pH and oxidation potential

Returning to the metastable equilibrium (equal-affinity) reference state, the equilibrium activities of the proteins are plotted in Fig. 5e as functions of $\log a_{H_2(aq)}$ at constant temperature and pH (corresponding to site 3) for a total activity of residues equal to one. Nowhere does the equilibrium chemical activity of the overall model protein for site 3 rise above all the others, which is consistent with Fig. 5b, but it is also apparent that the activity for this model protein maximizes at intermediate values of $\log a_{H_2(aq)}$. The relative instability of the model proteins for site 3 throughout the different classes of model proteins is apparent from the low frequency of predominance fields representing this site appearing in Fig. 6. It can also be seen in Fig. 5e that the overall model proteins representing sites 4 and 5 are similar to each other in terms of their relative stabilities.

To portray the effects of changing temperature, pH and $\log a_{H_2(aq)}$ simultaneously, all of these variables are projected along the x-axis ("distance") in Fig. 5f. The values of temperature and pH at any point along the distance axis are taken from the curves shown in Figs. 5a and c, and the values of $\log a_{H_2(aq)}$ are calculated using Eq. (24). Fig. 5f has the advantage that the relative stabilities of the model proteins are shown as a function of a spatial variable and can therefore be compared with the physical location of the sampling sites in the hot spring.

In order to visualize the relative stabilities of all of the groups of model proteins on a single figure, the equilibrium activities (a_i) of the proteins were transformed into equilibrium degrees of formation. The degree of formation of the ith model protein (α_i) is given by

$$\alpha_i = a_i / \sum_i a_i, \tag{26}$$

where the summation occurs over all of the model proteins in the calculation, which in this case is five (one for each sampling site). Since $\sum_i \alpha_i = 1$, the degree of formation of any protein can be visualized as a fraction of a bar of unit length. The equilibrium degrees of formation of the overall and classified model proteins are shown as a function of distance in Fig. 7. In this figure, the color code refers to the five sampling sites, and the height of the bars represents the equilibrium degree of formation of the indicated model protein. At any point along the distance axis, the bars are stacked with the most relatively stable model protein on the top. The locations and color codes of the sampling sites are indicated by the tick marks.

Fig. 7 permits a visual test of the overall goodness of fit of Eq. (24) to the geographical relationships of the sampling sites and also helps in identifying outliers. At the high-temperature end, the most stable model protein is usually that from site 1, and rarely from site 2 (phosphatase, periplasmic). At the low-temperature end, the most stable protein is usually that from site 4 or 5 (with approximately equal frequency) and only occasionally from sites 2 (transcription) or 3 (transposase). The only model proteins for site 3 that are the most relatively stable over any part of the combined chemical gradient are those for transposase (over the mid- to low-temperature range), oxidoreductase and phosphatase (only at moderate temperatures). The transition zone occurring toward the middle of the plots is sometimes associated with an increase in the relative stability of the model proteins for site 3 (e.g., transferase, synthase), but in many other cases the equilibrium degree of formation of the model proteins for site 3 minimizes relative to the other sites. The overall relative instability of the proteins from site 3 is can be attributed to constraints on their amino acid compositions that also account for the shifts in O/C, H/C, and S/C for these proteins from the overall trends between sites apparent in Fig. 3. A major exception is transposase, for which the model proteins from site 3 are relatively stable over much of the chemical gradient. As in the general case, the relative stabilities are conditioned by trends in amino acid composition that also affect elemental ratios, apparent in Fig. 4 as a high oxidation state of carbon of the transposase model proteins at sites 4 and 5.

Measurements of oxidation-reduction potential

Eq. (24) represents a proposal for the temperature dependence of the activity of hydrogen derived from the relative stabilities of the model proteins. It can be compared with a variety of other measurements that are indicators of redox conditions of the hydrothermal solution including results of field measurements of redox conditions made using an oxidation-reduction potential (ORP) probe.

ORP, temperature and pH readings obtained in Summer 2009 for Bison Pool (four years after the biofilm sampling for metagenomic analysis and acquisition of chemical data reported in Table 2) and other flowing hot springs in Yellowstone are listed in Table 6 and in Dataset S5. The ORP measurements were obtained at three sites at Bison Pool that approximated the original locations of sites 3, 4 and 5. ORP, pH and temperature measurements at higher temperatures were also obtained at Mound Spring, which is the official name a nearby hot spring in Sentinel Meadows with chemical features similar to Bison Pool [55].

Temperature and pH were measured in the field with hand-held temperature/conductivity (YSI, Yellow Springs, Ohio) and pH meters (WTW, Weilheim, Germany 300i pH meter with SenTix 41 pH electrode). Oxidation-reduction potential was measured using a high-temperature ORP probe (PI-M11-ORP-HT) rated to greater than 80°C and a Thermo Scientific pH/mV meter with a readout sensitivity of 1 mV, both acquired from Pulse Instruments (Van Nuys, CA). The ORP probe contains a silver-silver chloride (Ag/AgCl) reference electrode with saturated KCl solution. Before the field work, the ORP meter was calibrated in the laboratory at 25°C using a stock of Light's Solution (ferrous/ferric sulfate in sulfuric acid) supplied by Pulse Instruments. The stated potential of the solution is 468 ± 25 mV at 25°C vs. saturated KCl/AgCl electrode.

To convert the ORP readings (referenced to the Ag/AgCl electrode) to Eh (referenced to the standard hydrogen electrode, or SHE) we used

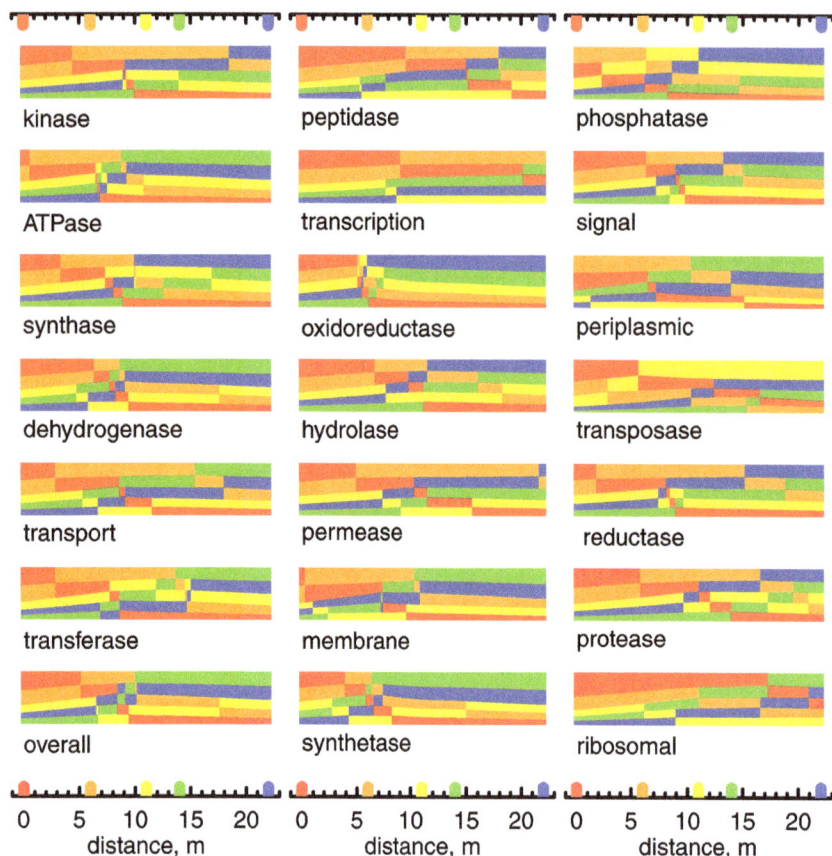

Figure 7. Equilibrium degrees of formation of model proteins in 21 classes. The locations of the sampling sites are indicated by the colored tick marks. The degrees of formation of the five model proteins in each class are shown by the heights of the color-coded bars. At any point on the distance axis, the bars are stacked in order of relative stability, so that the most stable (highest equilibrium degree of formation) is at the top.

$$Eh = ORP + E(Ag/AgCl), \qquad (27)$$

where E(Ag/AgCl) is the potential vs. SHE of the Ag/AgCl reference electrode. The potential of the reference electrode was calculated using an equation for Ag/AgCl with 1M KCl electrolyte [57]

$$E(Ag/AgCl, 1\,M\,KCl) = 0.23737 - 5.3783 \times 10^{-4} * T$$
$$- 2.3728 \times 10^{-6} * T^2 - 2.2671 \times 10^{-9} * (T + 273), \qquad (28)$$

where E is in volts and T is in degrees Celsius. The effect of differences between the saturated and 1M KCl Ag/AgCl electrodes is discussed further below.

Values of Eh calculated by combining field measurements of ORP with Eqs. (27) and (28) are shown in Fig. 8a. The hot springs are identified by the number codes given in Table 6, and the source pools are indicated by bold symbols. The values for Mound Spring and Bison Pool (points labeled "3" and "4" in the figure) increase with decreasing temperature and are similar to an unnamed hot spring in Sentinel Meadows for which data are available (points labeled "5" in the figure; the GPS locations of this and other springs are given in Table 6).

Equilibrium values of $\log a_{H_{2(aq)}}$ were calculated by combining Eh, pH and temperature. The values of Eh were first converted to pe (the negative of the logarithm of the activity of the electron) using [26]

$$pe = \frac{F}{2.303RT} Eh, \qquad (29)$$

where F denotes the Faraday constant. Then, $\log a_{H_{2(aq)}}$ was calculated from the law of mass action for

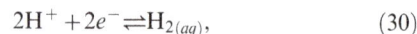

$$2H^+ + 2e^- \rightleftharpoons H_{2(aq)}, \qquad (30)$$

that is,

$$\log K = \log a_{H_{2(aq)}} + 2pH + 2pe. \qquad (31)$$

Values of $\log K$ calculated using the standard molal properties of the species in Reaction 30 (see Methods) were combined with pe and pH to generate the values of $\log a_{H_{2(aq)}}$ shown in Fig. 8b. These values correspond to equilibrium with both protons (activities constrained by pH measurements) and electrons (activities inferred from the ORP measurements, which does not capture the full spectrum of reactivity of electrons in the solution).

At 25°C the potential of the Ag/AgCl electrode (1M KCl) calculated using Eq. (28) is 0.222 V. In contrast, the potential of the Ag/AgCl electrode with saturated KCl, which might be a more appropriate choice for calculations given the specifications of the ORP probe used for the measurements, is about 0.197V [58], or about 0.025V lower. Eqs. 29 and 31 can be used to calculate

Table 6. Meter readings for selected hot springs and their outflows in Yellowstone National Park, Summer 2009.

Sample	Area	Spring	Latitude	Longitude	pH	T, °C	ORP, mV
090723G	Greater Obsidian Pool Area	1	0544526	4939786	5.12	79.0	27
090723K	Greater Obsidian Pool Area	1	0544541	4939799	5.38	57.6	98
090723F	Greater Obsidian Pool Area	2	0544482	4939773	4.21	68.3	185
090723E	Greater Obsidian Pool Area	2	0544497	4939806	4.21	53.0	183
090724PA	Sentinel Meadows	3	0511112	4934624	8.28	93.9	−258
090724OA	Sentinel Meadows	3	0511112	4934623	8.31	87.7	−227
090724NA	Sentinel Meadows	3	0511093	4934632	8.76	66.4	−98
090724O1	Sentinel Meadows	4	0510717	4935156	7.82	75.7	−55
090724P1	Sentinel Meadows	4	0510722	4935156	7.96	70.1	−58
090724Q1	Sentinel Meadows	4	0510723	4935156	8.06	66.2	−41
090724UA	Sentinel Meadows	5	0510846	4934731	7.84	87.3	−217
090724VA	Sentinel Meadows	5	0510842	4934733	8.07	71.3	−155
090728NA	Crater Hills	6	0541100	4944727	3.69	90.0	−50
090728-13	Crater Hills	6	0541100	4944727	3.50	51.6	274
090729DA	South of Sylvan Springs	7	0518409	4949162	5.84	79.5	−234
090729GA	South of Sylvan Springs	7	0518405	4949180	7.55	57.5	−130
090729RA	South of Sylvan Springs	7	0518395	4949185	7.92	44.9	−47
090729MA	South of Sylvan Springs	8	0518426	4949136	5.57	86.2	−248
090729HA	South of Sylvan Springs	8	0518426	4949144	6.17	73.3	−175
090729PA	South of Sylvan Springs	8	0518419	4949157	7.42	50.4	−42
090801HA	Heart Lake, Fissure Area	9	0538074	4905829	8.67	92.4	−373
090801I1	Heart Lake, Fissure Area	9	0538068	4905836	8.70	88.4	−351

The column labeled "spring" contains a unique number code for each hot spring and is used to label the points in Fig. 8. Named hot springs are Obsidian Pool (1), Mound Spring (3), Crater Hills Geyser (6). "Bison Pool" is number 4. Data for Bison Pool shown in this table were obtained in 2009 at three locations along the outflow channel. Latitude and longitude are the northing and easting, in meters, for the 12T grid zone for the Universal Transverse Mercator (UTM) projection using the WGS 84 datum. Supporting Dataset S5 includes the data in this table and estimated uncertainty in pH and ORP measurements.

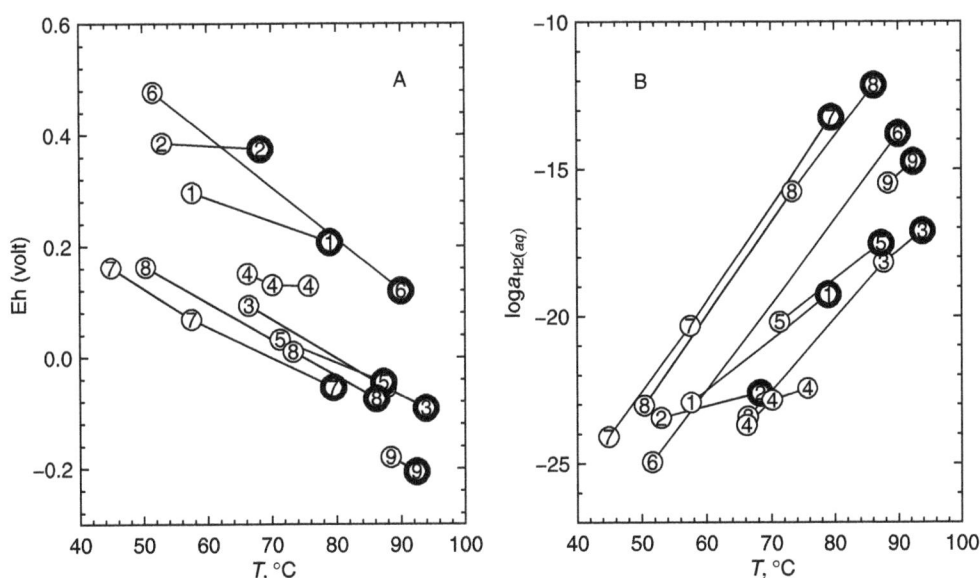

Figure 8. Calculated Eh and activity of hydrogen in hot springs. Eh and equilibrium $\log a_{H_2(aq)}$ as functions of temperature for various hot springs were calculated from measured values of pH and ORP listed in Table 6. The different hot springs are identified by numbers (see Table 6 for key), and bold symbols indicate the sources of the hot springs.

that a decrease of 0.025V at constant pH would increase the equilibrium $\log a_{H_{2(aq)}}$ by approximately 0.84 at 25°C and 0.68 at 100°C. Changes of this magnitude would not drastically affect the relative positions of the points and lines shown in Fig. 8a or the comparisons drawn further below. The difference between the potentials of the 1M and saturated KCl electrode is probably also comparable in size to the total uncertainty in the measurement of oxidation-reduction potential in reactive hydrothermal fluids (see for example Ref. [59]). Therefore, the values of E(Ag/AgCl) used to calculate Eh from ORP (Eq. 27) were taken from Eq. (28) without modification.

Although the water at Mound Spring had a higher pH than reported for Bison Pool (see Table 6) the equilibrium values of $\log a_{H_{2(aq)}}$ calculated using Reaction 30 for the two hot springs are close to each other at similar temperatures. The values of $\log a_{H_{2(aq)}}$ calculated in this manner for the outflow channels of Bison Pool and Mound Spring are lower than other hot springs at the same temperatures shown in Fig. 8b, but a more reduced fluid at higher temperatures seems to be the case for any of the measured hot springs.

Comparison of equilibrium activities of hydrogen

Concentrations of chemical species such as dissolved oxygen, and species with an element in different oxidation states (e.g. sulfide and sulfate) can be expressed on the $\log a_{H_{2(aq)}}$ redox scale using relationships derived from chemical equilibrium. Values of $\log a_{H_{2(aq)}}$ for equilibrium between dissolved sulfide (as HS^-) and sulfate (SO_4^{-2}) were obtained using the law of mass action for

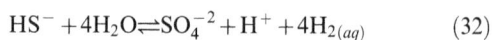

$$HS^- + 4H_2O \rightleftharpoons SO_4^{-2} + H^+ + 4H_{2(aq)} \qquad (32)$$

at the temperature of each sampling site. The law of mass action is the relationship (Eq. 4) between the equilibrium constant (K) and activity product (Q) of this reaction when the chemical affinity (A) of the reaction is zero (see Eq. 31 for an example). Activities of HS^- and SO_4^{-2} were taken to be equal to molalities of the species listed in Table 2, and values of $\log K$ were again obtained using standard molal thermodynamic properties of the species as a function of temperature calculated as described in the Methods.

Values of $\log a_{H_{2(aq)}}$ in equilibrium with dissolved oxygen were obtained using the law of mass action for

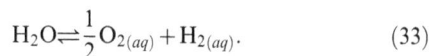

$$H_2O \rightleftharpoons \frac{1}{2}O_{2(aq)} + H_{2(aq)}. \qquad (33)$$

using activities of $O_{2(aq)}$ derived from the concentrations listed in Table 2.

Fig. 9 shows a comparison of the activities of hydrogen calculated using Eq. (24) and activities of hydrogen in equilibrium with measured oxidation-reduction potentials (ORP) and the oxygen and sulfur redox indicators described above. The temperature dependence of the equilibrium $\log a_{H_{2(aq)}}$ is strongest for the ORP measurements and weakest for the sulfur system. Among the three different redox indicators considered here (ORP, sulfide/sulfate, and dissolved oxygen) the lowest values of $\log a_{H_{2(aq)}}$ come from equilibrium with dissolved oxygen and the highest for equilibrium in the sulfide/sulfate reaction. All of the redox indicators have lower equilibrium values of $\log a_{H_{2(aq)}}$ than those calculated using Eq. (24).

These three proxies are limiting cases, indicating what the values of $\log a_{H_{2(aq)}}$ would be if each of the reactions dictated the actual hydrogen activity in the system. Therefore, the separation of the lines in Fig. 9 provides evidence that overall redox equilibrium

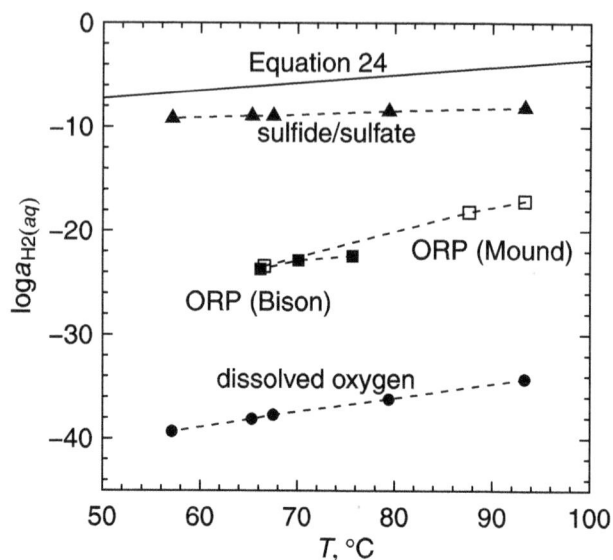

Figure 9. Comparison of different estimates of hydrogen activity as a function of temperature. Values shown for sulfide/sulfate and dissolved oxygen were computed using measured concentrations of those species listed in Table 2. Values shown for "Eh" were computed using pHs and ORP readings listed in Table 6. The solid line indicates values calculated using Eq. (24).

does not characterize the hot spring (see Ref. [55] for further evidence of redox disequilibria in hot springs). Nevertheless, the simplest explanation for the trends shown in Fig. 9 is that the states of various redox reactions in the water are all becoming more oxidized with decreasing temperature, and that the redox gradient derived from the relative stabilities of the proteins (Eq. 24) occurs at more reducing conditions than any of the inorganic oxidation-reduction reactions.

Discussion

By computing the standard Gibbs energies of overall reactions representing the formation of proteins, and using geochemical data on the gradients of temperature and chemistry, the relative stabilities of model proteins derived from metagenomic sequences in a hot spring could be calculated. It was possible to compute relative stabilities of the model proteins that reflect their overall spatial distribution in the system. An equation for hydrogen activity as a function of temperature was proposed, as a way of calibrating the model to maximize the correspondence between the geographical distributions of the sampling sites and the progression of relative stabilities of proteins.

There is a general increase in average oxidation state of carbon in the model proteins that parallels the rising redox potential of species in the water as it flows away, also cooling, from the hot spring. However, the average oxidation states of carbon in the different classes of model proteins are offset from each other, with membrane-associated proteins (including permeases, transport proteins and oxidoreductases) being more reduced. As noted above, hydrophobic amino acids tend to have lower oxidation states of carbon, so an increase in the number of hydrophobic amino acids can account for a decrease in average oxidation state of carbon in the proteins. An increased frequency of hydrophobic amino acids in proteins is a likely feature of adaptation to higher temperatures, as seen in the genomes of model organisms [60,61]. Our metagenomically based finding of a negative correlation between temperature and oxidation state of carbon is consistent with those results.

The oxidation states of carbon in the different classes of model proteins increase along the outflow channel of the hot spring, except for a slight decrease between sites 4 and 5. The apparent inversion in relative stabilities between sites 4 and 5 might be connected to additional hydrogen input, possibly from oxygenic photosynthesizers [62], as a byproduct of nitrogen fixation [63] or fermentation, and/or from secondary sources of reduced gases in the hot-spring system.

A distinctive feature of the chemical compositions of the proteins in the study area is the departure of site 3 from the general trends for H/C, O/C and S/C shown in Fig. 3, the infrequent appearance of the model proteins for site 3 in the diagrams in Fig. 6, and the generally low equilibrium degrees of formation of these proteins in Fig. 7. Therefore, the model proteins for site 3 are on average less stable, or have a greater energy of formation, than those from other sites. This result could imply that there are specific metabolic requirements of the organisms at site 3, near the photosynthetic fringe, that cause them to produce proteins that are relatively energetically demanding.

The calculations described here may help elucidate the redox gradients between and cellular interiors and their surroundings. The interiors of bacterial cells such as *Escherichia coli* are more reduced than the growth medium in laboratory experiments [64], but much remains unknown about the redox conditions of microbial interiors in high-temperature environments. Fig. 9 shows that effective values of $\log a_{H_{2(aq)}}$ in equilibrium with sulfur/sulfide, dissolved oxygen, and inferred Eh values, all measured on bulk water samples in the field, decrease as as a function of temperature (oxygen and Eh more so), but that the protein-equilibrium-based model equation for $\log a_{H_{2(aq)}}$ is more reduced than any of these proxies. This finding implies that the interiors of the microbial cells, where the proteins are mainly present, are more reduced than the environmental conditions at the temperatures in the hot spring. There is, however, not a general agreement about the redox conditions inside microbial cells; relatively oxidized conditions would help to account for the probable frequent occurrence of disulfide bonds in proteins in archaeal organisms (including some hyperthermophilic representatives) [65]. It would be useful to have data on subcellular redox indicators (e.g. oxidized and reduced forms of glutathione) at high temperatures to help resolve these questions.

Outlining the convergence of physical, chemical and biological forces that shape the information present in metagenomic sequences would benefit from the development of more sophisticated thermodynamic models and a tighter connection with phylogenetic approaches. For example, the construction of the model proteins could be based on identification of housekeeping genes that are conserved across phyla [66]. The use of only aligned sequences therein would help to eliminate some of the noise inherent in averaging shotgun sequence fragments, leading to a closer resolution of the differences between proteins from different environments.

Comparing values such as the average oxidation state of carbon with the phylogenetic relationships of gene families that have appeared in different redox conditions [67] might reveal correlations between chemical composition of proteins and evolutionary

constraints. Incorporating the energetics of protein-forming chemical reactions in such an analysis would permit even greater integration of available data on the organisms' environments. Comparative calculations of the energetics of overall protein formation reactions is conducive to integrative studies of microbial communities and environments because the energies depend on both molecular sequences and properties of the chemical environment. Extension of such an integrated thermodynamic framework could be a new way forward to quantifying the relationships between chemically distinct environments and their microbial communities.

Supporting Information

Dataset S1   Script for the CHNOSZ package (version 0.9–5) for the R software environment [42] demonstrating the relative stability calculation for the example described in the Methods.

Dataset S2   Script for the CHNOSZ package including code to produce all of the figures (except the map) and Table 4. The code also depends on the files in Dataset S4 (for Fig. 3), Dataset S3 (for Figs. 4, 5, 6, 7) and Dataset S5 (for Fig. 8). Instructions for running the code are provided in the comments at the top of the file.

Dataset S3   Amino acid compositions of model proteins. This file was produced using the "mkprot" function contained in the source code listing of Dataset S2.

Dataset S4   Results of bootstrap analysis of the elemental ratios and average oxidation number of carbon for all protein sequences at each site. This file was produced using the "boot.prep" function contained in the souce code listing of Dataset S2.

Dataset S5   Temperature, pH and oxidation-reduction potential measurements in hot springs and outflow channels, Summer 2009. Northing and easting (in meters) are listed for the Universal Transverse Mercator projection using the 12T grid zone; elevation is in meters.

Acknowledgments

Many people contributed to field work in Yellowstone and laboratory analysis that support the calculations described in this paper. We would like to acknowledge Jeff Havig, D'Arcy Meyer-Dombard and Jason Raymond for their work in collecting and preparing the biofilm samples for metagenomic analysis, Maggie Osburn, Brandon McLean and Nathan Schnebley for field geochemistry analysis, and Bob Osburn for preparing the map of Bison Pool. Jessie Havig collected temperature and pH data in the field and Natasha Zolotova gathered ion chromatography data in the lab. Thanks to Eric Boyd for his helpful comments on an earlier version of this manuscript and Kris Fecteau for assistance with analysis of the oxidation-reduction potential data.

Author Contributions

Conceived and designed the experiments: JMD ELS. Analyzed the data: JMD ELS. Wrote the paper: JMD ELS.

References

1. Acquisti C, Kleffe J, Collins S (2007) Oxygen content of transmembrane proteins over macroevolutionary time scales. Nature 445: 47–52.
2. Nakashima H, Fukuchi S, Nishikawa K (2003) Compositional changes in RNA, DNA and proteins for bacterial adaptation to higher and lower temperatures. J Biochem 133: 507–513.
3. Zeldovich KB, Berezovsky IN, Shakhnovich EI (2007) Protein and DNA sequence determinants of thermophilic adaptation. PLoS Comput Biol 3: 62–72.
4. Boyd ES, Hamilton TL, Spear JR, Lavin M, Peters JW (2010) [FeFe]-hydrogenase in Yellowstone National Park: evidence for dispersal limitation and phylogenetic niche conservatism. ISME J. pp 1–11.

5. Inskeep WP, Rusch DB, Jay ZJ, Herrgard MJ, Kozubal MA, et al. (2010) Metagenomes from hightemperature chemotrophic systems reveal geochemical controls on microbial community structure and function. PLoS ONE 5.
6. Akashi H, Gojobori T (2002) Metabolic efficiency and amino acid composition in the proteomes of *Escherichia coli* and *Bacillus subtilis*. Proc Natl Acad Sci U S A 99: 3695–3700.
7. Barton MD, Delneri D, Oliver SG, Rattray M, Bergman CM (2010) Evolutionary systems biology of amino acid biosynthetic cost in yeast. PLoS ONE 5: e11935.
8. Wagner A (2005) Energy constraints on the evolution of gene expression. Mol Biol Evol 22: 1365–1374.
9. Amend JP, Shock EL (1998) Energetics of amino acid synthesis in hydrothermal ecosystems. Science 281: 1659–1662.
10. Amend JP, Helgeson HC (2000) Calculation of the standard molal thermodynamic properties of aqueous biomolecules at elevated temperatures and pressures. II. Unfolded proteins. Biophys Chem 84: 105–136.
11. Dick JM, LaRowe DE, Helgeson HC (2006) Temperature, pressure, and electrochemical constraints on protein speciation: Group additivity calculation of the standard molal thermodynamic properties of ionized unfolded proteins. Biogeosciences 3: 311–336.
12. Havig JR, Raymond J, Meyer-Dombard D, Zolotova N, Shock E (2011) Merging isotopes and community genomics in a siliceous sinter-depositing hot spring. J Geophys Res - Biogeosciences 116: G01005.
13. Meyer-Dombard DR, Swingley W, Raymond J, Havig J, Shock EL, et al. (2011) Hydrothermal ecotones and streamer biofilm communities in the Lower Geyser Basin, Yellowstone National Park. Environ Microbiol, in press.
14. Buvet R (1983) General criteria for the fulfillment of redox reactions. In: Milazzo G, Blank M, eds. Bioelectrochemistry I: Biological Redox Reactions. New York: Plenum Press, volume 11 of Ettore *Majorana International Science Series*. pp 15–50.
15. Kroll JH, Donahue NM, Jimenez JL, Kessler SH, Canagaratna MR, et al. (2011) Carbon oxidation state as a metric for describing the chemistry of atmospheric organic aerosol. Nature Chemistry 948.
16. Masiello CA, Gallagher ME, Randerson JT, Deco RM, Chadwick OA (2008) Evaluating two experimental approaches for measuring ecosystem carbon oxidation state and oxidative ratio. J Geophy Res - Biogeosciences 113: G03010.
17. LaRowe DE, Van Cappellen P (2011) Degradation of natural organic matter: A thermodynamic analysis. Geochim Cosmochim Acta 75: 2030–2042.
18. Hendrickson JB, Cram DJ, Hammond GS (1970) Organic Chemistry. New York: McGraw-Hill, 3rd edition. 1279 p.
19. International Union of Pure and Applied Chemistry. IUPAC Compendium of Chemical Terminology -The Gold Book. online resource. URL http://goldbook.iupac.org/. Accessed 2011-01-26.
20. Calvert JG (1990) Glossary of atmospheric chemistry terms - (recommendations 1990). Pure Appl Chem 62: 2167–2219.
21. Kyte J, Doolittle RF (1982) A simple method for displaying the hydropathic character of a protein. J Mol Biol 157: 105–132.
22. Bowers TS, Jackson KJ, Helgeson HC (1984) Equilibrium Activity Diagrams for Coexisting Minerals and Aqueous Solutions at Pressures and Temperatures to 5 kb and 600°C. Heidelberg: Springer-Verlag. pp 397. URL http://www.worldcat.org/oclc/11133620.
23. De Donder Th (1927) L'Affinité. Paris: Gauthiers-Villars.
24. Kondepudi DK, Prigogine I (1998) Modern Thermodynamics: From Heat Engines to Dissipative Structures. New York: John Wiley & Sons. 486 p.
25. Helgeson HC, Kirkham DH, Flowers GC (1981) Theoretical prediction of the thermodynamic behavior of aqueous electrolytes at high pressures and temperatures. IV. Calculation of activity coefficients, osmotic coefficients, and apparent molal and standard and relative partial molal properties to 600°C and 5 Kb. Am J Sci 281: 1249–1516.
26. Drever JI (1997) The Geochemistry of Natural Waters. Upper Saddle RiverNew Jersey: Prentice Hall, 3rd edition.
27. Dick JM (2008) Calculation of the relative metastabilities of proteins using the CHNOSZ software package. Geochem Trans 9: 10.
28. Haney PJ, Badger JH, Buldak GL, Reich CI, Woese CR, et al. (1999) Thermal adaptation analyzed by comparison of protein sequences from mesophilic and extremely thermophilic *Methanococcus* species. Proc Natl Acad Sci U S A 96: 3578–3583.
29. Akca E, Claus H, Schultz N, Karbach G, Schlott B, et al. (2002) Genes and derived amino acid sequences of S-layer proteins from mesophilic, thermophilic, and extremely thermophilic methanococci. Extremophiles 6: 351–358.
30. Shock E, Canovas P (2010) The potential for abiotic organic synthesis and biosynthesis at seafloor hydrothermal systems. Geofluids 10: 161–192.
31. Spear JR, Walker JJ, McCollom TM, Pace NR (2005) Hydrogen and bioenergetics in the Yellowstone geothermal ecosystem. Proc Natl Acad Sci U S A 102: 2555–2560.
32. Hoehler TM, Alperin MJ, Albert DB, Martens CS (1998) Thermodynamic control on hydrogen concentrations in anoxic sediments. Geochim Cosmochim Acta 62: 1745–1756.
33. Tanger JC, IV, Helgeson HC (1988) Calculation of the thermodynamic and transport properties of aqueous species at high pressures and temperatures: Revised equations of state for the standard partial molal properties of ions and electrolytes. Am J Sci 288: 19–98.
34. Shock EL, Helgeson HC, Sverjensky DA (1989) Calculation of the thermodynamic and transport properties of aqueous species at high pressures and

temperatures: Standard partial molal properties of inorganic neutral species. Geochim Cosmochim Acta 53: 2157–2183.
35. Shock EL, Oelkers EH, Johnson JW, Sverjensky DA, Helgeson HC (1992) Calculation of the thermodynamic properties of aqueous species at high pressures and temperatures: Effective electrostatic radii, dissociation constants, and standard partial molal properties to 1000°C and 5 kbar. J Chem Soc, Faraday Trans 88: 803–826.
36. Levelt-Sengers JMH, Kamgarparsi B, Balfour FW, Sengers JV (1983) Thermodynamic properties of steam in the critical region. J Phys Chem Ref Data 12: 1–28.
37. Haar L, Gallagher JS, Kell GS (1984) NBS/NRC Steam Tables. Washington, D. C.: Hemisphere. 320 p.
38. Johnson JW, Norton D (1991) Critical phenomena in hydrothermal systems: state, thermodynamic, electrostatic, and transport properties of H2O in the critical region. Am J Sci 291: 541–648.
39. Johnson JW, Oelkers EH, Helgeson HC (1992) SUPCRT92: A software package for calculating the standard molal thermodynamic properties of minerals, gases, aqueous species, and reactions from 1 to 5000 bar and 0 to 1000°C. Comp Geosci 18: 899–947.
40. Shock EL, Helgeson HC (1988) Calculation of the thermodynamic and transport properties of aqueous species at high pressures and temperatures: Correlation algorithms for ionic species and equation of state predictions to 5 kb and 1000°C. Geochim Cosmochim Acta 52: 2009–2036.
41. Schulte MD, Shock EL, Wood RH (2001) The temperature dependence of the standard-state thermodynamic properties of aqueous nonelectrolytes. Geochim Cosmochim Acta 65: 3919–3930.
42. R Development Core Team (2011) R: A Language and Environment for Statistical Computing. R Foundation for Statistical Computing, Vienna, Austria. URL http://www.R-project.org.
43. Zimmerman SB, Minton AP (1993) Macromolecular crowding: Biochemical, biophysical, and physiological consequences. Annu Rev Biophys Biomolec Struct 22: 27–65.
44. Curtis RA, Prausnitz JM, Blanch HW (1998) Protein-protein and protein-salt interactions in aqueous protein solutions containing concentrated electrolytes. Biotechnol Bioeng 57: 11–21.
45. Engel T, Reid P (2006) Thermodynamics, Statistical Thermodynamics, and Kinetics. San Francisco: Benjamin Cummings. 589 p.
46. Nelson PG (1986) Treatment of chemical equilibrium without using thermodynamics or statistical mechanics. J Chem Ed 63: 852–853.
47. Pfeil W (1998) Protein Stability and Folding. Berlin: Springer-Verlag.
48. Privalov PL, Khechinashvili NN (1974) A thermodynamic approach to the problem of stabilization of globular protein structure: A calorimetric study. J Mol Biol 86: 665–684.
49. Meyer-Dombard DR, Shock EL, Amend JP (2005) Archaeal and bacterial communities in geochemically diverse hot springs of Yellowstone National Park, USA. Geobiology 3: 211–227.
50. Huber R, Eder W, Heldwein S, Wanner G, Huber H, et al. (1998) *Thermocrinis ruber* gen. nov., sp. nov., a pink-filament-forming hyperthermophilic bacterium isolated from Yellowstone National Park. Appl Environ Microbiol 64: 3576–3583.
51. Ruff-Roberts AL, Kuenen JG, Ward DM (1994) Distribution of cultivated and uncultivated cyanobacteria and *Chloroflexus*-like bacteria in hot spring microbial mats. Appl Environ Microbiol 60: 697–704.
52. Cox A, Shock EL, Havig JR (2011) The transition to microbial photosynthesis in hot spring ecosystems. Chem Geol 280: 344–351.
53. Markowitz VM, Ivanova NN, Szeto E, Palaniappan K, Chu K, et al. (2008) IMG/M: a data management and analysis system for metagenomes. Nucleic Acids Res 36: D534–D538.
54. McInerney MJ, Rohlin L, Mouttaki H, Kim U, Krupp RS, et al. (2007) The genome of *Syntrophus aciditrophicus*: Life at the thermodynamic limit of microbial growth. Proc Natl Acad Sci U S A 104: 7600–7605.
55. Shock E, Holland M, Meyer-Dombard D, Amend J, Osburn G, et al. (2010) Quantifying inorganic sources of geochemical energy in hydrothermal ecosystems, Yellowstone National Park, U.S.A. Geochim Cosmochim Acta 74: 4005–4043.
56. McCollom TM, Amend JP (2005) A thermodynamic assessment of energy requirements for biomass synthesis by chemolithoautotrophic micro-organisms in oxic and anoxic environments. Geobiology 3: 135–144.
57. Bard AJ, Parsons R, Jordan J (1985) Standard Potentials in Aqueous Solution. New York: M. Dekker.
58. Sawyer D, Roberts JL (1974) Experimental Electrochemistry for Chemists. New York: Wiley.
59. Oijerholm J, Forsberg S, Hermansson HP, Ullberg M (2009) Relation between the SHE and the internal Ag/AgCl reference electrode at high temperatures. J Electrochem Soc 156: P56–P61.
60. De Vendittis E, Castellano I, Cotugno R, Ruocco MR, Raimo G, et al. (2008) Adaptation of model proteins from cold to hot environments involves continuous and small adjustments of average parameters related to amino acid composition. J Theor Biol 250: 156–171.
61. McDonald JH (2001) Patterns of temperature adaptation in proteins from the bacteria *Deinococcus radiodurans* and *Thermus thermophilus*. Mol Biol Evol 18: 741–749.
62. Hoehler TM, Bebout BM, Des Marais DJ (2001) The role of microbial mats in the production of reduced gases on the early Earth. Nature 412: 324–327.

63. Lynch JM, Poole NJ (1979) Microbial Ecology: A Conceptual Approach. New York: John Wiley & Sons.
64. Hwang C, Sinskey AJ, Lodish HF (1992) Oxidized redox state of glutathione in the endoplasmic reticulum. Science 257: 1496–1502.
65. Mallick P, Boutz DR, Eisenberg D, Yeates TO (2002) Genomic evidence that the intracellular proteins of archaeal microbes contain disulfide bonds. Proc Natl Acad Sci U S A 99: 9679–9684.

66. Wu M, Eisen JA (2008) A simple, fast, and accurate method of phylogenomic inference. Genome Biol 9: 10.
67. Duval S, Ducluzeau AL, Nitschke W, Schoepp-Cothenet B (2008) Enzyme phylogenies as markers for the oxidation state of the environment: The case of respiratory arsenate reductase and related enzymes. BMC Evol Biol 8.

6

17β-Estradiol Alters Rat Type-II Alveolar Cell Recovery from High Levels of Ozone

Madeleine Chalfant, Karen K. Bernd*

Department of Biology, Davidson College, Davidson, North Carolina, United States of America

Abstract

Respiratory health is negatively impacted by exposure to ozone or to estrogens. Increasingly, individuals have simultaneous environmental exposure to both compounds. Characterizing the cellular responses stimulated by the combination of ozone and estrogens, therefore, is crucial to our complete understanding of the compounds' environmental health impacts. Our work introduces an alveolar cell culture model with defined media that provides evidence of ozone damage and determines sex hormones alter the cells' susceptibility to oxidative damage. Specifically, we investigated the individual and combined effects of environmentally relevant levels of ozone and 17β-estradiol on non-cancerous rat, type-II alveolar cells by examining biomarkers of cellular health and redox balance. The data reveal a complex role for 17β-estradiol in cellular recovery from 1 hr exposure to high ozone levels. At 0.5 hr post-ozone necrosis and inflammation markers show 17β-estradiol augments the detrimental effects of 350 ppb ozone, but after 24 hr of recovery, steroid treatment alters glutathione redox ratio and allows cellular proliferation.

Editor: Shama Ahmad, University of Colorado, Denver, United States of America

Funding: This work was funded by a Merck Foundation Undergraduate Research in Biochemistry Grant (to KB) and a Davidson Research Initiative Grant (to MC). The funders had no role in study design, data collection and analysis, decision to publish, or preparation of the manuscript.

Competing Interests: The authors have declared that no competing interests exist.

* E-mail: kabernd@davidson.edu

Introduction

Lungs encounter stressors, like ozone and estrogens, through simultaneous exposure to environmental and cellular sources including indoor and outdoor air, phytoestrogens and poly-aromatic hydrocarbons, and genetic makeup. Epidemiological analyses of the effects of exposure to higher ambient ozone concentrations have revealed a correlation with the incidence and severity of many lung pathologies including asthma [1], cancers [2], chronic obstructive pulmonary diseases (COPD), and pneumonia [3]. Cell-level studies of individuals with healthy or diseased lungs connect ozone exposure with acute and chronic pulmonary inflammation, with both types of inflammation postulated to be a part many lung disorders' pathogenesis [4–6]. When considering estrogens, compounds with both environmental routes of exposure and physiological differences due to genetic makeup and sex, there is agreement that the compounds alter lung pathogenesis. However, whether estrogens promote or inhibit disease remains in question ([7–9] *reviewed in* [10–12]). As indicated, previous research has focused on epidemiological or toxicological analysis of the individual effect of ozone or estrogen [13,14] but, in the body, lungs are exposed to them simultaneously. An understanding how the combination of ozone and estrogen alters pulmonary pathogenic processes not only could help assess health risks posed by environmental exposure to estrogen and endocrine disrupting chemicals but also delineate the different health risks ozone may pose to male and female subpopulations, with the potential to inform intervention and treatment efforts.

Estrogens' cellular role is known to be complex. Because many cell types defend against oxidative damage by up-regulating antioxidant levels [1], the presence of additional 17β-estradiol (E2), the reduced and biologically active form of estrogen, might be predicted to mitigate ozone-induced damage. However, additional products of E2 metabolism increase the complexity of the potential cellular outcomes induced by ozonolysis. E2 exposed to oxidation is broken down [15] and many E2 metabolites increase ROS [16], alter redox homeostasis, and may be involved in carcinogenesis [17]. Compounding this effect, oxidation-induced upregulation of E2 levels may, in turn, upregulate cyclooxygenase-2 (COX-2) and increase expression of the inflammation marker, prostaglandin E2 (PGE2; [18])

In lung epithelia, although a connection between E2 and increased PGE2 has been described, PGE2's function is not well understood. In alveoli increased PGE2 decreases apoptosis and suppresses fibroblast proliferation; suggesting increased PGE2 prevents idiopathic pulmonary fibrosis and promotes lung health [19]. However, PGE2 is also associated with tumorigenesis [20,21]. In type-II alveolar cells, because PGE2 alters the x_c^- system for cystine transport, thereby decreasing cysteine available for synthesis of the antioxidant glutathione, PGE2 may also increase sensitivity to ROS [22]. A mechanism of E2 metabolism increasing ROS levels and PGE2 synthesis- which further augments ROS and thereby reduces antioxidant capacity and inhibits apoptosis- is consistent with E2's association with carcinogenesis and underscores the need for further investigation of the combined effects of E2 and strong oxidants like ozone.

The effect of E2 on recovery from oxidative stress is context dependent. E2 can affect recovery through plasma membrane-associated estrogen receptors (ERs) resulting in immediate alteration of signaling cascades, or via classical nuclear estrogen receptors, a slower, longer-lasting mechanism changing transcription rates. For example, E2 can inhibit glutathione synthesis via interactions with plasma membrane ERs that increase cAMP [23,24]. However, E2 also increases expression of glutathione and enzymes in glutathione's biosynthetic pathway through nuclear ER-β in myocardial cells [25]. Because E2's effect on antioxidant levels is context dependent, research is needed determine how E2 alters non-cancerous cells' responses to oxidation.

Despite increased environmental exposure to ozone and estrogens and the individual correlations of each chemical to lung disease, the combined effect of ozone and estrogen on pulmonary health has not been examined in either whole animal or cell model systems. Due to the complexity of these variables' cellular effects, we developed a defined cell culture model focusing on the activation of cellular defense systems. Levels of well-established biomarkers were investigated to determine the effect of E2 and ozone, alone and in combination, on type-II alveolar cell health and redox homeostasis.

The survival of type-II alveolar cells during and after oxidative stress is critical to lung function. These cells participate in immune and inflammatory responses and, after lung injury, can proliferate and differentiate into type-I alveolar cells, the site of gas exchange [26]. We present a cell culture system using a non-cancerous, female rat type-II alveolar cell line (L2 cells; ATCC #CC-149) to characterize the combined effects of consistent exposure to physiologically relevant levels E2 (10 nM) and 1 hr exposure to Environmental Protection Agency (EPA)-defined 'very unhealthy' levels of ambient ozone (350 ppb). Separate recovery periods of 0.5 or 24 h hours allowed immediate and long-term responses to be evaluated. We assessed cellular health by determining relative levels of mitochondrial function, viability, necrosis, and apoptosis and by measuring the levels of total glutathione and glutathione disulfide (GSSG) and secreted PGE2.

Materials and Methods

Materials

Materials were obtained from the following suppliers: female rat non-cancerous type-II alveolar cells (CC-149, L2 cells) and fetal bovine serum (FBS), ATCC; low glucose DMEM, Hyclone. Phenol red-free low glucose DMEM, 17β-estradiol (E2) and 3,3′, 5-triiodo-L-thyronine sodium salt (T3), Sigma-Aldrich; charcoal-stripped FBS and 100X antibiotic-antimycotic, Invitrogen; Trypsin-EDTA .05%, VWR; Na-pyruvate, Cellgro; Hanks buffered saline solution (HBSS), Lonza; MTT, PGE2, and Apotox-Glo Triplex assay kits from Roche, Cayman Chemical, and Promega, respectively. Promega generously provided GSH/GSSG-Glo assay prior to public release.

Cell culture and O_3 exposure

L2 cells were cultured in a humidified atmosphere at 37°C, 5% CO_2 in low glucose DMEM, 10% FBS. Cells were seeded into either white bottom or clear bottom 96-well, tissue culture treated plates (Costar) at 10^4 cells/well. FBS contains uncharacterized levels of E2 and thyroid hormone (T3) and phenol red has been shown to have estrogen-like effects [34], therefore, after 18–24 hour attachment period, cells were washed (PBS) and *defined media* (phenol red-free, low glucose DMEM, 10% charcoal stripped FBS, 10^{-9} M T3) was added. Pretreatments (48 hours ±10 nM E2) occurred as indicated. To remove extracellular compounds that were oxidizable, cells were washed (PBS) and the media changed to HBSS ±10 nM E2 before gas exposure. Ozone was generated from O_2 via an Ozone Gas Generator (Pacific Ozone Technology) and diluted to indicated concentrations with sterile 5% CO_2/air. Exposure conditions included (2.5 L/min sterile 5%

CO_2/air) ±350 ppb O_3, 1 hr, 37°C. To isolate the effect of flowing air (itself a source of oxidation) versus non-flowing air, 'No-flow' (NF) samples, covered with parafilm, were included and used to normalize data as indicated. After gas exposure, cells were washed (PBS) and returned to defined media ±10 nM E2. Assays were performed 0.5-hour or 24-hour after gas exposure, as indicated.

Biomarker assays

Mitochondrial activity assay. Quadruplicate assays determining mitochondrial activity via reductase activity (MTT assay) were performed as per manufacturer instructions (Roche). Absorbance values were measured spectrophotometrically (Model 680 Microplate Reader; Bio-Rad) with background readings (Abs_{655nm}) subtracted from Abs_{600nm} readings. Data were normalized to the non-oxidized levels represented by average NF controls.

Viability, necrosis and apoptosis assays. Viability (GF-AFC cleavage), cytotoxicity (bis-AAF-R110 cleavage) and apoptosis (caspase 3/7 activity) were measured simultaneously via the Apotox-Glo^{TM} Triplex assay (Promega). To facilitate collection of fluorescent and luminescent data cells were seeded into white-bottom 96 well plates. Quadruplicate samples were treated as indicated and processed per manufacturer instructions. An FLx800 Microplate Fluorescence Reader (Bio-Tek Instruments Inc.) was used to measure both fluorescence ($420_{Ex}/485_{Em}$ and $485_{Ex}/528_{Em}$) and luminescence. Data were normalized to the non-oxidized levels represented by average NF samples for each subassay.

Assay of inflammation marker. Triplicate samples were seeded in clear-bottom 96 well plates, treated as indicated and PGE2 levels determined per manufacturer instructions (PGE2 Assay: Cayman Chemical Co.). Since fresh media was added after gas exposure, samples represent PGE2 secreted after oxidative stress. All samples were frozen (−80°C) immediately after collection and assays were performed within 2 weeks. PGE2 concentrations were calculated against concurrently run standards. The average concentration (pg/ml) is reported.

Redox state. Total glutathione and GSSG levels were used as a measure of the cells' redox state. Triplicate samples were prepared for each assay, treated as indicated, and processed per manufacturer instructions (*GSH/GSSG-Glo^{TM} Assay* Promega). Total glutathione and GSSG concentrations were calculated against concurrently run standards and average μM for each condition is reported.

Statistical analysis

Prior to norming, outliers were identified and removed from data sets by the Q-Test (90% confidence interval). Graphs present mean ± S.E.M. To compare the combined effects of ±350 ppb ozone and ±10 nM estrogen, pure model I two-way ANOVAs were performed followed by a Tukey HSD *post hoc* test (JMP statistical package, Cary, NC). $p \leq 0.05$ was considered significant.

Results and Discussion

Despite increasing environmental exposure to ozone and estrogens and correlations between each chemical and lung disease, no studies in either whole animal or cell model systems have reported the combined effects of ozone and estrogen on pulmonary health. In addition, while alveoli play a critical role in lung function much about the stress and recovery response of these cells remains unknown. Here we introduce L2 cells as an alveolar cell culture system suitable for determining the effects of environmental pollutants. We support use of the L2 cell system by showing that, consistent with data from other animal and culture systems, ozone has deleterious effects. However, unlike other systems used for ozone research, our L2 cell system has defined estrogen levels allowing dual analysis of ozone and hormonal influences. Our data indicated that estrogen plays a complex role in response to an oxidative stress event with differences between immediate and more long-term outcomes.

In order to remain close to conditions found in the environment we exposed alveolar type II cells (L2 cells) to physiologically relevant concentrations of ozone and estrogen. More specifically, we exposed L2 cells to the ozone level classified by the EPA as 'very unhealthy' (350 ppb ozone/1 hr). We recognize that an experimental design with alveolar cells directly exposed to ozone differs from the *in vivo* situation where inspired gases react with respiratory tract tissue before reaching the alveolus. Therefore, the effective ozone exposure that our +350 ppb ozone samples experience is greater than those found in an alveolus of a whole lung respiring in a 350 ppb ozone environment. Given that alveolar ozone concentrations *in vivo* are neither available nor part of the EPA exposure definitions, we note this limitation of our model and submit that, compared to cell and whole animal studies with exposure parameters of 1000 ppb+ozone [27–29], the exposure level in our system better models environmental and physiological conditions.

Cells were exposed to the biologically active form of estrogen, 17β-estradiol (E2). Reports indicate that exposures to some, but not all, concentrations of E2 increase cell growth rate [32]. Because E2-induced changes in growth rate would confound comparative analyses of biomarkers, prior to characterizing the combined effect of E2 and ozone on L2, we tested the effect of 0, 1, 10, and 100 nM E2 and determined that 10 and 100 nM E2 do not change L2 growth rate (data not shown). The fact that 10 nM E2 does not increase is consistent with data from other systems [32]. Because 10 nM E2 is more physiologically relevant than 100 nM E2, 10 nM E2 exposures were used in this study.

Within the body, E2 can alter recovery from oxidative stress via immediate mechanisms affecting signaling cascades through plasma membrane-associated estrogen receptors (ERs) and via slower, longer-lasting genomic mechanisms affecting transcription through classical nuclear ERs [23–25]. Therefore to capture data within the immediate and genomic response mechanisms, we characterized our model system by measuring biomarkers for cellular health, inflammation and oxidative stress at 0.5 hr and 24 hr after gas exposure, respectively. To examine the combined effect of E2 and ozone, we exposed L2 cells for 1 hour to all permutations of 0 nM or 10 nM E2 plus 0 ppb or 350 ppb ozone (i.e. ±E2 ±O$_3$) at a flow rate of 2.5 L/min.

First considering the 0.5 hr time point, exposure to 350 ppb ozone resulted in decreased viability and mitochondrial activity. Two-way ANOVA followed by Tukey HSD *post hoc* analysis revealed significant main effect of ozone, alone, on necrosis levels (p = 0.0113; Figure 1). 10 nM E2 treatment augmented ozone's effect and resulting in an additional statistically significant interaction effect and increase in necrosis (p = 0.0053; Figure 1).

Because ozone stimulates secretion of the inflammation marker, PGE2, and apoptotic enzyme activity and PGE2 levels show an inverse relationship in airway epithelia [30], we examined whether this relationship is maintained in L2 cells. Two-way ANOVA followed by Tukey HSD *post hoc* analysis revealed significant main effect of ozone on PGE2 secretion 0.5 hr post gas exposure (p = 0.0217) and an interaction effect between ozone and E2 resulting in an additional significant increase in PGE2 levels (p = 0.0295; Figure 2B). However, since neither ozone nor E2

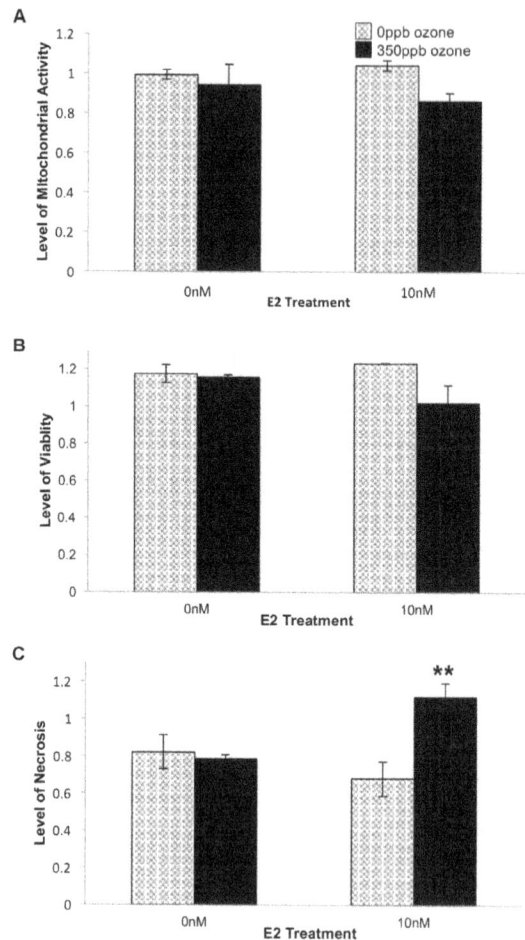

Figure 1. Effect of E2 and ozone (O$_3$) on relative levels of mitochondrial activity (A), viability (B) and necrosis (C) 0.5 hr after gas exposure. 10^4 L2 cells/well were treated with E2 and 350 ppb O$_3$ as indicated. After 0.5 hr recovery time, levels of mitochondrial activity (F = 1.7336, df = 3,15, p = 0.2133), viability (F = 2.6161, df = 3,12, p = 0.1152), and necrosis (F = 7.4798, df = 3,12, p = 0.0081) were determined. Values represent the mean of 3–4 replicates normalized to data from control cells (-E2, in non-flowing 5% CO$_2$/air), ±S.E.M. # p≤0.05 compared to 0 ppb O$_3$. ** p≤0.01 compared to the same E2 treatment group.

altered activity of the apoptotic enzymes, Caspase 3/7, alveolar cells responds differently than airway epithelia and did not show the inverse relationship between PGE2 levels and apoptotic enzyme activity (Figure 2A).

It has been suggested that an increase in PGE2 secretion may result in cell death through the generation of superoxide radicals [20,31]. Thus, the additional increase in PGE2 seen in E2- treated samples at 0.5 hr post oxidative stress could be caused by a concomitant increase in ROS that stressed the cells to the point of irrevocable damage and necrosis. However, an interpretation of irrevocable damage would predict that at 24 hr post gas exposure, cell cultures treated with E2 would have fewer cells and therefore secrete less PGE2, than counterparts that were not treated with estrogen. In contrast, the data show that no combination of E2 and ozone significantly altered PGE2 secretion at the 24 hr time point (Figure 3B). In fact, in both 350 ppb ozone and 5%CO2/air treatment groups, including 10 nM E2 resulted in an 11% increase in viability (p = 0.0009 Figure 4B).

A

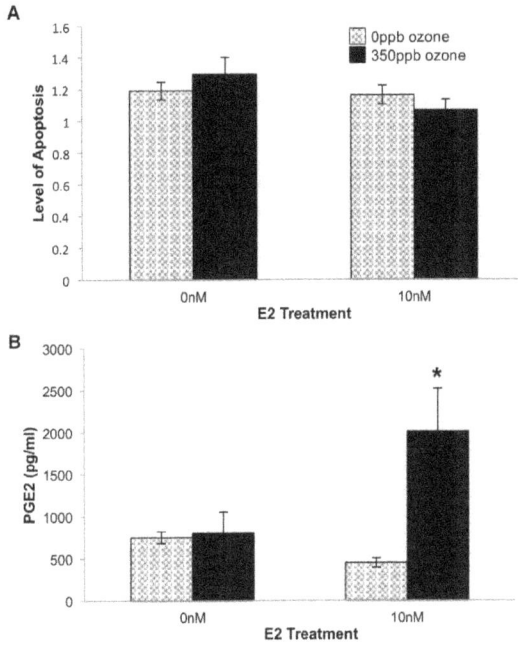

B

A

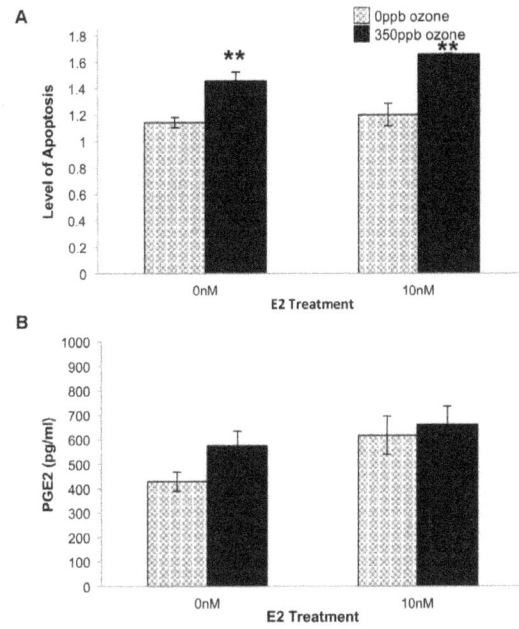

B

Figure 2. Effect of E2 and O₃ on relative levels of apoptosis (A) and PGE2 secretion (B) 0.5 hr after gas exposure. 10⁴ L2 cells/well were treated with E2 and O₃ as indicated. After 0.5 hr recovery time, levels of apoptosis ($F = 0.6948$, $df = 3,12$, $p = 0.5781$) were determined and media samples were collected from a separate set of cells to determine PGE2 secretion ($F = 5.8628$, $df = 3,11$, $p = 0.0203$). Level of apoptosis values represent the mean of 3–4 replicates normalized to data from control cells (-E2, in non-flowing 5% CO_2/air), ±S.E.M. PGE2 values represent the mean of 3 replicates, ±S.E.M. # $p \leq 0.05$ compared to 0 ppb O_3 group and * $p \leq 0.05$ compared to the same E2 treatment group.

Figure 3. Effect of E2 and O₃ on relative levels of apoptosis (A) and PGE2 secretion (B) 24 hr after gas exposure. 10⁴ L2 cells/well were treated with E2 and 350 ppb O_3 as indicated. After 24 hr, levels of apoptosis ($F = 12.0440$, $df = 3,13$, $p = 0.0012$) were determined and media samples were collected from a separate set of cells to determine PGE2 secretion ($F = 2.3803$, $df = 3,11$, $p = 0.1453$). Values of apoptosis represent the mean of 3–4 replicates normalized to data from control cells (-E2, in non-flowing 5% CO_2/air), ±S.E.M. ** $p \leq 0.01$ compared to 0 ppb O_3.

Previous work demonstrated that type-II alveolar cells repair the alveolar wall after damage [27]. Consequently, we expected the increase in viability to be accompanied by an increase in cell proliferation that would, in turn, raise the sample's mitochondrial activity. Instead we found that E2 had no significant effect on mitochondrial activity in our alveolar cell system (Figure 4A) making our findings more similar to those of Si and colleagues [32] who reported 10 nM E2 did not induce aortic endothelia proliferation.

Because the increased viability seen in E2-treated samples could be the consequence of a decrease in necrotic cell death, a decrease in apoptotic death, or both, we subjected cells to all combinations of ±E2 and ±O₃ and assayed necrosis and apoptosis biomarkers within the same sample. As predicted, samples treated with 10 nM E2 during and after 0 ppb ozone exposure (i.e. 2.5 L/min, 5% CO₂/air) showed a significant decrease in necrosis when compared to cells exposed to flowing gas but no E2 (p<0.05; Figure 4C). In addition, the +E2-O₃ samples exhibited significantly less necrosis than samples treated with E2+350 ppb ozone (p<0.05; Figure 4C). Comparing samples collected at the 0.5 hr and 24 hr post-gas recovery periods, necrosis in +E2+O₃ samples increased by only 5.51% while in all other exposure conditions necrosis increased by 30–32% between the two time points (Figure 4C and 1C). This suggests that at the 24 hr time point a majority of the necrosis seen in the samples treated with E2+O₃ reflects cell death that occurred immediately after gas exposure, rather than a significant increase in necrosis occurring *between* 0.5 and 24 hr of recovery. Examination of Caspase 3/7 activity levels

revealed that, independent of E2 treatment, cells exposed to 350 ppb ozone showed significantly greater levels of apoptosis than cells exposed to 0 ppb ozone (p = 0.0002; Figure 3A). Taken together, these data suggest that when 10 nM E2 is present during a recovery period that is long enough to include changes in gene expression (i.e. 24 hr) the steroid mitigates ozone- induced necrosis, but not ozone-induced apoptosis.

Differences in study design and culture conditions could account for differences between previous studies and our cell proliferation data. First, earlier studies approximated cell growth via markers that occur before cytokinesis [27,33] while our metrics required completion of cell division. Second, and perhaps more significantly, other studies, not focused on estrogens, used media containing phenol red, an estrogen mimic [34], and complete FBS containing undefined concentrations of E2. To decrease confounding media effects and better define E2 exposure levels in our system we cultured L2 cells in phenol-red free DMEM and 10% charcoal-stripped FBS. Thus, the increased cell proliferation in those other studies could be, in part, due to estrogenic effects of culture media. We find this hypothesis likely, as it is consistent with recent whole organism studies that report estrogen is responsible for some sex-specific differences in alveolar size [35] and is necessary for alveolar wall regeneration in mice [36]. Additionally, our initial feasibility studies showed that in the absence of any airflow 10 nM E2 treatment did not increase viability (Chalfant and Bernd, unpublished data). 10 nM E2 only increased viability when cells were under conditions more similar to those in the lung where low or high oxidative stress is present. These conditions are modeled in our system by 2.5 L/min, 5% CO2/air containing 0 ppb or 350 ppb ozone.

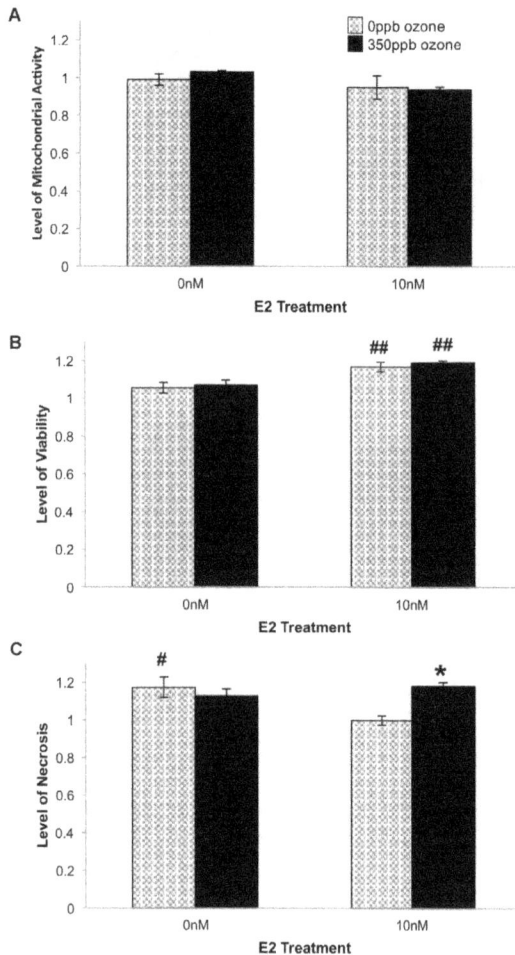

Figure 4. Effect of E2 and O$_3$ on relative levels of mitochondrial activity (A), viability (B) and necrosis (C) 24 hr after gas exposure. 10^4 L2 cells/well were treated with E2 and 350 ppb O$_3$ as indicated. After 24 hr, levels of mitochondrial activity (F = 1.4122, df = 3,15, p = 0.2872), viability (F = 7.2915, df = 3,13, p = 0.0071) and necrosis (F = 5.9990, df = 3,13, p = 0.0132) were determined. Values represent the mean of 3–4 replicates normalized to data from control cells (-E2, in non-flowing 5% CO$_2$/air), ±S.E.M.* p≤0.05 in the same E2 treatment group. # p≤0.05 and ## p≤0.01 compared to the same O$_3$ exposure.

The increase in viability seen in E2-treated cells could occur by two different mechanisms. First, because E2 is involved in alveolar wall repair [36], it could induce proliferation that compensates for cell death despite high intracellular ROS. Conversely, viable cell number could be maintained because E2 could decrease damage caused by cellular ROS, either directly by acting as an antioxidant, or indirectly by increasing expression of the antioxidant glutathione. To explore the effect of E2 on the expression of cellular antioxidants, we measured total glutathione and found no significant difference between any combination of E2 and ozone treatments (Figure 5A). We note that assay limitations preclude determining cell number within the actual test population, thus, conditions that decrease cell numbers but increase glutathione per cell could show no net change in total glutathione. However, our data indicate that E2 treatment *increases* the number of viable cells in +350 ppb O$_3$ conditions (Figure 4). Because we see no increase in glutathione in any condition tested, including those with increased viability, we are confident that E2's function in

+350 ppb O$_3$ conditions does not include upregulation of glutathione expression. Further investigation determined that ozone exposure (p = 0.0026) and E2 treatment (p<0.0001) significantly increased the amount of glutathione found in its oxidized form, GSSG (Figure 5B). In addition, together the two treatments synergistically increase GSSG levels in +E2+O$_3$ vs. -E2- O$_3$ controls (p = 0.0009). These data suggest that E2 increases ROS and enhances ozone-induced increases in ROS resulting in an altered glutathione redox ratio.

While E2 treatment increased cell viability after either low or high levels of oxidative stress, E2's effect on necrosis appears dependent upon the degree of oxidative stress in the system. E2-treated cells exposed to 0 ppb ozone had significantly lower levels of necrosis than those exposed to 0 ppb ozone without E2. Also, E2-treated cells exposed to 350 ppb O$_3$ showed comparable high levels of necrosis at 0.5 hr and 24 hr post gas exposure, suggesting that while E2 exacerbated the original oxidative insult, the hormone eventually mitigated further damage to the cells. While this result could be due to increased levels of apoptosis, given cell proliferation seen in these conditions, we find that explanation unlikely.

As mentioned earlier, E2 could mitigate necrosis through several different mechanisms. E2 could act directly as an antioxidant, decreasing intracellular ROS and thus allowing recovery [37]. However this hypothesis is unlikely because the raised level of GSSG seen in 10 nM E2, 350 ppb ozone conditions supports the presence of high ROS. Other mechanisms that warrant exploration in future studies include E2 reducing ATP depletion caused by oxidative stress [38] and enabling ATP secretion that, in lung epithelia, is known to decrease ozone induced necrosis and apoptosis [39]. Alternatively, several studies in non-lung cell types have shown that E2 treatment increases

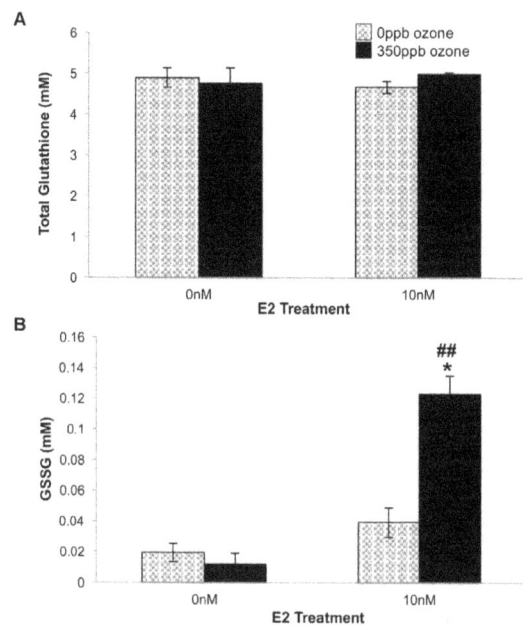

Figure 5. Effect of E2 and O$_3$ on total glutathione (A) and GSSG (B) 24 hr after gas exposure. 10^4 L2 cells/well were treated with E2 and 350 ppb O$_3$ as indicated. After 24 hr recovery time total glutathione (F = 0.3699, df = 3,11, p = 0.7770) and GSSG (F = 33.0667, df = 3,11, p<0.0001) levels were determined. Values represent the mean of 3 replicates, ±S.E.M. ** p≤0.01 relative to the same E2 treatment. ## p≤0.01 same O$_3$ exposure.

basal levels of heat shock proteins (HSPs), which could play a role in recovery [40–42]. These reports, combined with those showing ozone stimulates expression of HSPs and stress proteins in type-II alveolar cells [27,43], suggest that E2 treatment could enable cell survival by either increasing basal levels of HSPs or by further enhancing increases in HSP expression that ozone has induced.

The increase in cellular proliferation and reduction in cell death in the presence of increased ROS suggests that ozone exposure may induce E2 metabolism to one of its less understood metabolites. Both 2-hydroxyestradiol and 4-hydroxyestradiol increase cellular ROS and are known to induce DNA damage directly and via quinone – semi-quinone redox cycling [17]. Despite increasing intracellular ROS and DNA damage, 4-hydroxyestradiol also induces cell growth, providing a connection between E2 metabolism and cellular proliferation [16]. In our system E2 metabolism could be stimulated by tryptophan oxidized during ozone exposure. In mouse heptocarcinoma cells, AhR was activated both by oxidized tryptophan [44] and 2,3,7,8-tetrachlorodibenzo-p-dioxin (TCDD), the prototypical AhR ligand [45], resulting in increased expression of proteins involved in E2 metabolism [16]. In human bronchial epithelial cells, TCDD-AhR interactions were linked to induced expression of E2 metabolizing enzymes, decreased E2 and increased levels of its metabolites, 2-hydroxyestradiol and 4-hydroxyestradiol [46]. However while both metabolites are associated with increased ROS and cell growth, they also increase PGE2 secretion and decrease apoptosis [47]. Since, in our system, E2 treatment neither altered apoptosis

levels in ozone treatments nor affected PGE2 secretion, it is important to continue examining other potential mechanisms.

In summary, we present a novel alveolar type II cell culture model that uses defined media conditions allowing characterization of simultaneous exposure to estrogen and ozone. Using this culture model we provide evidence that ozone and E2 treatments alter alveolar type-II cellular health metrics, both independently and in concert with one another. Our data suggest ozone significantly decreases viability, immediately causing necrosis and eventually increasing apoptosis. E2 treatment augments some of ozone's deleterious effects, increasing PGE2 secretion and increasing GSSG levels 0.5 hr and 24 hr after ozone exposure, respectively. However, E2 mitigates ozone's other effects, resulting in increased viability 24 hr post gas exposure. Our research provides greater insight into cellular mechanisms involved in sex differences in lung diseases and the effects of ozone exposure. While these topics are far from being understood, it is clear that ozone causes pulmonary damage and that sex hormones alter susceptibility to oxidative damage. Because E2 levels in the body vary and individual's environmental exposure to estrogens is increasing, our work underscores the need for further research to determine the extent of these trends.

Author Contributions

Conceived and designed the experiments: MC KB. Performed the experiments: MC. Analyzed the data: MC KB. Contributed reagents/materials/analysis tools: KB. Wrote the paper: MC KB.

References

1. Nadadur SS, Costa DL, Slade R, Silbjoris R, Hatch GE (2005) Acute ozone-induced differential gene expression profiles in rat lung. Environ Health Perspect 113: 1717–1722.
2. Dhondt S, Beck C, Degraeuwe B, Lefebvre W, Kochan B, et al. (2012) Health impact assessment of air pollution using dynamic exposure profile: Implications for exposure and health impact estimates. Environ Impact Assess Rev 36: 42.
3. Medina-Ramon M, Zanobetti A, Schwartz J (2006) The effect of ozone and PM10 on hospital admissions for pneumonia and chronic obstructive pulmonary disease: A national multicity study. Am J Epidemiol 163: 579–588.
4. Park GY, Christman JW (2006) Involvement of cyclooxygenase-2 and prostaglandins in the molecular pathogenesis of inflammatory lung diseases. Am J Physiol Lung Cell Mol 209: 797–805.
5. Rahman I (1999) Inflammation and the regulation of glutathione level in lung epithelial cells. Antiox Redox Signal 1: 425.
6. Klaunig JE, Kamendulis LM, Hocevar BA (2010) Oxidative stress and oxidative damage in carcinogenesis. Toxicol Pathol 38: 96–109.
7. Dougherty SM, Mazhawidza W, Bohn AR, Robinson KA, Mattingly KA, et al. (2006) Gender difference in the activity but not expression of estrogen receptors α and β in human lung adenocarcinoma cells. Endocrine Related Cancer 13: 113.
8. Karlsson C, Helenius G, Fernandes O, Karlsson MG (2012) Oestrogen receptor beta in NSCLC - prevalence, proliferative influence, prognostic impact, and smoking. Acta Pathol Microbiol Immunol Scand 120: 451–458.
9. Cook MB, McGlynn KA, Devesa SS, Freedman ND, Anderson WF (2011) Sex disparities in cancer mortality and survival. Cancer Epidemiol Biomarkers 20: 1629–1637.
10. Verma MK, Miki Y, Sasano H (2011) Sex steroid receptors in human lung diseases. J Steroid Biochem Mol Biol 127: 216–222.
11. Dransfield MT, Washko GR, Foreman MG, Estepar RSJ, Reilly J, et al. (2007) Gender differences in the severity of CT emphysema in COPD. Chest 132: 464–470.
12. de Torres JP, Cote CG, Lopex MV, Casanova C, Diaz O, et al. (2009) Sex differences in mortality in patients with COPD. Eur Respir J 33: 528–535.
13. Diamanti-Kandarakis E, Bourguignon J, Guidice LC, Hauser R, Prins GS, et al. (2009) Endocrine-disrupting chemicals: An endocrine society scientific statement. Endocr Rev 30: 293.
14. Fucic A, Gamulin M, Ferencic Z, Rokotov DS, Katic J, et al. (2010) Lung cancer and environmental chemical exposure: A review of our current state of knowledge with reference to the role of hormones and hormone receptors as an increased risk factor for developing lung cancer in man. Toxicol Pathol 38: 869.
15. Sindhu RK, Kikkawa Y (1999) Superinduction of oxidized tryptophan-inducible cytochrome P450 1A1 by cycloheximide in hepa 1c1c7 cell. In Vitro and Mol Toxicol 12: 149–162.
16. Chang LW, Chang Y, Ho C, Tsai M, Lin P (2007) Increase of carcinogenic risk via enhancement of cyclooxygenase-2 expression and hyroxyestradiol accumu-

lation in human lung cells as a result of interaction between BaP and 17-beta estradiol. Carcinogenesis.
17. Roy D, Cai Q, Felty Q, Narayan S (2007) Estrogen-induced generation of reactive oxygen and nitrogen species, gene damage, and estrogen-dependent cancers. J Toxicol Environ Health 10: 235–257.
18. Ho C, Ling Y, Chang LW, Tsai M, Lin P (2008) 17-beta estradiol and hydroxyestradiols interact via the NF-kappa B pathway to elevate cyclooxygenase 2 expression and prostaglandin E2 secretion in human bronchial epithelial cells. Toxicol Sci 104: 294–302.
19. Maher TM, Evans IC, Bottoms SE, Mercer PF, Thorley AJ, et al. (2010) Diminished prostaglandin E2 contributes to the apoptosis paradox in idiopathic pulmonary fibrosis. American Journal of Critical Care Medicine 182: 73–82.
20. Greenhough A, Smartt HJM, Moore AE, Roberts HR, Williams AC, et al. (2009) The COX-2/PGE2 pathway: Key roles in the hallmarks of cancer and adaptation to the tumour microenvironment. Carcinogenesis 30: 377.
21. Klaunig JE, Kamendulis LM, Hocevar BA (2010) Oxidative stress and oxidative damage in carcinogenesis. Toxicol Pathol 38: 96–109.
22. van de Wetering JK, van Golde LM, Batenburg JJ (2004) Collectins: Players of the innate immune system. Eur J Biochem 271: 1229–1249.
23. Bjornstrom L, Sjoberg M (2005) Mechanisms of estrogen receptor signaling: Convergence of genomic and nongenomic actions on target genes. Mol Endocrinol 19: 833–842.
24. Lu SC, Kuhlenkamp J, Garcia-Ruiz C, Kaplowitz N (1991) Hormone-mediated down-regulation of hepatic glutathione synthesis in the rat. J Clin Invest 88: 260–269.
25. Urata Y, Ihara Y, Murata H, Goto S, Koji T, et al. (2006) 17β-estradiol protects against oxidative stress-induced cell death through the glutathione/glutaredoxin-dependent redox regulation of akt in myocardiac H9c2 cells. JBC 281: 13092–13102.
26. Wang G, Umstead TM, Phelps DS, Al-Mondhiry H, Floros J (2002) The effect of ozone exposure on the ability of human surfactant protein A variants to stimulate cytokine production. Environ Health Perspect 110: 79–84.
27. Wang J, Wang S, Manzer R, McConville G, Mason RJ (2006) Ozone induces oxidative stress in rat alveolar type II and type I-like cells. Free Radic Biol Med 40: 1914–1928.
28. Funabashi H, Shima M, Kuwaki T, Hiroshima K, Kuriyama T (2004) Effects of repeated ozone exposure on pulmonary function and bronchial responsiveness in mice sensitized with ovalbumin. Toxicol 204: 75–83.
29. Last JA, Gohil K, Mathrani VC, Kenyon NJ (2005) Systemic responses to inhaled ozone in mice: Cachexia and down-regulation of liver xenobiotic metabolizing genes. Toxicol Appl Pharmacol 208: 117–126.
30. Kafoury RM, Hernandez JM, Lasky JA, Toscano WA, Friedman M (2007) Activation of transcription factor IL-6 (NF-IL-6) and nuclear factor-kappa B by lipid ozonation productions is crucial to interleukin-8 gene expression in human airway epithelial cells. Environ Toxicol 22: 159–168.

31. Klaunig JE, Kamendulis LM, Hocevar BA (2010) Oxidative stress and oxidative damage in carcinogenesis. Toxicol Pathol 38: 96–109.

32. Si M, Al-Sharafi B, Lai C, Khardori P, Chang C, et al. (2001) Gender difference in cytoprotection induced by estrogen on female and male bovine aortic endothelial cells. Endocrine 15: 255–262.

33. Prokhorova S, Patel N, Laskin DL (1998) Regulation of alveolar macrophage and type II cell DNA synthesis: Effects of ozone inhalation. Am J Physiol Lung Cell Mol 275: LI200–LI207.

34. Welshons WV, Wolf MF, Murphy CS, Jordan VC (1988) Estrogenic activity of phenol red. Mol Cell Endocrinol 57: 169.

35. Carey MA, Card JW, Voltz JW, Germolec DR, Korach KS, et al. (2007) The impact of the sex and sex hormones on lung physiology and disease: Lessons from animal studies. Am J Physiol Lung Cell Mol 293: 272–278.

36. Massaro D, Clerch LB, DeCarlo Massaro G (2007) Estrogen receptor-alpha regulates pulmonary alveolar loss and regeneration in female mince: Morphometric and gene expression studies. Am J Physiol Lung Cell Mol 293: L222–L228.

37. Miyacuchi C, Muranaka S, Kanno T, Fujita H, Akiyama J, et al. (2004) 17β-estradiol suppresses ROS-induced apoptosis of CHO cells through inhibition of lipid peroxidation-couples membrane permeability transition. Physiol Chem Phys Med NMR 36: 21–35.

38. De Marinis E, Ascenzi P, Pellegrini M, Galluzzo P, Bulzomi P, et al. (2010) 17beta-estradiol - A new modulator of neuroglobin levels in neurons: Role in neuroprotection against H2O2 toxicity. Neurosignals 18: 223–235.

39. Ahmad S, Ahmad A, McConville G, Schneider BK, Allen CB, et al. (2005) Lung epithelial cells release ATP during ozone exposure: Signaling for cell survival. Free Radic Biol Med 39: 213–226.

40. Zhang Y, Champagne N, Beitel LK, Gooodyer CG, Trifiro M, et al. (2004) Estrogen and androgen protection of human neurons against intracellular amyloid beta toxicity through heat shock protein 70. J Neuroscience 24: 5315–5321.

41. Hamilton KL, Mbai FN, Gupta S, Knowlton AA (2004) Estrogen, heat shock proteins, and NF-kappaB in human vascular endothelium. Arterioscler Thromb Vasc Biol 24: 1628–1633.

42. Porter W, Wang F, Duan R, Qin C, Castro-Rivera E, et al. (2001) Transcriptional activaiton of heat shock protein 27 gene expression by 17beta-estradiol and modulation by antiestrogens and aryl hydrocarbon receptor agonists. J Mol Endocrinol 26: 31–42.

43. Kosmider B, Loader JE, Murphy RC, Mason RJ (2010) Apoptosis induced by ozone and oxysterols in human alveolar epithelial cells. Free Radic Biol Med 48: 1513.

44. Sindhu RK, Mitsuhashi M, Kikkawa Y (2000) Induction of cytochrome P-410 1A2 by oxidized tryptophan in hepa 1c1c7 cells. J Pharmacol Exp Ther 292: 1008–1014.

45. Singhal R, Shankar K, Badger TM, Ronis MJ (2008) Estrogenic status modulates aryl hydrocarbon receptor-mediated hepatic gene expression and carcinogenicity. Carcinogenesis 29: 227–236.

46. Lin P, Chang Y, Chen C, Yang W, Cheng Y, et al. (2004) A comparative study on the effects of 2,3,7,8-tetrachlorodibenzo-p-dioxin polychlorinated biphe-nyl126 and estrogen in human bronchial epithelial cells. Toxicol Appl Pharmacol 195: 83–91.

47. Ho C, Ling Y, Chang LW, Tsai M, Lin P (2008) 17-beta estradiol and hydroxyestradiols interact via the NF-kappa B pathway to elevate cyclooxygen-ase 2 expression and prostaglandin E2 secretion in human bronchial epithelial cells. Toxicol Sci 104: 294–302.

7

Health Risk Assessment of Inhalation Exposure to Formaldehyde and Benzene in Newly Remodeled Buildings, Beijing

Lihui Huang[1,2], Jinhan Mo[1,2,3], Jan Sundell[1], Zhihua Fan[4], Yinping Zhang[1,2,3]*

1 Institute of Built Environment, Department of Building Science, Tsinghua University, Beijing, China, 2 Key Laboratory of Eco Planning & Green Building, Ministry of Education (Tsinghua University), Beijing, China, 3 Built Environmental Test Center, Tsinghua University, Beijing, China, 4 Department of Environmental and Occupational Medicine, Robert Wood Johnson Medical School, Environmental and Occupational Health Sciences Institute, Rutgers University, Piscataway, New Jersey, United States of America

Abstract

Objective: To assess health risks associated with inhalation exposure to formaldehyde and benzene mainly emitted from building and decoration materials in newly remodeled indoor spaces in Beijing.

Methods: We tested the formaldehyde and benzene concentrations in indoor air of 410 dwellings and 451 offices remodeled within the past year, in which the occupants had health concerns about indoor air quality. To assess non-carcinogenic health risks, we compared the data to the health guidelines in China and USA, respectively. To assess carcinogenic health risks, we first modeled indoor personal exposure to formaldehyde and benzene using the concentration data, and then estimated the associated cancer risks by multiplying the indoor personal exposure by the Inhalation Unit Risk values (IURs) provided by the U.S. EPA Integrated Risk Information System (U.S. EPA IRIS) and the California Office of Environmental Health Hazard Assessment (OEHHA), respectively.

Results: (1) The indoor formaldehyde concentrations of 85% dwellings and 67% offices were above the acute Reference Exposure Level (REL) recommended by the OEHHA and the concentrations of all tested buildings were above the chronic REL recommended by the OEHHA; (2) The indoor benzene concentrations of 12% dwellings and 32% offices exceeded the reference concentration (RfC) recommended by the U.S. EPA IRIS; (3) The median cancer risks from indoor exposure to formaldehyde and benzene were 1,150 and 106 per million (based on U.S. EPA IRIS IURs), 531 and 394 per million (based on OEHHA IURs).

Conclusions: In the tested buildings, formaldehyde exposure may pose acute and chronic non-carcinogenic health risks to the occupants, whereas benzene exposure may pose chronic non-carcinogenic risks to the occupants. Exposure to both compounds is associated with significant carcinogenic risks. Improvement in ventilation, establishment of volatile organic compounds (VOCs) emission labeling systems for decorating and refurbishing materials are recommended to reduce indoor VOCs exposure.

Editor: Aditya Bhushan Pant, Indian Institute of Toxicology Reserach, India

Funding: The research was supported by Natural Science Foundation of China (51136002, 51076079) and Ministry of Science and Technology of China (2012BAJ02B-03). http://www.nsfc.gov.cn/Portal0/default166.htm. http://www.most.gov.cn/eng/. The funders had no role in study design, data collection and analysis, decision to publish, or preparation of the manuscript.

Competing Interests: The authors have declared that no competing interests exist.

* E-mail: zhangyp@tsinghua.edu.cn

Introduction

Indoor air quality (IAQ) is important for public health because most people spend over 80% of lifetime indoors [1–4]. Carbonyls and BTX (benzene, toluene and xylene), a subset of volatile organic compounds (VOCs), represent an important group of indoor air pollutants [5–8]. The emission sources of these compounds in indoor environment include building materials, decoration and renovation materials (e.g., vinyl floor and composite wood boards, adhesives, synthesized resins, paints, carpets, furniture) and consumer products (e.g., freshly dry cleaned clothes, mothball and deodorizers) [1,2,9–17]. Indoor carbonyls

can also be formed via ozonolysis of alkenes and terpenes [1,14,15,18,19]. Inhalation exposure to these compounds may result in a variety of acute and chronic adverse health effects [9,12,20,21] such as Sick Building Syndrome (SBS) symptoms [2,13], mucous membrane and lower respiratory irritation [1,12], neurologic effects [2,12], allergic effects [2,12,13], developmental and reproductive effects [12] as well as potential carcinogenic effects (e.g., lung cancer and childhood leukemia) [1,2,12,13].

During the economic boom in the past decades, China has experienced the largest industrialization and urbanization ever in human history [22]. The rapid economic growth and the dramatically increased household wealth result in a nationwide real estate boom. For example, more than 10 million square meters of newly built residential properties were sold each year in Beijing since 2000 (Beijing Statistical Yearbook 2000–2010). Accompanied with the real estate boom is a high demand for building decoration, renovation and refurbishment [23]. Therefore, the IAQ of newly remodeled buildings has become a major public concern in the cities in China. Exposure to formaldehyde in newly remodeled dwellings is suspected to be one of the main causes for the increased childhood leukemia incidence in Chinese mega-cities in recent years [12,13]. The unhealthy IAQ may also be one of the major causes for a 56% increase of lung cancer incidence in Beijing from 2000 to 2010 [24]. This dramatic increase cannot be fully explained by either ambient VOCs pollution or smoking. In fact, ambient concentrations of VOCs and carbonyls in Beijing have decreased in recent years due to implementation of vehicle emission regulation policy [25–27]; the smoking rate for Chinese adults has not significantly changed in the past 15 years [28]. It is challenging to evaluate the acute and chronic health risks from indoor inhalation exposure to VOCs in mega-cities such as Beijing, because it is lack of IAQ monitoring data. In particular, there are few studies on the health risk assessment of IAQ in newly remodeled buildings in China.

The objective of this study was to assess the health risks from inhalation exposure to formaldehyde and benzene in newly remodeled homes and offices in Beijing. The risk assessment was conducted based on the IAQ test results in this study. The reasons to select formaldehyde and benzene as the target VOCs are because they are the main indoor VOC pollutants regulated by GB/T 18883-2002 (the *Chinese National Indoor Air Quality Standard*) [29], and have been ranked top on the list of VOCs with cancer risk potency greater than 1 per million population in the U.S. [10,20,21].

Methods

Buildings Sampled

The formaldehyde and benzene concentrations in indoor air of 410 dwellings and 451 offices in Beijing were tested from July 2008 through September 2012 by Built Environment Test Center, Tsinghua University. The tested dwellings and offices were remodeled (i.e., renovated, decorated and/or refurbished) within the past year. The IAQ tests were requested by the occupants, who had health concerns on the IAQ. The specific reasons they requested such a test were not asked by the study team but likely included the following: uncomfortable odor, awareness of emissions of VOCs in newly remodeled buildings even without obvious odor, and general concern about the impact of VOC emissions on health, particularly families with vulnerable residents (e.g. infants, kids and elder people). These buildings were located in 13 different districts in Beijing (Figure 1), with 73% dwellings and 98% offices in Chaoyang, Haidian, Dongcheng and Xicheng Districts that are urban areas of Beijing. The field tests were conducted based upon permissions from the property owners, and authorized by Certification and Accreditation Administration of the People's Republic of China (CMA) and China National Accreditation Service for Conformity Assessment (CNAS).

All tested homes utilized natural ventilation. Split type air conditioners were used for cooling in the summer, except very limited luxury serviced apartments that used central air-conditioning. All homes utilized central heating in the winter. Central air-conditioning was utilized in about 65% of the tested office buildings, while about 35% of the tested office buildings utilized split type air conditioners. It is noted that windows are usually closed in the winter when central heating is used and in the summer when air conditioner is on. Thus, the air exchange rates (AERs) of dwellings in Beijing in the two seasons are expected to be lower than in spring and fall. Different from dwellings, most of the tested offices utilized central air conditioning, and these office rooms were in a closed environment. Therefore, the AERs of these "closed" offices are expected to be low as well. However, we acknowledge that the effects of ventilation on the indoor VOCs concentration cannot be evaluated in this study, since little data on AERs of the dwellings and offices in China is currently available.

Sample Collection and Chemical Analysis

Field tests were designed and implemented according to the *Chinese National Indoor Air Quality Standard* (GB/T 18883-2002). Prior to the tests, the occupants were asked to stop smoking, remove consumer products that could release VOCs (e.g. mothballs, cleaning products and air fresher). The impacts of human activities on IAQ were minimized as much as possible until completion of the tests. In addition, the occupants were asked to close doors and windows and turn off air conditioning (if they could) for 12–24 hours prior to the IAQ test. Therefore, the indoor environments were in an airtight state during the entire IAQ tests.

Collection of the air samples for formaldehyde and benzene analysis were conducted based on the *Chinese National Indoor Air Quality Standard* (GB/T 18883-2002). The formaldehyde air samples were analyzed using the *Methods for determination of formaldehyde in air of public places* (GB/T 18204.26-2000); whereas the benzene air samples were analyzed based on the *Ambient air-Determination of benzene and its analogies using sorbent adsorption thermal desorption and gas chromatography* (HJ 583-2010). Briefly, duplicate samples were collected for 45 minutes with a sampling rate of 200 mL/minute from the bedrooms in the dwellings. Additional rooms, such as living rooms, were sampled based on the size of the dwellings. Compounds were collected for 45 minutes with a sampling rate of 200 mL/minute from one to five locations in the tested offices, depending on the office size. Samplers were placed at about 1 m above the floor, located as centrally as possible given logistic constraints. Benzene was collected onto Tenax-TA sorbent bed and analyzed using thermodesorption-GC/MS (Series 6850; Agilent Technologies). Formaldehyde was absorbed by 3-methyl-2-benzothiazolinonehydrazone hydrochloride (MBTH) solution and analyzed using UV-VIS spectrometry at 630 nm. Details of the analysis methods are described elsewhere [29–31].

Data Analysis

The measurements obtained from multiple rooms in the dwellings were averaged for both formaldehyde and benzene. If multiple offices in one building were tested, they were regarded as different microenvironments with different occupants. Descriptive statistical analysis was performed for the concentrations of both species. Student T test was conducted to compare the formaldehyde and benzene concentrations between the dwellings and the offices. Since the data were not normally distributed, the lognormal transformed formaldehyde and benzene concentrations were used for analysis. All statistical analysis was conducted by SAS v9.0 (SAS Corporation, Cary, NC), and thereafter.

Health Risk Assessment

Non-carcinogenic health risk assessment. The non-carcinogenic health risks associated with IAQ of the tested buildings were evaluated in terms of the threshold mechanisms of toxic effects [32]. Quantitative risk characterization involves a simple calculation of a hazard index (HI) [32]

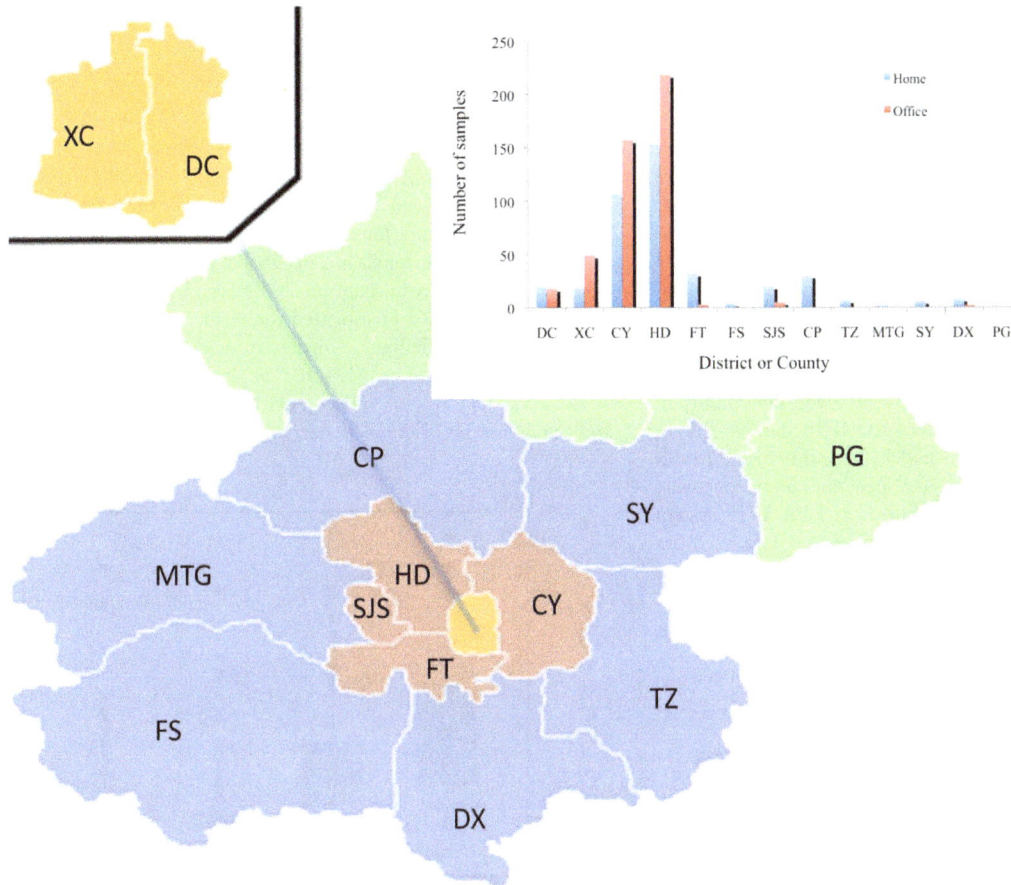

Figure 1. Distribution of the sampled buildings in 13 districts and counties of Beijing. Abbreviation in the figure: XC: Xicheng District, DC: Dongcheng District, HD: Haidian District, CY: Chaoyang District, FT: Fengtai District, FS: Fangshan District, SJS: Shijingshan District, CP: Changping District, TZ: Tongzhou District, MTG: Mengtougou District, SY: Shunyi District, DX: Daxing District, PG: Pinggu County.

$$HI = C_{exp}/RfC \qquad (1)$$

where C_{exp} represents inhalation exposure level of a given air toxic species and RfC represents a "threshold dose" of a given air toxic species. When HI is less than one, it may be inferred that such an exposure is unlikely at risk of toxicity or a given health problem, and vice versa [32].

We compared the indoor concentrations of formaldehyde and benzene to the reference concentrations (RfCs) defined by GB/T 18883-2002 (Chinese National Indoor Air Quality Standard), the RfCs for chronic inhalation exposure defined in the U.S. EPA Integrated Risk Information System (IRIS; U.S. EPA 2010), and the reference exposure levels (RELs) suggested by the California Office of Environmental Health Hazard Assessment (OEHHA; California Environmental Protection Agency, 2008). These reference values are summarized in the Supporting Information (Table S1).

Cancer risk assessment. Cancer risks posed by inhalation exposure to formaldehyde and benzene originate from the very exposure in both outdoor and indoor microenvironments, which include, but not limited to, ambient, home, office and transportation. In our study, we focused on cancer risks associated with indoor (i.e. home+office) exposure. For the assessment of cancer risks from indoor exposure to formaldehyde and benzene, it is necessary to obtain personal concentrations of the two compounds

in indoor microenvironments. Personal concentrations can be determined by two methods. The first method is personal exposure measurement, which was not feasible in this study due to many factors such as available human subjects and willingness of participation. The second method is to a) use computation modeling to simulate human's daily activity and air pollution in microenvironments; and b) derive the distribution of personal concentrations of the two compounds from the distributions of human activity and air toxics concentrations in indoor microenvironments [21,33]. Monte Carlo is a frequently used simulation method, as it can generate numerical distribution through repeated random sampling.

The model population in this study was adult males and females who live and work in newly remodeled buildings. We a) developed the distributions of indoor personal concentrations of formaldehyde and benzene using Monte Carlo Simulations in Crystal Ball (5,000 trials); b) calculated the cancer risks associated with the developed exposure to formaldehyde and benzene; c) compared the cancer risks to those reported in the literatures. Figure 2 illustrates the model framework. The indoor personal exposure to formaldehyde or benzene can be calculated using the following equation,

$$E = \sum_{i=1}^{N} (C_{i,home} \bullet T_{i,home} + C_{i,office} \bullet T_{i,office})/T \qquad (2)$$

where E is the indoor exposure to formaldehyde or benzene; $C_{i,home}$ is the concentration of formaldehyde or benzene for ith person at home; $C_{i,office}$ is the concentration of formaldehyde or benzene for ith person in office; $T_{i,home}$ is the time that ith person spends at home; $T_{i,office}$ is the time that ith person spends in office; T is total exposure time in all microenvironments, i.e. indoor and outdoor.

Distributions of the time that adult working males and females spend in home and office were taken from the *Time Use Patterns in China* [3]. Distributions of formaldehyde and benzene concentration were obtained using the concentration data in this study. We calculated risks by multiplying personal concentration by Inhalation Unit Cancer Risk values (IURs), which represent the excess number of cases per million people expected to develop cancer following lifetime (70 years) exposure to 1 $\mu g/m^3$ of a given agent [10,32]. Two sets of factor values were used given uncertainty in the toxicity estimates. One set is from the IRIS database (U.S. EPA 2010), and the values for formaldehyde and benzene in this system are 1.30×10^{-5} and 7.8×10^{-6} per million, respectively [21,32]. It is necessary to note that the U.S. EPA IRIS system provides two IURs for benzene. The other is 2.2×10^{-6} per million [21,32]. The higher IUR for benzene was used in our study to obtain a maximum estimate of cancer risk from benzene exposure. The other set is from the California OEHHA (California Environmental Protection Agency 2005), and the values for formaldehyde and benzene are 6.00×10^{-6} and 2.90×10^{-5} per million, respectively [21]. We disaggregated risks into home indoor and office indoor exposure parts.

Results and Discussion

Indoor Formaldehyde and Benzene Concentrations

The concentrations of indoor formaldehyde and benzene are illustrated in Figure 3. The concentrations of formaldehyde were 131 ± 90 (100) $\mu g/m^3$ (Mean $\pm$ SD (Median), N = 383) in the tested dwellings and 85 ± 56 (74) $\mu g/m^3$ (N = 406) in the tested offices. The benzene data is more highly skewed than the formaldehyde data. Benzene concentrations were 17 ± 16 (11) $\mu g/m^3$ (Mean $\pm$ SD (Median), N = 379) in the tested dwellings and 30 ± 34 (16) $\mu g/m^3$ (N = 375) in the tested offices. Lognormal

distribution is the best model that fits the formaldehyde and benzene data in our study. This is consistent with conventional concept on probability distribution of pollutant concentrations, which has been illustrated by Beaker Pouring Experiment [34]. The formaldehyde and benzene concentrations in this study are higher than the limited literature data for ordinary buildings (not recently remodeled) in China. Wang et al. [25] tested 3 homes in Beijing and found that the formaldehyde concentration ranged from 30 to 90 $\mu g/m^3$. Jiang and Zhang (2012) measured indoor concentrations of carbonyls in 22 offices of the academic buildings, and the formaldehyde concentrations were 22.6 ± 11.0 $\mu g/m^3$ [35]. Zhou et al. [36] investigated the residential indoor benzene concentration in Tianjin, a mega-city next to Beijing, and the values were 6.13 ± 7.58 (N = 10) in home environment and 1.38 ± 0.57 (N = 6) in office environment. The comparison indicated that emission of benzene, formaldehyde and potential

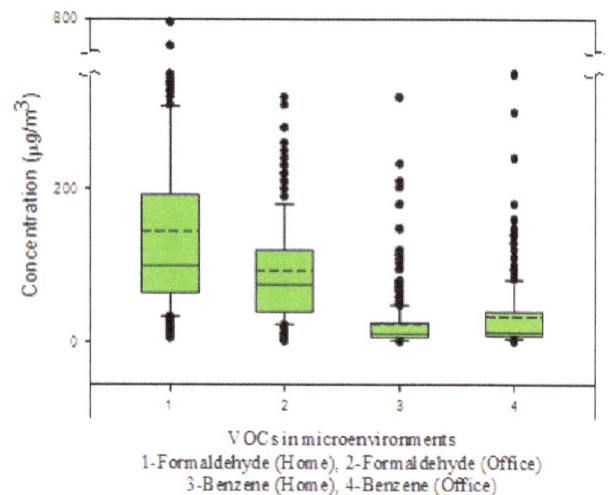

VOCs in microenvironments
1-Formaldehyde (Home), 2-Formaldehyde (Office)
3-Benzene (Home), 4-Benzene (Office)

Figure 3. The concentrations of indoor formaldehyde and benzene in newly remodeled buildings in Beijing.

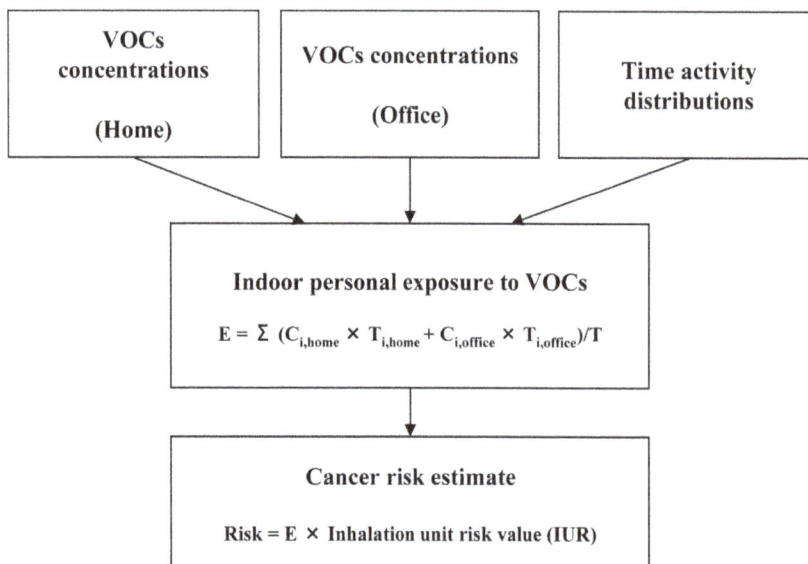

Figure 2. Diagram of personal exposure and cancer risk assessment model.

other VOCs from renovation and decoration materials led to the poor IAQ in the offices and dwelling investigated in the study.

The indoor concentrations of formaldehyde in our study were approximately two orders of magnitude higher than the outdoor formaldehyde concentrations in urban areas of Beijing, i.e. 7.8 ± 3.8, 8.8 ± 4.7 and 10.2 ± 4.2 $\mu g/m^3$ in 2008, 2009 and 2010, respectively [27]. This comparison indicated that formaldehyde emission from the indoor sources of newly remodeled buildings was much stronger than from outdoor sources. Given that urban residents in China spend more than 85% time indoor [3], exposure to indoor formaldehyde is expected to dominate personal formaldehyde exposure for urban residents who live and work in newly remodeled buildings. We also compared the indoor concentrations of formaldehyde by the two building types tested in the study. The concentrations of formaldehyde in the dwellings were significantly ($p<0.01$) higher than in the offices. This difference is due to the complex and intensive decoration and refurbishment that are very popular in Chinese dwellings [37]. Compared to relatively simple decoration in offices, the decoration in dwellings may result in greater formaldehyde emission from the decoration materials in the tested remodeled homes. It is interesting to find that the formaldehyde concentrations in newly remodeled dwellings in Beijing are similar to the formaldehyde concentrations in new homes in Korea and Japan, 134 $\mu g/m^3$ (mean, based on the measurements from 292 new homes) [38]. In Europe, the formaldehyde concentrations measured in newly remodeled homes are generally lower, e.g. 40 $\mu g/m^3$ (median) based on 367 measurements in newly prefabricated houses between 1996 and 2006 in Germany, and 37 $\mu g/m^3$ (median) in 36 newly remodeled Danish apartments [38].

Benzene is widely used as solvents and adhesives [1,13,15]. The use of benzene in those products were banned or restricted in the developed countries when its carcinogenic effects were confirmed [15]. Thus, outdoor sources, such as vehicle emission and gasoline station [1,20], generally make dominant contribution to benzene in indoor environment in the developed countries [10,20,23]. In our study, the indoor benzene concentrations, however, were noticeably higher than the ambient benzene concentrations in Beijing, 6.9 ± 6.7 and 9.2 ± 7.6 $\mu g/m^3$ in 2008 and 2009 [26]. This comparison indicated that benzene emission from indoor sources of newly remodeled buildings in Beijing, same as formaldehyde, was much higher than outdoor sources. It is thus suspected that some adhesives and solvents containing benzene were continued used in decoration and renovation. Contrary to formaldehyde, benzene concentrations in the tested offices were significantly ($p<0.01$) higher than in the tested dwellings. This difference suggests that the adhesives and solvents containing benzene may be more frequently used in office decoration and renovation than in home renovation.

Non-carcinogenic Health Risk Assessment

We compared the indoor concentrations of formaldehyde and benzene in the tested buildings to the health guidelines of China and USA. We note that the measured concentrations of the two compounds are for a 45-min period. As the concentrations were measured when the windows were closed and the impacts of human activities were minimized for 12–24 hours, the indoor concentrations of VOCs were expected to be stable, and thus our measurements were likely to be representative for the concentrations resulted from emission from decoration, renovation and refurbishment materials.

The RfCs of the *Chinese National Indoor Air Quality Standard* (GB/T 18883-2002) are 0.10 and 0.11 mg/m^3 for formaldehyde and benzene, respectively. Formaldehyde concentrations in the tested

buildings were above the RfC of GB/T 18883-2002 in 53% dwellings and 28% offices; whereas benzene concentrations were above the RfC of GB/T 18883-2002 in 2% offices and no dwellings had benzene concentration exceeding the guideline. The acute inhalation RELs of OEHHA were 55 and 1,300 $\mu g/m^3$ for formaldehyde and benzene, respectively. The levels of benzene in all of the dwellings and offices met the guidelines, while 85% dwellings and 67% offices had indoor formaldehyde concentrations exceeding the guidelines. The chronic non-carcinogenic RELs of OEHHA were 9 and 60 $\mu g/m^3$ for formaldehyde and benzene, respectively. The formaldehyde concentrations of all buildings were above the chronic RELs of OEHHA, whereas only 9.6% offices and no dwellings had benzene concentrations above the chronic RELs. The U.S. EPA IRIS system does not have chronic RfC for formaldehyde but benzene (30 $\mu g/m^3$). The benzene concentrations of 12% dwellings and 32% offices exceeded the chronic RfC (U.S. EPA IRIS) for benzene.

The comparison to health guidelines highlights concern on the IAQ of these newly remodeled buildings that were investigated in our study. The exposure levels of formaldehyde in most tested indoor environments were very high, especially the dwellings. According to GB/T 18883-2002, ~50% dwellings and ~30% offices had formaldehyde concentrations above the standard. Based on the acute REL of OEHHA, the formaldehyde exposure in ~80% dwellings and ~65% offices may trigger acute adverse health effects. These acute effects include, but not limited to, eye, throat, and respiratory irritation, tearing, sneezing, coughing, chest congestion, fever, heartburn, lethargy, loss of appetite, and even asthma attacks [1,12]. The scenario of chronic non-carcinogenic effects associated with formaldehyde exposure could be even worse. Based on the chronic REL of OEHHA, formaldehyde exposure in all tested dwellings and offices may result in chronic adverse health effects on the occupants. The chronic effects include headaches, dizziness, sleep disorders, memory loss, pulmonary function damage, pancytopenia and possible menstrual disorders of adult females [12]. The formaldehyde concentration could decrease and maintain stable at ~35% of the initial concentration 3 years after remodeling [39]; however, 99% dwellings and 95% offices may still have formaldehyde concentrations above the chronic REL recommended by OEHHA even with a ~65% decrease of exposure level with prolonged time.

The scenario of benzene is less severe than formaldehyde: only 2% offices did not meet the guideline set by GB/T 18883-2002. In terms of the stricter U.S. EPA IRIS guidelines, the benzene concentrations of all dwellings and offices met the guidelines of both OEHHA and U.S. EPA IRIS for acute adverse health risks while occupants have risks to develop chronic diseases in ~10% dwellings and ~30% offices. Therefore, the risks of developing chronic diseases are the major concern for benzene exposure in the tested dwellings and offices. The critical chronic effects of benzene exposure include decreased lymphocyte count, hematotoxicity and immunotoxicity (U.S. EPA IRIS).

Cancer Risk Assessment

In our study, the indoor formaldehyde concentrations of all dwellings and offices exceeded the inhalation risk level corresponding to cancer risk of 100 excess cases per million (8 $\mu g/m^3$, U.S. EPA IRIS). The indoor benzene concentrations of all dwellings and offices were above the inhalation risk level corresponding to cancer risk of 10 excess cases per million (1.3 $\mu g/m^3$, U.S. EPA IRIS), whereas 48% dwellings and 69% offices had indoor benzene concentrations above the inhalation risk level of 100 excess cases per million (13 $\mu g/m^3$, U.S. EPA IRIS). The comparison results indicated the potential cancer risks from exposure to formaldehyde and benzene. The cancer risks were quantitatively assessed based on modeled indoor personal exposure to the two compounds, which are presented below.

Indoor personal exposure to formaldehyde and benzene. The indoor personal exposure to formaldehyde and benzene were modeled using Monte Carlo Simulation in Crystal Ball. The exposure distributions are illustrated in the Supporting Information (Figures S1–S4). We note that the modeled personal concentrations of formaldehyde and benzene were the indoor (home+office) fraction of total personal exposure for each of the two compounds. As shown in Figure 4, the personal exposure of formaldehyde in dwellings and offices were 86 ± 52 (73) $\mu g/m^3$ (Mean $\pm$ SD (Median)) and 15 ± 9.1 (12) $\mu g/m^3$, respectively; while the personal concentrations of benzene in dwellings and offices were 11 ± 8.6 (8.3) $\mu g/m^3$ and 5.0 ± 4.7 (3.5) $\mu g/m^3$, respectively.

Cancer risk assessment. We calculated the cancer risks from inhalation exposure to formaldehyde and benzene in indoor environment using both U.S. EPA IRIS and OEHHA IURs. The descriptive statistical information of the cancer risk estimates is summarized in Table 1. Based on the OEHHA IURs, the median cancer risks were 531 and 394 excess cases per million for formaldehyde and benzene, respectively, if model individuals live and work in newly remodeled homes and offices over lifetime. Based on the U.S. EPA IRIS IURs, the median cancer risks from indoor exposure to formaldehyde and benzene are 1,150 and 106 excess cases per million, respectively. We disaggregated the cumulative risks into the risk attributed to exposure at home and the risk attributed to exposure in office. The residential microenvironment accounts for ~85% and ~70% of the cumulative risks from indoor exposure to formaldehyde and benzene, respectively.

The calculated cancer risks, to our knowledge, are the highest reported in scientific literature. The comparison is shown in Table 1. With regards to the cancer risks from inhalation exposure to formaldehyde and benzene, our values were ~5 times and ~3 times of Loh et al. [21] (Table 1). It is necessary to point out that the values yielded by the model calculation in Loh et al. [21] represent the cancer risks from baseline exposure in both indoor and outdoor microenvironments. Sax et al. [20] assessed the cancer risks from inhalation exposure to VOCs for non-smoking teenagers from non-smoking homes in New York City and Los Angeles. The assessment was based on the measured personal

concentrations of VOCs. The risks in Sax et al. [20] were about one fourth (formaldehyde) and one third (benzene) of the values in our study (Table 1). Again, the risks in Sax et al. [20] also represent the cancer risks from exposure in both indoor and outdoor environments. Zhou et al. [36] assessed the cancer risks based on the measured personal concentrations of BETX for 12 adults in Tianjin, China. The cancer risk (all microenvironments, indoor+outdoor) attributed to benzene in their study was lower than our value as well (Table 1). Note: only 5 participants in Tianjin Study renovated their apartments within the past year [36]. In addition, indoor concentrations of formaldehyde and benzene in the newly remodeled properties in our study were much higher than outdoor counterparts as previously discussed. Thus, it is not surprising that the cancer risks of indoor exposure were much greater than the values that were reported for outdoor exposure only in Beijing, i.e. 91.1 (formaldehyde) and 41.9 (benzene) per million [26].

It is necessary to note again that the concentrations of indoor formaldehyde and benzene will decrease with time after remodeling. Therefore, the risk values that we obtained through model calculation represent the cancer risks for the highest exposure scenario. Nonetheless, Ohura et al. [23] found that indoor benzene concentration in China dropped to ~25% of initial concentration and maintained stable 3 months to 1 year after remodeling, and Zhao et al. [39] suggested that indoor formaldehyde concentration in China dropped to ~35% of the initial concentration and maintained stable 3 years after remodeling. Assuming formaldehyde and benzene concentrations in the tested buildings in our study decreased to ~35% and ~25% of the reported values, respectively, the cancer risks associated with indoor exposure to formaldehyde and benzene would decrease to ~406 and ~27 excess cases per million (based on U.S. EPA IRIS 2010), respectively. These values, however, are still noticeably higher than those reported by researchers in the U.S. (Table 1). Our study results showed that cancer risks associated with baseline inhalation exposure to indoor formaldehyde and benzene in these Chinese buildings in our study may be significantly higher than the cancer risks associated with the baseline formaldehyde and benzene exposure in the U.S.

Uncertainties and Limitations

Our study assessed the health risks from indoor formaldehyde and benzene exposure in recently remodeled dwellings and offices

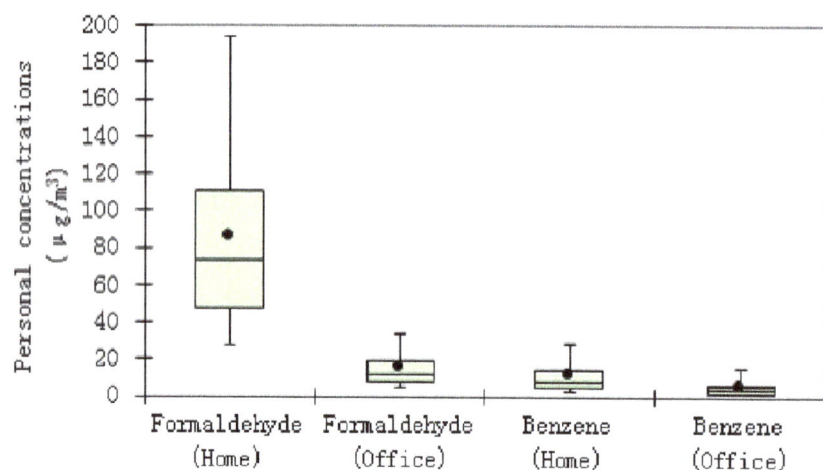

Figure 4. Personal concentrations of formaldehyde and benzene in dwellings and offices.

Table 1. The estimated cancer risks from indoor exposure to formaldehyde and benzene.

	City and country	Source of IURs[a]	Compound	Cancer risk (excess cases per 1 million population)			
				Mean	SD	Median	95%
Our study	Beijing, China	OEHHA[b]	Formaldehyde	604	314	531	1,260
			Benzene	463	285	394	1,030
		U.S. EPA IRIS[c]	Formaldehyde	1,340	771	1,160	2,880
			Benzene	131	96	107	304
Loh et al. [21][d]	USA	OEHHA	Formaldehyde	100[e]			
			Benzene	108[e]			
		U.S. EPA IRIS	Formaldehyde	240[e]			
			Benzene	30[e]			
Sax et al. [20][d]	New York City and Los Angeles, USA	U.S. EPA IRIS	Formaldehyde	205 (NYC), 258 (LA)[e]			
			Benzene	25 (NYC), 34 (LA)[e]			
Zhou et al. [36][f]	Tianjing, China	U.S. EPA IRIS	Benzene	~22[e]			

[a]IURs refers to Inhalation Unit Risk Values.
[b]OEHHA: California Office of Environmental Health Hazard Assessment.
[c]U.S. EPA IRIS: U.S. EPA Integrated Risk Information System.
[d]The cancer risk estimates from Loh et al. [21] and Sax et al. [20] represented the cancer risks associated with baseline exposure to formaldehyde and benzene in both indoor and outdoor microenvironments.
[e]Median cancer risk values are provided for Loh et al. [21], Sax et al. [20] and Zhou et al. [36].
[f]The cancer risk estimates from Zhou et al. [36] represented the cancer risks associated with exposure to benzene in both indoor and outdoor microenvironment.

in Beijing. Occupants of these buildings requested for the IAQ tests due to a variety of reasons, which included, but not limited to, uncomfortable odor, awareness of potential unhealthy IAQ in newly remodeled buildings and its impact on health. Therefore, the assessment results reflect the scenario of the newly remodeled buildings in Beijing, although the measurements were obtained from those requested for IAQ test.

Since the dwelling windows were closed 12–24 hours prior to field tests until completion, the ventilation rates were very low during the tests. As previously discussed, such an airtight state is prevalent in the winter and summer for Beijing residential spaces. This is because windows are usually closed when either central heating or air conditioning is on. Unlike the two seasons, Beijing residents frequently open windows for ventilation in the spring and fall. As a result, indoor exposure levels of formaldehyde and benzene and the associated health risks in newly remodeled dwellings may be lower in these two seasons. Nonetheless, using the summer and winter data can yield maximum estimate of health risks. Residents would actually benefit from regulation developed based on the maximum health risk estimate.

Conclusions and Recommendations

This study reports the formaldehyde and benzene levels in newly remodeled dwellings and offices in Beijing, in which the occupants have health concerns about IAQ. The concentration data and subsequent health risk assessment can help us understand the health risks associated with the IAQ of these buildings. Exposure to formaldehyde may pose both acute and chronic non-cancer risks to the occupants in the tested buildings, as 85% dwellings and 67% offices had concentrations over the acute REL of OEHHA and all dwellings and offices over the chronic REL of OEHHA. Exposure to benzene may pose chronic non-cancer risks to the occupants, as 12% dwellings and 32% offices had concentrations over the chronic RfC of US EPA IRIS. The median cancer risks (per million) of indoor exposure to formaldehyde and benzene were estimated to be 1,150 and 106 (based on

US EPA IRIS IURs), 531 and 394 (based on OEHHA IURs) if adult males and females work and live in the newly remodeled indoor environment over lifetime. Based on our assessment results, inhalation exposure to VOCs in newly remodeled buildings, which were mainly emitted from decoration and renovation materials, may trigger significant adverse health effects on occupants in China. Ventilation improvement is one of the potential strategies that can be considered to reduce indoor exposure to VOCs, especially for the airtight dwellings in Beijing. Another strategy would be the reduction of VOCs emission from building materials, decorating materials and furniture. For instance, the establishment of *Chinese indoor decorating and refurbishment materials and furniture VOCs emission labeling system* is recommended. This labeling system can serve as guidance for the consumers on selection of building and decoration materials.

There are still large knowledge gaps in the associations between indoor VOCs exposure and public health. For instance, current *Chinese National Indoor Air Quality Standard* (GB/T 18883-2002) does not involve some important carcinogenic VOCs species such as 1,4-dichlorobenzene, chloroform, 1,3-butadiene, acetaldehyde and tetrachloroethylene. Therefore, we recommend measuring more VOCs species in prospective IAQ monitoring campaigns and assessing associated health risks in China. We also recommend assessing the cancer risks from exposure to VOCs across various microenvironments and across various VOC species. Results of these research activities will provide a full scenario of Chinese environmental health that is associated with indoor VOCs pollution. These results will facilitate prioritization of air toxics for environmental regulation and pollution control, and ultimately protect public health in China.

Supporting Information

Figure S1 Distribution of personal exposure to formaldehyde in dwellings.

Figure S2   Distribution of personal exposure to formaldehyde in offices.

Figure S3   Distribution of personal exposure to benzene in dwellings.

Figure S4   Distribution of personal exposure to benzene in offices.

Table S1   Relevant guidelines and standards for indoor formaldehyde and benzene.

Author Contributions

Conceived and designed the experiments: LH JS. Performed the experiments: LH JM. Analyzed the data: LH. Contributed reagents/materials/analysis tools: JM YZ. Wrote the paper: LH. Revised the manuscript: ZF JS YZ.

References

1. Wu XM, Apte M, Maddalena R, Bennett D (2011) Volatile organic compounds in small- and medium-sized commercial buildings in California. Environ Sci Technol 45: 9075–9083.
2. Jones AP (1999) Indoor air quality and health. Atmo Environ 33: 4535–4564.
3. Department of Social, Science and Technology Statistics, National Bureau of Statistics (2010) Time use patterns in China. Beijing: China Statistics Press.
4. Wainman T, Zhang JF, Weschler C, Lioy P (2000) Ozone and limonene in indoor air: a source of submicron particle exposure. Environ Health Perspect 108: 1139–1145.
5. Massolo L, Rehwagen M, Porta A, Ronco A, Herbarth O, et al. (2009) Indoor-outdoor distribution and risk assessment of volatile organic compounds in the atmosphere of industrial and urban areas. Environ Toxicol DOI 10.1002/tox.20504.
6. Lee CW, Dai YT, Chien CH, Hsu DJ (2006) Characteristics and health impacts of volatile organic compounds in photocopy centers. Environ Res 100: 139–149.
7. Ramirez N, Cuadras A, Rovira E, Borrull F, Marce RM (2012) Chronic risk assessment of exposure to volatile organic compounds in the atmosphere near the largest mediterranean industrial site. Environ Int 39: 200–209.
8. Du Z, Mo J, Zhang Y, Li X, Xu Q (2013) Evaluation of a new passive sampler using hydrophobic zeolites as adsorbents for exposure measurement of indoor BTX. Analyst Methods 5: 3464–3472.
9. Dodson RE, Houseman EA, Levy J, Spengler J, Shine J, et al. (2007) Measured and modeled personal exposures to and risks from volatile organic compounds. Environ Sci Technol 41: 8498–8505.
10. Hun D, Siegel J, Morandi M, Stock T, Corsi R (2009) Cancer risk disparities between Hispanic and non-Hispanic white populations: the role of exposure to indoor air pollution. Environ Health Perspect 117: 1925–1931.
11. Xiong JY, Yao Y, Zhang YP (2011) C-History method: rapid measurement of the initial emittable concentration, diffusion and partition coefficients for formaldehyde and VOCs in building materials. Environ Sci Technol 45: 3584–3590.
12. Tang XJ, Bai Y, Duong A, Smith M, Li LY, et al. (2009) Formaldehyde in China: production, consumption, exposure levels, and health effects. Environ Int 35: 1210–1224.
13. Zhang JF, Smith K (2003) Indoor air pollution: a global health concern. British Med Bull 68: 209–225.
14. Weschler C (2004) Chemical reactions among indoor pollutants: what we've learned in the new millennium. Indoor air 14 (Suppl 7): 184–194.
15. Weschler C (2009) Changes in indoor pollutants since the 1950s. Atmos Environ 43: 153–169.
16. Xiong JY, Yan W, Zhang YP (2011) Variable volume loading method: a convenient and rapid method for measuring the initial emittable concentration and partition coefficient of formaldehyde and other aldehydes in building materials. Environ Sci Technol 45: 10111–10116.
17. Liu WW, Zhang YP, Yao Y, Li JG (2012) Indoor decorating and refurbishing materials and furniture volatile organic compounds emission labeling systems: a review. Chin Sci Bull 57: 2533–2543.
18. Finlayson-Pitts B, Pitts J (1986) Atmospheric chemistry: fundamentals and experimental techniques. New York City: John Wiley & Sons, Inc.
19. Calogirou A, Larsen B, Kotzias D (1999) Gas-phase terpene oxidation products: a review. Atmos Environ 33: 1423–1439.
20. Sax S, Bennett D, Chillrud S, Ross J, Kinney P, et al. (2006) A cancer risk assessment of inner-city teenagers living in New York City and Los Angeles. Environ Health Perspect 114: 1558–1566.
21. Loh M, Levy J, Spengler J, Houseman EA, Bennett D (2007) Ranking cancer risks of organic hazardous air pollutants in the United States. Environ Health Perspect 115: 1160–1168.
22. Zhang Y, Mo J, Weschler C (2013) Reducing health risks from indoor exposures in today's rapidly developing urban China. Environ Health Perspect 121: 751–755.
23. Ohura T, Amagai T, Shen XY, Li S, Zhang P, et al. (2009) Comparative study on indoor air quality in Japan and China: characteristics of residential indoor and outdoor VOCs. Atmos Environ 43: 6352–6359.
24. Beijing Municipal Health Bureau and Beijing Municipal People's Government (2012) Annual report of community and population health status in Beijing, China: 2011. Beijing: People's Medical Publishing House.
25. Wang YS, Ren XY, Ji DS, Zhang JQ, Sun J, et al. (2012) Characteristics of volatile organic compounds in the urban area of Beijing from 2000 to 2007. J Environ Sci 24: 95–101.
26. Zhang YJ, Mu YJ, Liu JF, Mellouki A (2012) Levels, Sources and health risks of carbonyls and BTEX In the ambient air of Beijing, China. J Environ Sci 24: 124–130.
27. Zhang YJ, Mu YJ, Liang P, Xu Z, Liu JF, et al. (2012) Atmospheric BTEX and carbonyls during summer seasons of 2008–2010 in Beijing. Atmos Environ 59: 186–191.
28. Chinese Center for Disease Control and Prevention; Available: http://www.notc.org.cn.
29. State Bureau of Quality and Technical Supervision, Ministry of Environmental Protection of the People's Republic of China, Ministry of Health of the People's Republic of China (2002) Indoor air quality standard. Beijing: China Standard Press.
30. State Bureau of Quality and Technical Supervision (2000) Methods for determination of formaldehyde in air of public places. Beijing: China Standard Press.
31. Ministry of Environmental Protection of the People's Republic of China (2010) Ambient air-determination of benzene and its analogies using sorbent adsorption thermal desorption and gas chromatography. Beijing: China Environmental Science Press.
32. Rodricks JV (2007) Calculated risks (second edition) Cambridge: Cambridge University Press.
33. Payne-Sturges D, Burke T, Breysse P, Diener-West M, Buckley T (2004) Personal exposure meets risk assessment: a comparison of measured and modelled exposures and risks in an urban community. Environ Health Perspect, 112(5): 589–598.
34. Ott W (1990) A physical explanation of the lognormality of pollutant concentrations. J air and waste management association 40: 1378–1383.
35. Jiang C, Zhang P (2012) Indoor carbonyl compounds in an academic building in Beijing, China: concentrations and influencing factors. Front. Environ. Sci. Engin. 6(2): 184–194.
36. Zhou J, You Y, Bai ZP, Hu YD, Zhang JF, et al. (2011) Health risk assessment of personal inhalation exposure to volatile organic compounds in Tianjin, China. Sci Total Environ 409: 452–459.
37. Yoshino H, Zhao JH, Yoshino Y, Kumagai K, Ni YY, et al. (2005) A study on indoor air quality of urban residential buildings in China. J Asian Archi Bild Enginr 4: 495–500.
38. Salthammer T, Mentese S, Marutzky R (2010) Formaldehyde in the indoor environment. Chem Rev 110: 2536–2572.
39. Zhao Y, Chen B, Guo YL, Peng FF, Zhao JL (2004) Indoor air environment of residential buildings in Dalian, China. Energy and Bild 36: 1235–1239.

8

Impact of Environmental Parameters on Marathon Running Performance

Nour El Helou[1,2,3]*, **Muriel Tafflet**[1,4], **Geoffroy Berthelot**[1,2], **Julien Tolaini**[1], **Andy Marc**[1,2], **Marion Guillaume**[1], **Christophe Hausswirth**[5], **Jean-François Toussaint**[1,2,6]

1 IRMES (bioMedical Research Institute of Sports Epidemiology), INSEP, Paris, France, 2 Université Paris Descartes, Sorbonne Paris Cité, Paris, France, 3 Faculté de Pharmacie, Département de Nutrition, Université Saint Joseph, Beirut, Lebanon, 4 INSERM, U970, Paris Cardiovascular Research Center – PARCC, Paris, France, 5 Research Department, INSEP, Paris, France, 6 Hôtel-Dieu Hospital, CIMS, AP-HP, Paris, France

Abstract

Purpose: The objectives of this study were to describe the distribution of all runners' performances in the largest marathons worldwide and to determine which environmental parameters have the maximal impact.

Methods: We analysed the results of six European (Paris, London, Berlin) and American (Boston, Chicago, New York) marathon races from 2001 to 2010 through 1,791,972 participants' performances (all finishers per year and race). Four environmental factors were gathered for each of the 60 races: temperature (°C), humidity (%), dew point (°C), and the atmospheric pressure at sea level (hPA); as well as the concentrations of four atmospheric pollutants: NO_2 – SO_2 – O_3 and PM_{10} ($\mu g.m^{-3}$).

Results: All performances per year and race are normally distributed with distribution parameters (mean and standard deviation) that differ according to environmental factors. Air temperature and performance are significantly correlated through a quadratic model. The optimal temperatures for maximal mean speed of all runners vary depending on the performance level. When temperature increases above these optima, running speed decreases and withdrawal rates increase. Ozone also impacts performance but its effect might be linked to temperature. The other environmental parameters do not have any significant impact.

Conclusions: The large amount of data analyzed and the model developed in this study highlight the major influence of air temperature above all other climatic parameter on human running capacity and adaptation to race conditions.

Editor: Alejandro Lucia, Universidad Europea de Madrid, Spain

Funding: The authors have no support or funding to report.

Competing Interests: The authors have declared that no competing interests exist.

* E-mail: nour.elhelou@insep.fr

Introduction

Like most phenotypic traits, athletic performance is multifactorial and influenced by genetic and environmental factors: exogenous factors contribute to the expression of the predisposing characteristics among best athletes [1,2]. The marathon is one of the most challenging endurance competitions; it is a mass participation race held under variable environmental conditions and temperatures sometimes vary widely from start to finish [3–5]. Warm weather during a marathon is detrimental for runners and is commonly referenced as limiting for thermoregulatory control [3,6]. More medical complaints of hyperthermia (internal temperature $\geq 39°C$) occur in warm weather events, while hypothermia (internal temperature $\leq 35°C$) sometimes occurs during cool weather events [3].

In addition, participating in an outdoor urban event exposes athletes to air pollution which raises concerns for both performance and health [7]. Runners could be at risk during competitions as they are subject to elevated ventilation rate and increased airflow velocity amplifying the dose of inhaled pollutants and carrying them deeper into the lungs [7–9]. They switch from nasal to mouth breathing, bypassing nasal filtration mechanisms for large particles. Both might increase the deleterious effects of pollutants on health and athletic performance [8,10]. Exposure to air pollution during exercise might be expected to impair an athlete's performance in endurance events lasting one hour or more [7,10].

The relationship between marathon performance decline and warmer air temperature has been well established. Vihma [6] and Ely et al. [11,12] found a progressive and quantifiable slowing of marathon performance as WBGT (Wet Bulb Globe Temperature) increases, for men and women of wide ranging abilities. Ely et al. [13] as well as Montain et al. [14] also found that cooler weather (5–10°C) was associated with better ability to maintain running velocity through a marathon race compared to warmer conditions especially by fastest runners; weather impacted pacing and the impact was dependent on finishing position. Marr and Ely [9] found significant correlations between the increase of WBGT and PM_{10}, and slower marathon performance of both men and women; but they did not find significant correlations with any other pollutant.

Previous studies have mostly analysed the performances of the top 3 males and females finishers as well as the 25th-, 100th-, and 300th- place finishers [11,13–16]. Here we targeted exhaustiveness and analysed the total number of finishers in order to quantify the effect of climate on the full range of runners.

The objectives of this study were 1) to analyse all levels of running performance by describing the distribution of all marathons finishers by race, year and gender; 2) to determine the impact of environmental parameters: on the distribution of all marathon runners' performance in men and women (first and last finishers, quantiles of distribution); and on the percentage of runners withdrawals. We then modelled the relation between running speed and air temperature to determine the optimal environmental conditions for achieving the best running performances, and to help, based on known environmental parameters, to predict the distribution and inform runners on possible outcomes of running at different ambient temperatures. We tested the hypothesis that all runners' performances distributions may be similar in all races, and may be similarly affected by temperature.

Methods

Data Collection

Marathon race results were obtained from six marathons included in the « IAAF Gold Labeled Road Races » and « World Marathon Majors »: Berlin, Boston, Chicago, London, New York and Paris. From 2001 to 2010 (available data are limited before 2001) the arrival times in hours: minutes: seconds, of all finishers were gathered for each race. These data are available in the public domain on the official internet website of each city marathon, and on marathon archives websites [17] and complementary data when needed from official sites of each race. Written and informed consent was therefore not required from individual athletes. The total number of collected performances was 1,791,972 for the 60 races (10 years × 6 marathons), including 1,791,071 performances for which the gender was known. We also gathered the total number of starters in order to calculate the number and the percentage of non-finishers (runner withdrawal) per race.

Hourly weather data corresponding to the race day, time span and location of the marathon were obtained from "weather underground website" [5]. Four climatic data were gathered for each of the 60 races: air temperature (°C), air humidity (%), dew-point temperatures (°C), and atmospheric pressure at sea level (hPA). Each of these parameters was averaged for the first 4 hours after the start of each race. Hourly air pollution data for the day, time span and location of each race were also obtained through the concentrations of three atmospheric pollutants: $NO_2 - SO_2 - O_3$ ($\mu g.m^{-3}$) from the Environmental Agency in each state (the Illinois Environmental Protection Agency for Chicago maratho'n, the Massachusetts Department of environmental Protection for Boston marathon and the New York State Department of Environmental Conservation for New York marathon), and the Environmental agency websites of the three European cities [18–20]. All pollutants values were averaged for the first 4 hours after the start of each race.

Concurrent measurements of air pollution for all ten race years (2001–2010) were only available for 3 pollutants, because air pollution monitoring sites typically measure only a subset of pollutants and may not have been operational in all years. In addition, particulate matters PM_{10} were collected in Paris and Berlin, but there were not enough measurements in the other four cities races days.

Data Analysis and selection

Men and women performances were analysed separately. For each race and each gender every year, we fitted the Normal and log-Normal distributions to the performances and tested the normality and log normality using the Kolmogorov-Smirnov D statistic. We rejected the null hypothesis that the sample is normally or log–normally distributed when p values <0.01.

The following statistics (performance levels) were determined for all runners' performances distribution of each race, every year and for each gender:

- the first percentile of the distribution (P1), representing the elite of each race.
- the winner.
- the last finisher.
- the first quartile of the distribution (Q1), representing the 25th percentile of best performers of the studied race.
- the median.
- the inter quartile range (IQR), representing the statistical dispersion, being equal to the difference between the third and first quartiles.

A Spearman correlation test was performed between each performance level and climate and air pollution parameters, in order to quantify the impact of weather and pollution on marathon performances. Spearman correlation tests were also performed between each environmental parameter. The year factor was not included because we previously demonstrated that for the past ten years, marathon performances were now progressing at a slower rate [21].

Temperature and running speed

We modelled the relation between running speed of each performance level for each gender and air temperature, using a second degree polynomial quadratic model, which seems appropriate to depict such physiological relations [22–24].

The second degree polynomial equation was applied to determine the optimal temperature at which maximal running speed is achieved for each level of performance for each gender, and then used to calculate the speed decrease associated with temperature increase and decrease above the optimum.

We similarly modelled the relation between air temperature and the percentage of runners' withdrawal.

All analyses were performed using the MATLAB and SAS software.

Results

The total numbers of starters and finishers of the 6 marathons increased over the 10 studied years (Figure 1). Marathons characteristics are described in supplementary data (Table S1). The race with the least number of finishers was Boston 2001 with 13381 finishers and the highest number was seen in New York 2010 with 44763 finishers.

Three marathons were held in April, the other three during fall. Air temperatures ranged from 1.7°C (Chicago 2009) to 25.2°C (Boston 2004) (Table 1).

Performance distribution

For all 60 studied races, the women and men's performance distributions were a good approximation of the "log normal" and "normal" distributions (p-values of Kolmogorov-Smirnov statistics ≥0.01).

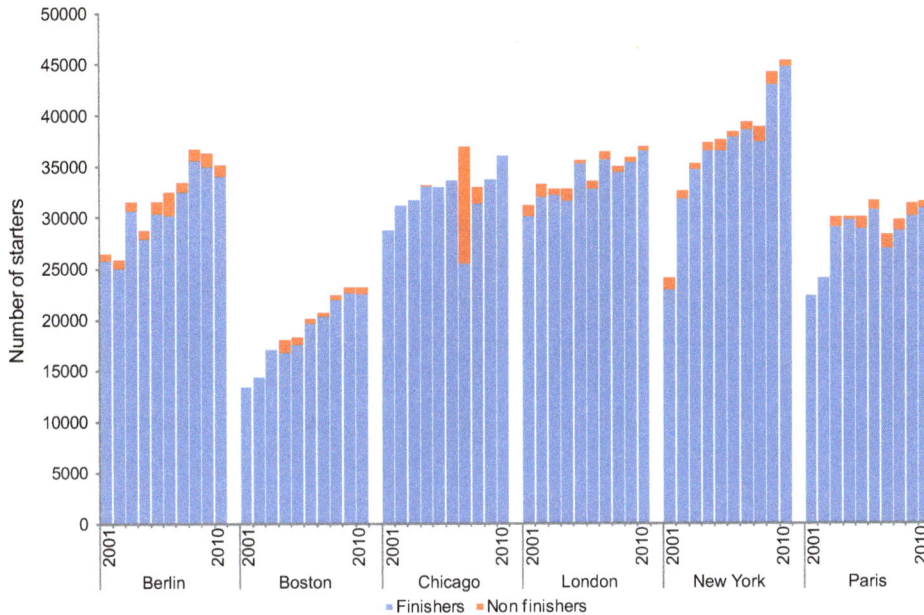

Figure 1. Number of starters and finishers by marathon and year (missing data points for Boston, Chicago and Paris marathons).

Figure 2 illustrates examples of 4 races' performances distribution fit: men's performances distribution of two races in Paris (2002: $T° = 7.6°C$; and 2007: $T° = 17.4°C$) and Chicago (2002: $T° = 5.4°C$; and 2007: $T° = 25°C$).

We notice a stable gap between male and female performances at all levels in all marathons, women being on average $10.3\% \pm 1.6\%$ (mean $\pm$ standard deviation) slower than men (Table S1); mean female winners are $9.9\% \pm 1.5\%$ slower than male winners, mean female median is $9.9\% \pm 1.6\%$ than male median, and mean female Q1 are $11.1\% \pm 1.5\%$ slower that male Q1.

Correlations

Spearman correlations results are displayed in Table 2, detailed correlations by marathon are available in supplementary data (Table S2).

The environmental parameter that had the most significant correlations with marathons performances was air temperature: it was significantly correlated with all performance levels in both male and female runners.

Humidity was the second parameter with a high impact on performance; it was significantly correlated with women's P1 and men's all performance levels.

The dew point and atmospheric pressure only had a slight influence ($p < 0.1$) in men's P1 and women's P1 respectively, and did not affect the other performance levels.

Concerning the atmospheric pollutants, NO_2 had the most significant correlation with performance: it was significantly correlated with Q1, IQR and the median for both genders. Sulfur dioxide (SO_2) was correlated with men's P1 ($p < 0.01$) and had a slight influence ($p < 0.1$) on men's Q1. Finally ozone (O_3) only had a slight influence ($p < 0.1$) on men's Q1. In the marathon by marathon analysis, ozone (O_3) had the most significant correlation with performance (Table S2): it was significantly correlated with all performance levels (P1, Q1, IQR and the median) of the Berlin and Boston (except men's IQR) marathon for both genders. It also affected Chicago (men's P1, Q1, and men's median), and New York (women's Q1) marathons.

Temperature and running speed

When temperature increased above an optimum, performance decreased. Figure 3 describes the relationship between marathons running speeds and air temperature, fit through a quadratic second degree polynomial curve for women's P1 and men's Q1 of all 60 races.

For each performance level the speed decrease associated with temperature increase and decrease is presented in supplementary data (Table S3).

For example the optimal temperature at which women's P1 maximal running speed was attained was $9.9°C$, and an increase of $1°C$ from this optimal temperature will result in a speed loss of 0.03%. The optimal temperatures to run at maximal speed for men and women, varied from $3.8°C$ to $9.9°C$ according to each level of performance (Table S3).

Warmer air temperatures were associated with higher percentages of runners' withdrawal during a race (Figure 4). After testing linear, quadratic, exponential and logarithmic fits, the quadratic equation was the best fit ($r^2 = 0.36$; $p < 0.0001$) for modelling the percentage of runners withdrawals associated with air temperature (Figure 4):

$$\%withdrawals = -0.59 \times t°C + 0.02 \times t°C^2 + 5.75$$

Discussion

Our study is the first to our knowledge to analyse the exhaustiveness of all marathon finishers' performances in the three major European (Berlin, Paris and London, which were not previously analysed) and three American marathons. Previous studies have mostly analysed American marathons including Chicago, Boston and New York that are analysed in the present paper [9,11–15], but they have only included the performances of the top 3 males and females finishers as well as the 25th-, 100th-, and 300th- place finishers [11,13–15]. In the present study we

Table 1. Average and range values of all weather and pollution parameters for the six marathons.

Marathon	Parameter	N	Mean	Std Dev	Minimum	Maximum
Berlin Run in September; Starts 9am	Temperature (°C)	10	14.9	3.2	11.3	21.3
	Dew Point (°C)	10	10.6	1.8	5.8	12.3
	Humidity (%)	10	78.0	14.5	55.0	98.5
	Atmospheric pressure (hPA)	10	1017.0	6.3	1003.0	1029.0
	NO_2 ($\mu g.m^{-3}$)	10	26.5	4.0	20.8	33.2
	O_3 ($\mu g.m^{-3}$)	10	41.0	17.3	21.2	81.8
	PM_{10} ($\mu g.m^{-3}$)	8	25.1	11.4	7.6	46.5
	SO_2 ($\mu g.m^{-3}$)	10	5.0	3.1	1.1	10.7
Boston Run in April; Starts 10am	Temperature (°C)	10	11.8	5.1	8.0	25.2
	Dew Point (°C)	10	3.9	3.8	−2.1	10.2
	Humidity (%)	10	62.6	19.9	28.3	91.0
	Atmospheric pressure (hPA)	10	1013.0	12.4	981.6	1029.0
	NO_2 ($\mu g.m^{-3}$)	10	29.3	10.3	14.6	50.5
	O_3 ($\mu g.m^{-3}$)	10	73.5	25.7	18.5	122.7
	PM_{10} ($\mu g.m^{-3}$)	0				
	SO_2 ($\mu g.m^{-3}$)	10	7.0	2.9	1.6	12.1
Chicago Run in October; Starts 7:30am	Temperature (°C)	10	12.1	7.5	1.7	25.0
	Dew Point (°C)	10	4.9	7.6	−5.9	19.0
	Humidity (%)	10	62.8	8.1	52.3	79.2
	Atmospheric pressure (hPA)	10	1022.0	6.4	1012.0	1031.0
	NO_2 ($\mu g.m^{-3}$)	10	27.9	13.0	9.7	52.0
	O_3 ($\mu g.m^{-3}$)	10	57.1	15.1	35.9	84.0
	PM_{10} ($\mu g.m^{-3}$)	2	26.7	11.6	15.3	38.0
	SO_2 ($\mu g.m^{-3}$)	9	6.5	3.1	2.1	12.4
London Run in April; Starts 9:30am	Temperature (°C)	10	12.4	3.2	9.5	19.1
	Dew Point (°C)	10	6.0	2.9	0.8	10.7
	Humidity (%)	10	66.9	16.7	42.9	86.1
	Atmospheric pressure (hPA)	10	1010.0	12.5	976.4	1020.0
	NO_2 ($\mu g.m^{-3}$)	10	44.8	14.5	22.8	72.2
	O_3 ($\mu g.m^{-3}$)	9	51.4	17.1	35.0	92.3
	PM_{10} ($\mu g.m^{-3}$)	2	27.8	14.5	13.7	41.9
	SO_2 ($\mu g.m^{-3}$)	10	4.5	2.8	0.0	8.8
New York Run in November; Starts 10am	Temperature (°C)	10	12.5	4.1	7.1	18.4
	Dew Point (°C)	10	2.3	6.4	−5.6	12.8
	Humidity (%)	10	51.1	12.1	36.5	79.8
	Atmospheric pressure (hPA)	10	1020.0	7.8	1009.0	1034.0
	NO_2 ($\mu g.m^{-3}$)	9	55.1	17.2	21.9	77.3
	O_3 ($\mu g.m^{-3}$)	10	32.6	12.3	11.1	53.8
	PM_{10} ($\mu g.m^{-3}$)	10	5.0	0.0	5.0	5.0
	SO_2 ($\mu g.m^{-3}$)	9	19.7	12.2	4.8	42.4
Paris Run in April; Starts 8:45am	Temperature (°C)	10	9.2	3.2	4.8	17.4
	Dew Point (°C)	10	4.2	4.1	−3.6	13.4
	Humidity (%)	10	72.4	10.1	45.9	85.4
	Atmospheric pressure (hPA)	10	1019.0	6.2	1005.0	1026.0
	NO_2 ($\mu g.m^{-3}$)	10	43.0	13.7	23.4	73.1

Table 1. Cont.

Marathon	Parameter	N	Mean	Std Dev	Minimum	Maximum
	O_3 ($\mu g.m^{-3}$)	10	66.9	9.8	55.2	82.1
	PM_{10} ($\mu g.m^{-3}$)	10	37.9	32.6	16.6	132.7
	SO_2 ($\mu g.m^{-3}$)	10	6.4	3.7	1.5	12.2

analysed the total number of finishers in order to exhaustively quantify the effect of climate on runners from all performance levels. Updating and extending earlier results, this study still concludes that the main environmental factor influencing marathon performance remains temperature. The pattern of performance reduction with increasing temperature is analogous in men and women, suggesting no apparent gender differences. In addition the mean gap between male and female performances is the same across all marathons and all performance levels (Table 1). This is consistent with our previous work that showed that the gender gap in athletic performance has been stable for more than 25 years, whatever the environmental conditions [25].

The more the temperature increases, the larger the decreases in running speeds (Table S3). This is supported by the increased percentage of runners' withdrawals when races were contested in very hot weather (Figure 4), and by the significant shift of the race's results through the whole range of performance distribution (Figure 2). The significant effect of air temperature on the median values (Table 2) also suggests that all runners' performances are similarly affected by an increase in air temperature, as seen in Figure 2 showing performances distribution of races in Paris and Chicago with different air temperatures: the significant shift of performance towards the right concerns all runners categories, from the elite to the less trained competitors. In addition the percentage of runner's withdrawals in Chicago 2007 was the

highest (30.74%) among all 60 studied races (Figure 1 and Figure 4). Roberts [26] reported that organisers tried to interrupt the race 3.5 h after the start. This was not successful as most of the finishers crossed the finish line much later (up to 7 h after the start); 66 runners were admitted to the hospital (12 intensive care cases with hydration disorders, heat shock syndromes and 1 death). During the 2004 Boston Marathon (T° = 22.5°C) more than 300 emergency medical calls were observed, consequently the race's start time changed from noon to 10 am in order to decrease heat stress and related casualties [26]. The 2007 London Marathon was hot by London standards (air T° = 19.1°C vs. an average of 11.6°C for the nine other years analysed in our study), 73 hospitalisations were recorded with 6 cases of severe electrolyte imbalance and one death, the total average time (all participants' average) was 17 min slower than usual. In contrast, the number of people treated in London 2008 in cool and rainy conditions (T° = 9.9°C), was 20% lower [26]. Our results showed that the percentage of runners' withdrawals from races significantly increases with increasing temperature (Figure 4). The acceptable upper limit for competition judged by the American College of Sports Medicine (ACSM) is a WBGT of 28°C, but it may not reflect the safety profile of unacclimatized, non-elite marathon runners [3,26–28]. Roberts [26] stated that marathons should not be allowed to start for non-elite racers at a WBGT of 20.5°C. Our results suggest that there is no threshold but a continuous process

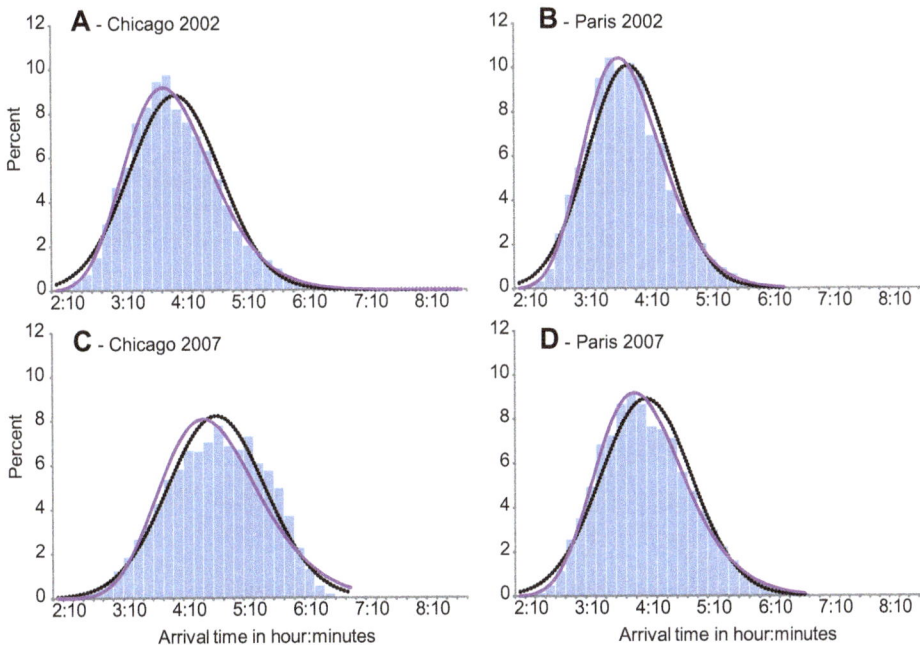

Figure 2. Distribution of performances: example of men's performances distribution for Chicago (in 2002: T°C = 5.4°C; and in 2007: T°C = 25°C); and Paris (in 2002: T°C = 7.6°C; and in 2007: T°C = 17.4°C).

Table 2. Spearman correlations results between all marathons performance levels and environmental parameters: \$ = $p<0.1$; * = $p<0.05$; ** = $p<0.01$; *** = $p<0.001$.

Parameter	Gender	P1	Median	Q1	IQR
Temperature	Women	0.31*	0.30*	0.35**	0.15
	Men	0.48***	0.40***	0.44***	0.25\$
Dew Point	Women	0.14	0.18	0.21	0.01
	Men	0.25\$	0.19	0.20	0.10
Humidity	Women	−0.3*	−0.16	−0.19	−0.21
	Men	−0.34**	−0.28*	−0.32*	−0.19
Atm. Pressure	Women	0.22\$	0.06	0.07	0.06
	Men	0.13	0.04	0.06	0.06
NO2	Women	0.11	0.40**	0.43***	0.33*
	Men	0.25\$	0.38**	0.35**	0.27*
O3	Women	0.01	−0.15	−0.11	−0.20
	Men	−0.05	−0.21	−0.24\$	−0.11
PM10	Women	0.08	0.15	0.25	0.03
	Men	0.10	0.10	0.09	0.16
SO2	Women	0.21	0.13	0.21	0.02
	Men	0.37**	0.20	0.25\$	0.04

P1: first percentile, Q1: first quartile, IQR: Inter Quartile Range.

on both side of an optimum: the larger the gap from the optimal temperature, the lower the tolerance and the higher the risk. In fact, in environments with high heat and humidity, not only is performance potentially compromised, but health is also at risk [29]; both are similarly affected. As soon as WBGT is higher than 13°C the rate of finish line medical encounters and on-course marathon dropouts begin to rise [26] as similarly seen in our study in Figure 4.

Warm weather enhances the risk of exercise induced hyperthermia; its first measurable impact is the reduction of physical performance [4,14,29–31] as it is detrimental for the cardiovascular, muscular and central nervous systems [32,33]. More recent work suggested that central fatigue develops before any elevation

in body temperature occurs: evidence supported that subjects would subconsciously reduce their velocity earlier after the start of an exercise in hot environment, when internal temperatures are still lower than levels associated with bodily harm. Exercise is thus homeostatically regulated by the decrease of exercise intensity (decrease of running performance and heat production) in order to prevent hyperthermia and related catastrophic failures [34,35]. On the other hand, cool weather is associated with an improved ability to maintain running velocity and power output as compared to warmer conditions, but very cold conditions also tend to reduce performance [29,36,37].

Among the studied races' winners, men's marathon world record was beaten in Berlin in 2007 and 2008 (Haile Gebrselassie in 02:03:59), as well as women's marathon world record, beaten in London 2003 (Paula Radcliffe in 02:15:25). The winners' speeds couldn't be affected in the same way than the other runners by air temperature and the other environmental parameters, because top performances can fluctuate from year to year due to numerous factors, such as prize money, race strategies, or overall competition [11]. Another explanation is that, in all of our 60 studied races, 89.5% of male winners were of African origin (57.9% from Kenya; 21.1% from Ethiopia; and 10.5% from Eritrea, Morocco and South Africa); as well as 54.5% of female winners (27.3% from Kenya and 27.3% from Ethiopia- data not shown). African runners might have an advantage over Caucasian athletes, possibly due to a unique combination of the main endurance factors such as maximal oxygen uptake, fractional utilization of VO_{2max} and running economy [38]. They might also perform better in warm environments as they are usually thinner than Caucasian runners (smaller size and body mass index) producing less heat with lower rates of heat storage [38–40]. Psychological factors may also play a role; some hypothesis suggested that regardless of the possible existence of physiological advantages in East African runners, belief that such differences exist may create a background that can have significant positive consequences on performance [41,42].

Genetics and training influence the tolerance for hyperthermia [4,38,43]. Acclimatisation involving repeated exposures to exercise in the heat also results in large improvements in the time to fatigue. Optimal thermoregulatory responses are observed in runners who have been acclimatized to heat and who avoid thirst before and during the race. Their best performances might be less influenced by temperature as winners had been more acclimatized to it [4,29,30,44]. The avoidance of thirst sensation rather than optimum

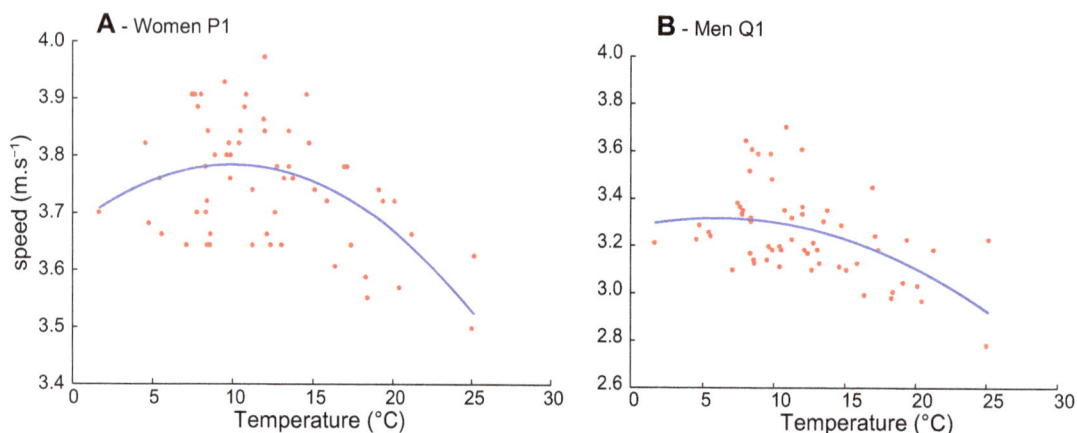

Figure 3. Quadratic second degree polynomial fit for Women's P1 running speeds vs. air temperature, $r^2=0.27$; $p<0.001$; max = 9.9°C. B) Men's Q1 running speeds vs. air temperature, $r^2=0.24$; $p<0.001$; max = 6°C.

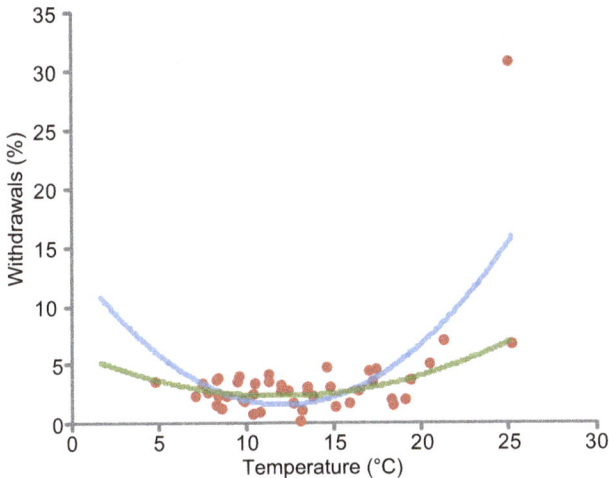

Figure 4. Relationship between air temperature and the percentage of runners' withdrawals, modeled with a quadratic fit (blue curve, $r^2 = 0.36$; $p<0.0001$). The green curve represents the quadratic fit without the maxima (Chicago 2007: 30.74% withdrawals at a race temperature of 25°C).

hydration prevents the decline in running performance [45]; contradicting the idea that dehydration associated with a body weight loss of 2% during an exercise will impair performance, recent studies reported that Haile Gebrselassie lost 10% of his body weight when he established his world record [45–47].

Previous studies suggested that the impact of weather on speed might depend on running ability, with faster runners being less limited than slower ones [6,13,14,29]. This could be attributable to a longer time of exposition to the environmental conditions of slower runners during the race [11]. Also, slower runners tend to run in closer proximity to other runners with clustering formation [48,49], which may cause more heat stress as compared with running solo [50]. These elements, however, are not supported after analyzing the full range of finisher's data; at a population level, temperature causes its full effect whatever the initial capacity. Differences in fitness relative to physiological potential may also contribute to differences in performance times and ability to cope with increasing heat stress [11,48,49].

There was a strong correlation of running speed with air temperature (Figure 3). The maximal average speeds were performed at an optimal temperature comprised between 3.8°C and 9.9°C depending on the performance level (Table S3); small increases in air temperatures caused marathon performances to decline in a predictable and quantifiable manner. On the other hand, large decreases in air temperatures under the optimum also reduce performances. These optimal temperatures found in the present study are comprised in the optimal temperature range of 5–10°C WBGT found in previous studies [14]; other studies stated that a weather of 10–12°C WBGT is the norm for fast field performance and reported a decrease of performance with increasing WBGT [12,27,51,52]. Best marathon times and most marathon world records were achieved in cool environmental temperatures (10–15°C) and have been run in the early morning during spring and fall [12]. Analysing Gebrselassie's performances in Berlin reveals that they follow the same trend, with both World Records obtained at the lowest temperatures (14°C in 2007 and 13°C in 2008, vs. 18°C in 2009 and 22°C in 2006 when he also won these two races without beating the world record).

The relationship between running speed and air temperature defined in our study (Figure 3) is similar to the relationship found

between mortality and air temperature (asymmetrical U-like pattern) in France defined by Laaidi et al [53], where mortality rates increase with the lowest and the highest temperatures. A "thermal optimum" occurs in between, where mortality rates are minimal [53]. The great influence that temperature has on performance is comparable to the influence it has on mortality, suggesting that both sports performance and mortality are thermodynamically regulated. This also emphasizes the utility of prevention programs, the assessment of public health impacts and acclimatization before participating in hot marathons [53]. Similar correlations were also found between temperature and swimming performance in juvenile southern catfish [22], and between increases in summer water temperature and elevated mortality rates of adult sockeye salmon [23]; suggesting that physiological adaptations to temperature, similarly occur in various taxons, but vary within specific limits that depend on species and will modify performances.

Air pollution and performance

The measured levels of pollution had no impact on performance, except for ozone (Table S2) and NO_2 (Table 2). Assessing the effect of any single air pollutant separately is not simple; it is not isolated in the inhaled air, but rather combined with other parameters. Therefore any possible influence might probably be due to a combination of components. In addition most marathons are held on Sunday mornings, when urban transport activity and its associated emissions are low, and photochemical reactions driven by solar radiation have not yet produced secondary pollutants such as ozone [9]. This is the most probable explanation to our results, confirming previous studies. Among the air pollutants analysed in the present study, ozone and NO_2 had the greatest effect on decreasing marathon performances (Table S2). Ozone concentrations on the ground increase linearly with air temperature [7,8,10]; thus the effect of ozone in our study may be mainly associated with the temperature effect, as seen in Berlin and Chicago. However ozone and other pollutants effects are known to be detrimental to exercise performance only when exposure is sufficiently high. Many studies showed no effect of air pollutants on sports performance [9]. Some of them showed that $PM_{2.5}$ and aerosol acidity were associated with acute decrements in pulmonary function, but these changes in pulmonary function were unlikely to result in clinical symptoms [54]. Others showed that chronic exposure to mixed pollutants during exercise may result in decreased lung function, or vascular dysfunction, and may compromise performance [55]. During the marathons studied here, concentrations of air pollutants never exceeded the limits set forth by national environmental agencies (US Environmental Protection Agency- EPA; AirParif; European Environmental Agency- EEA) or the levels known to alter lung function in laboratory situations [9].

Conclusions

Air temperature is the most important factor influencing marathon running performance for runners of all levels. It greatly influences the entire distribution of runners' performances as well as the percentage of withdrawals. Running speed at all levels is linked to temperature through a quadratic model. Any increase or decrease from the optimal temperature range will result in running speed decrease. Ozone also has an influence on performance but its effect might be linked to the temperature impact. The model developed in this study could be used for further predictions, in order to evaluate expected performance variations with changing weather conditions.

Supporting Information

Table S1   Time values of different descriptive statistics and their variability by marathon and gender. [1] Value of the described statistic for all performances of all year together, hour:min:sec [2] Standard deviation of the described statistic for all performances of each year, hour:min:sec [3] IQR: Inter Quartile Range.

Table S2   Spearman correlations results between each marathon performance levels and environmental parameters: $\$ = p<0.1$; $* = p<0.05$; $** = p<0.01$; $*** = p<0.001$. **P1: first percentile, Q1: first quartile, IQR: Inter Quartile Range.**

Table S3   Optimal temperatures for maximal running speeds of each level of performance, with speed losses associated with each temperature increase.

Acknowledgments

We thank the Centre National de Développement du Sport and the Ministry of Health, Youth and Sport. We thank INSEP teams for their full support. We thank Mrs Karine Schaal for carefully reviewing the manuscript.

Author Contributions

Conceived and designed the experiments: JFT NEH GB AM. Analyzed the data: JT GB AM NEH MG MT. Wrote the paper: NEH GB JFT. Reviewed the paper: CH JFT.

References

1. Lippi G, Favaloro EJ, Guidi GC (2008) The genetic basis of human athletic performance. Why are psychological components so often overlooked? J Physiol 586(Pt 12):3017; author reply 3019–3020.
2. Macarthur DG, North KN (2005) Genes and human elite athletic performance. Hum Genet 116(5): 331–339.
3. Cheuvront SN, Haymes EM (2001) Thermoregulation and marathon running, biological and environmental influences. Sports Med 31(10): 743–762.
4. Kenefick RW, Cheuvront SN, Sawka MN (2007) Thermoregulatory function during the marathon. Sports Med 37(4–5): 312–315.
5. Weather Underground website. Internet weather service. Weather data from each marathon race. Available: http://www.wunderground.com/history/. Accessed 2011 Mar, 30.
6. Vihma T (2010) Effects of weather on the performance of marathon runners. Int J Biometeorol 54(3): 297–306.
7. Shephard RJ (1984) Athletic performance and urban air pollution. Can Med Assoc J 131(2): 105–109.
8. Chimenti L, Morici G, Paterno A, Bonanno A, Vultaggio M, et al. (2009) Environmental conditions, air pollutants, and airway cells in runners: A longitudinal field study. J Sports Sci 27(9): 925–935.
9. Marr LC, Ely MR (2010) Effect of air pollution on marathon running performance. Med Sci Sports Exerc 42(3): 585–591.
10. Lippi G, Guidi GC, Maffulli N (2008) Air pollution and sports performance in Beijing. Int J Sports Med 29: 696–698.
11. Ely MR, Cheuvront SN, Roberts WO, Montain SJ (2007) Impact of weather on marathon-running performance. Med Sci Sports Exerc 39(3): 487–493.
12. Ely MR, Cheuvront SN, Montain SJ (2007) Neither cloud cover nor low solar loads are associated with fast marathon performance. Med Sci Sports Exerc 39(11): 2029–2035.
13. Ely MR, Martin DE, Cheuvront SN, Montain SJ (2008) Effect of ambient temperature on marathon pacing is dependent on runner ability. Med Sci Sports Exerc 40(9): 1675–1680.
14. Montain SJ, Ely MR, Cheuvront SN (2007) Marathon performance in thermally stressing conditions. Sports Med 37(4–5): 320–323.
15. Martin DE, Buoncristiani JF (1999) The effects of temperature on marathon runners' performance. Chance 12(4): 20–24.
16. Trapasso LM, Cooper JD (1989) Record performances at the Boston Marathon: biometeorological factors. Int J Biometeorol 33(4): 233–237.
17. Online worldwide athletic results database website. Marathons races results. Available: http://www.athlinks.com. Accessed 2011 Apr, 30.
18. AirParif website. Air pollution data for Paris retrieved (March – May 2009). Available: http://www.airparif.com.
19. Station Database of the Environmental Agency. Air pollution data for Berlin retrieved (June 24, 2009). Available: http://www.env-it.de/stationen/public/language.do;jsessionid = FB278996EE26B0351076A5D974C8BD04?language = en.
20. LondonAir website. Air pollution data for London retrieved (May 26, 2009). Available: http://www.londonair.org.uk/london/asp/default.asp.
21. Berthelot G, Tafflet M, El Helou N, Len S, Escolano S, et al. (2010) Athlete atypicity on the edge of human achievement: Performances stagnate after the last peak, in 1988. PLoS ONE, 5 (1), e8800: DOI: 10.1371/journal.pone.0008800.
22. Zeng LQ, Cao ZD, Fu SJ, Peng JL, Wang YX (2009) Effect of temperature on swimming performance in juvenile southern catfish (Silurus meridionalis). Comp Biochem Physiol A Mol Integr Physiol 153(2): 125–130.
23. Eliason EJ, Clark TD, Hague MJ, Hanson LM, Gallagher ZS, et al. (2011) Differences in thermal tolerance among sockeye salmon populations. Science 332(6025): 109–112.
24. Kirschbaum MUF, Watt MS (2011) Use of a process-based model to describe spatial variation in Pinus radiate productivity in New Zealand. Forest Ecology and Management 262: 1008–1019.
25. Thibault V, Guillaume M, Berthelot G, El Helou N, Schaal K, et al. (2010) Women and men in sport performance: the gender gap has not evolved since 1983. J Sports Sci Med 9: 214–223.
26. Roberts WO (2010) Determining a "do not start" temperature for a marathon on the basis of adverse outcomes. Med Sci Sports Exerc 42(2): 226–232.
27. Zhang S, Meng G, Wang Y, Li J (1992) Study of the relationships between weather conditions and the marathon race, and of meteorotropic effects on distance runners. Int J Biometeorol 36: 63–68.
28. Armstrong LE, Epstein Y, Greenleaf JE, Haymes EM, Hubbard RW, et al. (1996) American College of Sports Medicine position stand. Heat and cold illnesses during distance running. Med Sci Sports Exerc 28(12): i– x.
29. Maughan RJ, Watson P, Shirreffs SM (2007) Heat and cold, what does the environment do to the marathon runner? Sports Med 37(4–5): 396–399.
30. Hargreaves M (2008) Physiological limits to exercise performance in the heat. J Sci Med Sport 11(1): 66–71.
31. Walters TJ, Ryan KL, Tate LM, Mason PA (2000) Exercise in the heat is limited by a critical internal temperature. J Appl Physiol 89: 799–806.
32. Coyle EF (2007) Physiological regulation of marathon performance. Sports Med 37(4-5): 306–311.
33. González-Alonso J (2007) Hyperthermia Impairs Brain, Heart and Muscle Function in Exercising Humans. Sports Med 37(4–5): 371–373.
34. Tucker R, Rauch L, Harley YX, Noakes TD (2004) Impaired exercise performance in the heat is associated with an anticipatory reduction in skeletal muscle recruitment. Pflugers Arch 448(4): 422–430.
35. Tucker R, Marle T, Lambert EV, Noakes TD (2006) The rate of heat storage mediates an anticipatory reduction in exercise intensity during cycling at a fixed rating of perceived exertion. J Physiol 574(Pt 3): 905–915.
36. Nimmo M (2004) Exercise in the cold. J Sports Sci 22: 898–915.
37. Weller AS, Millard CE, Stroud MA, Greenhaff PL, Macdonald IA (1997) Physiological responses to a cold, wet, and windy environment during prolonged intermittent walking. Am J Physiol 272(1 Pt 2): R226–R233.
38. Larsen HB (2003) Kenyan dominance in distance running. Comp Biochem Physiol A Mol Integr Physiol 136(1): 161–170.
39. Marino FE, Lambert MI, Noakes TD (2004) Superior performance of African runners in warm humid but not in cool environmental conditions. J Appl Physiol 96: 124–130.
40. Marino FE, Mbambo Z, Kortekaas E, Wilson G, Lambert MI, et al. (2000) Advantages of smaller body mass during distance running in warm, humid environments. Pflügers Arch 441(2–3): 359–367.
41. Hamilton B (2000) East African running dominance: what is behind it? Br J Sports Med 34(5): 391–394.
42. Baker J, Horton S (2003) East African running dominance revisited: a role for stereotype threat? Br J Sports Med 37(6): 553–555.
43. Sawka MN, Young A (2006) Physiological systems and their responses to conditions of heat and cold. In: Tipton CM, ed. American College of Sports Medicine's Advanced exercise physiology. Philadelphia (PA): Lippincott Williams and Wilkins. 535–563 p.
44. Zouhal H, Groussard C, Vincent S, Jacob C, Abderrahman AB, et al. (2009) Athletic performance and weight changes during the "Marathon of Sands" in athletes well-trained in endurance. Int J Sports Med 30: 516–521.
45. Goulet ED (2011) Effect of exercise-induced dehydration on time-trial exercise performance: a meta-analysis. Br J Sports Med 45(14): 1149–1156.
46. Zouhal H, Groussard C, Minter G, Vincent S, Cretual A, et al. (2011) Inverse relationship between percentage body weight change and finishing time in 643 forty-two-kilometre marathon runners. Br J Sports Med 45(14): 1101–1105.
47. Beis LY, Wright-Whyte M, Fudge B, Noakes T, Pitsiladis YP (2012) Drinking Behaviors of Elite Male Runners During Marathon Competition. Clin J Sport Med. March [Epub ahead of print] doi: 10.1097/JSM.0b013e31824a55d7.
48. Alvarez-Ramirez J, Rodriguez E (2006) Scaling properties of marathon races. Physica A: Stat Mech Appl 365(2): 509–520.

49. Alvarez-Ramirez J, Rodriguez E, Dagduga L (2007) Time-correlations in marathon arrival sequences. Physica A: Stat Mech Appl 380: 447–454.
50. Dawson NJ, De Freitas CR, Mackey WJ, Young AA (1987) The stressful microclimate created by massed fun runners. Transactions of the Menzies Foundation 14: 41–44.
51. Galloway SDR, Maughan RJ (1997) Effects of ambient temperature on the capacity to perform prolonged cycle exercise in man. Med Sci Sports Exerc 29(9): 1240–2149.
52. Buoncristiani JF, Martin DE (1983) Factors affecting runners' marathon performance. Chance 6(4): 24–30.
53. Laaidi M, Laaidi K, Besancenot JP (2006) Temperature-related mortality in France, a comparison between regions with different climates from the perspective of global warming. Int J Biometeorol 51(2): 145–153.
54. Korrick SA, Neas LM, Dockery DW, Gold DR, Allen GA, et al. (1998) Effects of ozone and other pollutants on the pulmonary function of adult hickers. Environ Health Perspect 106: 93–99.
55. Rundell KW (2012) Effect of air pollution on athlete health and performance. Br J Sports Med. Epub ahead of print.

9

Early Life Ozone Exposure Results in Dysregulated Innate Immune Function and Altered microRNA Expression in Airway Epithelium

Candice C. Clay[1,9], Kinjal Maniar-Hew[1,9], Joan E. Gerriets[1], Theodore T. Wang[1], Edward M. Postlethwait[3], Michael J. Evans[1,2], Justin H. Fontaine[1], Lisa A. Miller[1,2]*

1 California National Primate Research Center, University of California Davis, Davis, California, United States of America, 2 Department of Anatomy, Physiology, and Cell Biology, School of Veterinary Medicine, University of California Davis, Davis, California, United States of America, 3 Department of Environmental Health Sciences, School of Public Health, University of Alabama, Birmingham, Alabama, United States of America

Abstract

Exposure to ozone has been associated with increased incidence of respiratory morbidity in humans; however the mechanism(s) behind the enhancement of susceptibility are unclear. We have previously reported that exposure to episodic ozone during postnatal development results in an attenuated peripheral blood cytokine response to lipopolysaccharide (LPS) that persists with maturity. As the lung is closely interfaced with the external environment, we hypothesized that the conducting airway epithelium of neonates may also be a target of immunomodulation by ozone. To test this hypothesis, we evaluated primary airway epithelial cell cultures derived from juvenile rhesus macaque monkeys with a prior history of episodic postnatal ozone exposure. Innate immune function was measured by expression of the proinflammatory cytokines IL-6 and IL-8 in primary cultures established following in vivo LPS challenge or, in response to in vitro LPS treatment. Postnatal ozone exposure resulted in significantly attenuated IL-6 mRNA and protein expression in primary cultures from juvenile animals; IL-8 mRNA was also significantly reduced. The effect of antecedent ozone exposure was modulated by in vivo LPS challenge, as primary cultures exhibited enhanced cytokine expression upon secondary in vitro LPS treatment. Assessment of potential IL-6-targeting microRNAs miR-149, miR-202, and miR-410 showed differential expression in primary cultures based upon animal exposure history. Functional assays revealed that miR-149 is capable of binding to the IL-6 3' UTR and decreasing IL-6 protein synthesis in airway epithelial cell lines. Cumulatively, our findings suggest that episodic ozone during early life contributes to the molecular programming of airway epithelium, such that memory from prior exposures is retained in the form of a dysregulated IL-6 and IL-8 response to LPS; differentially expressed microRNAs such as miR-149 may play a role in the persistent modulation of the epithelial innate immune response towards microbes in the mature lung.

Editor: Vladimir V. Kalinichenko, Cincinnati Children's Hospital Medical Center, United States of America

Funding: This work was supported by NIH ES011617, NIH ES000628, NIH HL081286, NIH HL097087, NIH OD011107, EPA STAR Grant 832947 and NIH T32 HL007013. The funders had no role in study design, data collection and analysis, decision to publish, or preparation of the manuscript.

Competing Interests: The authors have declared that no competing interests exist.

* E-mail: lmiller@ucdavis.edu

9 These authors contributed equally to this work.

Introduction

Ozone is a common inhaled air pollutant that is known to negatively impact respiratory health and contribute towards increased mortality in humans [1,2,3]. Epidemiologic studies have demonstrated a clear association between episodes of high ozone exposure and increased hospitalizations for respiratory illnesses such as exacerbations of asthma [4,5]. The interaction between ozone exposure and the pulmonary immune system is complex; both suppressive and stimulatory effects have been described [6,7]. Most experimental evidence supports a role for ozone in the enhancement of susceptibility to respiratory infections [8,9], with numerous rodent models exhibiting impaired pulmonary microbial clearance upon exposure [10,11,12]. The mechanisms by which ozone exposure compromises host defense in the lung are not well understood. Reduced bacterial clearance is primarily attributed to ozone-mediated impairment of macrophage phagocytosis [10], although modulation of adaptive immune responses following exposure in humans has also been reported [13,14].

Young children, with their immature mucosal immune system and limited host defense capacity, may be more sensitive to the immunomodulatory effects of ozone exposure [15,16]. In addition, several physiological parameters including differences in breathing patterns, ventilation rates and lung surface area per unit body weight can result in enhanced ozone exposure in children compared to adults [17]. The first year of life represents a highly dynamic phase for both the respiratory and mucosal immune systems. Many consider this a "window of susceptibility" for modulation by the environment [18,19], as postnatal irritant or toxicant exposure has the potential to permanently affect the growth trajectory and function of the respiratory system [20,21]. Given the challenges and ethical concerns involved in studying pediatric populations, our knowledge of infant pulmonary immunity is largely restricted to studies of neonatal laboratory animals. Because rodents and humans exhibit substantial differences in the postnatal maturation of both pulmonary and immune systems, it is critical to address the impact of environmental exposures on development of respiratory tract immunity in a primate species.

We have previously shown that ozone exposure of rhesus monkeys during the postnatal period of development results in altered immune cell composition in both peripheral blood and bronchoalveolar lavage, with significantly increased monocytes in both compartments that persisted with maturity [20]. Despite the higher monocyte frequency in animals with a history of ozone exposure, the peripheral blood and airway inflammatory response to inhaled lipopolysaccharide (LPS) was significantly attenuated. Furthermore, we demonstrated that the lasting effects of ozone were retained in the peripheral blood compartment, as *ex vivo* LPS treatment of peripheral blood mononuclear cells collected 6 months after ozone exposure showed significantly reduced proinflammatory cytokine responses. In this current study, we have focused on the long term impact of postnatal ozone exposure on the airway epithelium, using our previously described rhesus monkey model.

The epithelial cell of the conducting airways may play a significant role in the development of exposure-related effects, as it is architecturally and functionally poised to serve as a liaison between the external environment and the immune system. Acute inhalation of ozone has been shown to result in direct airway epithelial damage, with loss of ciliary function, enhanced permeability and impaired mucociliary clearance [7]. While the deleterious outcomes of ozone on airway epithelium have been documented, reports of ozone effects on epithelial cell immune responses are variable [22,23,24] and establishment of a persistent immunomodulatory phenotype in association with exposure has not been investigated. There is growing evidence to suggest that the respiratory tract develops a subtly unique inflammatory profile that is shaped by our prior exposures [25]. Although immunological memory is often attributed to the adaptive immune response, molecular programming of epithelial cell genes that likely impact innate immunity has been reported in asthmatics and cigarette smokers [26,27]. Epigenetics and transcriptional regulation by microRNAs may, in part, explain the causal role of environmental exposures in alteration of epithelial cell phenotypes [28].

Given that the molecular profile of the airway epithelial cell may vary in association with chronic lung disease, we hypothesized that postnatal ozone exposure intrinsically alters the ability of airway epithelial cells to respond to a microbial challenge, and that this may enhance susceptibility to respiratory infection with maturity. In the present study, we focused on the cellular response to LPS, as this molecule is ubiquitous in the environment and is recognized by the pathogen pattern receptor Toll-like receptor 4 (TLR4). We tested our hypothesis using primary airway epithelial cell cultures derived from juvenile monkeys that were exposed to episodic ozone during the first six months of life, followed by filtered air housing until one year of age. Both *in vivo* and *in vitro* approaches were used to address the question of whether postnatal air pollutant exposures persistently alter the ability of airway epithelial cells to synthesize proinflammatory cytokines in response to exogenous LPS.

Materials and Methods

Ethics Statement

All animal procedures were approved by the University of California, Davis, Institutional Animal Care and Use Committee (Protocol #06-12245). Care and housing of animals before, during, and after treatment complied with the provisions of the Institute of Laboratory Animal Resources and conforms to practices established by the Association for Assessment and Accreditation of Laboratory Animal Care International (AAALAC International).

Animal Exposure

Male rhesus macaque (*Macaca mulatta*) monkeys born at the California National Primate Research Center (CNPRC) were randomized into exposure groups based on birth order. Power projections at alpha level = 0.05 with POWERLIB202 software (University of North Carolina Chapel Hill) were used to calculate the minimum number of animals necessary in each group to provide statistically detectable differences based on antioxidant levels obtained in a pilot study. Animals were procedure naïve and weighed an average of 0.55 kg +/− 0.02 at the initiation of the study. Paired monkeys were housed in stainless steel cages with wire mesh bottoms within the CNPRC exposure chamber facility under high efficiency particulate air (HEPA) filtered conditions starting at 1–2 days following birth. Feeding and enrichment were provided according to the CNPRC standard operating procedure for nursery-reared infants including hand-feedings of human infant formula every 2 h until 5 weeks of age, when hanging bottles were provided. Water was available *ad libitum* and solid foods including monkey chow and fruit were provided twice per day starting at 2 weeks of age. For enrichment, a variety of swings, hanging perches and age-appropriate toys were offered. Temperature and humidity ranges were controlled along with 12 h light and dark cycles.

Details regarding the ozone exposure regimen for animals used in this study have been previously described [20]. In brief, animals (n = 8) received 11 successive cycles of ozone exposure from 1 month to 6 months of age (Figure 1). Animals were treated as an entire group within CNPRC exposure chambers. Each cycle consisted of 5 days of ozone (0.5 parts per million at 8 h/day) followed by 9 days of filtered air. Ozone was generated as previously described [20,29,30]. Oximetry and heart rate were monitored by CNPRC veterinary staff during exposure procedures, with no adverse effects reported. At completion of the ozone exposure regimen, all animals remained in filtered air housing for a period of 6 months. Control animals were nursery-raised indoors under HEPA filtered conditions until 1 year of age (n = 9). CNPRC nursery and inhalation and exposure staff monitored the well-being of all study animals with frequent health observations each day including assessment of appetite and body weight. All care is under the direct guidance of CNPRC veterinarians who set the criteria for modifying and/or terminating procedures to ameliorate suffering.

At 1 year of age, a subset of ozone or filtered air control juvenile animals received a single *in vivo* LPS aerosol challenge (filtered air n = 4, ozone n = 4). A dose of 25,000 endotoxin units in PBS (*E. coli* O26:B6; Sigma- Aldrich, St. Louis, MO) was administered via face mask in the morning ~24 h prior to necropsy. The same commercial lot of LPS was used for all animals and cultures in this study. For sample collection prior to necropsy, animals were anesthetized with an intramuscular dose of ketamine (10 mg/kg); the dose was adjusted as deemed necessary by attending the veterinarian. Animals were euthanized by intravenous overdose of sodium pentobarbitol IV (60 mg/kg). Lung tissue specimens collected at necropsy were immediately processed for primary cell culture and histology (see subsequent sections). These experiments were conducted over a 3 year period with filtered air controls

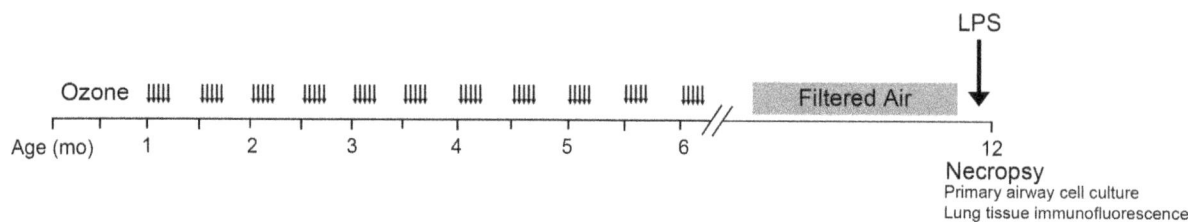

Figure 1. Experimental timeline for postnatal episodic ozone exposure and *in vivo* LPS challenge in juvenile monkeys. Starting at 1 month of age, infant rhesus monkeys (n = 8) were exposed to 11 cycles of ozone. Each cycle consisted of ozone exposure for 5 days (0.5 parts per million at 8 h/day for 6 months (small arrows)) followed by 9 days of filtered air. At the end of 11 exposure cycles, animals were housed in filtered air until 1 year of age. Control animals (n = 9) were housed in filtered air until 1 year of age. An aerosolized LPS challenge via face mask was given to the ozone-exposed and the filtered air-control animals at 1 year of age (n = 4 each group; marked by the large arrow in figure). All animals were necropsied within 24 h of *in vivo* LPS; lung tissues were collected for cell culture and histology.

completed in years 1 and 2, ozone exposure groups in year 2 and LPS groups in year 3.

Primary Airway Epithelial Cell Culture

Airway epithelial cells were isolated from trachea and large bronchi collected at necropsy using protease digestion as described in [30]. Isolated epithelial cells were plated at a density of 4×10^5 cells on 6.5 mm, 0.4 μm pore size transwell clear polyester membrane inserts (Corning, Corning, NY) coated with FNC coating mix (Athena Enzyme Systems, Baltimore, MD). Cells were first expanded in growth media (BEGM, Lonza, Walkersville, MD) ed with retinoic acid (50 nM Sigma-Aldrich Corp., St. Louis, MO) and upon confluency (approximately 2 weeks), cultured under air-liquid interface conditions for an additional seven days to achieve differentiation according to previously established methods [31]. Polarization of cultures was confirmed by measurement of transepithelial resistance using an STX2 Electrode (World Precision Instruments Inc., Sarasota, FL). Prior to *in vitro* LPS treatment, cultures underwent a complete media change, followed by addition of LPS in media supplemented with 0.5% fetal bovine serum (Invitrogen, Carlsbad, CA) on the apical surface (100 μL/well). After a one hour incubation with LPS, cultures were subsequently washed and incubated for 6 or 24 h with fresh culture media added to the apical surface. We have previously shown that this dose and LPS treatment protocol activates the toll-like receptor signaling pathway in airway epithelial cells as early as 3 h post-treatment and triggers IL-6 as well as IL-8 responses [30].

RNA Isolation and Analysis

RNA and microRNA were isolated from primary airway epithelial cell cultures using Trizol reagent (Invitrogen). cDNA was generated from total RNA using random hexamer primers and MultiScribe Reverse Transcriptase (Applied Biosystems, Foster City, CA). IL-6, IL-8, TLR4 and GAPDH mRNA was measured by Taqman Real-time PCR methods, using rhesus-specific primer-probe assays that map with 100% identity to the human homologue (Applied Biosystems) and detected using an Applied Biosystems PRISM 7900 Sequence Detection System. Progressive dilution of corresponding purified human cDNA plasmid constructs (Origene, Rockville, MD) for each gene was used to generate standard curves and allow absolute quantitation of copy number. Gene copy numbers are reported relative to GAPDH values.

Putative microRNAs (miR) that target the IL-6 mRNA 3'UTR were identified by the TargetScan 5.1 program which is based on previously-defined algorithms [32,33]. miR-149, miR-202 and miR-410 were selected based on conservation within the

mammalian species and a low total context score (< -0.1) for human IL-6 3' UTR (Supporting Data Figure 4). miR-let-7a was also included in the analysis as it is a known functional modulator of IL-6 expression [34]. The Taqman MicroRNA Reverse Transcription Kit was used for reverse transcription of hsa-miR-let7a, hsa-miR-149-5p, hsa-miR-202-3p and hsa-miR-410 with target-specific stem loop structure primers (Applied Biosystems). The small nuclear RNA, RNU-6b was amplified as an endogenous control. Analyses using specific microRNA probe sets and amplification reagents were performed on an ABI 7900 (Applied Biosystems). Results for miR data analyses were calculated by use of the deltaCT method, where deltaCT values = CT value of target miR - CT value of endogenous control RNU-6b.

Immunofluorescence Staining

Trachea specimens from necropsy were embedded in O.C.T. compound (Sakura Finetek, Torrance, CA) and immunostained as previously described [30] with the following modifications: sections were fixed in acetone, blocked in bovine serum albumin (Vector Laboratories, Burlingame, CA) and stained with a FITC-conjugated mouse anti-human IL-6 (clone MQ2-13A5; eBiosciences, San Diego, CA). A FITC-conjugated mouse IgG1 was used as an isotype control (eBiosciences). Both control and IL-6 antibodies were used at a concentration of 10 μg/ml with an overnight incubation.

Analysis of microRNA Function

The immortalized human bronchial airway epithelial cell line BEAS-2B S.6 [35] was transfected with 30 nM *mir*Vana mimics of hsa-miR-149-5p, hsa-miR-202-3p, hsa-miR-410 and the miR#1 negative control (lacking any mammalian target) using lipofectamine (Applied Biosystems). After 24 h, transfected cells were evaluated for microRNAs, IL-6 mRNA and IL-6 protein levels. For microRNA binding analysis, the papilloma virus–immortalized human bronchial epithelial cell line HBE [36] was transfected with a 3' UTR IL-6 gene target cloned upstream of a Firefly luciferase reporter in a dual luciferase reporter system (GeneCopoeia, Rockville, MD) using lipofectamine. HBE cells were co-transfected with the IL-6 3' UTR clone and 40 nM microRNA mimics, followed by evaluation of relative luciferase activity after 24 h. For transfection-normalization across samples, the firefly luciferase is reported relative to an internal control, consisting of renilla luciferase driven by a CMV promoter.

Cytokine ELISA

IL-6 and IL-8 protein concentration in apical supernatants collected from primary airway epithelial cell cultures were measured by ELISA Ready-SET-Go! Kits (eBioscience, San Diego, CA). The limit of detection for ELISA assays was 2 pg/mL (IL-6), and 4 pg/mL (IL-8).

Statistics

All data are reported as mean $+/-$ standard error (SE). Treatment and exposure differences were evaluated on log-transformed values using two-way ANOVA or Student's t-test when appropriate with GraphPad Prism software (GraphPad 5.0, La Jolla, CA). A p value of 0.05 or less was considered statistically significant.

Results

The Cytokine Response to LPS in Juvenile Airway Epithelial Cells is Altered by Postnatal Episodic Ozone Exposure

To determine if chronic ozone exposure during postnatal development results in persistent changes to innate immune function of airway epithelium, we established primary airway epithelial cell cultures from juvenile rhesus monkeys exposed as infants to episodic ozone. In addition to the effects of ozone, a subset of animals was also challenged *in vivo* with a single dose of aerosolized LPS (Figure 1). Primary cultures were evaluated for cytokine synthesis following *in vitro* treatment with LPS, focusing on IL-6 and IL-8 as measures of prototypic proinflammatory responses. A limited portion of the results from the filtered air control and filtered air + *in vivo* LPS animal groups has been previously reported as part of a manuscript examining the role of chronologic age on epithelial cell innate immune function [30]; the data are included here for exposure-related comparisons.

Airway epithelial cells cultured from animals with a history of postnatal ozone exposure exhibited altered synthesis of IL-6 in response to *in vitro* LPS treatment. Significantly reduced IL-6 mRNA ($p = 0.0002$) and protein ($p = 0.02$) levels were observed 24 h post- *in vitro* LPS treatment in airway epithelial cell cultures with a history of ozone exposure as compared to epithelial cells derived from filtered air control animals (Figure 2A–B; two-way ANOVA for *in vivo* exposure and *in vitro* LPS concentration). In epithelial cells from ozone-exposed animals that received a single *in vivo* LPS challenge (ozone + *in vivo* LPS), IL-6 mRNA levels were significantly greater ($p = 0.01$) than their corresponding filtered air + *in vivo* LPS controls and ozone-only animals (Figure 2C; two-way ANOVA for *in vivo* exposure and *in vitro* LPS concentration). The IL-6 protein response to *in vitro* LPS stimulation was not significantly different in the exposure groups that received the *in vivo* LPS challenge (Figure 2D), but there was a trend towards increased IL-6 protein relative to the ozone alone group (p = 0.055 by two-way ANOVA for *in vivo* exposure and *in vitro* LPS concentration). Significant exposure-dependent differences were also observed in the IL-6 mRNA, but not protein response at 6 h post- *in vitro* LPS treatment (Figure S1; $p = 0.01$ by two-way ANOVA for *in vivo* exposure and *in vitro* LPS concentration). Although exposure did not significantly affect IL-6 protein levels at the 6 h time point, *in vitro* LPS concentration was a significant source of variation in filtered air and ozone cultures (Figure S1B; $p = 0.04$).

The IL-8 response to *in vitro* LPS treatment was similar to IL-6, in that the epithelial cells from the ozone-only exposed animals exhibited attenuated IL-8 mRNA synthesis ($p = 0.02$) and cultures established from ozone + *in vivo* LPS animals showed the greatest IL-8 mRNA induction ($p = 0.002$) (Figure 3A, C; two-way ANOVA for *in vivo* exposure and *in vitro* LPS concentration). The IL-8 protein response was not significantly affected by exposure, although there was a trend towards increased IL-8 secretion in the ozone + *in vivo* LPS group relative to filtered air + *in vivo* LPS (Figure 3 B, D, p = 0.085 two-way ANOVA for *in vivo* exposure and *in vitro* LPS concentration). Similar exposure-dependent differences were observed in the IL-8 mRNA but not protein response at 6 h post- *in vitro* LPS treatment (Figure S2).

To determine if cytokine expression measured in primary airway epithelial cell cultures was reflective of airway epithelium *in vivo*, we conducted immunostaining of trachea cryosections obtained from the same animals used as the source of primary cell cultures (Figure 4). Focusing on IL-6, we detected positive immunofluorescence in animals that received *in vivo* LPS, including filtered air controls and those with a history of postnatal ozone exposure (Figure 4F, H). Little to no IL-6 staining was observed in animals that did not undergo *in vivo* LPS challenge (Figure 4B, D).

TLR4 recognizes LPS and mediates the subsequent cytokine response [37]. TLR4 has also been identified as an essential susceptibility gene for inflammatory and physiologic effects of ozone exposure in certain mouse strains [38,39,40]. To determine if our findings relative to postnatal ozone exposure were associated with altered TLR4 expression in airway epithelium, we measured TLR4 copy number in primary cultures and found no significant exposure-related differences (Figure S3).

Effect of Postnatal Ozone on microRNA Targets for IL-6 3' UTR

MicroRNAs are small (21–23 nucleotides) noncoding RNAs that regulate gene expression by affecting mRNA stability and protein synthesis through post-transcriptional modifications [41]. Recent expression profiling studies show that environmental pollutants can change the expression of microRNAs in the lung [42,43]. Given the discordant IL-6 mRNA and protein responses observed in airway epithelial cell cultures established from animals with a history of postnatal ozone and/or LPS exposure, we sought to determine if microRNAs that putatively target IL-6 were differentially expressed in association with antecedent ozone or LPS. miR-149, miR-202, and miR-410 were identified using TargetScan 5.1 (Figure S4), and miR-let7a expression was also measured as it is a known functional modulator of IL-6 expression [34].

Most of the evaluated microRNAs showed significant ozone/ *in vivo* LPS exposure- or *in vitro* LPS treatment-dependent expression differences in primary airway epithelial cell cultures (Figure 5). In untreated cultures (0 μg/ml), miR-149 showed the most significant difference in expression levels based upon prior exposure history, with epithelial cells from ozone + *in vivo* LPS animals having significantly reduced miR-149 as compared to all other exposure groups (Figure 5A, $p = 0.02$; two-way ANOVA for *in vivo* exposure and *in vitro* LPS treatment). *In vitro* LPS treatment (1 μg/ml) was also a significant source of variation for miR-149 expression ($p < 0.0001$). At 6 h post-LPS treatment *in vitro*, miR-149 expression significantly increased above baseline in epithelial cells from all exposure groups except filtered air controls that received an *in vivo* LPS challenge. Similar to miR-149, miR-202 expression at baseline was lowest in the ozone + *in vivo* LPS group, with exposure-dependent effects that were near statistical significance (Figure 5B; two-way ANOVA for *in vivo* exposure and *in vitro* LPS treatment; $p = 0.06$ for exposure and $p = 0.04$ for LPS). miR-202 decreased with *in vitro* LPS treatment compared to baseline values except in cultures with a history of ozone exposure ($p = 0.06$ for ozone alone, $p = 0.002$ for ozone + *in vivo* LPS). For miR-410, epithelial cell cultures derived from ozone-only animals showed the lowest constitutive expression levels but these were significantly increased with *in vitro* LPS treatment (Figure 5C; $p = 0.05$). In

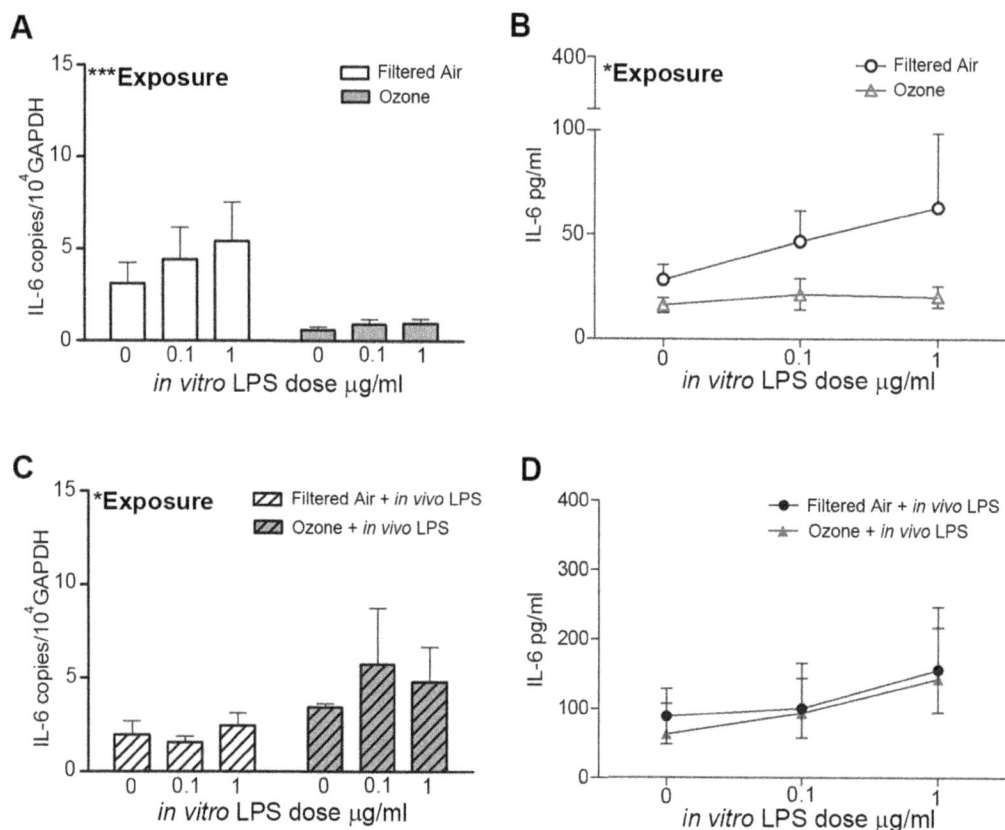

Figure 2. Effect of postnatal ozone exposure on IL-6 expression in juvenile monkey airway epithelial cell cultures following LPS treatment. Airway epithelial cells harvested from one-year-old juvenile rhesus monkeys with prior ozone and/or LPS exposure, as described in Figure 1 were cultured under air-liquid interface conditions and subsequently treated with increasing doses of LPS in vitro. Cultures were evaluated for IL-6 mRNA (A, C) and protein (B, D) expression at 24 h post-treatment. Results show the average +/− SE. *p<0.05, *** p<0.001 by two-way ANOVA comparing in vivo exposure and in vitro LPS concentration (n = 4 for each group except filtered air controls n = 5).

contrast, in vitro LPS treatment significantly reduced miR-410 expression in animals that were exposed to LPS in vivo (p<0.0001). In vitro LPS treatment but not prior in vivo exposures was a significant source of variation for miR-let-7a expression with significantly increased miR-let-7a levels in ozone-exposed animals following LPS treatment in vitro (Figure 5D; $p = 0.001$; two-way ANOVA for in vivo exposure and in vitro LPS treatment). Collectively, our data indicate that postnatal episodic ozone alters microRNA expression profiles in juvenile airway epithelial cells, both constitutively as well as in response to in vitro LPS treatment. Further, antecedent LPS (from in vivo challenge) also affects the microRNA response to secondary LPS treatment in airway epithelial cells.

Binding of microRNA Targets to the IL-6 3′ UTR and microRNA Modulation of IL-6 Expression

To evaluate the ability of putative IL-6 targeting microRNAs (miR-149, miR-202, miR-410) to modulate IL-6 mRNA and/or protein expression, the human bronchial epithelial cell line, BEAS-2B S.6 was transfected with microRNA mimics (Figure 6A). BEAS-2B S.6 were chosen for these functional experiments due to their ability to synthesize IL-6 both constitutively and in response to LPS treatment. At 24 h post-transfection, cells transfected with miR-149 showed minimal changes to IL-6 mRNA levels but had secreted significantly less IL-6 protein as compared to cells transfected with the negative microRNA control (Figure 6B–C,

$p = 0.004$). Comparatively, transfection with miR-202 or miR-410 did not significantly affect IL-6 mRNA or protein levels. To assess specificity of microRNA binding, the human bronchial epithelial cell line HBE (which expresses little to no endogenous IL-6) was co-transfected with microRNA mimics and an IL-6 gene target luciferase reporter plasmid (Figure 6D). Of the microRNA mimics tested, only miR-149 showed significant reduction of relative luciferase units ($p = 0.01$), indicating the ability of this microRNA to bind to the IL-6 3′ UTR. Relative luciferase units observed in HBE cells co-transfected with IL-6 gene target luciferase reporter plasmid and miR-202 or miR-410 were similar to transfection with the negative microRNA mimic.

Discussion

There is a very limited understanding of the mechanisms by which environmental perturbations during infancy affect pulmonary immunity later in adulthood, particularly in primate species. To address this deficiency, we investigated the immunomodulatory impact of early life air pollutant exposure using the rhesus monkey as an animal model of childhood development. We have previously shown that episodic exposure of infant monkeys to ozone followed by longitudinal evaluation as juveniles (at one year of age) resulted in attenuation of in vivo airway responses to LPS challenge, despite housing for 6 months exclusively in filtered air [20]. Furthermore, in vitro LPS treatment of peripheral blood mononuclear cells from juvenile animals with a history of postnatal

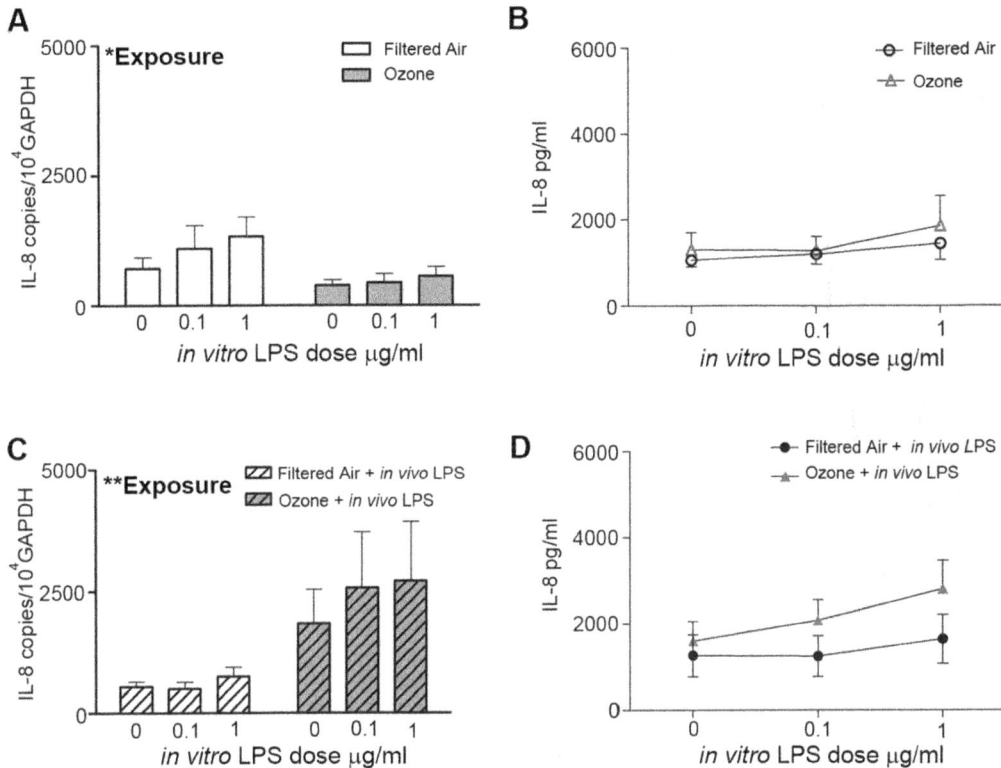

Figure 3. Effect of postnatal ozone exposure on IL-8 expression in juvenile monkey airway epithelial cell cultures following LPS treatment. Airway epithelial cells harvested from one-year-old juvenile rhesus monkeys with prior ozone and/or LPS exposure, as described in Figure 1 were cultured under air-liquid interface conditions and subsequently treated with increasing doses of LPS *in vitro*. Cultures were evaluated for IL-8 mRNA (A, C) and protein (B, D) expression at 24 h post-treatment. Results show the average +/− SE. *p<0.05, **p<0.01 by two-way ANOVA comparing *in vivo* exposure and *in vitro* LPS concentration (n = 4 for each group except filtered air controls n = 5).

ozone resulted in attenuation of both IL-6 and IL-8 protein synthesis, suggesting that air pollutant exposures may dampen systemic immune responses in an epigenetic fashion. In this current study, we have focused on the airway epithelium as an additional putative target for persistent immunomodulation by ozone. Using *in vitro* cultures of primary airway epithelial cells isolated from juvenile monkeys with a history of ozone exposure during postnatal development, we observed dysregulation of cytokine responses to *in vitro* LPS treatment and alterations in microRNA expression in association with exposure history.

Our finding of attenuated inflammatory responses to bacterial LPS in airway epithelium of rhesus monkeys previously exposed to ozone is consistent with published effects of ozone reducing host defense mechanisms during microbial infection of murine models [7,10,11]. However, it is important to emphasize that the animals evaluated in this study were housed under HEPA-filtered conditions for a period of six months after ozone exposure, indicating that the observed effects were not the result of recent exacerbation; rather, the functional change in epithelial cells from ozone-exposed infant airways was retained with increasing chronological age. We believe that our study is the first to report a persistently immunomodulated airway epithelial cell phenotype in response to a defined experimental exposure during postnatal development. The observed differential IL-6 and IL-8 mRNA expression in association with ozone highlights the inherent alterations that exposures have on epithelial cell immune function. Early life ozone exposure appears to reprogram the airway epithelium such that it synthesizes less IL-6 and IL-8 mRNA as

compared with filtered air control counterparts following treatment with LPS (Figure 2A, 3A). Interestingly, only IL-6 protein was significantly modulated by prior ozone exposure, whereas IL-8 protein secretion in response to LPS was not consistently affected (Figures 2B, 3B). The role of discordant mRNA and protein expression in airway epithelium is unclear but may be due to a delay in IL-8 protein translation relative to IL-6 that was not captured in our experimental time frame. Although our current study was limited to IL-6 and IL-8, we postulate that other proinflammatory cytokines, including TNF-α and IL-1β, may also be affected.

In addition to the impact of ozone on airway epithelium, we also observed that prior *in vivo* challenge with a single LPS aerosol significantly affected the cytokine response elicited by primary airway epithelial cell cultures. We have recently reported that chronological age and antecedent LPS exposure can have a significant effect on the ability of airway epithelium to respond to subsequent LPS exposure, resulting in a reduction of IL-6 mRNA but increased secretion of IL-6 protein [30]. Here, *in vivo* aerosol LPS exposure at one year of age in animals with prior ozone history elicited significantly higher IL-6 and IL-8 mRNA levels in primary airway epithelial cells constitutively and upon secondary LPS treatment *in vitro*, as compared with ozone alone counterparts (Figures 2C,3C). This finding indicates that the airway epithelial cell cytokine response is affected by both the type and sequence of exposures. Cytokine responses in our *in vitro* LPS epithelial cell cultures may reflect the timing and dose of the *in vivo* LPS challenge, as tracheal tissues were harvested from all animals

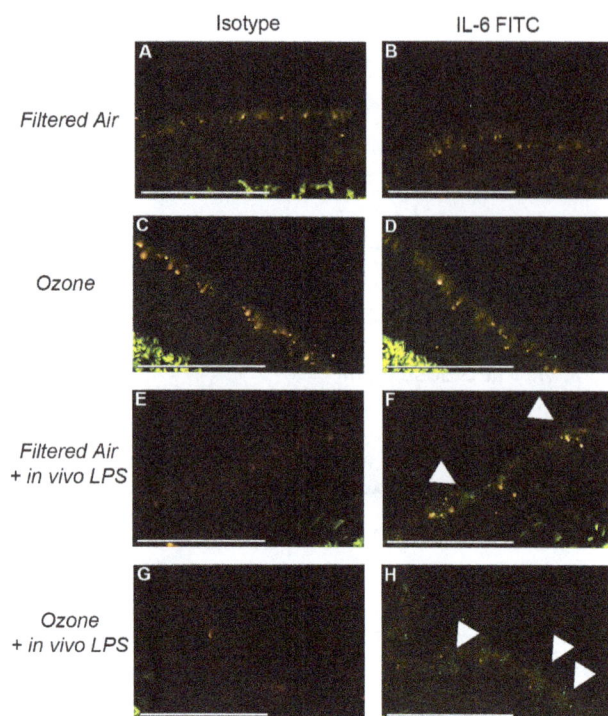

Figure 4. IL-6 immunofluorescence staining in juvenile monkey trachea following postnatal ozone and/or *in vivo* LPS. Trachea cryosections from a representative animal in each exposure group were stained with FITC-conjugated anti-human IL-6 mouse monoclonal antibody (B, D, F, and H). FITC-conjugated mouse IgG1 isotype controls are included for comparison (A, C, E, and G). Images were collected at 20× magnification. Scale bar = 100 μm. White arrows indicate IL-6 staining.

approximately 24 h after aerosolized LPS exposure, and our previous work shows that significant pulmonary inflammation lasts for at least 24 h after a single LPS dose is administered [20]. The observation that IL-6 protein was detected exclusively in trachea cryosections from animals that received inhaled LPS (Figure 4), suggests that the patterned cytokine response was established *in vivo*. While it is unclear why *in vitro* LPS treatment of airway epithelial cells derived from animals that only have postnatal ozone exposure does not recapitulate the effects of *in vivo* LPS, it may be speculated that interactions with other cells in the lung microenvironment (i.e. monocytes/macrophages) may impart a persistent cytokine phenotype in epithelium. Indeed, we have previously reported that postnatal ozone exposure resulted in elevated monocyte frequencies in bronchoalveolar lavage that persisted with maturity [20]. Taken together, our data emphasize the importance of studying the effects of co-exposures during this early life period given that environmental exposures likely involve multiple components which may act synergistically at a molecular level and result in distinct immunomodulation.

Accumulating evidence suggests that epigenetic mechanisms may play a role in persistent modification of host responses following air pollution exposure [28]. Recently, microRNAs have been identified as regulators of RNA responses in various biological processes including inflammatory signaling via TLRs [44]. In this study we focused our investigation on microRNAs that may regulate IL-6, based upon the identification of microRNAs that directly or indirectly affect IL-6 expression in other cell types [34,45,46,47,48,49,50]. miR-149 expression was

significantly dependent upon exposure, as airway epithelial cells derived from animals with a history of postnatal ozone + *in vivo* LPS exhibited reduced constitutive levels relative to other animal groups (Figure 5A). There was also a trend towards exposure-dependent differences in miR-202 expression in airway epithelium, with reduced levels in cultures derived from animals with prior ozone exposure (Figure 5B). Upon *in vitro* LPS treatment, miR-149, miR-410 and miR-let7a levels increased in epithelial cells from animal groups without *in vivo* LPS; *in vivo* LPS resulted in a comparatively attenuated microRNA response. In contrast, miR-202 expression levels were reduced in response to *in vitro* LPS treatment. In our study, differential expression of miR-149 and miR-let7a may contribute to the distinct IL-6 protein responses of airway epithelial cells based upon postnatal exposure history. The low IL-6 protein response in ozone and filtered air groups was associated with high LPS-induced levels of miR-149 and miR-let7a whereas the high IL-6 protein in the *in vivo* LPS-exposed groups corresponded to lower levels of these miRNAs. Induction of miRs -149 and -let7a in LPS-treated airway epithelial cell cultures may also explain why LPS concentration was a significant source of variation in IL-6 protein levels at 6 but not 24 h. Using microRNA mimics and 3′UTR binding assays, we also found that miRNA-149 was capable of regulating IL-6 protein expression in human airway epithelial cell lines, suggesting that this microRNA may be an important modulator of IL-6 protein expression following LPS exposure of airway epithelium. The current study was limited to microRNAs targeting IL-6 but it is likely that other microRNAs are differentially expressed due to early life environmental exposures and future studies will focus on identifying additional targets that may be impacted by ozone exposure.

A single microRNA is capable of controlling hundreds of different gene targets and individual genes are often targeted by many different microRNAs. Under steady-state conditions, microRNAs appear to act as only moderate regulators, however it is important to consider that microRNAs rarely function through a single gene target and it is the combined regulation of many different genes that determines their true functionality [51]. miR-149, which in our study appears to modulate IL-6 protein expression in airway epithelial cells, has also been shown to target several other diverse genes including E2F1 [52,53,54]. E2F1 is a transcriptional activator recruited by NF-kB upon TLR activation to control the LPS inducibility of proinflammatory cytokines [55]. In addition to IL-6, E2F1 may have also been modulated by miR-149 in our epithelial cell cultures, contributing to the observed downregulation of this proinflammatory cytokine. Several gene targets involved in innate immune responses have also been described for both miR-410 and miR-202; it has been speculated that these microRNAs play a role in repressing inflammatory reactions [56,57,58].

In conclusion, results from this study show that exposure to ozone in early life intrinsically alters the ability of airway epithelial cells to mount a robust proinflammatory IL-6 response to bacterial LPS challenge. Further, the expression of IL-6-targeting microRNAs is persistently affected by environmental exposures and likely contributes to the dysregulated cytokine response profiles. Because epidemiologic studies strongly support a window of immune susceptibility within the first year of life in humans, it will be important to evaluate other environmental insults for their ability to impose epigenetic reprogramming of airway immune responses.

Acknowledgments

The authors thank Sarah Davis, Paul-Michael Sosa, Sona Santos, Louise Olsen, and Brian Tarkington for technical support

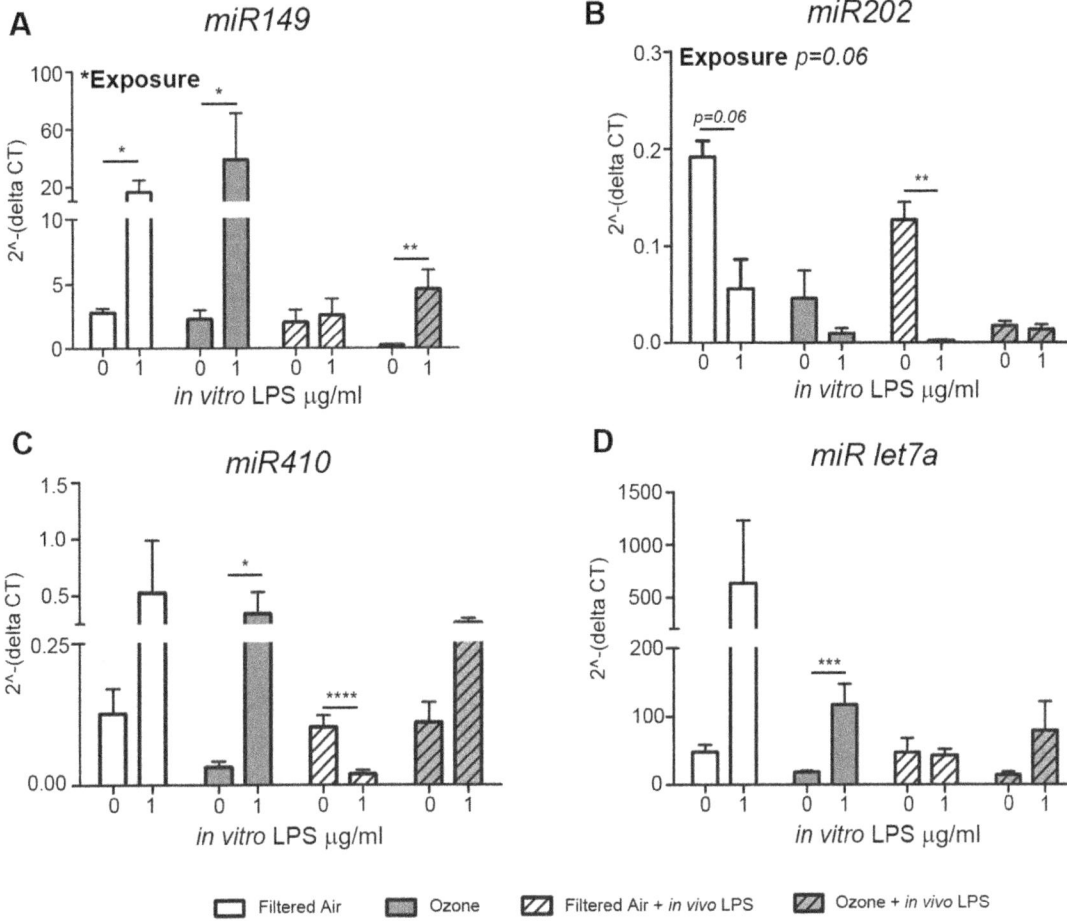

Figure 5. Effect of postnatal ozone exposure on miR-149, miR-202, miR-410 and miR-let7a expression in juvenile monkey airway epithelial cell cultures. Comparison of constitutive (0 µg/ml) and LPS-induced (1 µg/ml, 6 h post-treatment) microRNA expression in primary airway epithelial cell cultures derived from one-year-old juvenile rhesus monkeys with prior ozone and/or LPS exposure as described in Figure 1. (A) miRNA-149, (B) miRNA-202, (C) miRNA-410, (D) miRNA-let7a. Graphs show the average +/− SE of the 2-(delta CT) where the delta CT = (microRNA of interest − endogenous control microRNA (RNU-6b)). microRNAs were extracted from n = 3–5 animals per group. Results from Student's t-tests comparing microRNA expression at baseline and post- in vitro LPS treatment are shown with horizontal bars. *p<0.05, **p<0.005, ***p<0.0003.

during this project. We also thank Candace Burke and Carolyn Black for helpful discussions during the preparation of this manuscript.

Supporting Information

Figure S1   Effect of postnatal ozone on IL-6 expression in juvenile monkey airway epithelial cell cultures 6 h post-LPS treatment. Airway epithelial cells harvested from one-year-old juvenile rhesus monkeys with prior ozone and/or LPS exposure (as described in Figure 1) were cultured under air-liquid interface conditions and subsequently treated with increasing doses of LPS in vitro. IL-6 mRNA and protein expression was evaluated at 6 h post-treatment in filtered air and ozone cultures (A, B) as well as in filtered air + in vivo LPS and ozone + in vivo LPS cultures (C, D). Results show the average +/− SE. *p<0.05, **p<0.01, *** p<0.001 by two-way ANOVA comparing in vivo exposure and in vitro LPS concentration (n = 4–5 for each group).

Figure S2   Effect of postnatal ozone exposure on IL-8 expression in juvenile monkey airway epithelial cell cultures 6 h post-LPS treatment. Airway epithelial cells harvested from one-year-old, juvenile rhesus monkeys with prior ozone and/or LPS exposure (as described in Figure 1) were cultured under air-liquid interface conditions and subsequently treated with increasing doses of LPS in vitro. IL-8 mRNA and protein expression was evaluated at 6 h post-treatment in filtered air and ozone cultures (A, B) as well as in filtered air + LPS and ozone + in vivo LPS cultures (C, D). Results show the average +/− SE. *p<0.05, **p<0.01 by two-way ANOVA comparing in vivo exposure and in vitro LPS concentration (n = 4–5 for each group). (TIF)

Figure S3   Comparison of TLR4 expression in juvenile airway epithelial cells from different exposure groups. Constitutive TLR4 mRNA expression in primary airway epithelial cell cultures derived from juvenile rhesus monkeys exposed postnatally to filtered air, ozone, filtered air + in vivo LPS, or ozone + in vivo LPS. TLR4 copy number relative to GAPDH was determined by RT-PCR and calculated based on standard curves

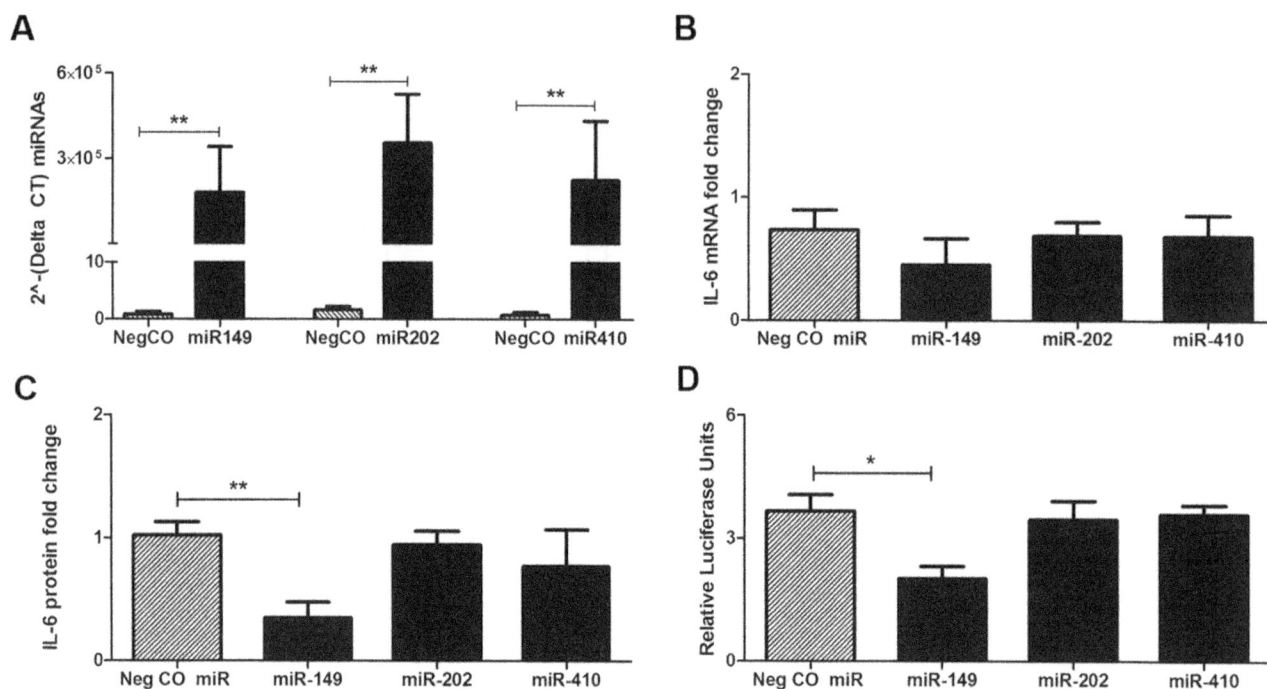

Figure 6. Assessment of microRNA regulation of IL-6 in human airway epithelial cell cultures. To evaluate the ability of the putative IL-6-targeting microRNAs to regulate expression of IL-6, the BEAS-2B cell line was transfected with a negative control (NegCO) microRNA or mimics for miR-149, miR-202 or miR-410. (A) Expression of microRNAs in BEAS-2B 24 h post-transfection with a NegCO microRNA or mimics for miR-149, miR-202, or miR-410. Values are reported as the average +/− SE of the $\hat{2}$-(delta CT) relative to the endogenous control microRNA RNU-6b. (B) IL-6 mRNA and (C) IL-6 protein expression was evaluated in transfected BEAS-2B cells after 24 h. Data are reported as the fold-change in IL-6 mRNA or IL-6 protein as compared to no mimic controls for each experiment. (D) Binding of microRNA mimics to IL-6 mRNA was tested in the HBE1 cell line with the IL-6 3′UTR cloned into a firefly/renilla luciferase plasmid reporter system. For transfection normalization across samples, the relative luciferase units are reported for firefly versus renilla luciferase. Data are from 4 separate experiments. *$p<0.05$, **$p<0.005$ by Student's t-test.

with the average +/−SE graphed for n = 3–5 per group. 1-way ANOVA for exposure-dependent differences showed no significance.

Figure S4 Identification of IL-6 targeting microRNAs. (A) A schematic of human IL-6 mRNA 3′UTR showing three potential binding sites for miR-149, miRlet-7a/miR-202 (overlap), and miR-410. Binding sites are indicated in gray by their position on the 3′UTR. (B) Context scores for rhesus and human microRNAs evaluated in this study are listed as determined by the TargetScan 5.1 program. The context score for each site is the sum of the (1) site-type contribution, (2) 3′ pairing contribution, (3) local AU contribution, and (4) position contribution, as described

in Grimson et al. [33]. A lower score indicates more favorable binding.

Author Contributions

Conceived and designed the experiments: CCC KMH EMP MJE LAM. Performed the experiments: CCC KMH TTW JEG JHF. Analyzed the data: CCC KMH TTW JEG LAM. Wrote the paper: CCC KMH LAM. Assay development: JHF.

References

1. Dockery DW, Pope CA, 3rd, Xu X, Spengler JD, Ware JH, et al. (1993) An association between air pollution and mortality in six U.S. cities. The New England journal of medicine 329: 1753–1759.
2. Bell ML, McDermott A, Zeger SL, Samet JM, Dominici F (2004) Ozone and short-term mortality in 95 US urban communities, 1987-2000. JAMA : the journal of the American Medical Association 292: 2372–2378.
3. Gryparis A, Forsberg B, Katsouyanni K, Analitis A, Touloumi G, et al. (2004) Acute effects of ozone on mortality from the "air pollution and health: a European approach" project. American journal of respiratory and critical care medicine 170: 1080–1087.
4. Burnett RT, Smith-Doiron M, Stieb D, Raizenne ME, Brook JR, et al. (2001) Association between ozone and hospitalization for acute respiratory diseases in children less than 2 years of age. American journal of epidemiology 153: 444–452.
5. Delfino RJ, Coate BD, Zeiger RS, Seltzer JM, Street DH, et al. (1996) Daily asthma severity in relation to personal ozone exposure and outdoor fungal spores. American journal of respiratory and critical care medicine 154: 633–641.

6. Jakab GJ, Spannhake EW, Canning BJ, Kleeberger SR, Gilmour MI (1995) The effects of ozone on immune function. Environmental health perspectives 103 Suppl 2: 77–89.
7. Al-Hegelan M, Tighe RM, Castillo C, Hollingsworth JW (2011) Ambient ozone and pulmonary innate immunity. Immunologic research 49: 173–191.
8. Medina-Ramon M, Zanobetti A, Schwartz J (2006) The effect of ozone and PM10 on hospital admissions for pneumonia and chronic obstructive pulmonary disease: a national multicity study. American journal of epidemiology 163: 579–588.
9. Kesic MJ, Meyer M, Bauer R, Jaspers I (2012) Exposure to ozone modulates human airway protease/antiprotease balance contributing to increased influenza A infection. PloS one 7: e35108.
10. Gilmour MI, Park P, Doerfler D, Selgrade MK (1993) Factors that influence the suppression of pulmonary antibacterial defenses in mice exposed to ozone. Experimental lung research 19: 299–314.

11. Goldstein E, Tyler WS, Hoeprich PD, Eagle C (1971) Ozone and the antibacterial defense mechanisms of the murine lung. Archives of internal medicine 127: 1099–1102.

12. Van Loveren H, Rombout PJ, Wagenaar SS, Walvoort HC, Vos JG (1988) Effects of ozone on the defense to a respiratory Listeria monocytogenes infection in the rat. Suppression of macrophage function and cellular immunity and aggravation of histopathology in lung and liver during infection. Toxicology and applied pharmacology 94: 374–393.

13. Devlin RB, McDonnell WF, Mann R, Becker S, House DE, et al. (1991) Exposure of humans to ambient levels of ozone for 6.6 hours causes cellular and biochemical changes in the lung. American journal of respiratory cell and molecular biology 4: 72–81.

14. Peterson ML, Harder S, Rummo N, House D (1978) Effect of ozone on leukocyte function in exposed human subjects. Environmental research 15: 485–493.

15. Holt PG, Jones CA (2000) The development of the immune system during pregnancy and early life. Allergy 55: 688–697.

16. PrabhuDas M, Adkins B, Gans H, King C, Levy O, et al. (2011) Challenges in infant immunity: implications for responses to infection and vaccines. Nat Immunol 12: 189–194.

17. Bateson TF, Schwartz J (2008) Children's response to air pollutants. Journal of toxicology and environmental health Part A 71: 238–243.

18. Pinkerton KE, Joad JP (2000) The mammalian respiratory system and critical windows of exposure for children's health. Environ Health Perspect 108 Suppl 3: 457–462.

19. Plopper CG, Fanucchi MV (2000) Do urban environmental pollutants exacerbate childhood lung diseases? Environ Health Perspect 108: A252–253.

20. Maniar-Hew K, Postlethwait EM, Fanucchi MV, Ballinger CA, Evans MJ, et al. (2010) Postnatal Episodic Ozone Results in Persistent Attenuation of Pulmonary and Peripheral Blood Responses to LPS Challenge. Am J Physiol Lung Cell Mol Physiol.

21. Murphy SR, Schelegle ES, Edwards PC, Miller LA, Hyde DM, et al. (2012) Postnatal exposure history and airways: oxidant stress responses in airway explants. American journal of respiratory cell and molecular biology 47: 815–823.

22. Devlin RB, McKinnon KP, Noah T, Becker S, Koren HS (1994) Ozone-induced release of cytokines and fibronectin by alveolar macrophages and airway epithelial cells. The American journal of physiology 266: L612–619.

23. Manzer R, Wang J, Nishina K, McConville G, Mason RJ (2006) Alveolar epithelial cells secrete chemokines in response to IL-1beta and lipopolysaccharide but not to ozone. American journal of respiratory cell and molecular biology 34: 158–166.

24. Jaspers I, Flescher E, Chen LC (1997) Ozone-induced IL-8 expression and transcription factor binding in respiratory epithelial cells. The American journal of physiology 272: L504–511.

25. Goulding J, Snelgrove R, Saldana J, Didierlaurent A, Cavanagh M, et al. (2007) Respiratory infections: do we ever recover? Proceedings of the American Thoracic Society 4: 618–625.

26. Beane J, Sebastiani P, Liu G, Brody JS, Lenburg ME, et al. (2007) Reversible and permanent effects of tobacco smoke exposure on airway epithelial gene expression. Genome biology 8: R201.

27. Bai TR, Knight DA (2005) Structural changes in the airways in asthma: observations and consequences. Clinical science 108: 463–477.

28. Silveyra P, Floros J (2012) Air pollution and epigenetics: effects on SP-A and innate host defence in the lung. Swiss medical weekly 142: w13579.

29. Schelegle ES, Miller LA, Gershwin LJ, Fanucchi MV, Van Winkle LS, et al. (2003) Repeated episodes of ozone inhalation amplifies the effects of allergen sensitization and inhalation on airway immune and structural development in Rhesus monkeys. Toxicol Appl Pharmacol 191: 74–85.

30. Maniar-Hew K, Clay CC, Postlethwait EM, Evans MJ, Fontaine JH, et al. (2013) Innate immune response to LPS in airway epithelium is dependent on chronological age and antecedent exposures. American journal of respiratory cell and molecular biology 49: 710–720.

31. Matsui H, Randell SH, Peretti SW, Davis CW, Boucher RC (1998) Coordinated clearance of periciliary liquid and mucus from airway surfaces. The Journal of clinical investigation 102: 1125–1131.

32. Lewis BP, Burge CB, Bartel DP (2005) Conserved seed pairing, often flanked by adenosines, indicates that thousands of human genes are microRNA targets. Cell 120: 15–20.

33. Grimson A, Farh KK, Johnston WK, Garrett-Engele P, Lim LP, et al. (2007) MicroRNA targeting specificity in mammals: determinants beyond seed pairing. Molecular cell 27: 91–105.

34. Iliopoulos D, Hirsch HA, Struhl K (2009) An epigenetic switch involving NF-kappaB, Lin28, Let-7 MicroRNA, and IL6 links inflammation to cell transformation. Cell 139: 693–706.

35. Ke Y, Reddel RR, Gerwin BI, Miyashita M, McMenamin M, et al. (1988) Human bronchial epithelial cells with integrated SV40 virus T antigen genes retain the ability to undergo squamous differentiation. Differentiation; research in biological diversity 38: 60–66.

36. Yankaskas JR, Haizlip JE, Conrad M, Koval D, Lazarowski E, et al. (1993) Papilloma virus immortalized tracheal epithelial cells retain a well-differentiated phenotype. The American journal of physiology 264: C1219–1230.

37. Lu YC, Yeh WC, Ohashi PS (2008) LPS/TLR4 signal transduction pathway. Cytokine 42: 145–151.

38. Hollingsworth JW, 2nd, Cook DN, Brass DM, Walker JK, Morgan DL, et al. (2004) The role of Toll-like receptor 4 in environmental airway injury in mice. American journal of respiratory and critical care medicine 170: 126–132.

39. Kleeberger SR, Levitt RC, Zhang LY, Longphre M, Harkema J, et al. (1997) Linkage analysis of susceptibility to ozone-induced lung inflammation in inbred mice. Nature genetics 17: 475–478.

40. Williams AS, Leung SY, Nath P, Khorasani NM, Bhavsar P, et al. (2007) Role of TLR2, TLR4, and MyD88 in murine ozone-induced airway hyperresponsiveness and neutrophilia. Journal of applied physiology 103: 1189–1195.

41. Bartel DP (2009) MicroRNAs: target recognition and regulatory functions. Cell 136: 215–233.

42. Jardim MJ, Fry RC, Jaspers I, Dailey L, Diaz-Sanchez D (2009) Disruption of microRNA expression in human airway cells by diesel exhaust particles is linked to tumorigenesis-associated pathways. Environmental health perspectives 117: 1745–1751.

43. Schembri F, Sridhar S, Perdomo C, Gustafson AM, Zhang X, et al. (2009) MicroRNAs as modulators of smoking-induced gene expression changes in human airway epithelium. Proceedings of the National Academy of Sciences of the United States of America 106: 2319–2324.

44. O'Neill LA, Sheedy FJ, McCoy CE (2011) MicroRNAs: the fine-tuners of Toll-like receptor signalling. Nature reviews Immunology 11: 163–175.

45. He Y, Sun X, Huang C, Long XR, Lin X, et al. (2013) MiR-146a Regulates IL-6 Production in Lipopolysaccharide-Induced RAW264.7 Macrophage Cells by Inhibiting Notch1. Inflammation.

46. Rossato M, Curtale G, Tamassia N, Castellucci M, Mori L, et al. (2012) IL-10-induced microRNA-187 negatively regulates TNF-alpha, IL-6, and IL-12p40 production in TLR4-stimulated monocytes. Proceedings of the National Academy of Sciences of the United States of America 109: E3101–3110.

47. Garg M, Potter JA, Abrahams VM (2013) Identification of microRNAs that regulate TLR2-mediated trophoblast apoptosis and inhibition of IL-6 mRNA. PloS one 8: e77249.

48. Yang X, Liang L, Zhang XF, Jia HL, Qin Y, et al. (2013) MicroRNA-26a suppresses tumor growth and metastasis of human hepatocellular carcinoma by targeting interleukin-6-Stat3 pathway. Hepatology 58: 158–170.

49. Xu Z, Xiao SB, Xu P, Xie Q, Cao L, et al. (2011) miR-365, a novel negative regulator of interleukin-6 gene expression, is cooperatively regulated by Sp1 and NF-kappaB. The Journal of biological chemistry 286: 21401–21412.

50. Sun Y, Varambally S, Maher CA, Cao Q, Chockley P, et al. (2011) Targeting of microRNA-142-3p in dendritic cells regulates endotoxin-induced mortality. Blood 117: 6172–6183.

51. van Rooij E (2011) The art of microRNA research. Circulation research 108: 219–234.

52. Wang Y, Zheng X, Zhang Z, Zhou J, Zhao G, et al. (2012) MicroRNA-149 inhibits proliferation and cell cycle progression through the targeting of ZBTB2 in human gastric cancer. PloS one 7: e41693.

53. Jin L, Hu WL, Jiang CC, Wang JX, Han CC, et al. (2011) MicroRNA-149*, a p53-responsive microRNA, functions as an oncogenic regulator in human melanoma. Proceedings of the National Academy of Sciences of the United States of America 108: 15840–15845.

54. Lin RJ, Lin YC, Yu AL (2010) miR-149* induces apoptosis by inhibiting Akt1 and E2F1 in human cancer cells. Molecular carcinogenesis 49: 719–727.

55. Lim CA, Yao F, Wong JJ, George J, Xu H, et al. (2007) Genome-wide mapping of RELA(p65) binding identifies E2F1 as a transcriptional activator recruited by NF-kappaB upon TLR4 activation. Molecular cell 27: 622–635.

56. Zhao Y, Zhao J, Mialki RK, Wei J, Spannhake EW, et al. (2013) Lipopolysaccharide-induced phosphorylation of c-Met tyrosine residue 1003 regulates c-Met intracellular trafficking and lung epithelial barrier function. American journal of physiology Lung cellular and molecular physiology 305: L56–63.

57. Chen L, Zhang J, Feng Y, Li R, Sun X, et al. (2012) MiR-410 regulates MET to influence the proliferation and invasion of glioma. The international journal of biochemistry & cell biology 44: 1711–1717.

58. Sweeney TE, Suliman HB, Hollingsworth JW, Piantadosi CA (2010) Differential regulation of the PGC family of genes in a mouse model of Staphylococcus aureus sepsis. PloS one 5: e11606.

10

The True Cost of Greenhouse Gas Emissions: Analysis of 1,000 Global Companies

Nagisa Ishinabe[1], Hidemichi Fujii[2], Shunsuke Managi[2,3]*

1 Department of Agricultural Economics, Purdue University, West Lafayette, Indiana, United States of America, **2** Graduate School of Environmental Studies, Tohoku University, Sendai, Japan, **3** Institute for Global Environmental Strategies, Hayama, Kanagawa, Japan

Abstract

This study elucidated the shadow price of greenhouse gas (GHG) emissions for 1,024 international companies worldwide that were surveyed from 15 industries in 37 major countries. Our results indicate that the shadow price of GHG at the firm level is much higher than indicated in previous studies. The higher shadow price was found in this study as a result of the use of Scope 3 GHG emissions data. The results of this research indicate that a firm would carry a high cost of GHG emissions if Scope 3 GHG emissions were the focus of the discussion of corporate social responsibility. In addition, such shadow prices were determined to differ substantially among countries, among sectors, and within sectors. Although a number of studies have calculated the shadow price of GHG emissions, these studies have employed country-level or industry-level data or a small sample of firm-level data in one country. This new data from a worldwide firm analysis of the shadow price of GHG emissions can play an important role in developing climate policy and promoting sustainable development.

Editor: Alejandro Raul Hernandez Montoya, Universidad Veracruzana, Mexico

Funding: This research was funded by a Grant-in-Aid for Japanese Ministry of the Environment conducted in 2011 and a Grant-in-Aid for Scientific Research from (B) in 2011 from the Japanese Ministry of Education, Culture, Sports, Science and Technology (MEXT). The funders had no role in study design, data collection and analysis, decision to publish, or preparation of the manuscript.

Competing Interests: The authors have declared that no competing interests exist.

* E-mail: managi.s@gmail.com

Introduction

Our study focuses on the true cost of emissions reduction at the firm level by computing the evidence-based shadow price of greenhouse gas (GHG) emissions by company, which is estimated by a production function approach using GHG data. The study covers 1,024 major companies of 17 industries in 37 countries over the period from 2002 to 2009. This survey allows us to observe the patterns of the shadow price of GHG emissions among firms and sectors over many years and to examine efficient and effective pathways to transform our socio-economic systems into sustainable systems.

To ameliorate the effects of climate change, each country in the world is currently advancing research on GHG reduction methods and the costs of such reductions [1–2]. Although many previous studies have focused on the shadow price of GHG, most of these studies are focused on future shadow prices at the national and global levels based on projected scenarios and do not actually calculate the current shadow price or calculate the shadow price at the company level[3]. Therefore, when regulations pertaining to the total quantity of GHG emissions are introduced, the degree of impact on each company in each country is uncertain. Consequently, it is difficult to allay the concerns of companies in any country as to whether they will suffer an adverse effect on their international competitiveness when carbon constraints are imposed. As a result, companies and countries are currently not assuming obligations to reduce the total quantity of GHG emissions. Therefore, this study determines the present shadow price and the effect of market competitiveness in terms of GHG

emissions at the company level as well as the degree of impact on each company under carbon constraints and on its international competitiveness when reductions in the total quantity of GHG emissions are introduced.

Methods

The shadow price that is calculated by a production function is equivalent to the revenue to be sacrificed when a company is forced to reduce one ton of GHG emissions. A lower shadow price signals that it is relatively less expensive to reduce GHGs. The shadow price captures the holistic price of GHG emissions for a company by considering both technological advancements and operational efficiencies (e.g., switching off lights and ensuring the optimal use of materials) by employing all firm-level data. As a result, the shadow price is distinct from technology-based abatement costs or GHG intensity, which captures the cost of only a single technology or the ratio of GHG emissions and sales.

The economic valuation method for handling environmentally undesirable outputs using the directional distance function (DDF) as a nonparametric approach was developed by [3–4]. Following [3], we can estimate q, the economic value of an environmentally undesirable output (shadow price), using the measure below.

We denote $x \in \Re_+^L$, $b \in \Re_+^R$ and $y \in \Re_+^M$ as the vectors of inputs, environmental output (or undesirable output), and market outputs (or desirable output), respectively, and we then define the production technology as follows:

$$P(x) = \{(x,y,b) : x \ can \ produce \ (y,b)\} \quad (1)$$

The inefficiency $D(x, y, b| \ g_x, g_y, g_b)$ of the production units in $P(x)$ for each of the firms in this study is defined with the distance from the production frontier consisting of the efficient production units as follows:

$$\vec{D}(x,y,b|g_x,g_y,g_b) = Sup\{\beta : (y+\beta g_y, b-\beta g_b) \in P(x-\beta g_x)\} \quad (2)$$

where g_x, g_y, and g_b denote the non-negative directional vectors of the input, the desirable output, and the undesirable output, respectively. From the above definition, equation (3) is determined to be valid.

$$(y,b) \in P(x) \ if \ and \ only \ if \ D(x,y,b|g_x.g_y.g_b) \geq 0 \quad (3)$$

Under a perfect competitive market, the prices of market goods, p, and the prices of undesirable goods, q, are assumed to be $p>0$ and $q<0$, respectively. If the aggregate economic value of the output for each production unit is given by $R(x, p, q)$, then the specific combination of y and b to maximize $py+qb$, $(y^*,b^*) \in P(x)$ for given prices of p and q exists in the production possibility set. Therefore, $R(x, p, q)$ can be expressed as follows:

$$R(x,p,q) = Max_{(y*,b*)}\{py^* + qb^*|(y^*,b^*) \in P(x)\} \quad (4)$$

Here, with equation (3), $R(x, p, q)$ becomes the following:

$$R(x,p,q) = Max_{(y*,b*)}\{py^* + qb^*|D(\cdot) \geq 0\} \quad (5)$$

where $D(x, y, b| \ g_x, g_y, g_b)$ is expressed as $D(\cdot)$. Given that equation (5) is formed for any pairs (y, b) $P(x)$, the relation $(y^*,b^*) = (y+D(\cdot)\times g_y, \ b+D(\cdot)\times g_b)$ can replace equation (5). Thus, $R(x,p,q)$ is obtained as follows:

$$R(x,p,q) = Max_{(y,b)}[p\{y+D(\cdot)g_y\}+q\{b+D(\cdot)g_b\}|D(\cdot)\geq 0]$$
$$\geq [p\{y+D(\cdot)g_y\}+q\{b+D(\cdot)g_b\}|D(\cdot)\geq 0] \quad (6)$$
$$= \{py+qb+(pg_y+qg_b) \times D(\cdot)|D(\cdot)\geq 0\}$$

Furthermore,

$$0 \leq D(\cdot) \leq \{R(x,p,q)-(py+qb)\}/(pg_y+qg_b) \quad (7)$$

Simultaneously, because $R(x, p, q)$ is function of p and q (hence, for any pairs $(y,b) \ \varepsilon \ P(x)$), a certain p and q exists to satisfy the following relationship:

$$D(\cdot) = Min_{(p,q)}[\{R(x,p,q)-(py+qb)\}/(pg_y+qg_b)] \quad (8)$$

Executing the partial differentiation for both sides of equation (8) with respect to b and y, equations (9) and (10), respectively, can be derived:

$$\frac{\partial D(\cdot)}{\partial b} = \frac{-q}{pg_y+qg_b} \geq 0 \quad (9)$$

$$\frac{\partial D(\cdot)}{\partial y} = \frac{-p}{pg_y+qg_b} \leq 0 \quad (10)$$

Equation (9) describes the extent of the increase in inefficiency $D(\cdot)$ while emitting an additional environmental undesirable output by one unit marginally. Similarly, equation (10) describes the extent of the decrease in inefficiency $D(\cdot)$, while increasing the additional market output by one unit marginally. Combining equations (9) and (10) then results in equation (11):

$$\frac{q}{p} = \frac{\partial D(\cdot)/\partial b}{\partial D(\cdot)/\partial y} \quad (11)$$

Therefore, by simply solving equation (11) with q, the economic value of the environmentally undesirable output is defined as follows:

$$q = p \times \frac{\partial D(\cdot)/\partial b}{\partial D(\cdot)/\partial y} \quad (12)$$

Here, the economic value of the market output can be normalized as $p = 1$ if the market output variable consists of monetary data; thus, the economic value q is regarded as the value of the environmentally undesirable good relative to the value of the market goods.

We can estimate the adjusted shadow price q^{adj}, the economic value of an environmentally undesirable output considering an inefficiency score, using the following measure:

$$q^{adj} = p \times \frac{\partial \vec{D}(x,y^*,b^*)/\partial b^*}{\partial \vec{D}(x,y^*,b^*)/\partial y^*} \times \frac{\sigma_b}{\sigma_y} \quad (13)$$

where (y^*, b^*) is the intersecting point on the frontier curve with the directional vector of an inefficient province. The inefficiency factors σ_b and σ_y are defined as follows:

$$\sigma_b = \frac{1}{1+\vec{D}(x,y,b)\frac{g_b}{b*}} \quad (14)$$

$$\sigma_y = \frac{1}{1-\vec{D}(x,y,b)\frac{g_y}{y*}} \quad (15)$$

Here, we set the production function using input x, undesirable output b, and desirable output y. We assume desirable and undesirable outputs under a null-joint hypothesis; a company cannot produce a desirable output without producing undesirable outputs.

Table 1. Shadow price by country.

	Country	Average Shadow Price (U.S. $)	Median Shadow Price	Shadow Price (<100$)	Shadow Price (100 $– 1,000$)	Shadow Price (1,001$ – 10,000$)	# obs.
1	UNITED STATES	9,809	3,340	10%	19%	46%	2,321
2	UNITED KINGDOM	13,544	7,335	6%	13%	41%	1,119
3	JAPAN	6,626	2,332	8%	27%	47%	689
4	FRANCE	16,686	8,697	7%	8%	37%	319
5	CHINA+HK+TAIWAN	8,453	4,316	14%	22%	38%	277
6	GERMANY	15,433	9,423	6%	11%	36%	223
7	SWITZERLAND	11,003	5,579	10%	8%	47%	155
8	SWEDEN	14,074	7,086	8%	14%	34%	152
9	ITALY	11,406	5,780	15%	7%	43%	137
10	SPAIN	12,482	3,146	22%	18%	30%	133
11	CANADA	2,120	469	11%	59%	24%	118
12	NETHERLANDS	9,917	3,536	3%	11%	50%	109
13	FINLAND	13,088	1,986	18%	18%	30%	98
14	NORWAY	2,955	612	12%	45%	33%	67
15	SOUTH KOREA	5,739	3,216	3%	36%	43%	67
16	DENMARK	7,093	5,640	12%	17%	41%	58
17	INDIA	19,279	18,259	13%	16%	16%	45
18	BRAZIL	13,707	10,107	5%	15%	29%	41
19	MALAYSIA	4,253	582	38%	18%	25%	40
20	IRELAND	11,596	8,704	16%	8%	29%	38
21	AUSTRALIA	2,080	878	30%	24%	41%	37
22	BELGIUM	6,904	2,817	5%	22%	46%	37
23	PORTUGAL	9,417	7,830	29%	6%	26%	35
24	GREECE	5,743	5,103	11%	4%	70%	27
25	SINGAPORE	6,610	4,115	0%	13%	67%	24
26	AUSTRIA	2,815	685	41%	9%	45%	22
27	THAILAND	69	18	86%	14%	0%	21
28	MEXICO	5,830	5,349	24%	6%	53%	17
29	INDONESIA	92	9	75%	25%	0%	8
30	LUXEMBOURG	5,660	3,189	25%	13%	38%	8
31	PHILIPPINES	1,265	1,100	0%	25%	75%	8
32	POLAND	519	570	25%	50%	25%	8
33	SOUTH AFRICA	113	98	50%	50%	0%	8
34	BERMUDA	319	240	17%	83%	0%	6
35	ISRAEL	3,496	3,291	0%	50%	50%	6

Table 1. Cont.

	Country	Average Shadow Price (U.S. $)	Median Shadow Price	Shadow Price (<100$)	Shadow Price (100 $– 1,000$)	Shadow Price (1,001$ – 10,000$)	# obs.
36	RUSSIA	22	25	100%	0%	0%	4
37	PAKISTAN	73	102	33%	67%	0%	3
	Average	91,358	90,649	10%	18%	42%	

$$(y,b) \in P(x); \quad b=0 \Rightarrow y=0 \qquad (16)$$

We also assume weak disposability. Weak disposability implies that the pollutant should not be considered freely disposable.

$$(y,b) \in P(x) \ and \ 0 \leq \beta \leq 1 \Rightarrow (\beta y, \beta b) \in P(x) \qquad (17)$$

Under the null-joint hypothesis and weak disposability, this directional distance function can be computed for firm k by solving the following optimization problem:

$$\vec{D}(x_k^l, y_k^m, b_k^r | g_{x^l}, g_{y^m}, g_{b^r}) = Maximize \ \beta_k \qquad (18)$$

$$\text{s.t.} \ \sum_{i=1}^{N} \lambda_i x_i^l \leq x_k^l + \beta_k g_{x^l} \quad l=1,\cdots,L \qquad (19)$$

$$\sum_{i=1}^{N} \lambda_i y_i^m \geq y_k^m + \beta_k g_{y^m} \quad m=1,\cdots,M \qquad (20)$$

$$\sum_{i=1}^{N} \lambda_i b_i^r = b_k^r + \beta_k g_{b^r} \quad r=1,\cdots,R \qquad (21)$$

$$\lambda_i \geq 0 \qquad i=1,\cdots,N \qquad (22)$$

where l, m, and r are the input, the desirable output, and the undesirable output, respectively; x is the input factor in the L × N input factor matrix; y is the desirable output in the M × N desirable output factor matrix; and b is the undesirable output factor in the R × N undesirable output matrix. In addition, g_x is the directional vector of the input factor, g_y is the directional vector of the desirable output factors, and g_b is the directional vector of the undesirable output factors. is the inefficiency score of the kth firm, and is the weight variable. To estimate the inefficiency score of all firms, the model must be independently applied N times for each firm. One objective of this study is to clarify the extent to which firms have improved their levels of productivity with respect to the CO_2 emissions under consideration. We set the directional vector as $(g_x, g_y, g_b) = (0, y^m, b^r)$ to estimate the productivity change by applying the Luenberger productivity indicator. Under this directional vector setting, we obtain the following equation.

Objective function

$$Max. \ \beta_k \qquad (23)$$

Restriction

$$\sum_{i=1}^{N} y_{q,i} \lambda_i \geq (1+\beta_k) y_{q,k} \quad q=revenue \qquad (24)$$

$$\sum_{i=1}^{N} x_{p,i} \lambda_i \leq x_{p,k} \quad p=cogs, capital \qquad (25)$$

Table 2. Shadow price by sector.

	Industry	Average Shadow Price (US$)	Median Shadow Price	Shadow Price (<100$)	Shadow Price (100$– 1,000$)	Shadow Price (1,001$– 10,000$)	# obs.
1	Utilities	245	46	60%	32%	8%	308
2	Construction & Materials	2,315	65	52%	13%	31%	311
3	Basic Resources	456	184	37%	52%	10%	434
4	Oil & Gas	1,673	582	13%	53%	31%	375
5	Chemicals	2,020	766	10%	50%	37%	254
6	Food & Beverages	4,457	3,269	3%	13%	80%	328
7	Automobiles & Parts	6,969	3,491	3%	22%	60%	146
8	Travel & Leisure	4,060	4,200	5%	20%	72%	289
9	Industrial Goods & Services	11,478	4,811	2%	17%	46%	1,614
10	Retail, Financial & Real Estate	9,620	5,602	0%	6%	65%	547
11	Personal & Household Goods	16,819	8,718	0%	4%	49%	434
12	Telecommunications	14,235	11,044	0%	5%	38%	208
13	Healthcare	13,666	11,605	1%	3%	41%	387
14	Media	20,341	17,833	1%	0%	24%	285
15	Technology	29,399	22,092	0%	3%	29%	565
	Average	119,018	118,606	10%	18%	42%	

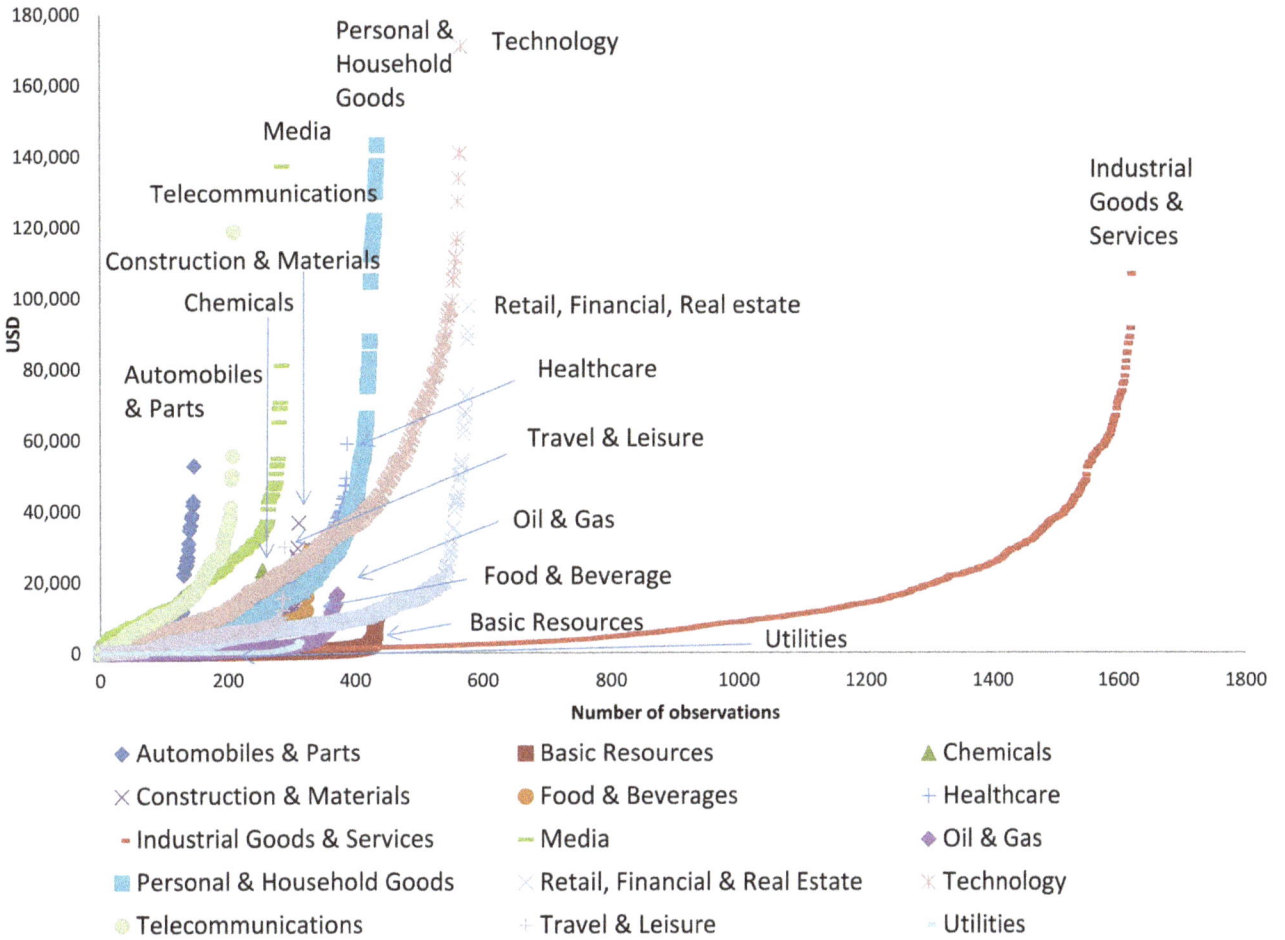

Figure 1. Shadow price by sector.

$$\sum_{i=1}^{N} b_{r,i}\lambda_i = (1-\beta_k)b_{r,k} \quad r = CO_2 \ emissions \qquad (26)$$

$$\sum_{i=1}^{N} \lambda_i \leq 1 \qquad (27)$$

$$\lambda_i \geq 0 \qquad (i=1,2,\cdots,k,\cdots,N) \qquad (28)$$

The model above estimates the inefficiency score by considering the extent to which a company can reduce its CO_2 emissions and increase its revenue without increasing the cogs and capital. β represents the inefficiency score, and β >0 indicates that firm k inefficiently discharges CO_2 emissions and loses revenue compared with the efficient firms that form the production frontier line curve.

Equation (27) representing the DDF model has applied the decreasing return-to-scale (DRS) assumption. The DDF model commonly requires the return-to-scale assumption. In this study, we apply the DRS assumption to avoid an infeasible calculation in the time series analysis. Note that assuming DRS does not

eliminate the possibility of infeasible linear programming (LP) problems when weak disposability is imposed on bad outputs. When we model bad outputs in LP, the potential for infeasible LP problems results from imposing weak disposability on the undesirable outputs and from specifying an LP problem that uses period t reference technology with observations from period t+1. The variable return-to-scale (VRS) model tends to yield infeasible results more often than the DRS model when computing the productivity change because the VRS has stronger restrictions for solving a linear program. We confirmed with our models that the calculation of productivity changes under the VRS is infeasible. Therefore, we applied the DRS model in this study.

This study used the GHG emissions data and sales revenues by industry provided by the Trucost, and financial data provided by the Factset, respectively. Therefore, this study used data from these two companies, which is considered to ensure the highest level of quality of the data available at present.

The Trucost's GHG emissions data is created by taking the quantity of GHG emissions disclosed by the businesses through environmental and financial reports supplemented by Trucost's own processes of verification and modification. When verifying the data, Trucost confirms whether each company reports irregular figures by comparing the reported data with calculations of the plausible quantities of GHG emissions based on the quantity of energy used by the business in question, as well as by comparison with the GHG emissions of other companies in the same industry.

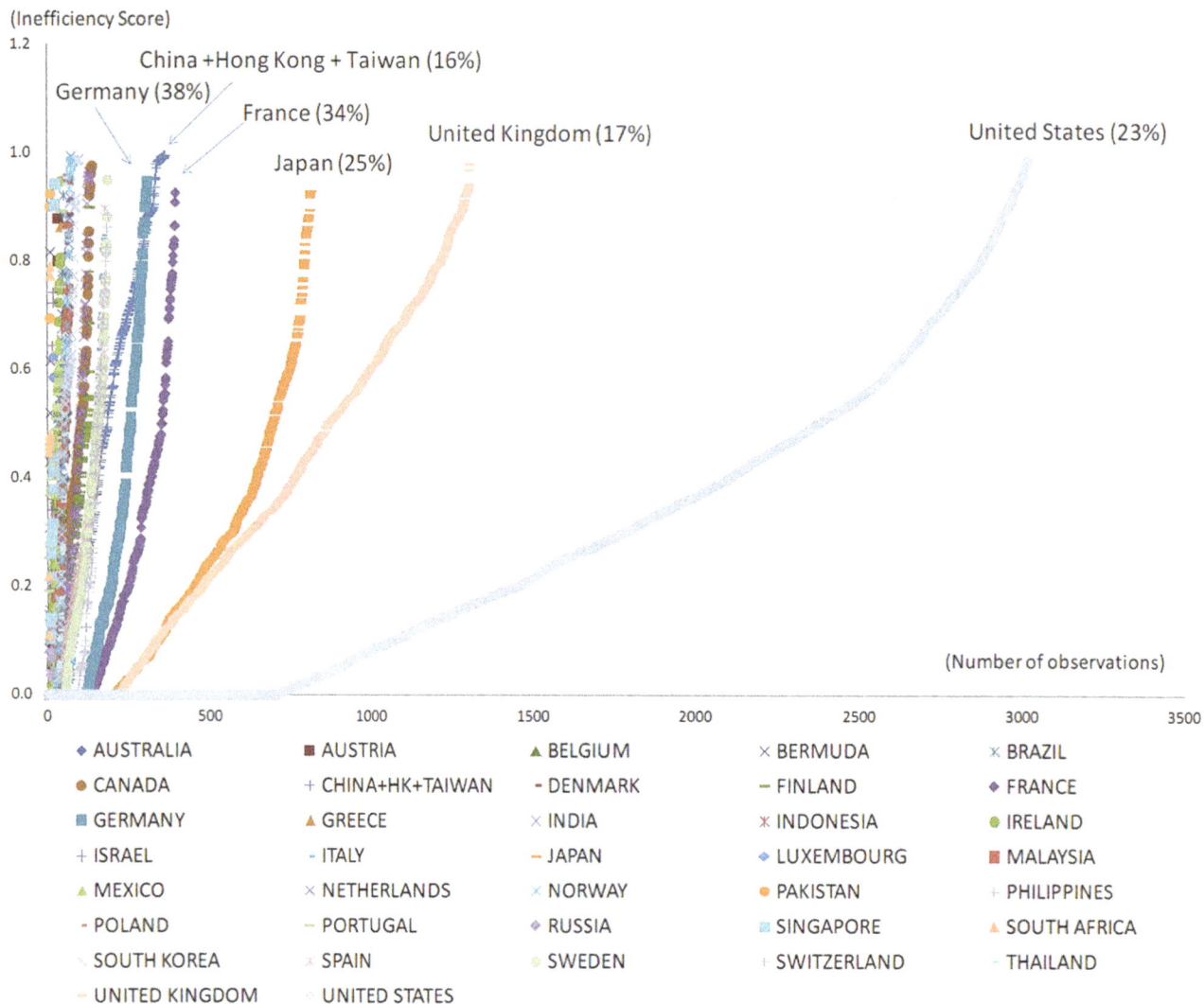

Figure 2. Inefficiency score by country.

Furthermore, if in this process a figure that could be considered irregular is discovered, interviews are held with representatives from the company concerned and corrections are made as required for the maintenance of the database. Since the data set is not a collection of unverified data released voluntarily by individual businesses, their data quality is better than data sets of Carbon Disclosure Project and similar sources, which is provided directly by the firms. At present, this is the most reliable database.

Regarding financial data, because this is reported by each company pursuant to the accounting standards of each country, and because each type of figure has undergone corporate auditing, reliable data that enables comparisons between companies can be obtained relatively easily. This study used the figures of the Factset, which is one of the largest business financial data firms in the world. Note that since accounting standards vary depending on the country, figures were used that were recompiled according to the Factset standard (a global integrated standard created by Factset). US dollars were adopted as the currency, and, after consolidating the exchange rates as of the end of the accounting year for each year, the figures were deflated to the year-2000 prices.

The research period covered in this study is in principle the eight-year period from 2002. The 17 industry sectors subject to analysis in this study are Automobiles & Parts, Basic Resources (including steel and paper industries), Chemicals, Oil & Gas, Utilities, Industrial Goods & Services, Construction & Materials, Personal & Household Goods, Telecommunications, Technology, Healthcare, Travel & Leisure, Retail, Food & Beverage, Financial, Real Estate, and Media services. For the classification of businesses into industry sectors, the ICB-Super Sector classification of the Financial Times of the UK and the London Stock Exchange which creates finance-related indexes, was referenced. In addition, retail, financial, and real estate services were integrated to ensure a sufficient quantity of data in each industry when running the model.

The data used comes under sales revenues, cost of goods sold (COGS), total assets, current assets, and GHG emissions (for the six gases listed for reduction in the Kyoto Protocol). For a figure for capital, fixed assets (total assets minus current assets) was chosen, and to express the (total) amount of labor and materials used, cost of goods sold was chosen. Cost of goods sold is the costs required to provide goods and services. For GHG emissions it was decided that all three scope categories would be included: Scope 1 (emissions in manufacturing processes), Scope 2 (emissions due to power usage among others), and Scope 3 (other emissions from commuting, business trips, and the supply chain).

This study applies the emission data for all the categories of Scope 1, 2, and 3 because of the following reasons; (1) using all three enables a more thorough assessment of the overall picture of each company's GHG emissions and (2) since this study covers a broad range of industries, including IT, Media, Healthcare, and financial institutions, rather than specifically the manufacturing industry, limiting the scope would increase the number of cases that do not reflect the actual state of the company. However, standards for the supply chain data in Scope 3 are currently being debated in each industry, so the figures were deemed unworkable for comparisons between companies at present and it was decided to use only those for GHG emitted during the use of airplanes, railroads among others.

Results

The average shadow price of GHG emissions for the observed sample of 1,024 companies in 37 countries worldwide is $10,414, and the median price is $4,189; these values are much higher than the corresponding values in any of the previous studies (Table 1). The reason for the higher shadow price is the use of the Scope 3 GHG emissions data in this study. This research implies that a firm would carry a high cost of GHG emissions if Scope 3 GHG emissions were the focus of discussions regarding corporate social responsibility. In fact, 10% of the observed companies may reduce emissions at a cost of $100 or less, approximately 30% at $1,000 or less, and 70% at $10,000 or less. Germany's median shadow price is the highest at a price of $9,423 among the six countries with more than 200 observations: the United States, the United Kingdom, Japan, France, and China (including Hong Kong and Taiwan). France follows Germany at a price of $8,697, with the United Kingdom at $7,335, the United States at $3,340, China at $4,316, and Japan at $2,332 (see Figure S1 and Figure S2).

Among these six countries, the proportion of companies that could reduce emissions at $100 or less is the largest in China at a rate of 14%, followed by the United States at 10%, but this order changes at shadow prices of $1,000 and $10,000 or less. At $1,000 or less, Japan's proportion becomes the second largest at a rate of 35%, slightly below China at 36%. At $10,000 or less, Japan's proportion becomes the largest at a rate of 83%, followed by the United States and China, both at 74%. Although the number of observations in emerging countries is limited, the study found that a majority of companies in certain emerging countries can reduce emissions relatively inexpensively. For instance, 63% of the observed companies in Indonesia can reduce emissions at $20 or less, followed by 52% in Thailand and 35% in Malaysia. These price gaps among different countries indicate the economic rationale for an emissions trading scheme and the importance of the Clean Development Mechanism (CDM) to achieve emissions reductions on a global level.

In addition, a large disparity was found to exist between the average and median shadow prices, suggesting that the average price in each country is being supported by a set of companies and that many companies can reduce emissions relatively inexpensively.

The shadow price is the lowest for utilities companies at a median price at $46, followed by construction and material companies at $65 and basic resource companies at $184; by contrast, the price is the highest in the technology sector at $22,092 (Table 2). Shadow prices are relatively low in heavy industries such as construction and materials at $65 and basic

resources at $184, and more than 50% of the observed companies in these industries can reduce emissions at a cost of $100 or less. Half of the observed companies in the oil and gas and chemicals industries can reduce emissions at a cost of $1,000 or less, whereas the shadow prices in the technology and media sectors are high–only a low percentage of the observed companies in these sectors can reduce emissions at a cost of $1,000 or less. If a cap on emissions were introduced at the same level in all industries, then those most readily affected would be industries with high shadow prices, such as technology and media, and those least affected would be industries with low shadow prices, such as utilities and basic resources.

Discussion

There are major disparities in the shadow price both among industries and within industries–some companies can reduce emissions at a relatively inexpensive price, whereas others cannot, even in the same industry (Figure 1). This disparity is particularly evident in industries such as technology, personal and household goods, and media. This finding supports the idea of a cap and trade emissions scheme because the variations in the shadow price for different companies suggests that it is economically efficient to trade carbon credits between companies with a low cost of emissions reductions and those with a high cost. The wide variation of shadow prices among companies within individual industries further indicates that an emissions trading scheme could be established not only among industries but also within industries; even if the system targets a single domestic sector, it is still effective. A shadow price that varies by company also assists in explaining which firms would become sellers and buyers of carbon credits and thus in determining the approximate quantities bought and sold in carbon markets. If the same level of carbon constraints (total quantity reduction) were to be imposed on all industries, then technology and other industries with high shadow prices would become buyers of carbon credits, and utilities and other industries with low shadow prices would become sellers. However, in reality, it is highly unlikely that the same level of emissions cap would be introduced for all industries, given their differing quantities of GHG emissions. As in the cap and trade system, it would be possible to estimate the tax burden on each business if an environmental tax were introduced.

Fig. 1 indicates that the shadow price of GHG emissions in the utility sector is much lower than in other sectors. The average shadow price of utility companies is $245 per ton of GHG, and the median is $46 per ton of GHG. The results reveal that 35% of the observed companies can reduce emissions at $20 per ton of GHG or less, 60% at $100 per ton of GHG or less, and 75% at $180 per ton of GHG or less. Furthermore, more than 40% of the observed companies in the United States can reduce emissions at $20 per ton of GHG or less. Moreover, a majority of companies in these five countries can reduce emissions at $100 per ton of GHG or less.

Fig. 2 shows the operating efficiency (inefficiency score) by country in terms of the GHG emissions for 1,024 companies in 37 countries worldwide during the eight-year period from 2002 through 2009. The six countries shown in the figure are the United States, the United Kingdom, Japan, France, China (including Hong Kong and Taiwan), and Germany, all of which had large sample sizes; the number in parentheses is the frontier percentage (the percentage of results that were assessable as having an inefficiency score of zero and as conducting efficient operations) for each country.

Each country has companies with a high level of operating efficiency (companies with low inefficiency scores) and companies with low operating efficiency (companies with high inefficiency scores) (see Fig. 2). Because the data sample from the companies in the United States is the largest, the curve for the United States swings far to the right, but the frontier percentage clearly indicates that the United States does not simply have an especially high concentration of top-quality companies that would constitute the frontier curve. In fact, quality companies exist in every country in the world, and although quality companies may be concentrated in the center of the sample, the existence of frontier companies in China at a similar percentage relative to those in the United Kingdom indicates that any misconception of emerging market businesses being synonymous with inefficient businesses should be eliminated. Moreover, frontier companies and inefficient companies are not distributed in a manner that indicates that companies in advanced countries are synonymous with highly efficient companies or that emerging market companies are synonymous with inefficient companies. Rather, inefficient companies exist in every country, which suggests that a general improvement in such companies is required to reduce the world's GHG emissions and to ensure continued economic growth (see Table S1 and Figure S3). To reduce GHG emissions in the world, policy measures are needed to advance operations, and such measures need to consider the GHG emissions of companies in specific countries or regions and the emissions in all countries.

Policy Implications

Disparities in market competitiveness originate from the medium- to long-term management efforts of businesses, and eliminating these disparities is time consuming because it requires improvements in operating technology.

The figures and tables above indicate that there are major disparities in the shadow price, both among industries and within industries, similar to the disparities observed in market competitiveness. This finding appears to have extremely important implications concerning the administration of environmental issues, especially for the design and introduction of emissions trading schemes, as the research findings support the introduction of emissions trading schemes from all perspectives. Emissions trading schemes allow for the flexible fulfillment of GHG reduction obligations by establishing limits (gaps) for GHG emissions and allowing trading within those emissions limits. By allowing companies and groups with varying GHG reduction costs

to trade emission rights according to their respective needs, inexpensive initiatives to reduce emissions will be selected efficiently, thus promoting the efficient reduction of emissions for society as a whole.

In this study, the shadow prices were found to vary greatly, both within and outside of industries. The disparities in shadow prices that were observed in this study clearly indicate that the introduction of emissions trading schemes would make it economically rational for companies with high shadow prices to trade emissions rights with companies with low shadow prices to promote the overall reduction of GHG emissions in an efficient manner in Japan and in other countries. Furthermore, the finding that shadow prices are widely divergent among individual companies, both across industries and within industries, indicates that emissions trading schemes could be established not only among industries but also within industries. In other words, emissions trading schemes could function meaningfully even if they target a single domestic industry.

Supporting Information

Figure S1   Shadow price of GHG emissions by country.

Figure S2   Mean and median value of the shadow price by region and by country.

Figure S3   Inefficiency score by region.

Table S1   Summary of inefficiency scores and the number of companies by country.

Text S1   Supporting Information.

Acknowledgments

Disclaimer: The results and conclusions of this article do not necessary represent the views of the funding agencies.

Author Contributions

Conceived and designed the experiments: NI HF SM. Performed the experiments: NI HF SM. Wrote the paper: NI HF SM.

References

1. Lobell DB, Schlenker W, Costa-Roberts J (2011) Climate Trends and Global Crop Production Since 1980. Science 333: 616–620.
2. IPCC (2007) Climate Change 2007: Impacts, Adapation and Vulnerability. Contribution of Working Group II to the Fourth Assessment Report of the Intergovernmental Panel on Climate Change. Parry ML, Canziani OF, Palutikof JP, van der Linden PJ, Hanson CE, editors. Cambridge, UK: Cambridge Univ. Press. 7–22
3. Lee JD, Park JB, Kim TY (2002) Estimation of the shadow prices of pollutants with production/environment inefficiency taken into account: a nonparametric directional distance function approach. J of Env. Manag. 64: 365–375.
4. Färe R, Grosskopf S, Noh D, Weber W (2005) Characteristics of a polluting technology: theory and practice. Journal of Econometrics 126: 469–492.

11

Respiratory Health before and after the Opening of a Road Traffic Tunnel: A Planned Evaluation

Christine T. Cowie[1,2]*, **Nectarios Rose**[1,3], **Wafaa Ezz**[1,2], **Wei Xuan**[1,2], **Adriana Cortes-Waterman**[1,2], **Elena Belousova**[1], **Brett G. Toelle**[1], **Vicky Sheppeard**[4], **Guy B. Marks**[1,2]

1 Woolcock Institute of Medical Research, Sydney, Australia, 2 Cooperative Research Centre for Asthma and Airways, Sydney, Australia, 3 NSW Health Department, Sydney, Australia, 4 Western Sydney and Nepean Blue Mountains Public Health Unit, Sydney, Australia

Abstract

Objective: The construction of a new road tunnel in Sydney, Australia, and concomitant reduction in traffic on a major road presented the opportunity to study the effects of this traffic intervention on respiratory health.

Methods: We made measurements in a cohort of residents in the year before the tunnel opened (2006) and in each of two years afterwards (2007–2008). Cohort members resided in one of four exposure zones, including a control zone. Each year, a respiratory questionnaire was administered (n = 2,978) and a panel sub-cohort (n = 380) performed spirometry once and recorded peak expiratory flow and symptoms twice daily for nine weeks.

Results: There was no consistent evidence of improvement in respiratory health in residents living along the bypassed main road, despite a reduction in traffic from 90,000 to 45,000 vpd. Residents living near tunnel feeder roads reported more upper respiratory symptoms in the survey but not in the panel sub-cohort. Residents living around the tunnel ventilation stack reported more upper and lower respiratory symptoms and had lower spirometric volumes after the tunnel opened. Air pollutant levels measured near the stack did not increase over the study period.

Conclusion: The finding of adverse health effects among residents living around the stack is unexpected and difficult to explain, but might be due to unmeasured pollutants or risk factors or an unrecognized pollutant source nearby. The lack of improvement in respiratory health among people living along the bypassed main road probably reflects a minimal change in exposure due to distance of residence from the road.

Editor: Stephania Ann Cormier, Louisiana State University Health Sciences Center, United States of America

Funding: The authors acknowledge funding support from the Cooperative Research Centre for Asthma & Airways and the NSW Health Department. The funders had no role in study design, data collection and analysis, decision to publish, or preparation of the manuscript, other than the contributions made by NR and VS as indicated under Competing Interests.

Competing Interests: The authors have read the journal's policy and have the following conflicts: NR was employed with the NSW Health Department on the Biostatistics Training Scheme and VS was employed in the NSW Health system, at the time the study was conducted. There are no other competing interests.

* E-mail: christine.cowie@sydney.edu.au

Introduction

Road tunnels are increasingly being constructed in major world cities to help alleviate traffic congestion. They are often located within densely populated areas, have high traffic volumes, including a substantial proportion of heavy duty vehicles, and pollutants are commonly vented to the external environment through stacks [1–3]. However, studies investigating the health impact of traffic-related air pollution (TRAP) on communities residing around road tunnels, or their ventilation stacks and portals, are scarce, probably due to difficulty in disentangling health effects associated with general traffic from effects attributable to emissions arising in the tunnels [4].

We are aware of only one study that has investigated health effects associated with exposure to road tunnel stack emissions [2]. This cross-sectional study, conducted after the tunnel was opened, found no difference in the prevalence of self-reported respiratory and irritant symptoms between three zones estimated to have varying exposure to the ventilation stack plume.

The Lane Cove Tunnel (LCT) is a 3.6 km long road tunnel that connects two motorways and is vented by two stacks located near each end of the tunnel. Modelling prior to tunnel construction predicted that areas around the tunnel feeder roads would experience increased exposure to nitrogen dioxide (NO_2) and particulate matter (PM), and areas adjacent to the bypassed main road would experience a decrease in exposure [5]. Modelling of stack emissions forecast little detectable impact on the surrounding area. The tunnel commenced operation on 25 March 2007, with an average of 43,446 vehicles per day (vpd) during May, 2007 increasing to 58,218 vpd by November, 2008, with 2.9% and 3.5% of vehicles respectively classified as heavy or articulated vehicles. Opening of the tunnel was accompanied by an approximate halving of vehicle numbers from 90,000 vpd along the bypassed main road [6].

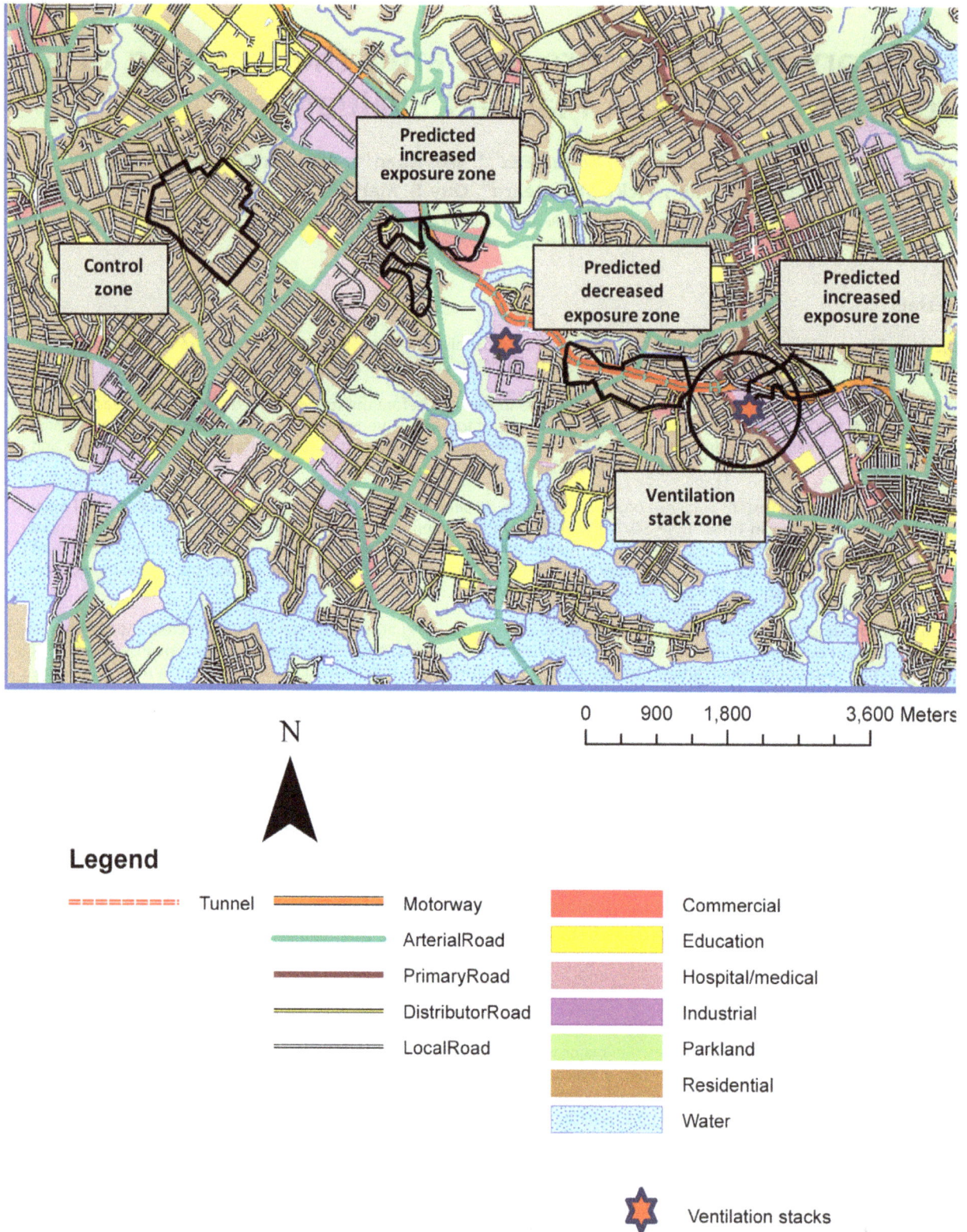

Figure 1. Map of study area showing the zones, tunnel, ventilation stacks and land use.

This study was commissioned by the NSW Health Department to inform public policy on road tunnels and their ventilation stacks and to opportunistically study respiratory health effects of a substantial traffic intervention. The principal hypothesis was that, after tunnel opening, adverse respiratory health outcomes would increase among residents near the tunnel feeder roads and

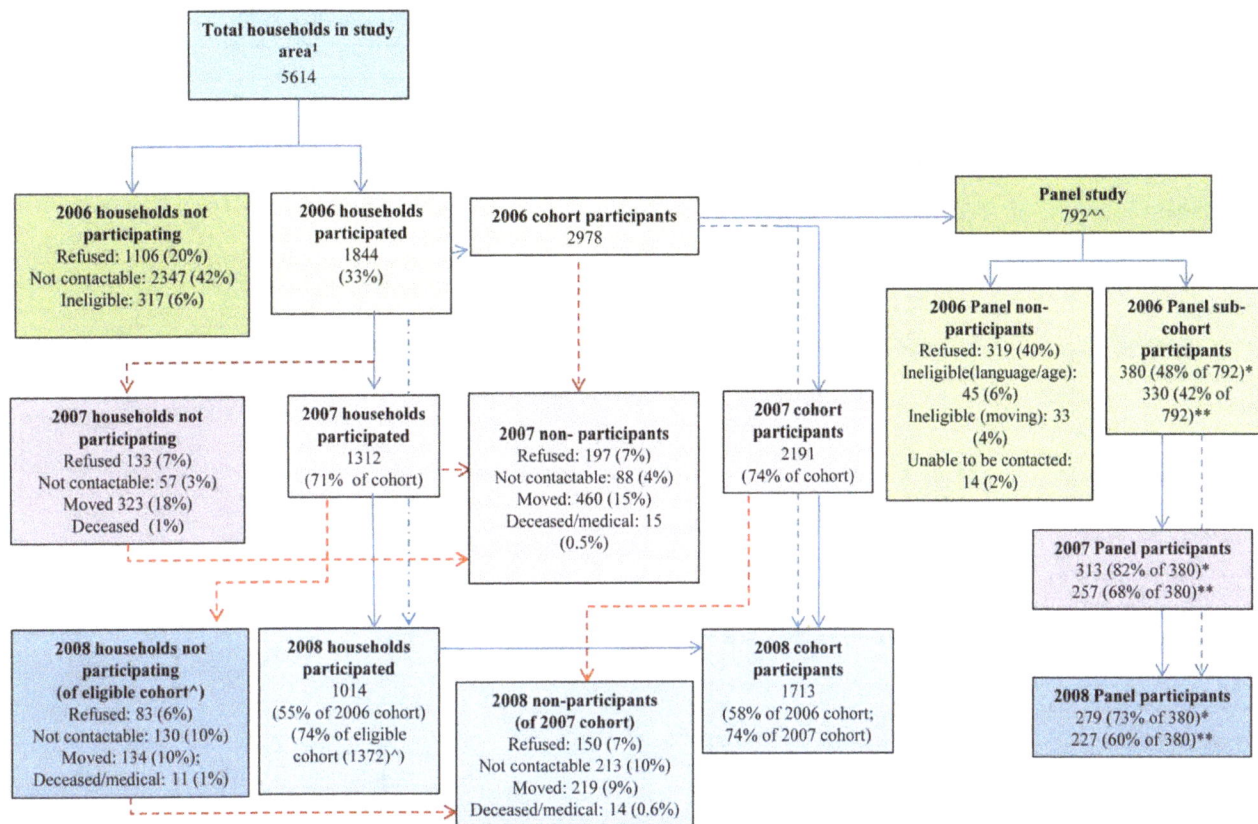

Figure 2. Participation flowchart. [1] Estimate derived from census of households within the study area. ^ Eligible cohort = 2007 cohort + additional households potentially eligible to be interviewed in 2008. ^^ Subjects approached to participate in panel study. * Subjects who took part in spirometry and eNO measurement. ** Subjects who took part in daily lung function and symptom diary. Blue arrow: Household and participant participation. Red arrow: Loss to follow-up.

decrease among residents along the bypassed main road, after adjustment for concurrent changes in a control zone. A subsidiary hypothesis was that people living around the ventilation stack would not experience a change in respiratory health status.

Methods

Ethics Statement

Approval for the study was obtained from the University of Sydney Human Research Ethics Committee (HREC). Recruitment of subjects occurred through a household doorknock survey. At the time of the home visit an introductory letter about the study was provided to an adult resident, the nature of the study was discussed, and consent to participate in the survey was sought. If consent was given, this adult was considered to be the "primary respondent" for the household, and was interviewed using a standardised questionnaire. Parents completed questionnaires for children less than 16 years, and children aged 16–18 years completed questionnaires with a parent/caregiver present. Self-completion questionnaires were left for other household adults not present at the time of the visit. Up to three follow-up phone calls were made for return of the self-completed questionnaires. Interviewers returned to households up to three times and left calling cards for households unable to be contacted. The household contact and interviewing process was documented using a standardised form. Participants in the panel sub-cohort

provided separate and additional written consent. The recruitment process for the whole cohort and sub-cohort was approved by the HREC.

Subject Recruitment and Exposure Assignment

We recruited a cohort of 2,978 participants during June-December 2006 (pre-tunnel opening) and followed them up during the same time of year in 2007 and 2008 (post-tunnel opening), within two weeks of their initial participation date. Participants were recruited by door-knocking all households within four exposure zones (Figure 1), defined as follows:

a) a zone along the bypassed main road (predicted 'decreased exposure zone');

b) a zone around the tunnel feeder roads (predicted 'increased exposure zone');

c) a zone of 650 m radius around the tunnel's eastern ventilation stack ('stack zone'); and,

d) a control zone.

The first two zones were defined using NO_2 contours from Cal3qhc and Calmet/Calpuff dispersion modelling (see Supporting Information, Appendix S1). The zone around the eastern stack was included to address community concern about stack emissions, and the radius of 650 m was chosen to ensure a sufficient sample size. The control zone was in a nearby suburb, away from the

Table 1. Symptom prevalence and other characteristics of the study cohort by zone and year (n = 2978).

	Reduced exposure zone			Increased exposure zone[a]			Eastern stack zone[a]			Control zone		
	2006	2007	2008	2006	2007	2008	2006	2007	2008	2006	2007	2008
n	1024	756	608	589	422	302	645	500	394	768	537	424
(% of 2006)		(74)	(59)		(72)	(51)		(78)	(61)		(70)	(55)
%												
Adult (age ≥18 yrs)	72	70	68	74	72	71	75	72	72	72	72	72
Female	56	57	58	53	54	57	54	53	54	53	54	54
Diagnosed asthma	18	20	20	18	24	21	16	19	21	15	17	19
Current asthma	11	13	13	12	14	12	9	12	12	9	10	9
Wheeze	15	18	15	14	19	18	11	20	17	10	13	13
Asthma medication	11	14	15	11	14	14	10	14	13	9	10	11
Inhaled corticosteroids	6	8	9	4	8	7	4	7	6	5	5	7
Cough	40	49	50	33	52	48	36	52	53	34	47	48
Lower resp. symptoms[b]	39	47	47	35	51	50	34	48	52	33	43	43
Severe lower resp. symptoms[b]	10	11	12	12	14	13	8	17	12	8	9	10
Upper resp. symptoms[b]	17	24	22	10	25	28	12	27	25	12	22	20
Mouth symptoms	14	17	18	9	21	21	11	24	22	13	17	18
Smoker [c]	7	6	6	12	10	9	11	9	7	12	13	10
Home environmental factors												
Unflued gas heater	25	28	33	12	21	25	18	25	27	20	24	29
Gas cooktop or oven	52	58	59	25	52	51	44	56	56	42	46	48
ETS[d] at home	7	3	4	14	9	10	8	5	5	8	5	7
Educational status												
Tertiary educated	61	65	64	46	46	44	62	61	61	44	42	36
High school/diploma[e]	31	28	28	43	38	41	31	34	31	43	38	46
Up to middle school[f]	8	8	8	11	16	15	7	6	7	13	19	18
Work status												
Paid Work	74	76	77	66	70	70	75	76	78	71	70	73

[a]Participants in overlapping area of increased exposure zone and stack zone, entire cohort (2006 = 48; 2007 = 24; 2008 = 15) contributed data to both zones.
[b]Composite variables derived from questionnaire data (see Supporting information, Table S1).
[c]Subjects <18 years assumed to be non-smokers.
[d]Environmental tobacco smoke.
[e]Included Diploma/TAFE, and participants who responded "Other".
[f]Included participants who refused to respond.

influence of the tunnel, and not expected to experience any change in TRAP.

Eligibility criteria included: all residents within the zones; ages 2–75 years; being at home for ≥1 hour during rush hours (7:00 am–9.30 am or 4.30 pm–7:00 pm); and having sufficient English. Subjects were not told which zone they resided in. Participants were interviewed using the adult or children's version of the questionnaire.

A panel sub-cohort was recruited sequentially from the entire cohort to keep a nine week diary of symptoms and lung function. We originally intended recruiting only children to the sub-cohort, but, because of the scheduled opening of the tunnel and slower than expected recruitment, we decided, in late 2006, to also recruit adults. We recruited adults and children into two approximately equal strata until the sub-cohort quota was filled.

Questionnaire

Questionnaires collected information on respiratory symptoms and diagnoses (wheeze, cough and asthma), eye, nose and throat symptoms, and use of medications, and on potential confounders and effect modifiers including smoking, environmental tobacco smoke (ETS) exposure, and use of gas heating and gas cooking in the home. The respiratory health questions were adapted from the ISAAC [7] and Belmont studies [8], with questions about eye, nose and throat symptoms from the M5East study [2]. We asked about symptoms experienced within the preceding three months (in contrast to 12 months for ISAAC) so that the recall period for 2007 did not extend into the pre-tunnel period. Questions on mouth symptoms in adults, which we did not expect to be associated with air pollution exposure, were used to assess reporting bias. We used questionnaire responses to construct composite variables for three primary outcomes: lower respiratory tract symptoms (LRS), severe LRS (SLRS), and upper respiratory tract symptoms (URS) (see Supporting information, Table S1).

Diary Panel Sub-cohort

In panel sub-cohort participants (n = 380) we performed allergen skin prick tests to define atopic status in 2007 and

Table 2. Lung function measures, and other characteristics of the diary panel sub-cohort, by zone and year (n = 380)[a].

	Reduced exposure zone			Increased exposure zone[b]			Eastern stack zone[b]			Control zone		
	2006	2007	2008	2006	2007	2008	2006	2007	2008	2006	2007	2008
n[c] (% of 2006)	93	85 (91)	79 (85)	98	77 (78)	70 (71)	97	85 (88)	74 (76)	98	71(72)	61(62)
%												
Adult	46	42	41	48	44	43	45	42	46	52	54	56
Female	54	52	53	51	52	47	56	54	54	45	46	44
Atopic		58	60		57	53		50	48		58	58
Mean (SD)												
FEV_1 pre-bd[d] (L)	2.6 (0.8)	2.7 (0.8)	2.8 (0.8)	2.5 (0.9)	2.6(0.9)	2.8 (0.9)	2.8 (0.8)	2.9 (0.8)	3.1(0.8)	2.6 (0.9)	2.7 (0.8)	2.7 (0.7)
FEV_1 post-bd (L)	2.7 (0.8)	2.8 (0.8)	2.9 (0.8)	2.6 (0.9)	2.7(0.9)	2.8 (0.9)	2.9 (0.9)	3.0 (0.8)	3.1(0.8)	2.7 (0.9)	2.8 (0.8)	2.8 (0.8)
FVC pre-bd (L)	3.2 (1.1)	3.4 (1.1)	3.5(1.1)	3.1 (1.2)	3.3(1.2)	3.4 (1.1)	3.3 (1.1)	3.6 (1.1)	3.8(1.1)	3.2 (1.1)	3.4 (1.0)	3.4 (0.9)
FVC post-bd (L)	3.3 (1.1)	3.4 (1.1)	3.5 (1.1)	3.1 (1.2)	3.3(1.2)	3.4 (1.1)	3.4 (1.2)	3.6 (1.1)	3.8(1.1)	3.2 (1.1)	3.4 (1.0)	3.4 (1.0)
eNO (ppb)[e]	9.7 (1.8)	17.8(2.0)	8.9 (2.3)	8.9 (1.9)	18.7(1.8)	9.7 (2.0)	9.2 (2.0)	16.3(1.7)	8.0(2.0)	8.9 (1.8)	17.9(1.8)	9.1 (1.8)
n[f] (% of 2006)	86	76 (88)	70 (81)	82	57(70)	52 (63)	85	69 (81)	62(73)	81	57 (70)	46 (57)
Mean (SD)												
PEF am (L/min)[g]	453 (106)	456 (94)	447 (98)	372 (78)	377 (72)	351 (83)	429(137)	420(131)	459(125)	382(110)	377(103)	379 (86)
PEF pm (L/min)[g]	450 (108)	462 (90)	453 (97)	377 (83)	386 (79)	355 (80)	420(136)	429(126)	460 (136)	396 (103)	378 (97)	379 (87)

[a]Each subject contributed up to 63 days of data in the diary panel.
[b]Participants in overlapping area of increased exposure zone and stack zone: diary panel cohort (2006 = 6(4); 2007 = 5(2); 2008 = 5(3)) contributed data to both zones.
[c]Participants with spirometry and eNO.
[d]bd-bronchodilator.
[e]Geometric mean and SD.
[f]Participants with PEF/symptom diary.
[g]PEF recorded three times twice daily at morning and evening. Highest of three measurements used for each session.

measured spirometric lung function (forced expiratory volume in one second (FEV_1) and forced vital capacity (FVC)) and exhaled nitric oxide (eNO) once each year (2006–2008) at a home visit. We calculated the ratio of FEV_1 as a proportion of FVC (FEV_1/FVC) as a measure of airflow obstruction. We also asked participants to record symptoms, medication use and peak expiratory flows (PEF) twice daily for nine weeks each year at the same time of year (see Supporting information, Appendix S2).

Air Quality Monitoring

Two elevated air quality monitoring stations located near the ventilation stacks measured oxides of nitrogen (NOx), nitric oxide (NO), NO_2, particulate matter less than 2.5 μm and less than 10 μm in diameter respectively ($PM_{2.5}$ and PM_{10}) and carbon monoxide (CO). Monitoring methods were identical to those previously described for fixed site monitoring in the study area [9].

Statistical Analysis

We compared the study respondents' characteristics to the local population using Australian Census 2006 data [10].

Study hypotheses were tested using a regression model. Main study factors were study year, exposure zone and their interaction. We tested the hypotheses by fitting a contrast statement to this interaction term. This approach modelled changes in symptoms and lung function between pre-tunnel year (2006) and each post-tunnel year (2007 and 2008) by zone, adjusted for the change in the control zone. Coefficients for contrasts were reported as mean differences for the continuous variables, or odds ratios for binary or multinomial variables.

To adjust for autocorrelation we fitted random intercepts for subjects and households. An additional random intercept for study year was fitted for PEF data, due to repeated measures each year. Some models with both subject and household random effects did not converge, and we re-fitted the model with random intercepts for subject only.

We fitted both baseline models and models adjusted for covariates. All models, including baseline models, included adjustment for changes in the control zone. For adult questionnaire outcomes we fitted a model adjusting additionally for mouth symptoms to assess potential reporting bias. Covariates incorporated in adjusted models were specified a priori and included: asthma diagnosis at baseline; smoking status; ETS at home; use of gas cooking (stove or oven); use of unflued gas heaters; education; and employment status. Covariates were retained in these models if they contributed significantly (P<0.05) to the model.

Baseline models for FEV_1, FVC and PEF were adjusted for height, age, age[2], gender and their interaction, and models for FEV_1/FVC were adjusted for age and gender and their interaction, based on models used in predictive equations [11]. Baseline models for eNO were adjusted for age and gender. The same covariates as incorporated for the questionnaire analysis were used and all covariates were retained in these models.

Continuous variables were tested using linear regression. The distribution of eNO was skewed so its values were log-transformed. Other continuous variables were analysed without transformation. Questionnaire symptoms were analysed as binary outcomes using logistic regression. For the individual daily panel symptom data, we attempted to fit a general linear mixed model incorporating random effects for subject and day with a logit link. As these models did not converge, we used summarised data for each subject, converting daily

Table 3. Symptom prevalence of the diary panel sub-cohort, by zone and year (n = 380)[a].

	Reduced exposure zone			Increased exposure zone[b]			Eastern stack zone[b]			Control zone		
	2006	2007	2008	2006	2007	2008	2006	2007	2008	2006	2007	2008
N [c] (% of 2006)	86	76 (88)	70 (81)	82	57 (70)	52 (63)	85	69 (81)	62 (73)	81	57 (70)	46 (57)
%												
Symptom days [d]												
At least once	99	96	94	96	93	98	95	97	94	94	91	98
10% of days	93	84	79	89	70	77	82	78	89	84	72	76
>20% of days	69	67	60	73	53	62	62	64	66	63	60	61
Day-time symptoms [d]												
At least once	79	72	70	84	70	69	67	65	63	63	58	65
>10% of days	52	43	53	63	42	48	49	46	48	41	40	43
>20% of days	36	33	37	39	26	37	33	32	31	23	25	30
Night-time symptoms [d]												
At least once	64	61	59	73	44	50	66	59	65	54	42	52
>10% of days	36	30	27	38	25	29	34	32	40	30	26	28
>20% of days	19	13	13	21	12	15	18	22	21	15	19	17
Wheeze[d]												
At least once	28	24	20	35	21	23	22	28	35	35	28	33
>10% of days	9	7	9	17	4	10	11	9	5	15	14	7
>20% of days	3	7	3	10	4	4	6	4	0	11	9	4
Bronchodilator use [d]												
At least once	12	16	10	21	11	12	6	7	8	12	12	13
>10% of days	3	7	6	12	7	10	2	3	2	6	11	4
>20% of days	2	1	1	5	5	4	0	0	0	5	9	4
Cough [d]												
At least once	71	68	60	77	53	56	66	64	58	57	46	63
>10% of days	43	33	39	48	32	35	40	38	45	36	32	37
>20% of days	26	18	23	32	12	21	21	23	26	20	25	30
Upper resp. symptoms [d]												
At least once	50	50	53	60	44	44	48	49	53	40	44	50
>10% of days	19	17	29	24	11	21	18	23	18	9	14	26
>20% of days	5	8	13	12	7	10	6	9	3	5	11	15
Lower resp symptoms[e][f]	37	44	33	26	57	51	38	32	44	31	38	39
Severe lower resp symp[e][f]	6	7	8	12	16	13	5	7	6	9	7	12
Upper resp. symptoms [e][f]	12	18	18	5	27	33	11	22	15	8	26	20
Mouth symptoms [f]	12	14	22	9	18	15	7	22	21	8	14	9

[a]Each subject contributed up to 63 days of data in the diary study.
[b]Participants in overlapping area of increased exposure zone and stack zone, diary panel cohort (2006 = 6(4); 2007 = 5(2); 2008 = 5(3)), contributed data to both zones. n = subjects in overlapping area with spirometry; (n) = subjects in overlapping area taking part in peak flow and symptom diary.
[c]Panel sub-cohort participants who took part in peak flow/symptom diary.
[d]Symptoms recorded by diary. % represents proportion of subjects experiencing those symptoms a) at least one day during the study period; b) >10% of days during the study period; c) >20% of days during the study period.
[e]Composite variables derived from variables measured in questionnaire (See Table S1, Supporting Information).
[f]Attributes measured by questionnaire.

symptom data to the proportion of days each year the subject reported symptoms, classified on an ordinal scale as: no days; 1–10%; 11–20%; and 21–100% of days. These data were analysed using ordinal regression with a cumulative logit link and a multinomial error distribution. The same covariates as specified previously were incorporated and retained in the diary symptom models.

Linear regression and multinomial regression models were fitted in Proc MIXED and Proc GLIMMIX, respectively using SAS 9.2. The logistic regression with two random effects used for the questionnaire analysis was fitted in MLwiN using the Markov Chain Monte Carlo (MCMC) estimation method, as these models did not converge when using numerical methods such as adaptive quadrature.

Table 4. Odds ratios for symptoms reported by questionnaire in 2007 and in 2008 relative to 2006 adjusted for change in the control zone.

Outcomes measured by questionnaire		2007 vs 2006		2008 vs 2006	
		Odds Ratio (95% CI)	P	Odds Ratio (95% CI)	P
Upper respiratory symptoms					
Reduced exp zone	Baseline model	0.7 (0.5, 1.2)	0.2	0.8 (0.5, 1.3)	0.3
	Adjusted	0.7 (0.4, 1.2)[a]	0.2	0.8 (0.5, 1.3)[a]	0.3
Increased exp zone	Baseline model	2.0 (1.2, 3.6)	0.02	3.2 (1.7, 6.1)	<0.001
	Adjusted	1.9 (1.0, 3.3)[a]	0.04	3.0 (1.6, 5.6)[a]	<0.001
Stack zone	Baseline model	1.8 (1.1, 3.2)	0.03	1.8 (1.0, 3.2)	0.05
	Adjusted	1.8 (1.1, 3.1)[a]	0.03	1.7 (1.0, 3.0)[a]	0.07
Lower respiratory symptoms					
Reduced exp zone	Baseline model	0.9 (0.6, 1.3)	0.7	0.9 (0.6, 1.4)	0.7
	Adjusted	0.9 (0.6, 1.3)[b]	0.6	0.9 (0.6, 1.4)[c]	0.6
Increased exp zone	Baseline model	1.5 (1.0, 2.2)	0.07	1.4 (0.8, 2.2)	0.2
	Adjusted	1.5 (1.0, 2.2)[b]	0.08	1.4 (0.8, 2.2)[c]	0.2
Stack zone	Baseline model	1.2 (0.8, 1.7)	0.5	1.6 (1.0, 2.4)	0.05
	Adjusted	1.2 (0.8, 1.7)[b]	0.4	1.6 (1.0, 2.5)[c]	0.04
Severe lower respiratory symptoms					
Reduced exp zone	Baseline model	1.1 (0.6, 2.0)	0.9	1.0 (0.5, 1.9)	1.0
	Adjusted	1.1 (0.6, 2.0)[d]	0.9	1.0 (0.5, 2.0)[d]	0.9
Increased exp zone	Baseline model	1.2 (0.6, 2.4)	0.7	1.0 (0.5, 2.1)	1.0
	Adjusted	1.1 (0.5, 2.2)[d]	0.8	1.0 (0.5, 2.0)[d]	0.9
Stack zone	Baseline model	3.0 (1.5, 6.0)	0.002	1.4 (0.7, 2.8)	0.4
	Adjusted	3.2 (1.6, 6.5)[d]	0.001	1.6 (0.8, 3.3)[d]	0.2
Wheeze in last 3 months					
Reduced exp zone	Baseline model	0.9 (0.5, 1.5)	0.7	0.9 (0.5, 1.5)	0.7
	Adjusted	0.9 (0.5, 1.6)[e]	0.8	0.7 (0.3, 1.2)[f]	0.2
Increased exp zone	Baseline model	1.1 (0.6, 2.0)	0.8	1.1 (0.6, 2.0)	0.8
	Adjusted	1.2 (0.6, 2.2)[e]	0.6	1.1 (0.5, 2.4)[f]	0.7
Stack zone	Baseline model	1.8 (1.0, 3.3)	0.06	1.8 (1.0, 3.3)	0.06
	Adjusted	2.0 (1.1, 3.6)[e]	0.03	1.5 (0.7, 3.0)[f]	0.3
Cough in last 3 months					
Reduced exp zone	Baseline model	0.8 (0.6, 1.1)	0.2	0.9 (0.6, 1.3)	0.4
	Adjusted	0.8 (0.6, 1.1)[g]	0.2	0.9 (0.6, 1.3)[h]	0.5
Increased exp zone	Baseline model	1.4 (0.9, 2.0)	0.1	1.1 (0.7, 1.8)	0.6
	Adjusted	1.5 (1.0, 2.2)[g]	0.06	1.2 (0.8, 1.9)[h]	0.4
Stack zone	Baseline model	1.1 (0.8, 1.7)	0.5	1.2 (0.8, 1.9)	0.3
	Adjusted	1.1 (0.8, 1.7)[g]	0.5	1.3 (0.8, 1.9)[h]	0.3

[a]Adjusted for: asthma, gas cooker or oven.
[b]Adjusted for: age, age^2, gender, asthma, smoker, unflued gas heater.
[c]Adjusted for: age, age^2, asthma, smoker, unflued gas heater.
[d]Adjusted for: age, age^2, asthma, smoker, ETS at home.
[e]Adjusted for: asthma, unflued gas heater.
[f]Adjusted for: age, age^2, asthma, unflued gas heater.
[g]Adjusted for: age, age^2, asthma, smoker, gas cooker or oven.
[h]Adjusted for: age, age^2, gender, asthma, smoker, gas cooker or oven.

We used an autoregressive model to test whether there was a difference in air quality at the elevated monitoring sites between 2006 and later years, adjusting for changes in regional air quality measured at three Sydney sites by the NSW Office of Environment and Heritage (OEH). Analysis was conducted in SAS 9.2 using the AUTOREG procedure.

Results

We identified 5,614 households in all study zones in 2006 (Figure 2). Overall household participation rate at baseline was 33%, varying from 25% in the stack zone to 38% in the control zone (see Supporting information, Table S2). 20% of households

Table 5. Differences in spirometry and peak flow and odds ratios for eNO, in 2007 and 2008 relative to 2006, adjusted for change in the control zone (n = 380).

Lung function outcomes		2007 vs 2006		2008 vs 2006	
		Difference (95% CI)	P	Difference (95% CI)	P
FEV1-pre bronchodilator (L) [a] [b]					
Reduced exp zone	Baseline model	−0.002 (−0.06, 0.06)	1.0	−0.01 (−0.08, 0.07)	0.9
	Adjusted	−0.002 (−0.06, 0.06)	1.0	−0.001 (−0.08, 0.08)	1.0
Increased exp zone	Baseline model	−0.02 (−0.08, 0.04)	0.5	−0.07 (−0.15, 0.00)	0.06
	Adjusted	−0.02 (−0.08, 0.04)	0.5	−0.07 (−0.15, 0.01)	0.09
Stack zone	Baseline model	0.01 (−0.05, 0.07)	0.8	−0.002 (−0.08, 0.07)	1.0
	Adjusted	0.01 (−0.04, 0.07)	0.6	0.004 (−0.07, 0.08)	0.9
FEV1-post bronchodilator (L) [a] [b]					
Reduced exp zone	Baseline model	−0.03 (−0.12, 0.05)	0.5	−0.05 (−0.15, 0.05)	0.4
	Adjusted	−0.03 (−0.12, 0.06)	0.5	−0.05 (−0.15, 0.06)	0.4
Increased exp zone	Baseline model	0.02 (−0.07, 0.10)	0.7	−0.06 (−0.17, 0.04)	0.2
	Adjusted	0.01 (−0.07, 0.10)	0.8	−0.06 (−0.17, 0.04)	0.2
Stack zone	Baseline model	−0.08 (−0.17, 0.00)	0.06	−0.06 (−0.16, 0.05)	0.3
	Adjusted	−0.08 (−0.16, 0.01)	0.07	−0.05 (−0.16, 0.05)	0.3
FEV1/FVC-pre bronchodilator [b] [c]					
Reduced exp zone	Baseline model	0.01 (−0.01, 0.03)	0.2	0.006 (−0.01, 0.03)	0.6
	Adjusted	0.01 (−0.01, 0.03)	0.2	0.005 (−0.01, 0.02)	0.6
Increased exp zone	Baseline model	0.002 (−0.02, 0.02)	0.8	−0.002 (−0.02, 0.02)	0.8
	Adjusted	0.002 (−0.02, 0.02)	0.8	−0.003 (−0.02, 0.02)	0.8
Stack zone	Baseline model	0.004 (−0.01, 0.02)	0.7	0.004 (−0.01, 0.02)	0.7
	Adjusted	0.004 (−0.01, 0.02)	0.7	0.005 (−0.01, 0.02)	0.6
FEV1/FVC-post bronchodilator [b] [c]					
Reduced exp zone	Baseline model	0.009 (−0.00, 0.02)	0.2	0.003 (−0.01, 0.02)	0.6
	Adjusted	0.009 (−0.00, 0.02)	0.2	0.003 (−0.01, 0.02)	0.6
Increased exp zone	Baseline model	0.00 (−0.01, 0.01)	1.0	−0.009 (−0.02, 0.00)	0.2
	Adjusted	−0.001 (−0.01, 0.01)	0.9	−0.009 (−0.02, 0.00)	0.1
Stack zone	Baseline model	0.005 (−0.01, 0.02)	0.5	0.009 (−0.00, 0.02)	0.2
	Adjusted	0.005 (−0.01, 0.02)	0.5	0.009 (−0.00, 0.02)	0.1
FVC-pre bronchodilator (L) [a] [b]					
Reduced exp zone	Baseline model	−0.04 (−0.12, 0.05)	0.4	−0.02 (−0.12, 0.08)	0.7
	Adjusted	−0.04 (−0.12, 0.05)	0.4	−0.02 (−0.12, 0.08)	0.8
Increased exp zone	Baseline model	−0.03 (−0.11, 0.05)	0.5	−0.09 (−0.19, 0.01)	0.09
	Adjusted	−0.03 (−0.11, 0.05)	0.5	−0.08 (−0.18, 0.02)	0.1
Stack zone	Baseline model	0.001 (−0.08, 0.08)	1.0	−0.02 (−0.11, 0.08)	0.8
	Adjusted	0.004 (−0.08, 0.08)	0.9	−0.01 (−0.11, 0.09)	0.8
FVC-post bronchodilator (L) [a] [b]					
Reduced exp zone	Baseline model	−0.05 (−0.15, 0.05)	0.4	−0.05 (−0.16, 0.07)	0.4
	Adjusted	−0.04 (−0.14, 0.06)	0.4	−0.05 (−0.16, 0.07)	0.4
Increased exp zone	Baseline model	0.01 (−0.09, 0.11)	0.8	−0.05 (−0.16, 0.06)	0.4
	Adjusted	0.01 (−0.09, 0.11)	0.8	−0.05 (−0.16, 0.07)	0.4
Stack zone	Baseline model	−0.11 (−0.21, −0.01)	0.03	−0.09 (−0.20, 0.02)	0.1
	Adjusted	−0.10 (−0.20, −0.01)	0.04	−0.09 (−0.20, 0.02)	0.1
Peak flow (L/min) morning [a] [b] [d]					
Reduced exp zone	Baseline model	−4.76 (−13.63, 4.11)	0.3	−5.74 (−16.90, 5.43)	0.3
	Adjusted	−5.20 (−14.40, 4.01)	0.3	−6.62 (−18.03, 4.79)	0.3
Increased exp zone	Baseline model	−2.56 (−11.85, 6.71)	0.6	−5.93 (−17.52, 5.65)	0.3
	Adjusted	−2.52 (−12.00, 6.95)	0.6	−4.56 (−16.46, 7.34)	0.5

Table 5. Cont.

Lung function outcomes		2007 vs 2006		2008 vs 2006	
		Difference (95% CI)	P	Difference (95% CI)	P
Stack zone	Baseline model	−9.69 (−18.66, −0.72)	0.03	4.31 (−6.83, 15.45)	0.5
	Adjusted	−9.77 (−18.94, −0.60)	0.04	4.64 (−6.70, 15.97)	0.4
Peak flow (L/min) evening [a] [b] [d]					
Reduced exp zone	Baseline model	−5.32 (−14.32, 3.67)	0.3	−6.93 (−18.39, 4.53)	0.2
	Adjusted	−5.29 (−14.61, 4.02)	0.3	−7.65 (−19.35, 4.06)	0.2
Increased exp zone	Baseline model	−0.83 (−10.24, 8.58)	0.9	−3.71 (−15.59, 8.18)	0.5
	Adjusted	−0.74 (−10.34, 8.85)	0.9	−1.82 (−14.02, 10.39)	0.8
Stack zone	Baseline model	−8.16 (−17.25, 0.94)	0.08	2.82 (−8.62, 14.25)	0.6
	Adjusted	−7.90 (−17.19, 1.38)	0.1	3.24 (−8.38, 14.86)	0.6
Exhaled nitric oxide (eNO) [b] [e]		Ratio (95% CI)		Ratio (95% CI)	
Reduced exp zone	Baseline model	0.94 (0.75, 1.19)	0.6	0.89 (0.71, 1.12)	0.3
	Adjusted	0.93 (0.74, 1.17)	0.5	0.92 (0.73, 1.14)	0.4
Increased exp zone	Baseline model	1.05 (0.83, 1.32)	0.7	1.08 (0.86, 1.35)	0.5
	Adjusted	1.00 (0.79, 1.26)	1.0	1.09 (0.87, 1.36)	0.4
Stack zone	Baseline model	0.90 (0.72, 1.13)	0.4	0.90 (0.73, 1.13)	0.4
	Adjusted	0.91 (0.73, 1.14)	0.4	0.93 (0.75, 1.15)	0.5

[a]Baseline models adjusted for: age; age^2; height; gender; gender*age; gender*age^2; gender*height.
[b]All adjusted models additionally adjusted for: diagnosed asthma at baseline; smoking; gas cooker or oven; unflued gas heating; ETS; education and employment status.
[c]Baseline models adjusted for: age, gender, gender*age.
[d]PEF recorded three times twice daily at morning and evening.
[e]Adjusted for age and gender.

refused to participate, 6% were ineligible and no contact could be made with 42% of households. Survey respondents had a similar distribution of demographic characteristics to the general population in the study area [10].

Of 2,978 cohort members who completed the 2006 questionnaire, 2,191 (74%) provided data in 2007 and 1,713 (58%) in 2008 (Figure 2). Loss to follow up was lower in the stack zone than the control zone (odds ratio (OR) 0.73, 95% CI 0.58–0.92 for 2007 and OR 0.82, 95% CI 0.67–1.01 for 2008) and higher in the reduced exposure zone than the control zone in 2008 (OR 1.28, 95% CI 1.04–1.57).

Baseline Data and Description of Study Group Over Time

Proportions of adults and females in the entire cohort did not vary substantially between zones or over the study period (Table 1). The prevalence of reporting wheeze, cough, diagnosed and current asthma, and use of asthma medication was similar across all zones at baseline (Table 1). The reduced exposure zone and the stack zone had more university-educated people and people in paid work, compared with the increased exposure zone and control zone. The proportion of participants using unflued gas heaters and gas cooktops/ovens increased in all zones over the study period. The prevalence of diagnosed asthma, cough, LRS, URS, and mouth symptoms increased in 2007 and 2008 in all zones.

Supporting information, Table S3, presents baseline data, reported in 2006, for the cohort that participated in each year of the study. Remaining members of the cohort in 2007 and in 2008 were similar in most baseline characteristics to the initial cohort, suggesting loss-to-follow-up is unlikely to have biased comparisons between years.

Of subjects asked to participate in the panel sub-cohort, 40% refused, 4% intended moving residence, and almost 6% were ineligible due to language or age constraints (Figure 2). The sub-cohort included more children than the entire cohort (Table 2), and the proportion of subjects with atopy was relatively high across all zones (50–58% in 2007). Spirometric measurements were similar across all zones in all years, although values for FEV_1 and FVC were slightly higher in subjects residing within the stack zone (Table 2). Participants in the stack zone and the reduced exposure zone had higher PEF readings in all years than participants from the other two zones. eNO readings were higher in 2007 than in 2006 and 2008, across all zones. Over the study period, participants in the stack zone reported increased prevalence of wheeze (at least once) and slightly more symptomatic days (>20% of days), whereas in other zones participants reported a decrease in these symptoms (Table 3). There was some year-to-year variation in the prevalence and frequency of diary symptoms, within each zone.

Supporting information, Table S4, presents baseline data, reported in 2006, for the panel sub-cohort that participated in each year. Remaining participants in 2007 and 2008 were similar in most baseline characteristics to the initial cohort except that, in all zones other than the increased exposure zone, those remaining had reported substantially more cough at baseline. Also, in the increased exposure zone and the stack zone, the remaining 2008 sub-cohort had reported slightly less "current asthma" and "asthma medication use" at baseline than the initial sub-cohort and those remaining in 2007. This suggests limited potential for selection bias due to loss-to-follow up in this sub-cohort.

Table 6. Odds ratios for symptoms reported in the diary panel sub-cohort in 2007 and in 2008 relative to 2006, adjusted for change in the control zone, (n = 380).

Symptoms measured by diary		2007 vs 2006		2008 vs 2006	
		Odds Ratio[a] (95% CI)	P	Odds Ratio[a] (95% CI)	P
Day symptoms					
Reduced exp zone	Baseline model	0.7 (0.2, 1.7)	0.4	0.6 (0.2, 1.6)	0.3
	Adjusted[b]	0.6 (0.2, 1.6)	0.3	0.5 (0.2, 1.4)	0.2
Increased exp zone	Baseline model	0.4 (0.1, 1.1)	0.06	0.4 (0.1, 1.3)	0.1
	Adjusted	0.3 (0.1, 0.9)	0.04	0.4 (0.1, 1.1)	0.09
Stack zone	Baseline model	1.1 (0.4, 2.9)	0.9	0.5 (0.2, 1.4)	0.2
	Adjusted	1.1 (0.4, 2.8)	0.9	0.4 (0.2, 1.3)	0.1
Night symptoms					
Reduced exp zone	Baseline model	1.0 (0.4, 2.7)	1.0	0.6 (0.2, 1.6)	0.3
	Adjusted	1.0 (0.4, 2.8)	1.0	0.5 (0.2, 1.4)	0.2
Increased exp zone	Baseline model	0.4 (0.2, 1.2)	0.1	0.5 (0.2, 1.4)	0.2
	Adjusted	0.5 (0.2, 1.3)	0.1	0.4 (0.1, 1.2)	0.1
Stack zone	Baseline model	1.4 (0.5, 3.8)	0.5	1.1 (0.4, 3.0)	0.9
	Adjusted	1.5 (0.5, 4.0)	0.5	1.0 (0.4, 2.8)	1.0
Symptom days					
Reduced exp zone	Baseline model	1.2 (0.4, 3.6)	0.7	0.6 (0.2, 2.0)	0.4
	Adjusted	1.1 (0.4, 3.4)	0.8	0.6 (0.2, 1.9)	0.4
Increased exp zone	Baseline model	0.5 (0.1, 1.4)	0.2	0.8 (0.3, 2.6)	0.7
	Adjusted	0.4 (0.1, 1.2)	0.1	0.6 (0.2, 2.0)	0.4
Stack zone	Baseline model	2.4 (0.8, 7.1)	0.1	1.8 (0.6, 5.6)	0.3
	Adjusted	2.5 (0.8, 7.5)	0.1	1.6 (0.5, 4.9)	0.4
Upper respiratory symptoms					
Reduced exp zone	Baseline model	0.6 (0.2, 1.6)	0.3	0.5 (0.2, 1.4)	0.2
	Adjusted	0.5 (0.2, 1.5)	0.2	0.4 (0.1, 1.2)	0.1
Increased exp zone	Baseline model	0.2 (0.07, 0.6)	0.005	0.1 (0.04, 0.4)	0.001
	Adjusted	0.2 (0.07, 0.6)	0.005	0.1 (0.04, 0.4)	<0.001
Stack zone	Baseline model	0.8 (0.3, 2.3)	0.7	0.3 (0.1, 0.9)	0.03
	Adjusted	0.8 (0.3, 2.2)	0.6	0.3 (0.1, 0.8)	0.02
Wheeze					
Reduced exp zone	Baseline model	1.2 (0.3, 4.0)	0.8	0.9 (0.3, 3.5)	0.9
	Adjusted	1.4 (0.4, 4.7)	0.6	1.1 (0.3, 4.0)	0.9
Increased exp zone	Baseline model	0.5 (0.1, 2.0)	0.3	1.2 (0.3, 4.6)	0.8
	Adjusted	0.5 (0.2, 1.9)	0.3	1.1 (0.3, 4.2)	0.9
Stack zone	Baseline model	2.2 (0.6, 7.5)	0.2	2.9 (0.8, 10.5)	0.1
	Adjusted	2.6 (0.8, 8.6)	0.1	3.3 (0.9, 11.6)	0.06
Cough					
Reduced exp zone	Baseline model	0.9 (0.3, 2.3)	0.8	0.4 (0.1, 1.1)	0.07
	Adjusted	0.8 (0.3, 2.2)	0.7	0.3 (0.1, 0.9)	0.03
Increased exp zone	Baseline model	0.4 (0.1, 1.1)	0.08	0.3 (0.1, 0.9)	0.03
	Adjusted	0.4 (0.1, 1.1)	0.06	0.2 (0.1, 0.7)	0.01
Stack zone	Baseline model	1.5 (0.5, 4.1)	0.4	0.6 (0.2, 1.8)	0.4
	Adjusted	1.5 (0.5, 4.1)	0.4	0.5 (0.2, 1.5)	0.2

[a]OR-odds of higher values on the ordinal scale (symptoms experienced: 0 days; 1–10% of days;10–20% of days; and 21–100% of days) compared with lower values, assuming proportional odds.
[b]Adjusted for: age; gender; diagnosis of asthma at baseline; smoker; ETS; gas stove or oven; unflued gas heating; education and employment status.

Table 7. Mean pollutant concentrations by year and difference in pollutant concentration from post-tunnel year compared to pre-tunnel year.

Site & pollutant	Mean pollutant levels (SD)		Difference in pollutant levels[a]	
	Pre-tunnel year[a]	First post-tunnel year[b]	Unadjusted estimate (95% CI)	Adjusted estimate (95% CI)[c]
Elevated site near eastern ventilation stack				
NO_2 ppb	16.7 (7.0)	15.7 (5.3)	−2.55 (−4.43, −0.66)	−1.00 (−2.21, 0.22)
NOx ppb	30.7 (17.8)	27.1 (14.1)	−6.33 (−10.97, −1.70)	−2.10 (−3.84, −0.37)
PM_{10} $\mu g/m^3$	20.2 (8.1)	16.1 (6.8)	−4.02 (−6.24, −1.80)	−1.74 (−3.04, −0.43)
$PM_{2.5}$ $\mu g/m^3$	6.4 (3.6)	5.3 (3.2)	−1.13 (−2.38, 0.12)	−0.26 (−0.65, 0.12)
Elevated site near western ventilation stack				
NO_2 ppb	13.4 (5.6)	12.1 (4.5)	−1.96 (−3.08, −0.84)	−1.07 (−1.68, −0.45)
NOx ppb	26.1 (19.9)	20.6 (14.6)	−6.05 (−10.47, −1.62)	−2.79 (−4.46, −1.11)
PM_{10} $\mu g/m^3$	18.7 (7.4)	16.2 (7.0)	−2.39 (−4.22, −0.56)	−0.46 (−1.94, 1.03)
$PM_{2.5}$ $\mu g/m^3$	6.2 (3.5)	5.3 (3.3)	−1.02 (−1.98, −0.06)	−0.19 (−0.65, 0.27)

[a]Difference between pre-tunnel and first post-tunnel years, estimated by autoregressive analysis.
[b]Pre-tunnel is 25 March 06–24 March 07 and first post-tunnel year is 25 March 2007–24 March 2008.
Data for only one post-tunnel year available.
[c]Adjusted for change in air quality at the three Sydney regional sites (averaged).

Questionnaire Symptom Analysis

Relative to 2006, the odds for URS in 2007 and in 2008 were significantly increased in the increased exposure zone and the stack zone, after adjusting for change in the control zone (Table 4). The odds for LRS and SLRS were significantly increased in 2008 and 2007 respectively, in the stack zone. The odds ratios for symptoms in the reduced exposure zone were generally at or below 1.0 in 2007 and 2008, although were not statistically significant. Adjusting for significant covariates did not change the estimates substantially (Table 4). Similarly, adjusting for mouth symptoms in the analysis of adult questionnaire responses ((model 2), made little difference to the effect estimates (model 1) (see Supporting information, Table S5).

Diary Analysis

In the panel sub-cohort, mean post-bronchodilator FVC and morning PEF were significantly reduced among stack zone participants in 2007, compared with 2006, after adjusting for change in the control zone (Table 5). This decrease was present, although not significant, for evening PEF. There was much less evidence of a similar change between 2006 and 2008. Although there were decreases in FEV_1 and FVC (pre-bronchodilator) in the increased exposure zone in 2008, they were not significant. There were no significant changes in lung function, airflow obstruction (FEV_1/FVC) or eNO in the other exposure zones.

In contrast to the questionnaire findings for the entire cohort, odds for symptoms recorded by panel sub-cohort participants were generally lower in all three exposure zones for both post-tunnel years, after adjustment for the control zone (Table 6). No significant increases in symptom prevalence were observed in any zone. Odds ratios for symptom days and wheeze in the stack zone were substantially greater than 1 but the confidence intervals (CIs) around these estimates were very broad. Adjustment for covariates did not materially alter the findings.

After adjusting for changes in regional air quality, monitors at the elevated sites near the ventilation stacks recorded significant decreases in PM_{10}, NOx and NO_2 in 2007 compared to 2006 (Table 7). Similarly, we previously showed using data from four additional ground level monitors that after adjusting for changes in

regional air quality and meteorological conditions, PM_{10}, NOx and NO_2 concentrations decreased in the eastern part of the study area after the tunnel opened [9].

Discussion

We did not observe consistent evidence of an improvement in respiratory health among residents living near a major road which experienced a traffic volume reduction of 50% after opening of a road tunnel. Among residents living adjacent to tunnel feeder roads there was increased reporting of upper respiratory tract symptoms in the questionnaire survey but not in the sub-cohort who recorded symptoms in a diary. The most consistently observed change was in residents living within a 650 m radius of the eastern ventilation stack. After the tunnel opened these subjects reported more upper and lower respiratory tract symptoms and had evidence of lower spirometric function, particularly lower FVC. Significant adverse effects were seen only for the first year of tunnel use (2007) and not for the second year (2008), except for questionnaire reported lower respiratory symptoms.

The strengths of this study are the cohort intervention design, in which measurements were made in the same individuals before and after commencement of operation of the tunnel, the inclusion of both subjective and objective measures of respiratory health, adjustment for coincident trends in a control site and for time-varying confounders, and testing for reporting bias.

Apart from the stack zone findings, findings from the questionnaire survey of the entire cohort and the panel sub-cohort were not consistent. This is perplexing but points to limitations of questions and diary cards that rely on self-report. Using a summary annual measure of symptoms for each subject, rather than daily symptom scores, in the sub-cohort analysis resulted in reduced statistical power, possibly contributing to inconsistency in results, but does not explain odds ratios being in the opposite direction.

Recall or other reporting bias is a potential weakness of self-reporting, which we sought to address by asking adults about mouth symptoms. The reported prevalence of mouth symptoms increased in all zones and was greatest in the increased exposure zone and the stack zone, a pattern similar for wheeze and LRS, which might be suggestive of information bias. However, when we included mouth symptoms as a covariate in the analysis, the observed effects for other symptoms were not attenuated. Hence, we conclude that the observed associations are unlikely to be explained by reporting bias.

The previous health investigation on a different tunnel stack [2] did not find an association between modelled annual average NOx concentrations representing ground level exposures from stack emissions and self-reported eye, nose and throat symptoms assessed by telephone interview. That study did not assess other respiratory symptoms or include lung function measures, thus making it difficult to compare with findings from our study. A few traffic intervention studies have reported improvements in respiratory health outcomes including wheeze and peak flow [12], hospitalisations for asthma [13], emergency department presentations [14], eNO [15], and nose and eye symptoms [16], after redirection or closure of roads or implementation of traffic restrictions during major events. These health benefits were demonstrated despite somewhat modest reductions in air pollutant concentrations, aside from major reductions seen during the Beijing Olympic Games. However, other studies have not found an effect [17]. For this local traffic intervention we did not see substantial benefits in health outcomes along the bypassed main road.

The most likely explanation for the lack of change in respiratory outcomes where this was expected is that the traffic intervention did not result in substantial changes in TRAP for most subjects. Although NO_2 concentrations decreased adjacent to the bypassed main road, these changes were smaller than predicted, probably due to lower than predicted traffic volume through the tunnel [9]. Furthermore, changes in NO_2 decreased markedly with increasing distance from the affected roads [18], which is consistent with previous evidence of a rapid decay in NO_2 concentrations with distance from major roads [19,20].

The findings of this study suggest an adverse health effect associated with living in the stack zone. This is based on lower and upper respiratory tract symptoms and wheeze recorded by questionnaire in the entire cohort, lung function analysis of panel participants, and (albeit not significantly) diary symptom analysis. These findings related primarily, but not exclusively to 2007, the first year of tunnel operation. While the confidence intervals for the diary symptom scores were wide, the outcomes are consistent with other effects reported for the stack zone. These findings are unexpected, given that neither the modelled stack emissions nor measured NO_2, PM_{10} or $PM_{2.5}$ at the elevated monitors, showed any evidence of increased concentration after the tunnel commenced operation.

Bias of some kind as an explanation for positive associations cannot be ruled out in observational studies, although, as previously discussed, we believe reporting bias is an unlikely explanation. While bias due to differential loss to follow-up is possible, there was no evidence that the remaining cohort had a higher prevalence of diagnosed asthma, current asthma or wheeze at baseline compared to the complete cohort or that smoking, education or work status changed appreciably (see Supporting information, Tables S3 and S4). This suggests a lack of any major selective factors underlying loss to follow-up. Another possibility is that the findings are due to unmeasured confounding factors.

We considered whether the adverse effects in the stack zone could be due to exposure to tunnel portal emissions due to its location near the tunnel's eastern portal, an area with potential to experience an increase in pollutant concentrations [4]. However,

the LCT was designed not to allow for portal emissions. Furthermore, the residents in the 'increased exposure zone', also located near the tunnel portal, did not show the same pattern of adverse health effects. The observed effects in the stack zone might have been due to TRAP associated with a highway which traverses the zone, however, highway traffic decreased after 2006 [9].

Another possible explanation for the stack zone findings is an alternative pollutant source located in the industrial area near the zone. While the local government authority indicated that land zoning did not change in the industrial estate during the study timeframe [21], it does not keep records of processes undertaken on individual properties. Thus we cannot rule out this possibility but also have no direct evidence for it.

The adverse findings in the stack zone are also not attributable to any measured pollutants emitted from the eastern ventilation stack, as discussed above. In addition, previous analysis of air quality data showed a decrease in NO_2, PM_{10} or $PM_{2.5}$ in the eastern section of the study area, over the study period [9]. Furthermore, the choice of a 650 m buffer for the stack zone was based on sample size considerations, rather than on any modelled or measured pollutant profiles.

It is possible, though, that the adverse findings are due to pollutants or size fractions other than those measured. Measurements of PM_{10} probably do not adequately reflect the particulate content of emissions in tunnels [4] and some studies show that PM_{10} and $PM_{2.5}$ do not represent TRAP as well as other pollutants such as NOx, NO_2, black carbon or elemental carbon [22,23]. Smaller size PM fractions that is, PM_1 or $PM_{0.1}$ (ultrafine particles (UFP)) are of lower mass, are present in far greater numbers, and, due to their oxidative potential, larger surface areas and ability to attach to other compounds, are considered to pose greater toxicity risks than PM_{10} [24–26]. Furthermore, elevated levels of UFP have been observed inside tunnels [27].

The finding of adverse respiratory health effects among residents living in the eastern stack zone is unexpected and difficult to explain, and as pollutant concentrations decreased in the immediate area, we are unable to attribute the effects to the ventilation stack. However, the findings imply the need to investigate the emission profile of the stack and other pollutant sources in the area that might explain the findings. The lack of improvement in respiratory health among people living along the bypassed main road is disappointing but probably reflects the short distance from the bypassed road within which exposure reduction might result in measurable health improvement.

Supporting Information

Table S1 Composite variables for LRS, SLRS and URS.

Table S2 Response rates for household participation and participant description at recruitment (2006), and comparison with general population (2006 Census).

Table S3 Baseline (2006) characteristics of the cohort members who participated, by zone and study year.

Table S4 Baseline (2006) characteristics of the diary panel sub-cohort who participated, by zone and year (n = 380).

Table S5 Odds ratios for symptoms reported by questionnaire in 2007 and in 2008, compared to 2006,

without (model 1) and with (model 2) adjustment for potential reporting bias (adult subjects only).

Appendix S1 Definition and choice of zones.

Appendix S2 Measurement of lung function, eNO, atopy and sample size calculations.

Acknowledgments

We are grateful to the participants in this study. We acknowledge advice received from Professor Bert Brunekreef. We acknowledge the assistance of

research staff: Kitty Ng, Angelica Ordonez, Paula Garai, Adrian Ferrero, and Jennifer Manning. The study was overseen by a Scientific Advisory Committee which provided advice on study design, methodology, analysis and interpretation, and consisted of Professor Bruce Armstrong, Professor Judy Simpson, Professor Bin Jalaludin, Dr Peter Lewis, and Dr Gavin Fisher.

Author Contributions

Conceived and designed the experiments: CTC VS ACW GBM. Performed the experiments: CTC WE ACW BGT GBM. Analyzed the data: CTC NR WX EB GBM. Wrote the paper: CTC NR WE WX ACW EB BGT VS GBM.

References

1. Bartonova A, Clench-Aas J, Gram F, Gronskei KE, Guerreiro C, et al. (1999) Air pollution exposure monitoring and estimation. Part V. Traffic exposure in adults. Journal of Environmental Monitoring 1: 337–340.
2. Capon A, Sheppeard V, Irvine K, Jalaludin B, Staff M, et al. (2008) Investigating health effects in a community surrounding a road tunnel stack - a cross sectional study. Environmental Health 7: 10.
3. Kuykendall JR, Shaw SL, Paustenbach D, Fehling K, Kacew S, et al. (2009) Chemicals present in automobile traffic tunnels and the possible community health hazards: A review of the literature. Inhalation Toxicology 21: 747–792.
4. National Health and Medical Research Council website. Air Quality In and Around Traffic Tunnels. Commonwealth of Australia 2008; ISBN online 1864964510 Available: http://wwwnhmrcgovau/guidelines/publications/eh42 Accessed 2011 Feb 25.
5. RTA (2001) Lane Cove Tunnel and Associated Road Improvements. Environmental Impact Statement. Working Paper Nine. Air Quality and Health Risk. Roads & Traffic Authority, NSW Government.
6. RTA (2010) Post Implementation Review. M7 Motorway, Cross City Tunnel and Lane Cove Tunnel. Roads & Traffic Authority (NSW), Motorways Project Branch.
7. ISAAC (1998) Worldwide variation in the prevalence of symptoms of asthma, allergic rhinoconjunctivitis, and atopic eczema: ISAAC. Lancet 351: 1225–1232.
8. Toelle B, Ng K, Belousova E, Salome C, Peat J, et al. (2004) The prevalence of asthma and allergy in schoolchildren in Belmont, Australia: three cross sectional surveys over 20 years. British Medical Journal 328: 386–387.
9. Cowie CT, Rose N, Gillett R, Walter S, Marks GB (2012) Redistribution of traffic related air pollution associated with a new road tunnel. Environmental Science & Technology 46: 2918–2927.
10. Australian Bureau of Statistics website. Census of Population and Housing, 2006. Available: http://www.abs.gov.au/census. Accessed 2008 Sep 17.
11. Hankinson J, Odencrantz J, Fedan K (1999) Spirometric Reference Values from a Sample of the General U.S. Population. American Journal of Respiratory & Critical Care Medicine 159: 179–187.
12. MacNeill SJ, Goddard F, Pitman R, Tharme S, Cullinan P (2009) Childhood peak flow and the Oxford Transport Strategy. Thorax 64: 651–656.
13. Lee JT, Son JY, Cho YS (2007) Benefits of mitigated ambient air quality due to transportation control on childhood asthma hospitalization during the 2002 summer Asian games in Busan, Korea. J Air Waste Manag Assoc 57: 968–973.
14. Li Y, Wang W, Kan H, Xu X, Chen B (2010) Air quality and outpatient visits for asthma in adults during the 2008 Summer Olympic Games in Beijing. Science of the Total Environment 408: 1226–1227.
15. Lin W, Huang W, Zhu T, Hu M, Brunekreef B, et al. (2011) Acute Respiratory Inflammation in Children and Black Carbon in Ambient Air before and during the 2008 Beijing Olympics. Environ Health Perspect 119: 1507–1512.
16. Burr ML, Karani G, Davies B, Holmes BA, Williams KL (2004) Effects on respiratory health of a reduction in air pollution from vehicle exhaust emissions. Occup Environ Med 61: 212–218.
17. Peel J, Klein M, Flanders W, Mulholland J, Tolbert P (2010) Impact of Improved Air Quality During the 1996 Summer Olympic Games in Atlanta on Multipple Cardiovascular and Respiratory Outcomes. HEI Research Report 148. Health Effects Institute, Boston, MA.
18. Rose N, Cowie C, Gillett R, Marks GB (2011) Validation of a Spatiotemporal Land Use Regression Model Incorporating Fixed Site Monitors. Environmental Science & Technology 45: 294–299.
19. Karner AA, Eisinger DS, Niemeier DA (2010) Near-Roadway Air Quality: Synthesizing the Findings from Real-World Data. Environmental Science & Technology 44: 5334–5344.
20. Zhou Y, Levy JI (2007) Factors influencing the spatial extent of mobile source air pollution impacts: a meta-analysis. BMC Public Health 7: 89.
21. WCC (2010) Willoughby, City Council. Personal communication, 2 December 2010.
22. Fischer PH, Hoek G, van Reeuwijk H, Briggs DJ, Lebret E, et al. (2000) Traffic-related differences in outdoor and indoor concentrations of particles and volatile organic compounds in Amsterdam. Atmospheric Environment 34: 3713–3722.
23. Roorda-Knape MC, Janssen NAH, De Hartog JJ, Van Vliet PHN, Harssema H, et al. (1998) Air pollution from traffic in city districts near major motorways. Atmospheric Environment 32: 1921–1930.
24. Lodovici M, Bigagli E (2011) Oxidative stress and air pollution exposure. Journal of Toxicology 2011, Article ID 487074, 9 pages.
25. Brook RD (2008) Cardiovascular effects of air pollution. Clinical Science 115: 175–187.
26. Mazzoli-Rocha F, Fernandes S, Einicker-Lamas M, Zin W (2010) Roles of oxidative stress in signaling and inflammation induced by particulate matter. Cell Biology and Toxicology 26: 481–498.
27. Knibbs LD, de Dear RJ, Morawska L, Mengersen KL (2009) On-road ultrafine particle concentration in the M5 East road tunnel, Sydney, Australia. Atmospheric Environment 43: 3510–3519.

12

Activity Change in Response to Bad Air Quality, National Health and Nutrition Examination Survey, 2007–2010

Ellen M. Wells[1,2]*, Dorr G. Dearborn[1], Leila W. Jackson[2]

1 Department of Environmental Health Sciences, Case Western Reserve University School of Medicine, Cleveland, Ohio, United States of America, 2 Department of Epidemiology and Biostatistics, Case Western Reserve University School of Medicine, Cleveland, Ohio, United States of America

Abstract

Air pollution contributes to poor respiratory and cardiovascular health. Susceptible individuals may be advised to mitigate effects of air pollution through actions such as reducing outdoor physical activity on days with high pollution. Our analysis identifies the extent to which susceptible individuals changed activities due to bad air quality. This cross-sectional study included 10,898 adults from the National Health and Nutrition Examination Survey (NHANES) 2007–2010. Participants reported if they did something differently when air quality was bad. Susceptible categories included respiratory conditions, cardiovascular conditions and older age (≥65 years). Analyses accounted for complex survey design; logistic regression models controlled for gender, race, education, smoking, and body mass index. 1305 individuals reported doing something differently (12.0%, 95% confidence interval (CI): 10.9, 13.1). This percentage was 14.2% (95% CI: 11.6, 16.8), 25.1% (95% CI: 21.7, 28.6), and 15.5% (95% CI: 12.2, 18.9) among older adults, those with a respiratory condition, and those with a cardiovascular condition, respectively. In adjusted regression models the following were significantly more likely to have changed activity compared to those who did not belong to any susceptible group: respiratory conditions (adjusted odds ratio (aOR): 2.61, 95% CI: 2.03, 3.35); respiratory and cardiovascular conditions (aOR: 4.36, 95% CI: 2.47, 7.69); respiratory conditions and older age (aOR: 3.83; 95% CI: 2.47, 5.96); or all three groups (aOR: 3.52; 95% CI: (2.33, 5.32). Having cardiovascular conditions alone was not statistically significant. Some individuals, especially those with a respiratory condition, reported changing activities due to poor air quality. However, efforts should continue to educate the public about air quality and health.

Editor: Mauricio Rojas, University of Pittsburgh, United States of America

Funding: EMW and DGD were supported by the Swetland Center for Environmental Health. The funders had no role in study design, data collection and analysis, decision to publish, or preparation of the manuscript.

Competing Interests: The authors have read the journal's policy and have the following competing interests: EMW is a former member of the Greater Cleveland Clean Air Campaign and DGD is a former member of the American Thoracic Society's Environmental Health Policy Committee and is currently a co-chair of the Greater Cleveland Asthma Coalition.

* E-mail: ellen.wells@case.edu

Introduction

Ample evidence links ambient air pollution with decreased health outcomes. Fine particulate matter and ozone have been linked with increased respiratory and cardiovascular morbidity and mortality [1–3]. Air pollution exposure is a major public health concern: using air quality data from 2005, Fann et al. estimated 130,000 deaths in the United States were related to particulate matter exposure and 4700 deaths were related to ozone exposure [4].

Moreover, reductions in these pollutants have been shown to reduce rates of related conditions [5,6]. Reduced exposure to air pollutants may occur regionally via reduced emissions; however, individuals may also reduce their personal exposures through actions such as shortening the duration and intensity of outdoor activities on days with elevated air pollution. For example, Langrish et al. showed in an open randomized case-crossover trial that wearing a mask while walking outdoors was associated with improved cardiovascular function among patients with coronary heart disease [7].

The United States Environmental Protection Agency (US EPA) recommends that individuals susceptible to the health effects of air pollution take precautions at lower concentrations than the general population, as susceptible individuals may be affected at these concentrations [8]. Susceptible populations include infants and children, the elderly, and those with respiratory or cardiovascular health conditions [9–13]. Physicians are also urged to encourage their patients, especially those in susceptible groups, to limit their exposure to ambient air on days with elevated air pollution [14–17].

The objective of this study was to determine the extent to which individuals take action to limit their exposure to ambient air pollution based on knowledge of poor air quality, and whether those in a susceptible population were more likely to do so. Susceptible groups considered in the present analysis include those with a self-reported respiratory condition (asthma, emphysema, and chronic bronchitis), those with a self-reported cardiovascular condition (congestive heart failure, coronary heart disease, angina, heart attack or stroke) and the elderly (≥65 years of age).

Materials and Methods

Study Population

This is a cross-sectional study using data from the 2007–2010 National Health and Nutrition Examination Survey (NHANES) conducted by the National Center for Health Statistics (NCHS) within the United States Centers for Disease Control and Prevention (CDC) [18]. Additional details about NHANES are available at http://www.cdc.gov/nchs/nhanes.htm. Participants provided written informed consent, and the study was operated under approval from the NCHS Research Ethics Review Board. This analysis had approval from the Case Western Reserve University Institutional Review Board.

Inclusion criteria included participation in both the household interview and the health examination for NHANES 2007–2010 (n = 20,015). Those less than 20 years of age (n = 8249) were excluded because key self-reported medical conditions were only asked among those 20 years or older. Other individuals were excluded due to missing data on key variables including activity change due to air quality (n = 2); educational status (n = 20); body mass index (BMI) (n = 159); or blood cotinine (n = 687). A total of 10,898 individuals were included in analyses.

Activity Change

Within the air quality questionnaire, activity change was determined from the question "During the past 12 months, when you thought or were informed air quality was bad, did you do anything differently?" [18]. Respondents could answer "yes", "no" or "never thought/not informed about air quality"; hereafter referred to as "not informed" for brevity. For analyses, those in the not informed category were included with those who said no, as these individuals did not change their activities based upon knowledge about air quality. The supporting information includes a sensitivity analysis, where models were also constructed when excluding individuals in the not informed category.

Individuals who responded yes to the air quality question were then asked if they made any of the following changes: wore a mask; spent less time outdoors; avoided roads that have heavy traffic; did less strenuous activities; took medication; closed windows of your house; drove your car less; canceled outdoor activities; exercised indoors instead of outside; used buses, trains or subways; or other. During data editing, a new category, "used or changed air filter or air cleaner", was created for responses that mentioned doing so. In a second sensitivity analysis, we created additional models where activity change was limited to those who identified making changes which could have resulted in reducing their exposure to ambient air pollution or reducing the severity of health effects resulting from their exposure. These included activities which may reduce exposure (spent less time outdoors, closed windows of your house, canceled outdoor activities, exercised indoors instead of outside, wore a mask, did less strenuous activities) and activities which may reduce the impact of exposure (took medication).

Susceptible Groups

Three groups susceptible to the health effects of air pollution were considered in analyses: 1) the elderly; 2) those with a respiratory condition; and 3) those with a cardiovascular condition. The elderly were defined as those 65 years of age or older. Respiratory condition was based on self-report of at least one of the following from the medical conditions questionnaire: current asthma, emphysema, or current bronchitis. Similarly, cardiovascular condition was based on self-report of at least one of the following from the medical conditions questionnaire: congestive heart failure, coronary heart disease, angina, heart attack or stroke. As there is substantial overlap among these three variables, we created a susceptible category variable which classified participants according to their status for each of the three susceptible groups.

Additional Covariates

The demographics questionnaire provided data on age, education level and race/ethnicity. Age was categorized into those 20–34, 35–49, 50–64 and 65 and older for descriptive statistics. Education was categorized as less than high school, high school, some college or a 2-year degree, and a 4-year degree or higher. Race/ethnicity was categorized as Hispanic, non-Hispanic white, non-Hispanic black, and other or mixed race.

Serum cotinine levels were measured as an indicator of smoking status. Whole blood specimens were collected by trained medical staff and analyzed for cotinine using dilution-high performance liquid chromatography/atmospheric pressure chemical ionization tandem mass spectrometry. Limit of detection was 0.015 ng/mL. Smoking status was based on serum cotinine levels, where <1 ng/mL indicates a nonsmoker, 1–10 ng/mL indicates a passive smoker and >10 ng/mL indicates an active smoker [19].

Height and weight were was obtained by trained health technicians during the health examination and used to calculate BMI (kg/m^2). BMI was categorized using standard definitions ($<18.5 \ kg/m^2$ = underweight; 18.5–$24.9 \ kg/m^2$ = normal weight; 25–$29.9 \ kg/m^2$ = overweight and $\geq 30 \ kg/m^2$ = obese) [20]; the underweight category was then combined with the normal weight category due to the small number of underweight individuals (population weighted proportion: 1.5%).

Data Analysis

All data were analyzed using Stata 11.2 (College Station, TX). Appropriate survey weights were used to account for the complex design and non-response. Sampling error was estimated using the Taylor series linearized method. Reported sample sizes are those of the study; however, proportions, percentages and odds ratios are all population-based estimates.

Descriptive statistics were used to assess the frequencies and distribution of individual variables, particularly in relationship to activity change and susceptible groups. Pearson's chi-square test was used to evaluate differences across groups. The association between activity change (dependent variable) and susceptibility category (independent variable) was created using logistic regression models. Covariates were considered for inclusion in the model based on a priori hypotheses and strength of the relationship with activity change in bivariate analyses. The final logistic regression model was adjusted for gender, educational level, race/ethnicity, smoking and body mass index.

Results

Population demographics are presented in Table 1. Mean age of the study population was 46.9 years (95% confidence interval (CI): 46.3, 47.6); 17.0% (95% CI: 15.9, 18.1) were at least 65 years old. Slightly more than half of the participants were women (51.8%, 95% CI: 50.9, 52.6), and the majority (69.3%, 95% CI: 64.3, 74.4) were non-Hispanic white. Mean body mass index was 28.6 kg/m^2 (95% CI: 28.4, 28.8). Serum cotinine levels were highly right skewed, with a geometric mean of 0.36 ng/mL (95% CI: 0.30, 0.44). The range of cotinine was from below the limit of detection to 1438 ng/mL. Active smokers comprised 25.8% (95% CI: 23.9, 27.6) of the population.

Current asthma was more prevalent than emphysema or current bronchitis (7.4% versus 1.8% and 2.5%, respectively)

Table 1. Population characteristics by activity change status, NHANES 2007–2010, N = 10,898.

Characteristic		Changed activity		Did not change activity		Not informed about air quality	
		N[a]	Percent (95% CI)[b]	N[a]	Percent (95% CI)[b]	N[a]	Percent (95% CI)[b]
Entire population		1305	12.0 (10.9, 13.1)	8895	81.5 (79.5, 83.5)	698	6.5 (4.5, 8.6)
Age[c]	20–34 years	230	8.3 (6.6, 9.9)	2244	85.6 (85.6, 88.6)	2635	6.2 (4.0, 8.3)
	35–49 years	336	12.3 (10.5, 14.0)	2329	80.9 (78.2, 83.6)	2845	6.8 (4.4, 9.3)
	50–64 years	398	14.2 (12.0, 16.4)	2186	79.7 (76.7, 82.6)	2752	6.1 (3.7, 8.5)
	≥65 years	341	14.2 (11.6, 16.8)	2136	79.7 (76.5, 82.6)	2666	7.0 (4.3, 9.8)
Gender[c]	Male	513	9.3 (8.2, 10.5)	4453	84.4 (82.2, 86.7)	335	6.3 (4.1, 8.5)
	Female	792	14.5(13.0, 15.9)	4442	78.8 (76.6, 80.9)	363	6.7 (4.7, 8.8)
Education[c]	Less than high school	243	7.9 (6.5, 9.3)	2737	84.4 (81.8, 87.0)	246	7.8 (4.8, 10.7)
	High school	281	10.6 (8.8, 12.4)	2181	84.5 (82.7, 86.4)	124	4.9 (3.4, 6.4)
	Some college or 2-year degree	436	13.3 (11.7, 15.0)	2315	79.5 (82.7, 86.4)	188	7.2 (4.6, 9.9)
	4-year college degree or higher	345	14.8 (12.1, 17.5)	1662	78.9 (75.9, 81.9)	140	6.3 (4.0, 8.6)
Race/ethnicity[c]	Non-Hipanic white	659	12.3 (10.8, 13.7)	4222	80.9 (78.6, 83.2)	367	6.9 (4.4, 9.3)
	Non-Hispanic black	293	14.2 (10.4, 17.9)	1637	83.8 (79.7, 87.9)	43	2.0 (1.2, 2.8)
	Hispanic	276	8.0 (6.0, 10.0)	2616	83.8 (79.5, 88.0)	249	8.2 (4.2, 12.2)
	Other/mixed	77	14.0 (10.3, 17.7)	420	79.5 (74.0, 85.0)	39	6.5 (3.1, 9.9)
Serum cotinine[c]	<1 ng/mL	958	12.8 (11.5, 14.0)	6115	80.2 (78.1, 82.3)	538	7.1 (4.9, 9.2)
	1 to 10 ng/mL	50	10.2 (6.2, 14.2)	393	81.8 (76.5, 87.2)	30	8.0 (4.2, 11.7)
	>10 ng/mL	297	10.2 (8.7, 11.8)	2387	85.0 (82.7, 87.2)	130	4.8 (2.9, 6.8)
Body mass index	<25 kg/m²	354	11.5 (10.1, 12.8)	2552	82.1 (79.1, 85.0)	193	6.5 (4.2, 8.7)
	25–29 kg/m²	414	11.6 (9.9, 13.3)	3083	81.6 (79.6, 83.7)	247	6.8 (4.7, 8.9)
	≥30 kg/m²	537	12.9 (11.2, 14.6)	3260	80.8 (78.5, 83.2)	258	6.3 (4.0, 8.6)
Respiratory condition[c]	No	1018	10.6 (9.6, 11.6)	8120	82.8 (80.7, 85.0)	637	6.6 (4.5, 8.7)
	Yes	287	25.1 (21.7, 28.6)	775	69.1 (65.1, 73.0)	61	5.8 (3.1, 8.5)
Cardiovascular condition[c]	No	1121	11.7 (10.6, 12.8)	7971	81.8 (79.6, 84.0)	618	6.5 (4.5, 8.6)
	Yes	184	15.5 (12.2, 18.9)	924	78.0 (74.5, 81.4)	80	6.5 (3.7, 9.3)

NHANES = National Health and Nutrition Examination Survey; CI = confidence interval.
[a]Unweighted sample N.
[b]Percents are corrected for survey design, are row percents, and may not sum to 100 due to rounding.
[c]Significant (p<0.05) Pearson's chi-squared test corrected for survey design, comparing the characteristic to activity change.

and 9.7% (95% CI: 8.4%, 10.9%) had at least one respiratory condition. The proportion with cardiovascular conditions ranged from 2.0% (angina) to 3.3% (heart attack), with 8.1% (95% CI: 7.2, 9.1) having at least one cardiovascular condition. Additional data regarding health outcomes is included in Table S1.

A total of 1305 (12.0%, 95% CI: 10.9, 13.1) individuals responded that they did something differently due to bad air quality (Table 1). Among those who reported changing an activity, the most commonly reported activities changed were spending less time outdoors (69.4% of those who had changed an activity) and closing the windows of your house (25.5%) (Table S2).

In bivariate comparisons, those more likely to change activities were older, women, those with more education, those with a respiratory condition, and those with a cardiovascular condition. Hispanics and active smokers were less likely to have changed activities.

Twenty-seven percent of the population belongs to at least one of the three susceptible groups (Table 2). The most common susceptible categories were a) those only >65 years old (10.9%), b) those only with a respiratory condition (6.4%), and c) those ≥65 with a cardiovascular condition (3.8%); all other susceptible group categories contained less than two percent of the study population.

Table 2. Prevalence of individuals susceptible to health effects from air pollution, NHANES 2007–2010, N = 10,898.

Susceptible category	N[a]	Percent (95% CI)[a]
None	7135	73.0 (71.4, 74.6)
Respiratory only	642	6.4 (5.5, 7.2)
Cardiovascular only	319	2.6 (2.3, 2.9)
≥65 years only	1713	10.9 (10.1, 11.8)
Respiratory and cardiovascular	136	1.0 (0.7, 1.3)
Respiratory and ≥65 years	220	1.6 (1.3, 1.8)
Cardiovascular and ≥65 years	608	3.8 (3.3, 4.3)
All three groups	125	0.7 (0.5, 0.9)

NHANES = National Health and Nutrition Examination Survey; CI = confidence interval.
[a]N is the unweighted sample N; percents are corrected for survey design and may not sum to 100 due to rounding.

Logistic regression model results are presented in Table 3, Table S3, and Figure S1. In adjusted regression models, there were significantly increased odds of having changed activity for those with a respiratory condition, either alone or in combination with another susceptible condition. The strongest relationships were among those with a cardiovascular and respiratory condition (adjusted odds ratio (aOR): 4.36, 95% CI: 2.47, 7.69) and those ≥65 years old with a respiratory condition (aOR: 3.83, 95% CI: 2.47, 5.96). Results were similar in the sensitivity analyses exploring the relationship when excluding the population not informed about air quality or limiting those who changed activities to those who changed an activity which may reduce exposure or the impact of exposure (Tables S3, S4).

Discussion

This research demonstrates that some individuals, particularly those with respiratory disease, changed activities in response to bad air quality. Our results are consistent with the limited body of literature exploring activity change in response to poor air quality. Wen et al. looked at change in outdoor activity based on individual perception of air quality as well as awareness of medical alerts using data from the 2005 Behavioral Risk Factor Surveillance System [21]. They found that 12.0% of those without asthma and 25.6% of those with lifetime asthma changed activities based on personal perceptions of bad air quality. Semenza et al. report 10–15% of respondents changed activities due to poor air quality based on a telephone survey during July-September 2005 in Houston, Texas and Portland, Oregon [22]. In our analysis, 12.0% of the study population changed activities due to bad air quality, and 25% of those with a respiratory condition changed activities (Table 1).

There was no association between having a cardiovascular condition or being at least 65 years old, without also having a respiratory condition, and changing activities due to bad air quality. Although the cause for this cannot be definitively determined from this study, it is possible that individuals with a respiratory condition are more likely to understand the connection between ambient air pollution and their personal health, as decreased respiratory function may be easier to detect in comparison with decreased cardiovascular function. In a survey of five United Kingdom neighborhoods, Howel et al. found between

82–89% of respondents thought asthma was related to air pollution and 69–78% of respondents thought bronchitis was related to air pollution; additionally, having the condition in question meant a person was more likely to perceive air pollution as affecting it [23]. In contrast, in a survey among patients of a cardiology outpatient clinic in Michigan, only 43% were aware that air pollution negatively affects the heart, and only 8.2% of patients had ever discussed health risks from outdoor air pollution with their doctors [24]. Wen and colleagues noted that advice from a professional, such as a physician, had an impact on the percentage of individuals changing activity [21]; this suggests that current recommendations that physicians discuss outdoor air pollution with cardiology patients [14,17] may be effective in increasing activity change.

It is important to recognize that the results presented here are specific to those who changed activities on days with bad air quality; not merely those who were aware of or concerned about air quality. In order to change an activity due to bad air quality one must have planned an activity that could be changed on a day which had high ambient air pollution. Therefore the proportion of individuals who changed activities may be an underestimate of the proportion of individuals who would potentially have made changes if they had planned activities that were suitable to change. These results should be interpreted accordingly.

This study has a few limitations. The analysis relies on self-reported data for medical conditions. As a result, the prevalence of those with a respiratory or cardiovascular condition are likely to be underestimated, as not all specific respiratory or cardiovascular illnesses were included in the respiratory or cardiovascular condition variables, and it is possible that some individuals with these illnesses are undiagnosed. If this were a substantial factor in analysis, it would have served to weaken our ability to detect a difference in changing actions due to air quality between these groups. Recall bias related to reporting of activity change within the past 12 months may also be a concern. If recall bias were present in this analysis, it could have increased the strength of the association between being in a susceptible group and changing activity. Furthermore, given that this is a cross-sectional survey, we are unable to definitively establish that individuals were part of a susceptible group prior to changing activity due to bad air quality.

Local media may share information on local air quality alerts to their audience; however, the extent to which this was carried out during the study period may have varied considerably. Individuals may have also obtained information about local air quality directly from federal websites [8]. Wen and colleagues demonstrated that media alerts significantly contributed to an individual's changing outdoor activity [21]. A limitation of the current analysis is that we were unable to assess the impact of the frequency and awareness of media alerts on individuals' knowledge of air quality or change in activities, as these data were not collected within NHANES.

This study also has several strengths. It has a large sample size, which allows for comparisons among several demographic and susceptible groups. Additionally, NHANES is based on a representative sample of the United States population; therefore the proportions and odds ratios from this study are representative of the United States population.

Table 3. Odds ratio (95% confidence interval)[a] for changing activity due to poor air quality by susceptible category, NHANES 2007–2010, N = 10,898.

Susceptible category	Unadjusted	Adjusted[b]
None (referent)		
Respiratory only	2.64 (2.06, 3.37)	2.61 (2.03, 3.35)
Cardiovascular only	1.16 (0.76, 1.77)	1.33 (0.86, 2.04)
≥65 years only	1.20 (0.93, 1.54)	1.22 (0.95, 1.57)
Respiratory and cardiovascular	4.06 (2.31, 7.15)	4.36 (2.47, 7.69)
Respiratory and ≥65 years	3.64 (2.35, 5.64)	3.83 (2.47, 5.96)
Cardiovascular and ≥65 years	1.23 (0.78, 1.91)	1.38 (0.89, 2.13)
All three groups	2.80 (1.94, 4.04)	3.52 (2.33, 5.32)

NHANES = National Health and Nutrition Examination Survey.
[a]The model incorporates complex survey design and survey weights.
[b]Model adjusted for gender, education, race/ethnicity, smoking status (based on serum cotinine), and body mass index category.

Conclusions

This analysis demonstrates that those with a respiratory condition are more likely to change activities based on poor air quality; however, more can be done by health and public health professionals to encourage persons susceptible to the effects of air pollution to make changes that will minimize their exposure to air pollution.

Supporting Information

Figure S1 Odds ratio (95% confidence interval) for changing activities, by susceptible group category. NHANES = National Health and Nutrition Examination Survey; aOR = adjusted odds ratio; 95% CI = 95% confidence interval. Probability of changing activity due to poor air quality by susceptible category, based on the adjusted regression model from Table 3 (main article); NHANES 2007–2010, N = 10,898. Model adjusted for gender, education, race/ethnicity, smoking status, and body mass index.

Table S1   Self reported respiratory and cardiovascular conditions by activity change status, NHANES 2007–2010.

Table S2   Population distribution by type of activity changed, among those who changed at least one activity, NHANES 2007–2010, N = 1305.

Table S3 Odds ratio (95% confidence interval) for adjusted models predicting changing activity due to bad air quality, comparing the population with and without those who had no knowledge about air quality, NHANES 2007–2010.
(PDF)

Table S4 Odds ratio (95% confidence interval) for adjusted models predicting changing activities that are related to reducing exposure to or health impact from bad air quality, comparing the population with and without those who had no knowledge about air quality, NHANES 2007–2010.

Author Contributions

Conceived and designed the experiments: EMW LWJ. Analyzed the data: EMW. Wrote the paper: EMW. Interpretation of data and results: EMW DGD LWJ. Edited and critically revised manuscript: EMW DGD LWJ. Approval of final version: EMW DGD LWJ.

References

1. Bell ML, McDermott A, Zeger SL, Samet JM, Dominici F (2004) Ozone and short-term mortality in 95 US urban communities, 1987–2000. JAMA 292: 2372–2378.
2. Dominici F, Peng RD, Bell ML, Pham L, McDermott A, et al. (2006) Fine particulate air pollution and hospital admission for cardiovascular and respiratory diseases. JAMA 295: 1127–1134.
3. Peel JL, Tolbert PE, Klein M, Metzger KB, Flanders WD, et al. (2005) Ambient air pollution and respiratory emergency department visits. Epidemiology 16: 164–174.
4. Fann N, Lamson AD, Anenberg SC, Wesson K, Risley D, et al. (2012) Estimating the national public health burden associated with exposure to ambient PM2.5 and ozone. Risk Anal 32: 81–95.
5. Friedman MS, Powell KE, Hutwagner L, Graham LM, Teague WG (2001) Impact of changes in transportation and commuting behaviors during the 1996 Summer Olympic Games in Atlanta on air quality and childhood asthma. JAMA 285: 897–905.
6. Laden F, Schwartz J, Speizer FE, Dockery DW (2006) Reduction in fine particulate air pollution and mortality: Extended follow-up of the Harvard Six Cities study. Am J Respir Crit Care Med 173: 667–672.
7. Langrish JP, Li X, Wang S, Lee MM, Barnes GD, et al. (2012) Reducing personal exposure to particulate air pollution improves cardiovascular health in patients with coronary heart disease. Environ Health Perspect 120: 367–372.
8. US Environmental Protection Agency, National Oceanic and Atmospheric Adminstration, National Aeronautics and Space Administration, National Park Service, National Association of Clean Air Agencies, et al. (2011) AIRNow. US Environmental Protection Agency,.
9. Bateson TF, Schwartz J (2004) Who is sensitive to the effects of particulate air pollution on mortality? A case-crossover analysis of effect modifiers. Epidemiology 15: 143–149.
10. Medina-Ramon M, Schwartz J (2008) Who is more vulnerable to die from ozone air pollution? Epidemiology 19: 672–679.
11. Pope CA 3rd (2000) Epidemiology of fine particulate air pollution and human health: biologic mechanisms and who's at risk? Environ Health Perspect 108 Suppl 4: 713–723.
12. Sacks JD, Stanek LW, Luben TJ, Johns DO, Buckley BJ, et al. (2011) Particulate matter-induced health effects: who is susceptible? Environ Health Perspect 119: 446–454.
13. Stafoggia M, Forastiere F, Faustini A, Biggeri A, Bisanti L, et al. (2010) Susceptibility factors to ozone-related mortality: a population-based case-crossover analysis. Am J Respir Crit Care Med 182: 376–384.
14. Abelsohn A, Stieb DM (2011) Health effects of outdoor air pollution: approach to counseling patients using the Air Quality Health Index. Can Fam Physician 57: 881–887, e280–887.
15. Kelly FJ, Fuller GW, Walton HA, Fussell JC (2012) Monitoring air pollution: use of early warning systems for public health. Respirology 17: 7–19.
16. Laumbach RJ (2010) Outdoor air pollutants and patient health. Am Fam Physician 81: 175–180.
17. Shofer S, Chen TM, Gokhale J, Kuschner WG (2007) Outdoor air pollution: counseling and exposure risk reduction. Am J Med Sci 333: 257–260.
18. Centers for Disease Control and Prevention C (2007–2008) National Health and Nutrition Examination Survey. Hyattsville, MD: U.S. Department of Health and Human Services, Centers for Disease Control and Prevention, National Center for Health Statistics.
19. Hukkanen J, Jacob P 3rd, Benowitz NL (2005) Metabolism and disposition kinetics of nicotine. Pharmacol Rev 57: 79–115.
20. National Heart Lung and Blood Institute (1998) Clinical guidelines on the identification, evaluation and treatment of overweight and obesity in adults: the evidence report. Washington, D.C.: National Institutes of Health.
21. Wen XJ, Balluz L, Mokdad A (2009) Association between media alerts of air quality index and change of outdoor activity among adult asthma in six states, BRFSS, 2005. J Community Health 34: 40–46.
22. Semenza JC, Wilson DJ, Parra J, Bontempo BD, Hart M, et al. (2008) Public perception and behavior change in relationship to hot weather and air pollution. Environ Res 107: 401–411.
23. Howel D, Moffatt S, Bush J, Dunn CE, Prince H (2003) Public views on the links between air pollution and health in Northeast England. Environ Res 91: 163–171.
24. Nowka MR, Bard RL, Rubenfire M, Jackson EA, Brook RD (2011) Patient awareness of the risks for heart disease posed by air pollution. Prog Cardiovasc Dis 53: 379–384.

13

Whole-Tree Water Use Efficiency Is Decreased by Ambient Ozone and Not Affected by O₃-Induced Stomatal Sluggishness

Yasutomo Hoshika[1], Kenji Omasa[1], Elena Paoletti[2]*

1 Graduate School of Agricultural and Life Sciences, The University of Tokyo, Tokyo, Japan, **2** Institute of Plant Protection, National Research Council, Sesto Fiorentino, Florence, Italy

Abstract

Steady-state and dynamic gas exchange responses to ozone visible injury were investigated in an ozone-sensitive poplar clone under field conditions. The results were translated into whole tree water loss and carbon assimilation by comparing trees exposed to ambient ozone and trees treated with the ozone-protectant ethylenediurea (EDU). Steady-state stomatal conductance and photosynthesis linearly decreased with increasing ozone visible injury. Dynamic responses simulated by severing of a leaf revealed that stomatal sluggishness increased until a threshold of 5% injury and was then fairly constant. Sluggishness resulted from longer time to respond to the closing signal and slower rate of closing. Changes in photosynthesis were driven by the dynamics of stomata. Whole-tree carbon assimilation and water loss were lower in trees exposed to ambient O₃ than in trees protected by EDU, both under steady-state and dynamic conditions. Although stomatal sluggishness is expected to increase water loss, lower stomatal conductance and premature leaf shedding of injured leaves aggravated O₃ effects on whole tree carbon gain, while compensating for water loss. On average, WUE of trees exposed to ambient ozone was 2–4% lower than that of EDU-protected control trees in September and 6–8% lower in October.

Editor: Carl J. Bernacchi, University of Illinois, United States of America

Funding: JSPS Fellowships for Young Scientists. The funders had no role in study design, data collection and analysis, decision to publish, or preparation of the manuscript.

Competing Interests: The authors have declared that no competing interests exist.

* E-mail: e.paoletti@ipp.cnr.it

Introduction

Tropospheric ozone (O₃) is an important phytotoxic air pollutant and is also recognized as a significant greenhouse gas [1]. Tropospheric O₃ level has been continuously increasing since the first direct measurements in 1874 and its atmospheric concentration is now twice or more than in the pre-industrial age in the northern hemisphere [2–4]. Phytotoxic nature of O₃ has been well known for decades [5–12]. Ozone concentrations recorded in rural areas are higher than those in the city [13] and thus O₃ is now considered as the air pollutant with the highest damage potential to forests [14].

As the penetration of O₃ through the cuticle can be considered as negligible [15], uptake through the stomata is a crucial factor for assessing the adverse effect of O₃ on plants [16–20]. However, effects of O₃ on stomatal responses are not straightforward, as both reductions and sluggish responses have been reported [21,22]. Reductions of stomatal conductance occur when measurements are carried out under steady-state conditions [23]. Sluggishness has been reported during dynamic stomatal responses to fluctuating photosynthetic photon flux density (PPFD) [22,24–27], vapor pressure deficit (VPD) [27], and severe water stress imposed by severing a leaf [26,28–30]. Sluggish stomatal control over transpiration may increase water loss. Plants live in a fluctuating environment. A fast gas exchange response to rapid changes in the environmental stimuli is the key for successful plant

adaptation and competition [31]. Because of climate change, forest ability of water control and carbon sequestration under O₃ pollution is of rising importance [14].

Scalar and conceptual uncertainties still limit the current understanding of the basic physiological mechanisms that underline responses of forests to O₃ [32]. The scalar uncertainties are due to transfer of results from seedlings in controlled environments to mature trees in the field, while the conceptual uncertainties are due to contrasting results about whole-tree water use responses to ambient O₃ [32–35]. In contrast, there is a general agreement about O₃ exposure as a factor of reduced tree carbon sequestration and biomass [36], although the results usually come from steady-state measurements of photosynthesis.

Ozone visible injury of leaves may be used as a clear and easily quantifiable proxy of O₃ foliar damage and is the only method to assess O₃ damage in the field [37]. Ozone visible injury has been investigated in many European and North American tree and herbaceous species, and partly validated under controlled conditions [38,39]. There are few reports of relationship between stomatal conductance and O₃ visible injury. After onset of O₃ visible injury, significant reductions in steady-state leaf gas exchange were recorded for tree species in chamber experiments [40–42]. Omasa et al. (1981) did not report any correlation between visible injury and stomatal O₃ uptake in a leaf [43]. Dynamic stomatal response was slower in injured leaves (20%

injury) compared to control leaves (0% injury) for manna ash (*Fraxinus ornus* L.) [28].

Our main objectives were to improve our knowledge of steady-state and dynamic stomatal response to O_3 visible injury in adult trees in the field, and to assess whole-tree water loss and carbon assimilation under ambient O_3 impacts. Measurements were carried out in an O_3-sensitive poplar clone (Oxford, *Populus maximoviczii* Henry × *berolinensis* Dippel) [44]. The amount of leaf injury per tree was experimentally manipulated by applying the O_3-protectant ethylenediurea (EDU, N-[2-(2-oxo-1-imidazolidi-nyl)ethyl]-N'-phenylurea). EDU *per se* does not affect gas exchange [45] and has been widely used to prevent O_3 visible injury and determine O_3 effects in many plant species [39,45–47].

Materials and Methods

Experimental Site and Plant Material

The study was carried out in an experimental field site located in central Italy (Antella: 43°44' N, 11°16' E, 50 m a.s.l., 14.7°C as mean annual temperature and 1233 mm as total annual precipitation in 2010). Forty root cuttings of the O_3-sensitive Oxford clone were planted in two lines in 2007. Every week over the growing seasons 2008–2010, each tree was irrigated with 1 to 2 L of water (WAT, control line) or 450 ppm EDU solution (EDU, treated line), according to the successful application of EDU as soil drench to adult trees [48]. In 2010, the mean tree height was 2.9 m, and the mean stem diameter at breast height was 19 mm. Soil moisture was measured in the root layer (30 cm depth) by EC5 sensors equipped with an EM5b data logger (Decagon Devices, Pullman WA, USA). On average, soil moisture was $21.1\pm0.2\%$ during the gas exchange measurements (September-October) and $24.5\pm0.1\%$ during the growing season (April to October). The values were between field capacity (25.5%) and wilting point (17.5%) for this type of soil, i.e. sandy clay loam. Air temperature, relative humidity and precipitation were recorded by a 110-WS-16 modular weather station (NovaLynx corp., Auburn CA, USA). Average vapor pressure deficit during daylight hours and total precipitation were 1.02 kPa and 197 mm in September to October and 1.42 kPa and 625 mm from April to October, respectively. Ozone concentrations were continuously recorded at canopy height (2.0 m) by an O_3 monitor (Mod. 202, 2B Technologies, Boulder CO, USA). The AOT40 value (accumulated exposure above a threshold concentration of 40 ppb during daylight hours) during the growing season (April to October) was 25.8 ppm·h and the maximum hourly O_3 concentration reached 118 ppb.

Assessment of Ozone Visible Injury

Ozone visible injury occurred as dark stippling on the upper leaf surface since early September 2010. The injury was identified as O_3-like because it was missing in shaded leaves and more severe in older than in younger leaves [38]. The symptoms were similar to those caused by ambient O_3 in *Populus nigra* [42]. In September (22th to 28th) and October (23th to 28th), all 9502 leaves from five trees per treatment (WAT and EDU) were counted and assigned to 5%-step injury classes by the same two observers. Photoguides quantifying visible injury (0~100%) by image analysis processing were used [38,39]. Pest, pathogen and mechanical injury occurred in both EDU and WAT trees and was assessed to be <5% of total leaves. Leaves for measurements of gas exchange showed O_3 visible injury only and were evaluated on a 1%-step basis.

Measurement of Steady-state and Dynamic Gas Exchange

Fully expanded sun leaves (medium size) with visible injury from 0% to 50% at set positions from the terminal shoot (5th to 16th) of WAT trees were measured in clear sky days of September and October 2010 between 10:00 and 15:00 CET. Preliminary measurements did not show significant differences in gas exchange of healthy leaves, i.e. without visible ozone injury, at those set positions. Gas exchange was measured with a portable infra-red gas analyzer (CIRAS-2 PPSystems, Herts, UK), equipped with a 2.5 cm^2 leaf cuvette which controlled leaf temperature (24°C), leaf-to-air vapour pressure deficit (1.0 kPa), saturating light (1800 µmol m^{-2} s^{-1}) and CO_2 concentration (375 ppm). Steady-state light-saturated photosynthesis (A_{max}), stomatal conductance to water vapor (g_s) and transpiration were measured in 41 leaves from WAT trees.

Dynamic measurements were carried out for 21 leaves from WAT trees. When both g_s and A_{max} reached equilibrium under constant light at 1800 µmol m^{-2} s^{-1}, the leaf petiole was severed with a sharp scalpel, similar to the methodology in Paoletti (2005) [26]. The data were logged at 1 min intervals for 30 min after severing. As the absolute value of g_s and A_{max} varied among individual leaves, relative g_s and A_{max} were expressed as a percentage of the average of the last 5 points at equilibrium, i.e., just before leaf severing. The following parameters were estimated based on fittings of two linear lines to minimize the root mean square error between measured and predicted values for g_s or A (Figure 1A): range of relative g_s decrease at 30 min after severing, Δg_s; time to start g_s decrease, T_{resp} (g_s); rate of g_s decrease at 30 min from severing, Slope(g_s) = $\Delta g_s/(30- T_{resp}$ (g_s)); range of relative A_{max} decrease at 30 min after severing, ΔA; time to start A_{max} decrease, T_{resp} (A); rate of A_{max} decrease at 30 min from severing, Slope(A) = $\Delta A/(30- T_{resp}$ (A)).

After measurements, the leaf area was measured by means of a leaf area meter (AM300, ADC, Herts, UK) for assessing a relationship between leaf size and the variation of g_s in single leaves. We hypothesized that the water content of a leaf may depend on leaf size and affect g_s response.

Tree Level Modeling

To assess effects of O_3 visible injury on leaf gas exchange at tree level, we constructed a simple model to scale up from single–leaf steady-state and dynamic gas exchange. The model was applied to the five trees per treatment (WAT and EDU) whose leaves were counted and assigned to a 5%-step visible injury class. Steady-state leaf water loss and photosynthesis at tree level, i.e. W_{loss}: mol H_2O tree^{-1} s^{-1}, and A_{tree}: µmol CO_2 tree^{-1} s^{-1}, were estimated as follows:

$$W_{loss} = \sum \left(T_{r\_inj} \cdot LA \cdot N_{inj} \right) \qquad (1)$$

$$A_{tree} = \sum \left(A_{max\_inj} \cdot LA \cdot N_{inj} \right) \qquad (2)$$

where T_{r_inj} and A_{max_inj} are transpiration rate (mmol m^{-2} s^{-1}) and photosynthesis (µmol m^{-2} s^{-1}), respectively, at 1800 µmol m^{-2} s^{-1} constant light for leaves showing O_3 visible injury. N_{inj} is the number of leaves in each 5%-step injury class. LA is the average leaf area per leaf (0.003 m^2 leaf^{-1}), calculated from subsamples of 30 randomly collected leaves per tree.

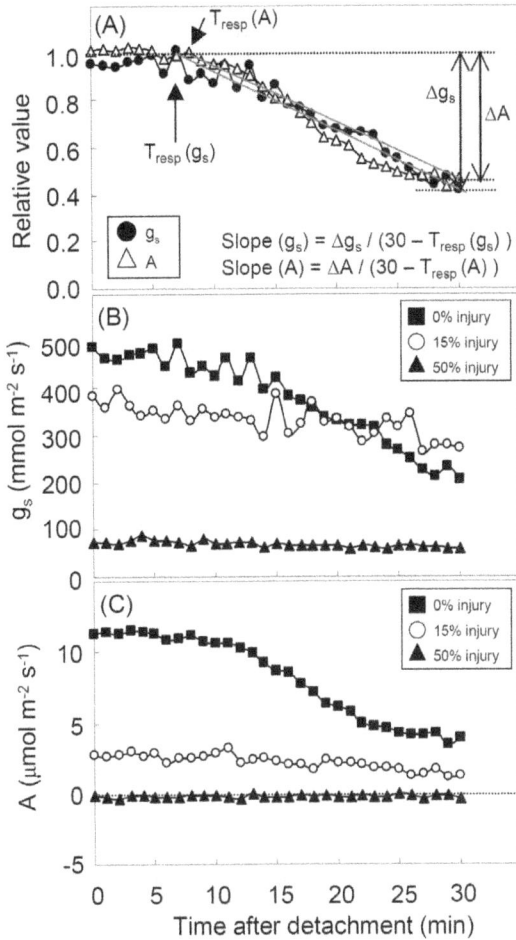

Figure 1. Examples of dynamic response of g_s and A_{max} after detachment of the leaf (A: calculation of the dynamic parameters in a leaf with 0% visible injury, B: time courses of absolute values in g_s, C: time courses of absolute values in A). Δg_s and ΔA show the range of g_s and A_{max} variation, respectively, over 30 min from the leaf severing. T_{resp} (g_s) and T_{resp}(A) show the time to start decrease of g_s and A_{max}, respectively. Slope(g_s) and Slope(A) show the rate of decrease for g_s and A_{max}, respectively, over 30 min.

Whole-tree leaf water loss and carbon assimilation under the severe water stress simulated by severing a leaf (W_{loss_st}: mol H_2O tree^{-1} s^{-1}, and A_{tree_st}: μmol CO_2 tree^{-1} s^{-1}) were estimated by the following equations:

$$W_{loss\_st} = \sum \left(\overline{T_{r\_inj}} \cdot LA \cdot N_{inj} \right) \qquad (3)$$

$$A_{tree\_st} = \sum \left(\overline{A_{max\_inj}} \cdot LA \cdot N_{inj} \right) \qquad (4)$$

where $\overline{T_{r\_inj}}$ is the average transpiration rate (mmol m^{-2} s^{-1}) and $\overline{A_{max\_inj}}$ is the average photosynthesis (μmol m^{-2} s^{-1}) at 1800 μmol m^{-2} s^{-1} constant light during the 30 min after severing a leaf with O_3 visible injury.

Statistical Analysis

Effects of O_3 visible injury on steady-state leaf gas exchange and dynamic responses after severing a leaf were tested with a regression analysis. Correlation between variables of dynamic stomatal response was tested. Two-way analysis of variance (ANOVA) was used to assess the effects of measuring month and EDU treatments on number of leaves, ozone visible injury and gas exchange parameters at whole tree level. Differences among means were tested by Tukey HSD test. Percents were arcsine square root transformed prior to analysis. Data were checked for normal distribution (Kolmogorov-Smirnov D test) and homogeneity of variance (Levene's test). Results were considered significant at p<0.05. All statistical analyses were performed with STATISTICA software (6.0, StatSoft Inc., Tulsa, OK, USA).

Results

Number of Leaves and Ozone Visible Injury

In September, EDU trees had 83% more leaves per tree than WAT trees (Figure 2A). In October, leaf abscission had progressed faster in WAT trees (-36% of leaves relative to September) than in EDU trees (-15%), resulting in EDU trees showing significantly more leaves ($+144\%$) than WAT trees. The percentage of injured leaves ($>5\%$ of visible injury) was significantly higher in WAT trees than in EDU trees in both September and October (Figure 2B). In October, the percentage of injured leaves was 3.13 and 7 times higher than in September in WAT and EDU trees, respectively.

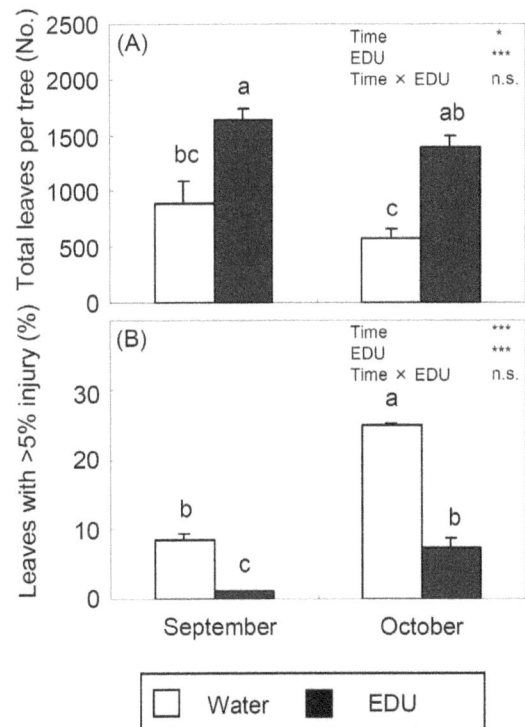

Figure 2. Total number of leaves (A) and percentage of ozone injured leaves (more than 5% of injured surface) (B) per tree (+SE) (WAT: water treated plants; EDU: EDU treated plants). * and *** denote significance at the 5% and 0.1% level, respectively; n.s. indicates no significance. Different letters above the bars indicate significant differences among bars (Tukey HSD test, P<0.05, n = 5 trees).

Steady-state g_s and A_{max}

Steady-state leaf gas exchange for both g_s and A_{max} decreased with increasing O_3 visible injury (Figure 1, 3). In healthy leaves (0% injury), g_s was 400 to 800 mmol m^{-2} s^{-1} whereas it was less than 100 mmol m^{-2} s^{-1} in leaves with 50% injury (Figure 3A). Leaves with higher injury (>50% injury) were tested but did not show a measurable g_s. In control leaves, A_{max} was 5 to 15 µmol m^{-2} s^{-1}, and dropped to around 0 µmol m^{-2} s^{-1} in leaves with more than 35% injury (Figure 3B).

Variation of g_s and A_{max} after Detachment of a Leaf

After detachment of a leaf, two phases of gas exchange response were observed (Figure 1A): no response until T_{resp} and then a linear decrease. The magnitude of change in g_s at 30 min after severing a leaf (Δg_s) did not depend on leaf size (data not shown: $R^2 = 0.02$, p = 0.537) and thus on the total water content of a leaf.

Figure 4 shows the relationships between O_3 visible injury and dynamic response of g_s and A_{max}. Δg_s showed a non-linear response to O_3 visible injury (Figure 4A). It sharply decreased from 45–60% in healthy leaves (0% injury) to 15–30% in leaves with >5% visible injury. Slope(g_s) sharply decreased from 2.5–3.2% min^{-1} in healthy leaves (0% injury) to 0.8–1.8% min^{-1} in leaves with >5% visible injury, and did not vary in leaves with 5–50% of injury (Figure 4B). The response time to start stomatal closing (T_{resp} (g_s)) was linearly correlated to O_3 visible injury (Figure 4C). T_{resp} (g_s) increased from about 10 min in healthy leaves to >13 min in leaves with >20% injury. The magnitude of decrease in photosynthetic rate (ΔA) sharply decreased from about 55% in healthy leaves to about 25% in leaves with >5% visible injury (Figure 4D). Slope(A) sharply decreased from about 3.3% min^{-1} in healthy leaves to about 1.6% min^{-1} in leaves with >5% visible injury (Figure 4E). There was a linear relationship between the response time to start decrease of photosynthesis (T_{resp} (A)) and O_3 visible injury (Figure 4F). T_{resp} (A) increased from 5–13 min in

healthy leaves to 25 min in a leaf with 50% injury. Table 1 shows correlation between the A_{max} and g_s variables obtained from dynamic response after severing of a leaf. The magnitude of change in A_{max} (ΔA) increased with increasing Δg_s. The rate of reduction in A_{max}, i.e. Slope(A), was positively correlated with Slope(g_s). The response times to start decrease of A_{max} and g_s, i.e., T_{resp}, were not significantly correlated, although they showed a statistical tendency to a positive correlation (p<0.1).

Carbon Assimilation and Water Loss at Tree Level

In September, A_{tree} and W_{loss} were significantly lower in WAT trees, being half of the values in EDU trees (Figure 5A–B). In October, the difference between WAT and EDU trees became even larger. Whole-tree water use efficiency (A_{tree}/W_{loss}) at steady-state was significantly higher in EDU trees than in WAT trees both in September and October (Figure 5C). WUE decreased over time, but the decrease was larger in WAT (−6%) than in EDU trees (−2%), resulting in a significant Time x EDU interaction. Both in September and October, both A_{tree_st} and W_{loss_st}, i.e. whole-tree carbon assimilation and water loss under the simulated severe water stress, were significantly lower in WAT trees (Figure 6A–B), similarly to the results from steady-state measurements (Figure 5A–B). Whole-tree instantaneous water use efficiency, expressed as A_{tree_st}/W_{loss_st}, was significantly higher in EDU trees than in WAT trees both in September and October (Figure 6C). Again, the decrease of WUE_st over time was larger in WAT (−8%) than in EDU trees (−3%).

Discussion

According to previous reports in different species [28,40–42], the steady-state measurements indicated that g_s and A_{max} linearly decreased with increasing leaf visible injury in the O_3-sensitive Oxford clone (Figure 3). A_{max} dropped to around 0 µmol m^{-2} s^{-1} in leaves with more than 35% injury and g_s was not measurable in leaves with more than 50% injury. In a previous field study, leaves of manna ash with 20% visible injury showed a 33% reduction in g_s and A_{max} relative to healthy leaves (measurement in September) [28]. The result of the present study showed a larger reduction in g_s (about 39%) and A_{max} (about 54%) of 20% injured leaves, suggesting effects of O_3 visible injury on gas exchange are species-specific. Paoletti et al. (2009a) suggested that the modifications of stomatal conductance in O_3 injured leaves were driven by the structural alterations found in the mesophyll rather than by structural changes in stomata or other epidermal cells [28]. Omasa et al. (1981) suggested that stomatal opening in leaves with O_3 visible injury varied with changes in the pressure balance between guard cells and epidermal cells caused by the water-soaking of epidermal cells [43]. The most likely changes, however, are due to photosynthetic impairment [21,49].

When analyzing dynamic g_s response to severing of a leaf, stomata of injured leaves were shown to be slower than those of healthy leaves in responding to the closing signal (T_{resp} (g_s)) and in the rate of closing (Slope(g_s)) (Figure 4B–C). These combined effects translated in a lower ability of injured leaves to close stomata, i.e. in a lower Δg_s than healthy leaves, resulting in a sluggish stomatal control over water loss. In a previous study, Paoletti et al. (2009a) also reported a slower response of stomata to severing in leaves of manna ash with O_3 visible injury [28], even though only leaves with 0% and 20% injury were compared. Here, we compared leaves with a range of O_3 visible injury, i.e. from no injury (control) until a measurable g_s was recorded (~50% injury) and showed that Δg_s decreased sharply above 5% injury and did not change any more (Figure 4A).

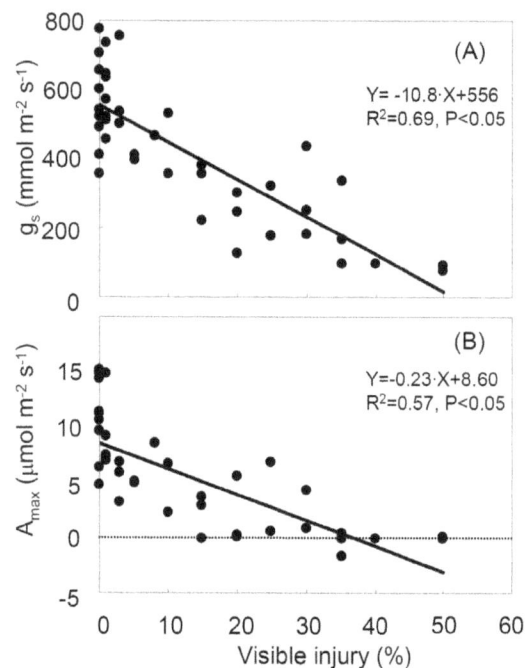

Figure 3. Relationships between steady-state leaf gas exchange (A: stomatal conductance (g_s), B: light-saturated photosynthesis (A_{max})) and visible ozone foliar injury.

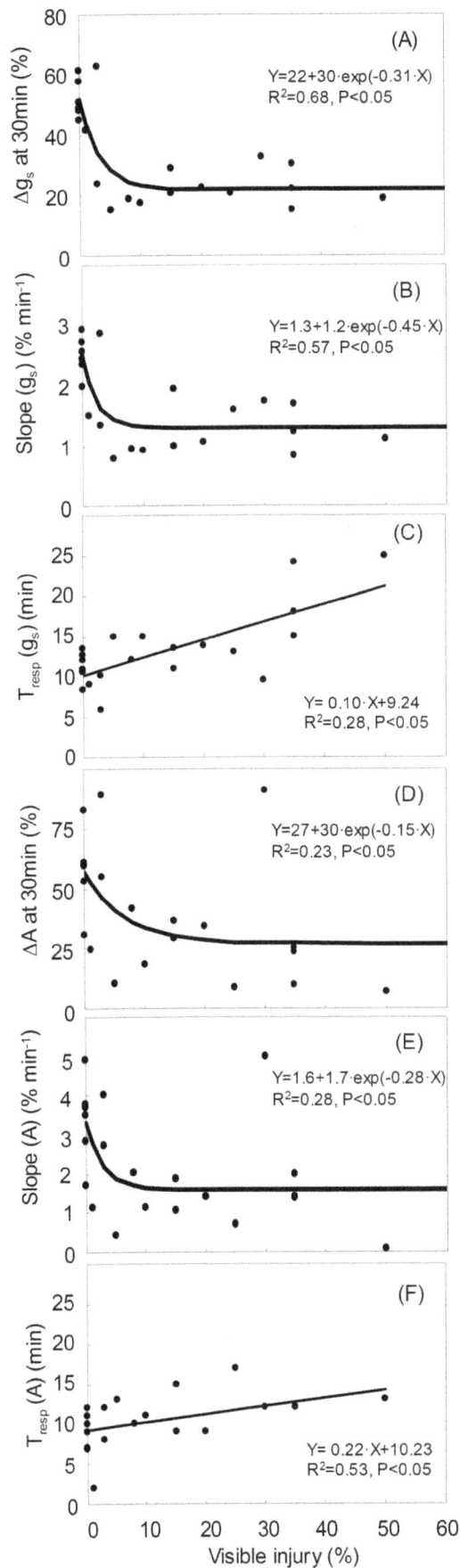

Figure 4. Relationships between visible ozone foliar injury and dynamic response of stomatal conductance (g_s) and photosynthesis (A_{max}) over 30 min after leaf severing (A: Δg_s at 30 min; B: Slope(g_s); C: T_{resp} (g_s); D: ΔA; E: Slope(A); F: T_{resp} (A)).

Literature results highlight several mechanisms by which O_3 may induce sluggishness. Omasa (1990) reported a slight increase in permeability of epidermal cell membranes and alteration of the osmotic pressure after O_3 exposure, that may modulate a balance in turgor between guard and subsidiary cells [50]. Vahisalu et al. (2010) found that Ca^{2+}-dependent signaling and O_3-induced stomatal movements were independent, and highlighted a temporary desensitization of the guard cells due to blocking of the K^+ channels [51]. Another cause of sluggishness may be O_3-induced lower rates of transpiration in which leaves take longer to perceive the same change in water status following petiole excision [26,28–30] or light variation [22,26]. All the above mechanisms, however, cannot explain the non-linear response of Δg_s to visible injury observed in the present study. Ozone may also delay stomatal responses by stimulating ethylene production and reducing stomatal sensitivity to ABA [52]. Ethylene production is known to increase with increasing O_3 visible injury [53,54]. In tomato plants, concentration of ACC (1-aminocyclopropane-1-carboxylic acid), a precursor of ethylene, increased when visible injury reached 5% and remained constant until the maximum injury recorded in the experiment, i.e. 35% [55]. A sharp rise of ethylene emission as soon as visible injury reaches 5% and a constant emission over this threshold would explain why Δg_s decreased sharply above 5% injury and did not change any more when injury was >5% (Figure 4A). Tuomainen et al. (1997) also showed that ethylene emission from detached leaves was enhanced fourfold in ozone-treated plants, while no changes were observed in control leaves that were similarly cut at the petiole [55].

Sluggish A_{max} responses with increasing O_3 visible injury were also found in the measurements of dynamic leaf gas exchange (Figure 4D–F). The response of A_{max} was similar to that of g_s after severing a leaf (Figure 1), i.e. no response until T_{resp} and then a linear decrease during stomatal closure. Although the response time to start reduction of A_{max} was not significantly correlated with the response time to closing stomata, the magnitude and rate of reduction in A_{max} were linearly correlated to those of stomatal closure (Table 1). Heber et al. (1986) showed that photosynthetic rate decreased following stomatal closure after severing of a leaf [56]. Slightly shorter $T_{resp}(g_s)$ than $T_{resp}(A)$ confirmed that the reduction of A_{max} was mediated by the response of g_s. The slower reduction of A_{max} in injured leaves than in healthy leaves would increase carbon assimilation under water stress conditions and may be interpreted as a feedback mechanism to maximize

Table 1. Correlation between A_{max} vs. g_s variables obtained during the dynamic response to severing of a leaf (Δ: magnitude of change in A_{max} and g_s over 30 min from the leaf severing; T_{resp}: time to start decrease in A_{max} and g_s after severing a leaf; Slope: rate of A_{max} and g_s decrease).

Parameter	Pearson coefficient	Level of significance
Δ	0.626	0.002**
Slope	0.622	0.003**
T_{resp}	0.371	0.098 n.s.

**denotes the significance at 1% level; n.s. indicates no significant correlation.

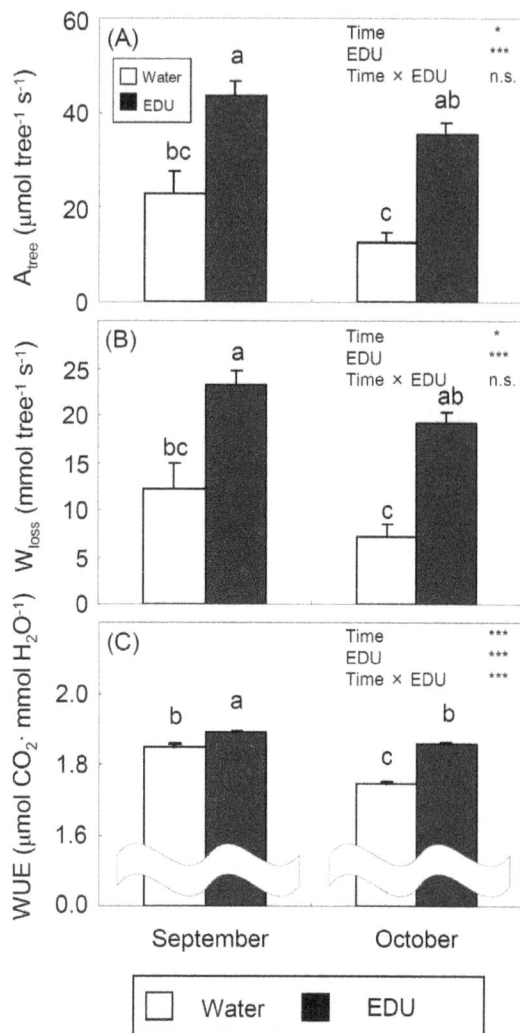

Figure 5. Estimated steady-state carbon assimilation (A: A_{tree}), water loss (B: W_{loss}) and instantaneous water use efficiency expressed as A_{tree}/W_{loss} (C: WUE) at tree level (+SE) (WAT: water treated plants; EDU: EDU treated plants). * and *** denote significance at the 5% and 0.1% level, respectively; n.s. indicates no significance. Different letters above the bars indicate significant differences among bars (Tukey HSD test, P<0.05, n = 5 trees).

Figure 6. Estimated carbon assimilation (A: A_{tree_st}), water loss (B: W_{loss_st}) and instantaneous water use efficiency as A_{tree_st}/W_{loss_st} (C: WUE_st) at tree level under severe water stress imposed by severing a leaf (+SE) (WAT: water treated plants; EDU: EDU treated plants). * and *** denote significance at the 5% and 0.1% level, respectively; n.s. indicates no significance. Different letters above the bars indicate significant differences among bars (Tukey HSD test, P<0.05, n = 5 trees).

photosynthesis under stress. However, severe O_3 visible injury (>35%) shifted carbon sink to source because A_{max} was <0 μmol m^{-2} s^{-1} (Figure 3B).

At whole-tree level, the total carbon assimilation (A_{tree}) and water loss (W_{loss}) assessed under steady-state conditions were significantly lower in WAT trees exposed to ambient ozone than in EDU-protected trees in both September and October (Figure 5A–B). Such O_3-induced reduction of photosynthesis and water loss was in agreement with meta-analysis results [36]. Dynamic and steady-state whole-tree WUEs showed a similar seasonal trend. WUE was significantly higher in EDU trees than in WAT trees, both in September and October and both when assessed under steady-state and dynamic conditions (Figure 5C and 6C). On average, WUE of trees exposed to ambient ozone was 2–4% lower than that of EDU-protected control trees in September and 6–8% lower in October. The decrease of tree-level WUE over time, in fact, was larger in WAT than in EDU trees, confirming the

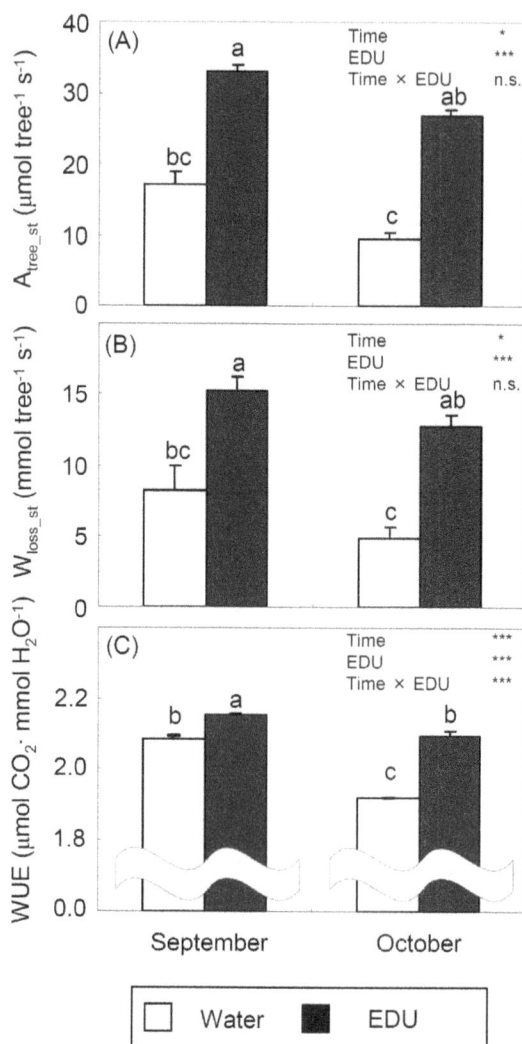

frequently reported decrease in leaf-level WUE in O_3-exposed plants [33] and O_3-injured leaves [41]. Also whole-tree dynamic carbon assimilation (A_{tree_st}) and water loss (W_{loss_st}) were significantly lower in WAT trees than in EDU-protected trees (Figure 6A–B). In contrast, ozone-induced stomatal sluggishness would be expected to increase whole-tree water loss. This response, however, was balanced by lower gas exchange (Figure 3) and premature shedding of injured leaves. After the onset of O_3 visible injury in early September, ozone visible injury increased quickly (Figure 2B). In parallel, leaf abscission also progressed (Figure 2A), so that both whole-tree water loss and carbon assimilation were reduced. However, McLaughlin et al. (2007) reported that ambient O_3 spikes significantly increased water loss of trees, as assessed from sap-flow measurements, suggesting that ozone-induced aberrations in the stomatal dynamics may differ depending on the species and the environmental conditions [32].

Conclusions

One of the topical subjects in the assessment of O_3 risk to forests is scaling up from leaf level to the stand and landscape level [4]. Further improvement of our understanding about stomatal responses to ambient O_3 can be regarded as an essential factor in modelling and predicting forest responses to both O_3 and climate [21]. Occurrence of O_3 visible injury resulted in loss of stomatal control for water loss, but was compensated by lower stomatal conductance and premature leaf shedding. The resulting decline in whole tree ability of transpiring and sequestering atmospheric carbon is a significant effect of ambient ozone pollution.

Stomata play a crucial role in regulating plant gas exchange with the atmosphere, including O_3 uptake [16–20]. Surface O_3 concentrations are continuously increasing [4]. The climate change brings about the risk of drought and flooding [1]. The results of this study contribute new knowledge about water control and carbon sequestration of trees under ambient O_3 exposure and suggest that the effects of O_3–induced stomatal sluggishness on the whole-tree carbon and water balance are negligible.

Acknowledgments

Prof. William J. Manning and Dr. Marcus Schaub are greatly acknowledged for providing EDU and poplar cuttings, respectively.

Author Contributions

Conceived and designed the experiments: EP KO. Performed the experiments: YH EP. Analyzed the data: YH EP. Contributed reagents/materials/analysis tools: EP. Wrote the paper: YH EP.

References

1. Bytnerowicz A, Omasa K, Paoletti E (2007) Integrated effects of air pollution and climate change on forests: A northern hemisphere perspective. Environ Pollut 147: 438–445.
2. Volz A, Kley D (1988) Evaluation of the Montsouris series of ozone measurements made in the nineteenth century. Nature 332: 240–242.
3. Akimoto H (2003) Global Air Quality and Pollution. Science 302: 1716–1719.
4. Paoletti E (2007) Ozone impacts on forests. CAB Reviews: Perspectives in Agriculture, Veterinary Science, Nutrition and Natural Resources 2 (No. 68), 13p.
5. NIES (National Institute for Environmental Studies) (1980) Studies on effects of air pollutants on plants and mechanisms of phytotoxicity. Res Rep Natl Inst Environ Stud Jap, Japan. 265 p.
6. NIES (1984) Studies on effects of air pollutant mixtures on plants. Part 1 & 2. Res Rep Natl Inst Environ Stud Jap, Japan. 163p & 155p.
7. Koziol MJ, Whatley FR (1984) Gaseous air pollutants and plant metabolism. Butterworths, London. 466 p.
8. Schulte-Hostede S, Darrall NM, Blank LW, Wellburn AR (1987) Air pollution and plant metabolism. London: Elsevier. 381p.
9. Yunus M, Iqbal M (1996) Plant response to air pollution. Chichester: Wiley. 545p.
10. Sandermann H, Wellburn AR, Heath RL (1997) Forest decline and ozone: a comparison of controlled chamber and field experiments. Berlin: Springer-Verlag. 400p.
11. De Kok LJ, Stulen I (1998) Responses of plant metabolism to air pollution and global change. Leiden: Backhuys. 519p.
12. Omasa K, Saji H, Youssefian S, Kondo K (2002) Air Pollution and Plant Biotechnology. Tokyo: Springer-Verlag. 455p.
13. Gregg JW, Jones CG, Dawson TE (2003) Urbanization effects on tree growth in the vicinity of New York City. Nature 424: 183–187.
14. Serengil Y, Augustaitis A, Bytnerowicz A, Grulke N, Kozovitz AR, et al. (2011) Adaptation of forest ecosystems to air pollution and climate change: a global assessment on research priorities. iForest 4: 44–48.
15. Kerstiens G, Lendzian KJ (1989) Interactions between ozone and plant cuticles. 1. Ozone deposition and permeability. New Phytol 112: 13–19.
16. Omasa K, Abo F, Natori T, Totsuka T (1979) Studies of air pollutant sorption by plants. (II) Sorption under fumigation with NO_2, O_3 or NO_2+O_3. J Agric Meteorol 35: 77–83 (in Japanese with English summary).
17. UNECE (2004) Mapping critical levels for vegetation. Chapter 3 Manual on Methodologies and Criteria for Modelling and Mapping Critical Loads and Levels and Air Pollution Effects, Risks and Trends. Berlin: Umweltbundesamt. 52p.
18. Grulke NE, Paoletti E, Heath RL (2007a) Comparison of calculated and measured foliar O_3 flux in crop and forest species. Environ Pollut 146: 640–647.
19. Paoletti E, Manning WJ (2007) Toward a biologically significant and usable standard for ozone that will also protect plants. Environ Pollut 150: 85–95.
20. Cieslik S, Omasa K, Paoletti E (2009) Why and how terrestrial plants exchange gases with air. Plant Biol 11: 24–34.
21. Paoletti E, Grulke NE (2005) Does living in elevated CO_2 ameliorate tree response to ozone? A review on stomatal responses. Environ Pollut 137: 483–493.
22. Paoletti E, Grulke NE (2010) Ozone exposure and stomatal sluggishness in different plant physiognomic classes. Environ Pollut 158: 2664–2671.
23. Wittig VE, Ainsworth EA, Long SP (2007) To what extent do current and projected increases in surface ozone affect photosynthesis and stomatal conductance of trees? A meta-analytic review of the last 3 decades of experiments. Plant Cell Environ 30: 1150–1162.
24. Reich PB, Lassoie JP (1984) Effects of low level O_3 exposure on leaf diffusive conductance and water-use efficiency in hybrid poplar. Plant Cell Environ 7: 661–668.
25. Reiling K, Davison AW (1995) Effects of ozone on stomatal conductance and photosynthesis in populations of *Plantago major* L. New Phytol 129: 587–594.
26. Paoletti E (2005) Ozone slows stomatal response to light and leaf wounding in a Mediterranean evergreen broadlieaf, *Arbutus unedo*. Environ Pollut 134: 439-445.
27. Grulke NE, Neufeld HS, Davison AW, Chappelka A (2007b) Stomatal behavior of O_3-sensitive and -tolerant cutleaf coneflower (*Rudbeckia laciniata* var. *digitata*) Great Smoky Mountain National Park. New Phytol 173: 100–109.
28. Paoletti E, Contran N, Bernasconi P, Gunthardt-Goerg MS, Vollenweider P (2009a) Structural and physiological responses to ozone in Manna ash (*Fraxinus ornus* L.) leaves of seedlings and mature trees under controlled and ambient conditions. Sci Tot Environ 407: 1631–1643.
29. Mills G, Hayes F, Wilkinson S, Davies WJ (2009) Chronic exposure to increasing background ozone impairs stomatal functioning in grassland species. Glob Chan Biol 15: 1522–1533.
30. Hoshika Y, Omasa K, Paoletti E (2012) Both ozone exposure and soil water stress are able to induce stomatal sluggishness. Environ Exp Bot doi:10.1016/j.envexpbot.2011.12.004.
31. Tinoco-Ojanguren C, Pearcy RW (1993) Stomatal dynamics and its importance to carbon gain in two rain forest Piper species. II. Stomatal vesus biochemical limitations during photosynthetic induction. Oecol 94: 395–402.
32. McLaughlin SB, Nosal M, Wullschleger SD, Sun G (2007) Interactive effects of ozone and climate on tree growth and water use in a southern Appalachian forest in the USA. New Phytol 174: 109–124.
33. Mansfield TA (1998) Stomata and plant water relations: does air pollution create problems? Environ Pollut 101: 1–11.
34. Robinson MF, Heath J, Mansfield TA (1998) Disturbances in stomatal behaviour caused by air pollutants. J Exp Bot 49: 461–470.
35. Wieser G, Matyssek R, Then C, Cieslik S, Paoletti E, et al. (2008) Upscaling ozone flux in forests from leaf to landscape. Ital J Agron 1: 35–41.
36. Wittig VE, Ainsworth EA, Naidu SL, Karnosky DF, Long SP (2009) Quantifying the impact of current and future tropospheric ozone on tree biomass, growth, physiology and biochemistry: a quantitative meta-analysis. Glob Chan Biol 15: 396–424.
37. Heath RL (2008) Modification of the biochemical pathways of plants induced by ozone: What are the varied routes to change? Environ Pollut 155: 453–463.
38. Innes JL, Skelly JM, Schaub M (2001) Ozone and broadleaved species. A guide to the identification of ozone-induced foliar injury. Bern: Paul Haupt Verlag. 136p.
39. Paoletti E, Ferrara AM, Calatayud V, Cervero J, Giannetti F, et al. (2009b) Deciduous shrubs for ozone bioindication: *Hibiscus syriacus* as an example. Environ Pollut 157: 865–870.
40. Zhang J, Ferdinand JA, VanderHeyden DJ, Skelly JM, Innes JL (2000) Variation in gas exchange within native plant species of Switzerland and relationships with ozone injury: an open-top experiment. Environ Pollut 113: 117–185.
41. Paoletti E, Nali C, Lorenzini G (2004) Gas exchange and ozone visible injury in Mediterranean evergreen broadleaved seedlings. In: Kinnunen H, Huttunen S, editors. Proceedings of the Meeting Forest Under Changing Climate, Enhanced UV and Air Pollution. Oulu: Dept Biology, Thule Institute, Univ. Oulu. 85–101.
42. Novak K, Schaub M, Fuhrer J, Skelly JM, Hug C, et al. (2005) Seasonal trends in reduced leaf gas exchange and ozone-induced foliar injury in three ozone sensitive woody plant species. Environ Pollut 136: 33–45.
43. Omasa K, Hashimoto Y, Aiga I (1981) A quantitative analysis of the relationships between O_3 absorption and its acute effects on plant leaves using image instrumentation. Environ Control Biol 19: 85–92.
44. Marzuoli R, Gerosa G, Desotgiu R, Bussotti F, Ballarin-Denti A (2009) Ozone fluxes and foliar injury development in the ozone-sensitive poplar clone Oxford (*Populus maximowiczii* × *Populus berolinensis*): a dose–response analysis. Tree Physiol 29: 67–76.
45. Paoletti E, Contran N, Manning WJ, Ferrara AM (2009c) Use of the antiozonant ethylenediurea (EDU) in Italy: verification of the effects of ambient ozone on crop plants and trees and investigation of EDU's mode of action. Environ Pollut 157: 1453–1460.

46. Paoletti E, Manning WJ, Spaziani F, Tagliaferro F (2007) Gravitational infusion of ethylenediurea (EDU) into trunks protected adult European ash trees (*Fraxinus excelsior* L.) from foliar ozone injury. Environ Pollut 145: 869–873.

47. Manning WJ, Paoletti E, Sandermann H Jr, Ernst D (2011) Ethylenediurea (EDU): A research tool for assessment and verification of the effects of ground level ozone on plants under natural conditions. Environ Pollut 159: 3283–3293.

48. Paoletti E, Manning WJ, Ferrara AM, Tagliaferro F (2011) Soil drench of ethylenediurea (EDU) protects sensitive trees from ozone injury. iForest 4: 66–68.

49. Pell EJ, Eckardt N, Enyedi AJ (1992) Timing of ozone stress and resulting status of ribulose bisphosphate carboxylase/oxygenase and associated net photosynthesis. New Phytol 120: 397–405.

50. Omasa K (1990) Study on changes in stomata and their surroundings cells using a nondestructive light microscope system: responses to air pollutants. J Agric Meteorol 45: 251–257 (in Japanese with English summary).

51. Vahisalu T, Puzorjova I, Brosche M, Valk E, Lepiku M, et al. (2010) Ozone-triggered rapid stomatal response involves the production of reactive oxygen species, and is controlled by SLAC1 and OST1. Plant J 62: 442–453.

52. Wilkinson S, Davies W (2010) Drought, ozone, ABA and ethylene: new insights from cell to plant community. Plant Cell Environ 33: 510–525.

53. Tingey DT, Standley C, Field RW (1976) Stress ethylene evolution: a measure of ozone effects on plants. Atmos Environ 10: 969–974.

54. Samuel MA, Walia A, Mansfield SD, Ellis BE (2005) Overexpression of SIPK in tobacco enhances ozone-induced ethylene formation and blocks ozone-induced SA accumulation. J Exp Bot 56: 2195–2201.

55. Tuomainen J, Betz C, Kangasjärvi J, Ernst D, Yin Z, et al. (1997) Ozone induction of ethylene emission in tomato plants: regulation by differential accumulation of transcripts for the biosynthetic enzymes. Plant J 12: 1151–1162.

56. Heber U, Neimanis S, Lange OL (1986) Stomatal aperture, photosynthesis and water fluxes in mesophyll cells as affected by the abscission of leaves. Simultaneous measurements of gas exchange, light scattering and chlorophyll fluorescence. Planta 167: 554–562.

14

High Environmental Ozone Levels Lead to Enhanced Allergenicity of Birch Pollen

Isabelle Beck[1], Susanne Jochner[2], Stefanie Gilles[1,4], Mareike McIntyre[3], Jeroen T. M. Buters[1], Carsten Schmidt-Weber[1], Heidrun Behrendt[1,4], Johannes Ring[3], Annette Menzel[2], Claudia Traidl-Hoffmann[1,3,4]*

1 ZAUM – Center of Allergy & Environment, Member of the German Center for Lung Research (DZL), Technische Universität München/Helmholtz Center, Munich, Germany, 2 Department of Ecology and Ecosystem Management, Ecoclimatology, Technische Universität München, Freising, Germany, 3 Department of Dermatology and Allergy, Technische Universität München, Munich, Germany, 4 Christine-Kühne-Center for Allergy Research and Education (CK Care), Davos, Switzerland

Abstract

Background: Evidence is compelling for a positive correlation between climate change, urbanisation and prevalence of allergic sensitisation and diseases. The reason for this association is not clear to date. Some data point to a pro-allergenic effect of anthropogenic factors on susceptible individuals.

Objectives: To evaluate the impact of urbanisation and climate change on pollen allergenicity.

Methods: Catkins were sampled from birch trees from different sites across the greater area of Munich, pollen were isolated and an urbanisation index, NO_2 and ozone exposure were determined. To estimate pollen allergenicity, allergen content and pollen-associated lipid mediators were measured in aqueous pollen extracts. Immune stimulatory and modulatory capacity of pollen was assessed by neutrophil migration assays and the potential of pollen to inhibit dendritic cell interleukin-12 response. *In vivo* allergenicity was assessed by skin prick tests.

Results: The study revealed ozone as a prominent environmental factor influencing the allergenicity of birch pollen. Enhanced allergenicity, as assessed in skin prick tests, was mirrored by enhanced allergen content. Beyond that, ozone induced changes in lipid composition and chemotactic and immune modulatory potential of the pollen. Higher ozone-exposed pollen was characterised by less immune modulatory but higher immune stimulatory potential.

Conclusion: It is likely that future climate change along with increasing urbanisation will lead to rising ozone concentrations in the next decades. Our study indicates that ozone is a crucial factor leading to clinically relevant enhanced allergenicity of birch pollen. Thus, with increasing temperatures and increasing ozone levels, also symptoms of pollen allergic patients may increase further.

Editor: Simon Patrick Hogan, Cincinnati Children's Hospital Medical Center, University of Cincinnati College of Medicine, United States of America

Funding: The study was supported by a grant of the German Research Council (Deutsche Forschungsgemeinschaft, DFG, TR467/5-1 and ME 179/3-1) (CTH and AM), the HWP grant from of the Technische Universität München (SG), and CK Care, Christine Kühne Center for Allergy Research and Education. The funders had no role in study design, data collection and analysis, decision to publish, or preparation of the manuscript.

Competing Interests: Bet v 1- specific antibodies MAK 2E10G6G7 and 4B10D10F8 were kindly provided by Allergopharma.

* E-mail: claudia.traidl-hoffmann@lrz.tu-muenchen.de

Introduction

Epidemiological studies show an increasing trend in allergies, leading to a major health problem. Reasons discussed for this trend include a westernized life style with diminished immune stimulation [1] and anthropogenic air pollution [2,3]. Particularly, irritant gases and diesel exhaust particles have been shown to exert adjuvant or aggravating effects on sensitisation and elicitation phases of allergic immune responses [4,5]. As underlying mechanisms, effects on cells of the immune system as well as epithelial barrier disruption are discussed [6]. However, pollutants in ambient air do not only impact humans but also the allergen-carrier itself, i.e. the plant and its pollen. Therefore, the question arises whether the observed increase in allergic diseases in the western world might in part be explained by modified allergenicity of pollen caused by urbanisation and paralleled climate change. These environmental changes — higher temperature, in combination with higher concentrations of specific anthropogenic pollutants — lead to higher tropospheric ozone concentrations. In this scenario UV-radiation delivers the energy for ozone generation, but besides this, higher temperatures can also lead to an increase in ozone formation promoted by emission of highly reactive hydrocarbons from vegetation and evaporation processes [7]. Climate extremes are often observed in cities, which function as heat islands and can be regarded as a mirror of future climate [8]. However, urbanisation is not only characterised by higher temperatures, but also by higher levels of pollutants like particulate matter, carbon dioxide (CO_2) or nitrogen dioxide (NO_2). In this respect, it has to be considered that ozone does not show the same distribution as other pollutants [7,9]. Ozone is a secondary pollutant whose formation underlies complex interactions, depending on the presence of precursors, degrading substances, temperature and UV-radiation. The main precursors are nitrogen oxides (NOx) and volatile organic compounds (VOCs). Especially NOx can be found at high concentrations in urban areas. NO, in turn, rapidly degrades the ozone generated in urban areas, especially during night-time. In rural areas, in contrast, we observe an accumulation of ozone due to lower NO levels and higher biogenic emissions [7]. Some studies already addressed the question of how pollutants affect the allergen carrier, showing that single pollutants strongly differ in their effects [10,11]. The current study expands these observations by analyzing one of the most relevant allergen producers – the birch tree – in its natural environment. Hereby, we work with real exposure conditions to analyse relevant factors under natural circumstances. We take into account the parallel occurrence of different environmental factors as well as mechanisms of plant adaptation. Recent studies showed that pollen do not only release allergens, but also non-allergenic compounds such as pollen-associated lipid mediators (PALMs) (reviewed in [12]), which have been shown to exert immune modulatory and stimulatory effects. Allergenicity, thus, was evaluated in a holistic approach also taking adjuvant factors into account.

This study aimed at understanding how long-term increases in urbanisation and concomitant increases in pollution might influence pollen allergenicity, and how this might translate into immune cell activation and symptoms of pollen-allergic patients.

Materials and Methods

Ethic Statement

The ethical committee of the Technical University of Munich approved the study, and volunteers were enrolled after written informed consent. The study was carried out on private land and the owners of the land gave permission to conduct the study on these sites.

Sampling of Birch Pollen

Catkins were collected from birch trees located in the greater area of Munich (n = 40; see Fig. S1). Sampling of catkins took place during the birch flowering season in spring 2010. Their developmental stages were assessed by an adopted and extended code of the BBCH (Biologische **B**undesanstalt, **B**undessortenamt

und **Ch**emische Industrie) [13]. The code included 12 different developmental stages, starting with winter rest (BBCH 50) and ending with end of flowering (BBCH 69). The stages of collection were BBCH 60 (single catkins sporadically emit pollen) and BBCH 61 (10% of the catkins emit pollen). After collection, catkins were air-dried, counted and weighed and pollen was extracted by sieving.

Preparation of Aqueous Pollen Extracts for Bet v 1, LTB$_4$ and PGE$_2$ ELISA

Aqueous pollen extracts (APEs) were prepared in 0.1 M NH$_4$HCO$_3$, pH 8.1, as previously described [14]. For skin prick tests, APEs were prepared as described in [15]. APE concentrations given refer to the extraction of a given amount of pollen per mL of buffer before centrifugation (e. g. 1 mg pollen was extracted in 1 mL buffer), but do not refer to actual protein concentrations in the extracts.

Bet v 1 ELISA

Bet v 1 levels were determined by sandwich ELISA as described in Buters et al. [14]. Bet v 1-specific antibodies MAK 2E10G6G7 and 4B10D10F8 were kindly provided by Joachim Ganzer, Allergopharma, Hamburg, Germany.

PGE$_2$ and LTB$_4$ ELISA

Concentrations of the eicosanoid-like PALMs in APEs were measured by commercially available enzyme immunoassays for prostaglandin E$_2$ and leukotriene B$_4$ (GE Healthcare, Germany) according to the supplier's protocol.

Passive Sampling Method for Ozone and Nitrogen Dioxide (NO$_2$) Determination

Passive samplers for ozone were provided and analysed by PASSAM AG, Männedorf, Switzerland. Passive sampling for ozone and NO$_2$ were done in parallel. Passive sampling was carried out during the one-week period from May 11th to 18th 2010, i.e. 2 weeks after the last catkins were sampled. Measurements for all 40 birch trees were done in the same time frame and directly at the tree. It is supposed that the traffic volume over the period of 1 week is a representative mean for this location. This data was used to characterize the tree for a relative ozone exposure.

The nitrogen dioxide concentration was measured at the 40 sites according to Palmes' principle [16]. Briefly, stainless steel

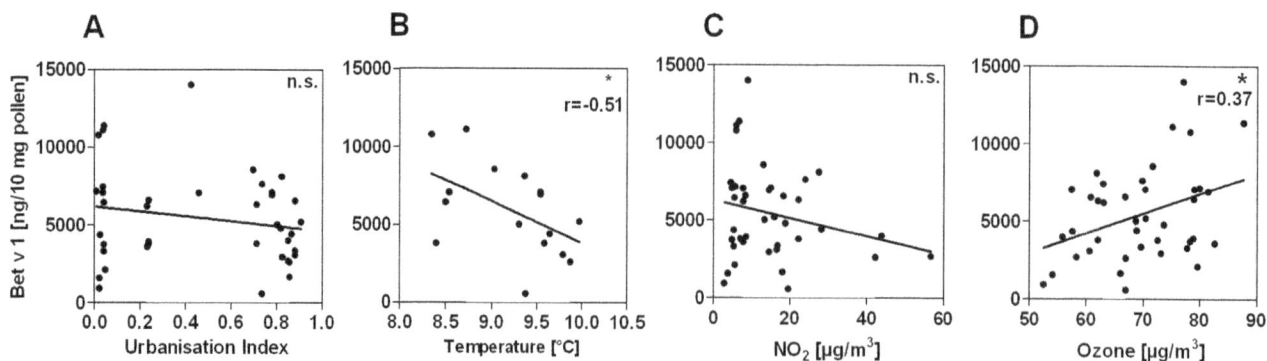

Figure 1. Bet v 1 content of birch pollen in trees against different, urbanisation-related environmental conditions. No significant correlation could be observed between Bet v 1 content and urbanisation index (A; n=40) or NO$_2$ concentration (B; n=40). Bet v 1 showed a significant and negative correlation with temperature (C; n=16) and was positively correlated with site-specific ozone levels (D; n=40). *: p<0.05.

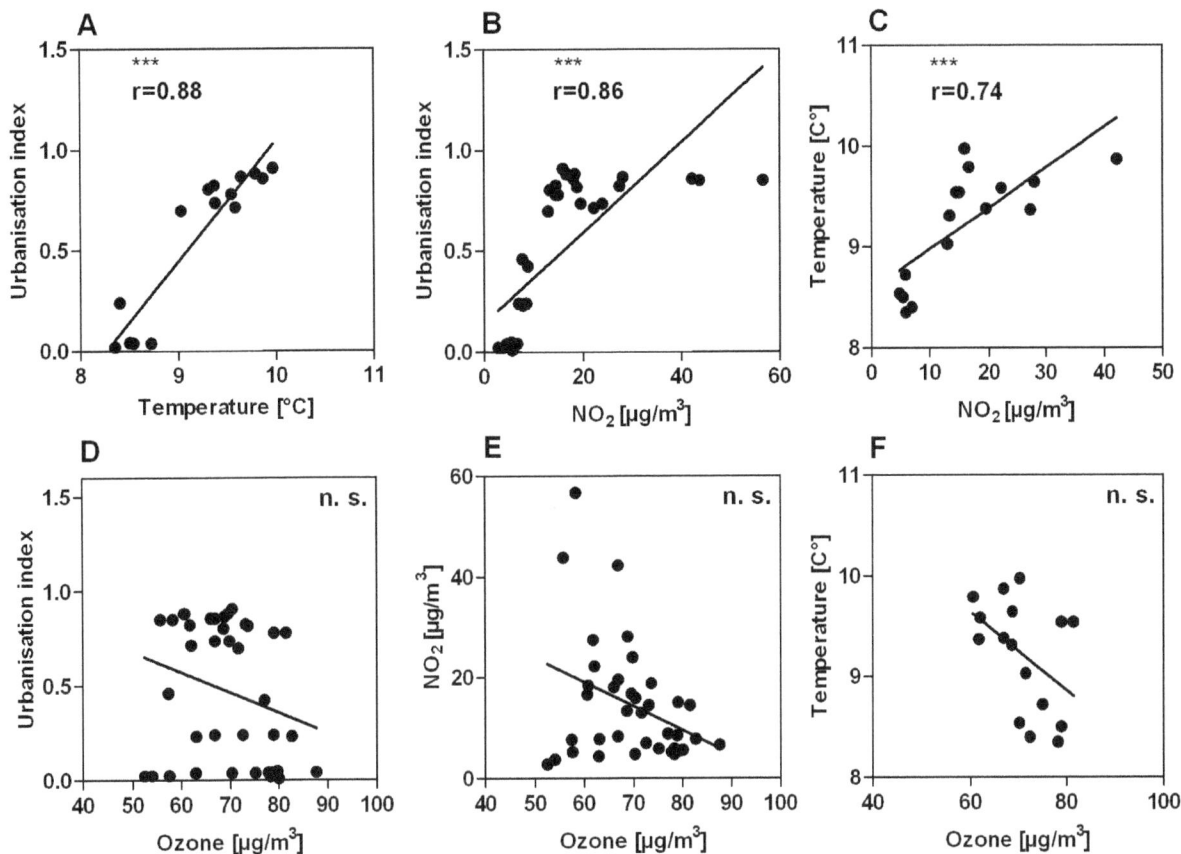

Figure 2. Scatter plots of different environmental parameters. A significant correlation was observed between the parameters urbanisation index (UI) and temperature (A; n = 16), (B) UI and NO_2 concentration (n = 40) and (C) temperature and NO_2 concentration (n = 16). Ozone was not correlated with either (D) UI (n = 40), (E) NO_2 (n = 40) or (F) temperature (n = 16); ***: p<0.001.

meshes were immersed in a triethanolamine-aceton mixture and were air-dried for 10 minutes. Three coated meshes were brought into an air-tight tube. NO_2 binds to the coated meshes by forming a triethanolamine-NO_2-complex. NO_2 adsorption was determined photometrically after one week of exposure.

Calculation of Urbanisation Index

An urbanisation index (UI) based on CORINE Land Cover 2000 data (European Environment Agency 2000) was calculated using ArcGIS 9.3. This index reflects the proportion of predefined built up areas (e.g. continuous and discontinuous urban fabric, industrial or commercial units) within a radius of 2 km and thus can vary between 0 and 1; i.e. from a low (UI = 0) to a high (UI = 1) degree of urbanisation [17].

Temperature Measurements

15 birch trees were provided with data recording devices measuring air temperature (HOBO U23-001, Onset Computer Corporation, Southern MA, USA). The devices were fixed in a radiation shield at the northern side of the trees in 3 m height. Air temperatures were recorded every 10 minutes, and daily temperature means were calculated. Temperature data were acquired between 1st July 2009 and 5th May 2010. The mean temperature for this period for each location was calculated.

Blood Donors

Healthy, non-atopic blood donors without a history of allergic diseases were tested by RAST for sensitisation against common allergens including birch allergens. All subjects were tested negative and total IgE was <100 IU/ml. Volunteers did not take any medication for at least 15 days before blood sampling.

Isolation and Culture of Monocyte-derived Dendritic Cells

Monocyte-derived dendritic cells (moDCs) were cultured from human peripheral blood monocytes as described by Gilles et al. [18]. Immature DCs were harvested on day 5 followed by stimulation with LPS (100 ng/ml) plus APEs from the highest (pollen[O3high]: mean ozone = 85 $\mu g/m^3$, n = 2) and lowest (pollen[O3high]: mean ozone = 54 $\mu g/m^3$, n = 2) ozone-exposed birch trees included in the study (mean ozone: 54 $\mu g/m^3$ versus 85 $\mu g/m^3$). After 24 h, supernatants were collected, and IL-12p70 release was measured by ELISA (BD Pharmingen, Heidelberg, Germany). The IL-12 response to APEs alone was measured exemplary in four APEs and virtually no IL-12 was detected (unstimulated: 18.7 pg/ml; APEs: median (min.-max.): 15.4 (9.9–20.8) pg/ml, n = 4) (data not shown). Viability of the cells after 24 h of culture was tested by propidium iodide staining and subsequent FACS analysis (see Fig. S4, supplementary material). Viability was not decreased by any of the conditions.

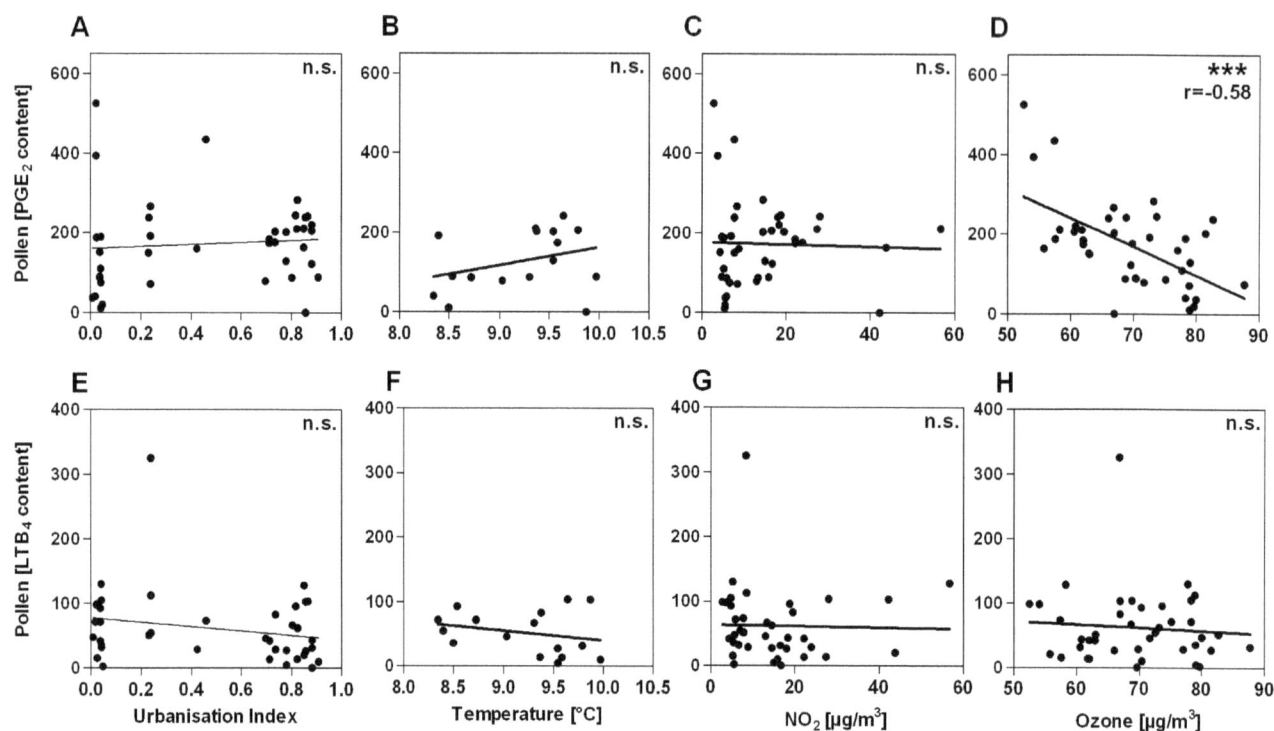

Figure 3. Content of PALMs in pollen samples plotted against different urbanisation-related environmental conditions. No correlation was seen between $PALM_{PGE2}$ and UI (A; n = 40), temperature (B; n = 16) or NO_2 concentration (C; n = 40). A significant association of high ozone concentrations and low $PALM_{PGE2}$ contents was observed (D; n = 40). $PALM_{LTB4}$ did not show any significant correlation. Neither UI (E; n = 40), nor temperature (F; n = 16), NO_2- (G; n = 40) and ozone (H; n = 40) were related to the content of $PALM_{LTB4}$. ***: p<0.001.

Neutrophil Migration Assays

The chemotactic activity of APEs from the highest (pollen^{O3high}: mean ozone = 85 $\mu g/m^3$) and lowest (pollen^{O3high}: mean ozone = 54 $\mu g/m^3$) ozone-exposed birch trees included in the study was evaluated by measuring neutrophil migration through a 5 μm pore polycarbonate membrane (ChemoTx Disposable Chemotaxis System, NeuroProbe). Neutrophils were isolated from peripheral blood as described by Traidl-Hoffmann et al. [19]. APEs were pipetted into the bottom chambers and neutrophils (1×10^6 cells/ml) were added to the top of the membrane. After 1 h of incubation the cell suspension was removed and cells that had transmigrated into the lower chamber were recovered and counted with a FACSCalibur (Becton Dickinson, Heidelberg, Germany).

Skin Prick Tests

Birch allergic patients (n = 5) with a specific IgE >0.35 kU/l were pricked on their forearms with APE (10 mg/mL) prepared of pollen samples from the highest (pollen^{O3high}: mean ozone = 85 $\mu g/m^3$) and lowest (pollen^{O3high}: mean ozone = 54 $\mu g/m^3$) ozone-exposed birch trees of the study. Each sample was pricked in replicates of four (proximally and distally on the same arm, right and left arm), and the mean wheal and flare sizes were calculated out of these four measurements after 15 min.

Statistics

Unpaired t-test was used for Gaussian distributed samples to determine statistically significant differences between groups. For non-Gaussian populations, the non-parametric Mann-Whitney U test was applied. The correlation coefficients (r) and 95% confidence intervals were calculated using the Pearson's approach

for Gaussian distributions and the Spearman's approach for non-parametric correlation. Additionally, we applied linear multiple regression analyses based on stepwise variable selection to test which environmental factor is most important for pollen allergenicity. To analyse cell assays, the area under the curve (AUC) was determined and the Wilcoxon matched-pairs signed-ranks test was applied. Prick tests were also analysed by the Wilcoxon matched-pairs signed-ranks test. P values of <0.05 were considered significant (*). **: p<0.01, ***: p<0.001. Statistics were done with Graph Pad Prism 5, San Diego, CA, USA.

Results

Correlation Analysis of Urbanisation Related Factors and Relationship to Pollen Allergenic Potential

No significant correlation was detected between Bet v 1 content of the pollen specimens and urbanisation index (UI) or NO_2 concentration (Fig. 1A, C; n = 40). However, a negative correlation could be observed for Bet v 1 content and temperature (Fig. 1B; r = -0.51; p = 0.042; n = 16). In contrast, ozone was positively correlated with Bet v 1 content. (r = 0.37; p = 0.017; Fig. 1D; n = 40).

Sites of birch trees were analyzed for environmental parameters. By correlation analyses the relationship of the determined parameters was investigated. A highly significant and positive correlation of the variables UI (n = 40), temperature (n = 16) and NO_2-concentration (n = 40) could be observed (Fig. 2 A, B, C). In contrast, ozone (n = 40) was not statistically associated with any of these parameters (Fig. 2 D, E, F).

To investigate whether ozone or temperature is the factor influencing pollen Bet v 1 content, we applied multiple regression analyses based on stepwise variable selection. Indeed, the role of the independent variable ozone for predicting Bet v 1 content was superior, since temperature has been excluded in the linear model, yielding in an R^2 of 27.6% (data not shown).

To determine immune stimulatory and modulatory potential of the birch pollen specimens, we analysed the content of PALMs in the different pollen extracts. Prostaglandin E_2-like PALMs, termed $PALM_{PGE2}$, harbour the immune modulatory PALMs [20,21]. No statistically significant correlation was observed between $PALM_{PGE2}$ contents and UI (n = 40), temperature (n = 16) or ambient NO_2 (n = 40) (Fig. 3A, B, C). Instead, $PALM_{PGE2}$ content was negatively correlated with ambient ozone (n = 40) levels (r = -0.58; p<0.0001; Fig. 3D). The LTB_4-like PALMs harbour the chemotactic, immune stimulatory PALMs [19]. $PALM_{LTB4}$ contents were not correlated with either UI (n = 40), temperature (n = 16), NO_2 (n = 40) or ozone (n = 40) levels (Fig. 3E, F, G, H).

Immune-stimulatory and Immune-regulatory Capacity of Pollen Samples

To assess whether the altered lipid compositions related to higher and lower ozone concentrations go along with differences in immune stimulatory and –modulatory potential, pollen from differentially ozone-exposed birch trees (mean ozone: 85 µg/m^3 (pollen^{O3high}) versus 54 µg/m^3 (pollen^{O3low})) were chosen and analysed for induction of neutrophil chemotaxis and modulation of dendritic cell (moDC) cytokine secretion. Migration to pollen^{O3high} and pollen^{O3low} was tested in 3 concentrations (Fig. S4) and the AUC was calculated. Pollen^{O3high} induced significantly stronger neutrophil chemotaxis than pollen^{O3low}. Neutrophil migration towards pollen extracts (median AUC (min.-max.)) were 3181 (2214–4429) for pollen^{O3high} versus 2327 (1839–4095) for pollen^{O3low}; p = 0.03 (Fig. 4A). To test the immune modulatory capacity of differentially ozone-exposed pollen, moDCs were stimulated with LPS in the presence and absence of APE, and IL-12 was measured in the supernatants (LPS induced IL-12 response: median (min.-max.): 14000 (8928–14500) pg/ml). APEs were applied in 3 concentrations (Fig. S3) and AUC was calculated. As shown in Fig. 4B, pollen^{O3high} were less potent inhibitors of the moDC's IL-12 response than pollen^{O3low}. IL-12 release of dendritic cells (% of LPS; median AUC (min-max)) after stimulation with APE was 127.7 (67.3–227.0) for pollen^{O3high} versus 71.0 (52.8–187.7) for pollen^{O3low}; p = 0.02.

To test for clinical relevance of enhanced Bet v 1 levels, birch pollen allergic patients were subjected to skin prick tests with APEs prepared from differently exposed pollen. Wheal and flare sizes were significantly larger when patients were pricked with APEs prepared from pollen^{O3high}. Wheal sizes (in mm^2; median (min.-max.)) were 21.3 (9.5–33.8) for pollen^{O3low} and 13.8 (4.7–22.5) for pollen^{O3high}; p = 0.02. Flare sizes (in mm^2; median (min.-max.)) were 256.3 (12.4–343.8) for pollen^{O3low} and 353.1 (43.8–381.3) for pollen^{O3high}; p = 0.005 (Fig. 5A, B). This is in line with higher Bet v 1 content of pollen^{O3high}.

Discussion

The present study reveals ozone exposure as important factor enhancing the allergenicity of birch pollen, with clinical relevance for susceptible individuals. Besides a positive relationship of the Bet v 1 content with ozone, we observed a negative association with temperature. Results of studies analysing the influence of temperature on Bet v 1 expression, however, are inconsistent. While Ahlholm et al. [22] reported a positive effect of high temperatures on Bet v 1 content, Helander et al. [23] indicated a

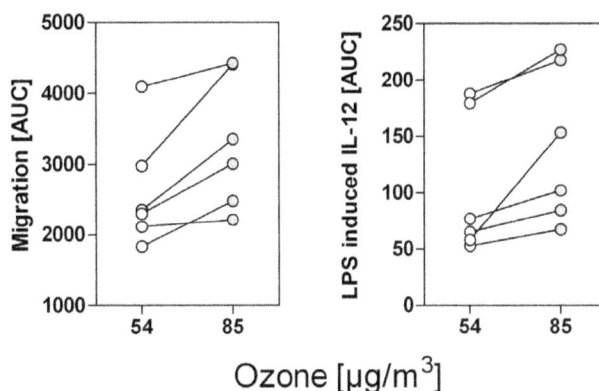

Figure 4. Immune stimulatory versus immune modulatory potential of pollen samples from higher- and lower-ozone-exposed birch trees. Aqueous extracts (APEs) of birch pollen sampled from high and low ozone exposed trees were chosen for neutrophil migration assays and stimulation of monocyte derived dendritic cells. APEs were applied in 3 concentrations and the AUC was calculated. Higher ozone-exposed pollen induced stronger neutrophil chemotaxis compared to pollen samples from lower ozone–exposed trees (**A**). In contrast, birch pollen from lower ozone-exposed trees were more potent in inhibiting the LPS-induced release of IL-12p70 from human monocyte-derived dendritic cells (**B**). APEs were prepared from birch pollen sampled from higher ozone-exposed trees (n = 2; mean ozone: 85 µg/m^3) and from lower ozone-exposed trees (n = 2; mean ozone: 54 µg/m^3). All APEs were tested in n = 3 patients. *: p<0.05 (Wilcoxon matched-pairs signed-ranks test).

negative association. The effect of temperature on the expression of Bet v 1 was also investigated by Tashpulatov et al. [24], using transgenic tobacco plants carrying a Bet v 1a-pomoter-reporter gene fusion. This study showed that warm temperatures positively regulate the activity of the Bet v 1 promoter. Our data indicate the opposite as we observed a negative correlation of temperature with Bet v 1. However, temperature and UI as well as NO_2 concentration were highly and significantly related, while UI and NO_2 did not correlate with Bet v 1. The number of entities for NO_2 and UI were 40, while merely 16 data points for temperature were available. Considering the strong relationship of the parameters UI and NO_2 with temperature it can be assumed that the correlation of temperature and Bet v 1 is observed by chance. Moreover when having a detailed look at the interrelationship of the environmental parameters of this study we also observed a non-significant negative association of ozone and temperature (Fig. 2F; p = 0.07; r = -0,46). This seems to be contradictory as ozone formation is associated with UV radiation and temperature [25]. However, ozone formation also depends on the presence of precursors (e.g. biogenic emissions) and degrading substances (e.g. NO). Rural areas are associated with lower temperatures. At the same time, the composition of ozone forming and -degrading factors in rural areas favor ozone accumulation. This might explain why we observe a negative correlation of ozone and temperature.

It might be hypothesized that a positive influence of ozone on Bet v 1 and the relationship of ozone and temperature implied a negative correlation of temperature and Bet v 1.

To confirm our suggestion, we applied multiple regression analyses based on stepwise variable selection. Indeed, the role of the independent variable ozone for predicting Bet v 1 content was superior, since temperature has been excluded in the linear model (data not shown).

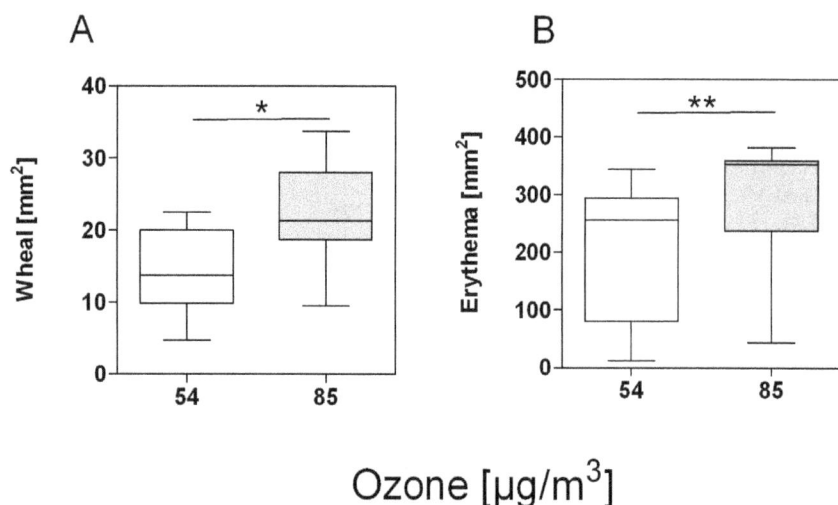

Figure 5. Cutaneous immune response towards pollen from higher and lower ozone-exposed birch trees. APEs were prepared from pollen sampled from higher ozone-exposed trees (n = 2; mean ozone: 85 $\mu g/m^3$) and from lower ozone-exposed trees (n = 2; mean ozone: 54 $\mu g/m^3$). All APEs were tested in n = 5 patients. Higher ozone-exposed pollen induced larger wheals (**A**) and flares (**B**) in skin prick tests compared to lower ozone-exposed pollen. *: $p < 0.05$; **: $p < 0.01$ (Wilcoxon matched-pairs signed-ranks test).

Besides, Tashpulatov et al. [24] showed that abscisic acid, a stress- and development-related plant hormone, positively regulates the activity of the Bet v 1a promoter. This goes in line with our study showing that ozone, a major stress factor for plants, is associated with a higher Bet v 1 expression. Besides, also former studies in ragweed and grass species gave evidence that ozone impacts on the allergen transcript and content [26,27]. In these studies, plants were exposed to defined ozone concentrations. In our study, trees were subjected to pollen sampling in their natural environment and under natural exposure conditions.

In a holistic approach analysing the allergenicity of birch pollen, we showed that elevated ozone exposure of birch trees was not only associated with increased allergen content in pollen but also with an altered composition of adjuvant PALMs. No relation of $PALM_{LTB4}$ and $PALM_{PGE2}$ content and the degree of urbanisation (UI), temperature or NO_2 concentration was observed. This seems to be in contrast to a recent study [28]. However, this discrepancy is most likely explained by the fact that in the present study, pollen sampling was carried out at defined, distinct maturation stages. As demonstrated in Fig. S2, the content of PALMs in pollen grains differs profoundly depending on the maturation stage of pollen, as does the Bet v 1 content, confirming former results [29].

Notably, $PALM_{PGE2}$ was significantly negatively associated to ozone concentrations. In our in vitro assays, pollen^{O3high} were significantly more chemotactic for neutrophils than extracts of pollen^{O3low}. Since $PALM_{LTB4}$ did not significantly differ in differently exposed pollen we hypothesize that other, up to now unknown, substances besides $PALM_{LTB4}$ account for this effect. Moreover, Armstrong [30] could show that PGE_2 is able to inhibit the chemotaxis of neutrophils in a concentration dependent manner. Consequently, lower concentrations of $PALM_{PGE2}$ in pollen^{O3high} could also have contributed to enhanced neutrophil chemotaxis. In contrast, APEs prepared from pollen^{O3low} were significantly more efficient in inhibiting dendritic cell IL-12 secretion, in line with higher levels of immune modulatory $PALM_{PGE2}$. Among the immune modulatory PALMs are plant isoprostanes identified as E_1-phytoprostanes. E_1-phytoprostanes inhibit the IL-12 response in maturing DCs [31,32], finally

licensing DCs to differentiate naïve $CD4^+$ T cells into Th2 cells [32]. Transferring our data to allergy mechanisms, we hypothesize that higher $PALM_{PGE2}$ concentrations in low-ozone areas might facilitate *de novo* sensitisation by providing Th2 promoting signals. Higher Bet v 1 concentration and less anti-inflammatory PALMs in pollen^{O3high} might in turn lead to pronounced allergic symptoms in already sensitized individuals.

In summary, urbanisation-related, anthropogenic environmental factors can influence birch trees to produce pollen with altered allergenic potential. Our study emphasizes the correlation of ozone exposure to the pollen content of allergen and non-allergenic, adjuvant factors. It is likely that future climate change with more frequent and intense warm spells [33] as well as increases in urbanisation and anthropogenic air pollutants such as NOx will further enhance the local accumulation of tropospheric ozone. As indicated by this study, ozone might – apart from direct adverse impacts on human health – lead to increased allergic symptoms *via* its impact on the allergen carrier.

Supporting Information

Figure S1   Locations of pollen sampling. Birch pollen were sampled during the birch flowering season of 2010. Red dots represent urban trees, green dots rural trees. Background: CORINE Land Cover 2000 (EEA 2000), major classes: red = urban fabric, green = forest and pastures, yellow = arable land, blue = rivers, lakes (see www.eea.europa.eu/themes/landuse/interactive/clc-download for a complete legend).

Figure S2   Catkin maturation and allergenic potential of pollen. A: Catkins of different maturation stages were collected at different time points from the same trees (BBCH-Code 51–52: n = 1; BBCH-Code 55–65: n = 5) and classified according to a BBCH-Code. Pollen were isolated from the catkins and aqueous pollen extracts were prepared. APEs were then analyzed for the presence of Bet v 1 and PALMs. The content of Bet v 1 peaked at maturation stages 60–61. Inversely to Bet v 1, levels of $PALM_{LTB4}$ and $PALM_{PGE2}$ were high in pollen from immature catkins and decreased during maturation. A concentration minimum of

PALMs corresponded to a maximum in Bet v 1. BBCH-Code: 52: catkins increase in length and show green expansion cracks; 55: enhanced expansion cracks through further increase in length; 60: first catkins emit pollen (sporadically); 61: beginning of flowering: few catkins emit pollen; 65: full flowering: more than 50% of the catkins emit pollen; 67: flowering finishing: just a few catkins still emit pollen.

Figure S3 Immune stimulatory versus immune modulatory potential of high versus low ozone-exposed pollen samples. Aqueous extracts (APEs) of birch pollen sampled from high and low ozone exposed trees were chosen for neutrophil migration assays and stimulation of monocyte derived dendritic cells. APEs were applied in 3 concentrations. Higher ozone-exposed pollen induced stronger neutrophil chemotaxis compared to pollen samples from lower ozone–exposed trees (**A**). In contrast, birch pollen from lower ozone-exposed trees were more potent in inhibiting the LPS-induced release of IL-12p70 from human monocyte-derived dendritic cells (**B**). APEs were prepared from birch pollen sampled from higher ozone-exposed trees (n = 2; mean ozone: 85 µg/m^3) and from lower ozone-exposed trees

(n = 2; mean ozone: 54 µg/m^3). All APEs were tested in n = 3 patients. *: p<0.05 (Wilcoxon matched-pairs signed-ranks test).

Figure S4 Viability of moDCs after stimulation with LPS plus APEs from high- and low ozone-exposed pollen. Viability of monocyte-derived dendritic cells (moDcs) after 24 h of stimulation with LPS (100 ng/ml) and APEs (1, 3, 10 mg/ml) was tested by propidium iodide staining and subsequent FACS analysis. APEs were prepared from birch pollen sampled from higher ozone-exposed trees (n = 2; mean ozone: 85 µg/m^3) and from lower ozone-exposed trees (n = 2; mean ozone: 54 µg/m^3). All APEs were tested in n = 3 patients.

Author Contributions

Conceived and designed the experiments: CTH AM. Performed the experiments: IB SG MM. Analyzed the data: IB SJ. Contributed reagents/materials/analysis tools: SJ JB CSW HB JR. Wrote the paper: IB HB SG CTH.

References

1. Strachan DP (1999) Lifestyle and atopy. Lancet 353: 1457–1458.
2. Behrendt H, Friedrichs KH, Kramer U, Hitzfeld B, Becker WM, et al. (1995) The role of indoor and outdoor air pollution in allergic diseases. Progress in Allergy and Clinical Immunology Volume 3, Stockholm 3: 83–89.
3. Kramer U, Koch T, Ranft U, Ring J, Behrendt H (2000) Traffic-related air pollution is associated with atopy in children living in urban areas. Epidemiology 11: 64–70.
4. Diaz-Sanchez D, Riedl M (2005) Diesel effects on human health: a question of stress? American journal of physiology Lung cellular and molecular physiology 289: L722–723.
5. Heinrich J, Wichmann HE (2004) Traffic related pollutants in Europe and their effect on allergic disease. Current opinion in allergy and clinical immunology 4: 341–348.
6. Ring J, Eberlein-Koenig B, Behrendt H (2001) Environmental pollution and allergy. Ann Allergy Asthma Immunol 87: 2–6.
7. Sillman S (1999) The relation between ozone, NOx and hydrocarbons in urban and polluted rural environments. Atmospheric Environment 33.
8. Ziska LH, Gebhard DE, Frenz DA, Faulkner S, Singer BD, et al. (2003) Cities as harbingers of climate change: common ragweed, urbanization, and public health. The Journal of allergy and clinical immunology 111: 290–295.
9. Health Effects Institute B, MA (2010) HEI Panel on the Health Effects of Traffic-Related Air Pollution. 2010. Traffic-Related Air Pollution: A Critical Review of the Literature on Emissions, Exposure, and Health Effects. HEI Special Report 17.
10. Behrendt H, Becker WM, Fritzsche C, Sliwa-Tomczok W, Tomczok J, et al. (1997) Air pollution and allergy: experimental studies on modulation of allergen release from pollen by air pollutants. Int Arch Allergy Immunol 113: 69–74.
11. Darbah JN, Kubiske ME, Nelson N, Oksanen E, Vaapavuori E, et al. (2007) Impacts of elevated atmospheric CO2 and O3 on paper birch (Betula papyrifera): reproductive fitness. ScientificWorldJournal 7 Suppl 1: 240–246.
12. Gilles S, Behrendt H, Ring J, Traidl-Hoffmann C (2012) The Pollen Enigma: Modulation of the Allergic Immune Response by Non-Allergenic, Pollen-Derived Compounds. Current pharmaceutical design.
13. Meier U (2001) Entwicklungsstadien mono- und dikotyler Pflanzen. BBCH-Monography. In: Forstwirtschaft BBfLu, editor. Berlin, Braunschweig. 165.
14. Buters JTM, Kasche A, Weichenmeier I, Schober W, Klaus S, et al. (2008) Year-to-year variation in release of Bet v1 allergen from birch pollen: Evidence for geographical differences between west and south Germany. International Archives of Allergy and Immunology 145: 122–130.
15. Gilles S, Fekete A, Zhang X, Beck I, Blume C, et al. (2011) Pollen metabolome analysis reveals adenosine as a major regulator of dendritic cell-primed T(H) cell responses. J Allergy Clin Immunol 127: 454–461 e451–459.
16. Palmes ED, Gunnison AF, Dimattio J, Tomczyk C (1976) Personal Sampler for Nitrogen-Dioxide. American Industrial Hygiene Association Journal 37: 570–577.
17. Jochner SC, Beck I, Behrendt H, Traidl-Hoffmann C, Menzel A (2011) Effects of extreme spring temperatures on urban phenology and pollen production: a case study in Munich and Ingolstadt. Climate Research 49: 101–112.
18. Gilles S, Mariani V, Bryce M, Mueller MJ, Ring J, et al. (2009) Pollen-derived E1-phytoprostanes signal via PPAR-gamma and NF-kappaB-dependent mechanisms. J Immunol 182: 6653–6658.
19. Traidl-Hoffmann C, Kasche A, Jakob T, Huger M, Plotz S, et al. (2002) Lipid mediators from pollen act as chemoattractants and activators of polymorphonuclear granulocytes. Journal of Allergy and Clinical Immunology 109: 831–838.
20. Traidl-Hoffmann C, Jakob T, Behrendt H (2009) Determinants of allergenicity. J Allergy Clin Immunol 123: 558–566.
21. Behrendt H, Kasche A, von Eschenbach CE, Risse U, Huss-Marp J, et al. (2001) Secretion of proinflammatory eicosanoid-like substances precedes allergen release from pollen grains in the initiation of allergic sensitization. International Archives of Allergy and Immunology 124: 121–125.
22. Hjelmroos M, Schumacher MJ, van Hage-Hamsten M (1995) Heterogeneity of pollen proteins within individual. Betula pendula trees. International Archives of Allergy and Applied Immunology 108.
23. Helander ML, Savolainen J, Ahlholm J (1997) Effects of air pollution and other environmental factors on birch pollen allergens. Allergy 52: 1207–1214.
24. Tashpulatov AS, Clement P, Akimcheva SA, Belogradova KA, Barinova I, et al. (2004) A model system to study the environment-dependent expression of the Bet v 1a gene encoding the major birch pollen allergen. Int Arch Allergy Immunol 134: 1–9.
25. Stathopoulou E, Mihalakakou G, Santamouris M, Bagiorgas HS (2008) On the impact of temperature on tropospheric ozone concentration levels in urban environments. J Earth Syst Sci 117: 227–236.
26. Eckl-Dorna J, Klein B, Reichenauer TG, Niederberger V, Valenta R (2010) Exposure of rye (Secale cereale) cultivars to elevated ozone levels increases the allergen content in pollen. J Allergy Clin Immunol 126: 1315–1317.
27. Kanter U, Heller W, Durner J, Winkler JB, Engel M, et al. Molecular and Immunological Characterization of Ragweed (Ambrosia artemisiifolia L.) Pollen after Exposure of the Plants to Elevated Ozone over a Whole Growing Season. PLoS One 8: e61518.
28. Behrendt H, Kasche A, Traidl C, Plotz S, Huss-Marp J, et al. (2002) Pollen grains contain and release not only allergens, but also eicosanoid-like substances with neutrophil chemotactic activity: A new step in the initiation of allergic sensitization? New Trends in Allergy V: 3–8.
29. Buters JT, Weichenmeier I, Ochs S, Pusch G, Kreyling W, et al. (2010) The allergen Bet v 1 in fractions of ambient air deviates from birch pollen counts. Allergy 65: 850–858.
30. Armstrong RA (1995) Investigation of the inhibitory effects of PGE2 and selective EP agonists on chemotaxis of human neutrophils. Br J Pharmacol 116: 2903–2908.
31. Gilles S, Mariani V, Bryce M, Mueller MJ, Ring J, et al. (2009) Pollen allergens do not come alone: pollen associated lipid mediators (PALMS) shift the human immune systems towards a T(H)2-dominated response. Allergy Asthma Clin Immunol 5: 3.
32. Traidl-Hoffmann C, Mariani V, Hochrein H, Karg K, Wagner H, et al. (2005) Pollen-associated phytoprostanes inhibit dendritic cell interleukin-12 production and augment T helper type 2 cell polarization. J Exp Med 201: 627–636.
33. Schar C, Vidale PL, Luthi D, Frei C, Haberli C, et al. (2004) The role of increasing temperature variability in European summer heatwaves. Nature 427: 332–336.

15

Gaseous Elemental Mercury (GEM) Emissions from Snow Surfaces in Northern New York

J. Alexander Maxwell[1], Thomas M. Holsen[2]*, Sumona Mondal[3]

1 Institute for a Sustainable Environment, Clarkson University, Potsdam, New York, United States of America, **2** Department of Civil and Environmental Engineering, Clarkson University, Potsdam, New York, United States of America, **3** Department of Mathematics, Clarkson University, Potsdam, New York, United States of America

Abstract

Snow surface-to-air exchange of gaseous elemental mercury (GEM) was measured using a modified Teflon fluorinated ethylene propylene (FEP) dynamic flux chamber (DFC) in a remote, open site in Potsdam, New York. Sampling was conducted during the winter months of 2011. The inlet and outlet of the DFC were coupled with a Tekran Model 2537A mercury (Hg) vapor analyzer using a Tekran Model 1110 two port synchronized sampler. The surface GEM flux ranged from -4.47 ng m^{-2} hr^{-1} to 9.89 ng m^{-2} hr^{-1}. For most sample periods, daytime GEM flux was strongly correlated with solar radiation. The average nighttime GEM flux was slightly negative and was not well correlated with any of the measured meteorological variables. Preliminary, empirical models were developed to estimate GEM emissions from snow surfaces in northern New York. These models suggest that most, if not all, of the Hg deposited with and to snow is reemitted to the atmosphere.

Editor: Stephen J. Johnson, University of Kansas, United States of America

Funding: New York State Energy Research and Development Authority (NYSERDA http://www.nyserda.ny.gov/) financially supported this research (Charles Driscoll, Syracuse University PI). It has not been subject to the NYSERDA's peer and policy review and, therefore, does not necessarily reflect the views of NYSERDA and no official endorsement should be inferred. The funders had no role in study design, data collection and analysis, decision to publish, or preparation of the manuscript.

Competing Interests: The authors have declared that no competing interests exist.

* E-mail: holsen@clarkson.edu

Introduction

Hg is a potent neurotoxin and regulated by the U.S. EPA [1], European Union Restriction of Hazardous Substances Directive (RoHS) [2], and other government agencies worldwide as a hazardous pollutant. In the form of monomethylmercury (MeHg) it can adversely impact the development and health of both humans and wildlife [3]. Gaseous elemental mercury (GEM) is emitted into the atmosphere from both natural and anthropogenic sources, and has an atmospheric residence time of 0.5–2 years, allowing it to be transported over great distances [4–6]. Anthropogenic sources can also emit Hg in the form of gaseous oxidized Hg (GOM) and particulate bound Hg (PBM), which have shorter atmospheric lifetimes on the order of days to weeks [4]. GOM is fairly soluble in water, thus allowing it to be readily deposited to terrestrial surfaces through wet deposition, including snow [4–6]. The Hg deposited with snow is then either quickly revolatilized back into the atmosphere or incorporated into the snowpack. Newly deposited Hg has been shown to preferentially revolatilize, depending on the deposition surface, in a process known as prompt recycling [7].

While the role of snow surfaces in Hg cycling has been widely studied in arctic regions [5,8–11], much less is known about its importance in more temperate climates [12–14]. Hg is deposited to snowpacks through both wet (snow) and dry deposition. Once deposited on the snowpack surface, it has been shown that >50% of the Hg deposited is reemitted within the first 24 hours [8,12]. This process is believed to be governed by photoinduced reduction of GOM to GEM. Hg in the snowpack is mainly found in the form

of GOM dissolved in snow grains, while <1% remains trapped in the interstitial air as GEM [8]. Hg concentrations are known to decrease with depth [12] with the higher concentrations up to 1.5 ng m^{-3} (GEM) remaining on the surface [8].

In the arctic, the snow surface-to-air flux of Hg is mainly the result of a diurnal pattern of GEM production in the interstitial air near the surface of the snowpack during the daytime ($\sim$15–50 ng m^{-2} hr^{-1}), with little contribution from deeper snow layers [15,16]. However, internal production of GEM increases slightly with higher temperatures and snowmelt [8]. Since this process has not been well studied in temperate climates, measurements of snow surface-to-air fluxes were made over the 2011 winter season in Potsdam, NY.

Materials and Methods

Site Description, Methods, and Materials

Flux measurements were conducted at an open field site located at the Potsdam Municipal Airport (Damon Field) in Potsdam, NY (44°40.41N, −74°57.06′W) near the Clarkson University Observatory. This site remains largely undisturbed throughout the year and has served as a background site for the New York State particulate matter (PM) monitoring network. Sampling periods were determined based on access to the site and snow conditions. Special considerations were made to ensure that the chamber was never buried in snow and that all inlets and outlets remained above the snow during sampling. Measurements were conducted on a concrete slab, isolating the snowpack from the soil surface.

Concentrations of GEM were measured using a DFC with a method previously described in Choi & Holsen (2009). Briefly, the ambient sampling line (inlet) and chamber sampling line (outlet) of the DFC (described below) were coupled with a Tekran Model 2537A Hg vapor analyzer operated at room temperature in a field shed (Tekran Corporation, Inc., Toronto, Ontario, Canada) using a Tekran Model 1110 two-port synchronized sampler. The Tekran 1110 unit allowed for alternating five minute sampling pairs to be made between the inlet and outlet sample lines every 20 minutes (trap A inlet, trap B inlet; trap A outlet, trap B outlet). During inlet sampling, outlet air is bypassed at the same 1 L min^{-1} flow rate as the Tekran Model 2537A to maintain a constant turnover time (TOT) of 0.78 minutes and an optimized flushing flow rate (FFR) of 5 L min^{-1}[17] through the flux chamber. The inlet and outlet openings were placed next to each other at the same height, roughly 2 cm above the snow surface. Four, 1 cm diameter holes were evenly distributed around the perimeter of the chamber wall to insure the chamber was well-mixed. Although a standard method for the use of DFCs does not exist and this method has not been used in other snow studies, this sampling approach is similar to methods used in past studies over soil surfaces [17–19]. The 5 L min^{-1} FFR and 0.78 minute TOT are also similar to those used in a study by Eckley, et al. (2010).

Modified Teflon fluorinated ethylene propylene (FEP) chambers were used in the study. The modified Teflon chamber was constructed using a polycarbonate (PC) chamber frame and thin, 25 μm thick Teflon FEP film (CS Hyde Company, Lake Villa, IL) to cover the top and side windows (Figure 1). In previous studies [20,21], Teflon film was shown to allow better UV permeability, up to 85±11% of light for wavelengths between 260 and 970 nm.

Figure 1. Modified Teflon Fluorinated Ethylene Propylene (FEP) Chamber with Polycarbonate (PC) Frame and 25 μm Teflon FEP Film Top and Side Windows.

Each DFC had a chamber volume of 3.9 L with a 18 cm diameter opening covering an area of approximately 254 cm^2 of the snow surface.

Manual spike Hg recovery tests were conducted at the start of each sampling period by injecting 20 μL of Hg at roughly 13.23 pg μL^{-1} (20°C) or approximately 0.26 ng into an operating chamber using a calibrated (ANSI/NCSL Z540-1-1994) Hamilton Digital Syringe (Hamilton Company, Reno, NV) and a Tekran Model 2505 Hg vapor calibration unit (Tekran Corporation, Inc., Toronto, Ontario, Canada). The recorded Hg concentrations after each manual spike test were roughly 9 ng m^{-3}, on the same order as the average daytime Hg concentrations around 2 ng m^{-3}. The recovery was 97.5±3.8%. Flow rates were calibrated using a Bios Definer 220 volumetric flow meter (Bios International Corporation, Butler, NJ) at the beginning of each sampling period.

Prior to all field measurements, the Tekran Model 2537A was calibrated with an internal permeation source to ensure acceptable response factors (>6,000,000) and that the concentration difference between the inlet and outlet samples was less than 5%. In addition, all soda-lime traps and 0.2 μm polytetrafluoroethylene (PTFE) membrane filters were replaced at the start of each sampling period.

Meteorological data was collected using a weather station (Vantage Pro 2 Weather Station, Davis Instruments, Hayward, CA) located 1–2 m away from the chamber. The weather station measured ambient air temperature (°C), relative humidity (%), and solar radiation (W m^{-2}) at a 10 minute time resolution.

Sampling Analysis and Calculations

The GEM flux from the snow under the chamber was calculated using the following mass balance equation:

$$F = (C_{outlet} - C_{inlet}) \times (Q/A) \qquad (1)$$

Where F is flux (ng GEM m^{-2} h^{-1}); C_{outlet} and C_{inlet} are the concentrations of GEM (ng GEM m^{-3}) at the outlet and inlet, respectively; Q is the FFR (m^3 h^{-1}) through the chamber; and A is the surface area (m^2) of the snow exposed in the chamber. When fluxes were negative (-), Hg was being deposited on the snow surface, and when fluxes were positive (+) Hg was being emitted from the snow surface. All flux data was then smoothed using a Savitzky-Golay smoothing filter [22], (Eqn 2), to account for random error/noise while also preserving the quantitative information and trends.

$$F_4^* = [(-2 \times F_1) + (3 \times F_2) + (6 \times F_3) + (7 \times F_4) + (6 \times F_5) + (3 \times F_6) + (-2 \times F_7)]/21 \qquad (2)$$

Where F_4^* is the smoothed flux (ng GEM m^{-2} h^{-1}), F_{1-7} are the range of measured abscissa flux values (ng GEM m^{-2} h^{-1}), and 21 is the normalizing factor.

Histograms of the GEM flux and the three meteorological predictor variables (temperature, solar radiation, and relative humidity) showed that none of the variables were normally distributed. Figure 2 provides histograms and residual plots of daytime GEM flux when compared to solar radiation. Similar plots were constructed for each individual variable, temperature, solar radiation, and relative humidity. Shapiro-Wilk normality tests [23] were then employed to confirm that the data deviated from normality. Non-parametric Pearson product-moment tests

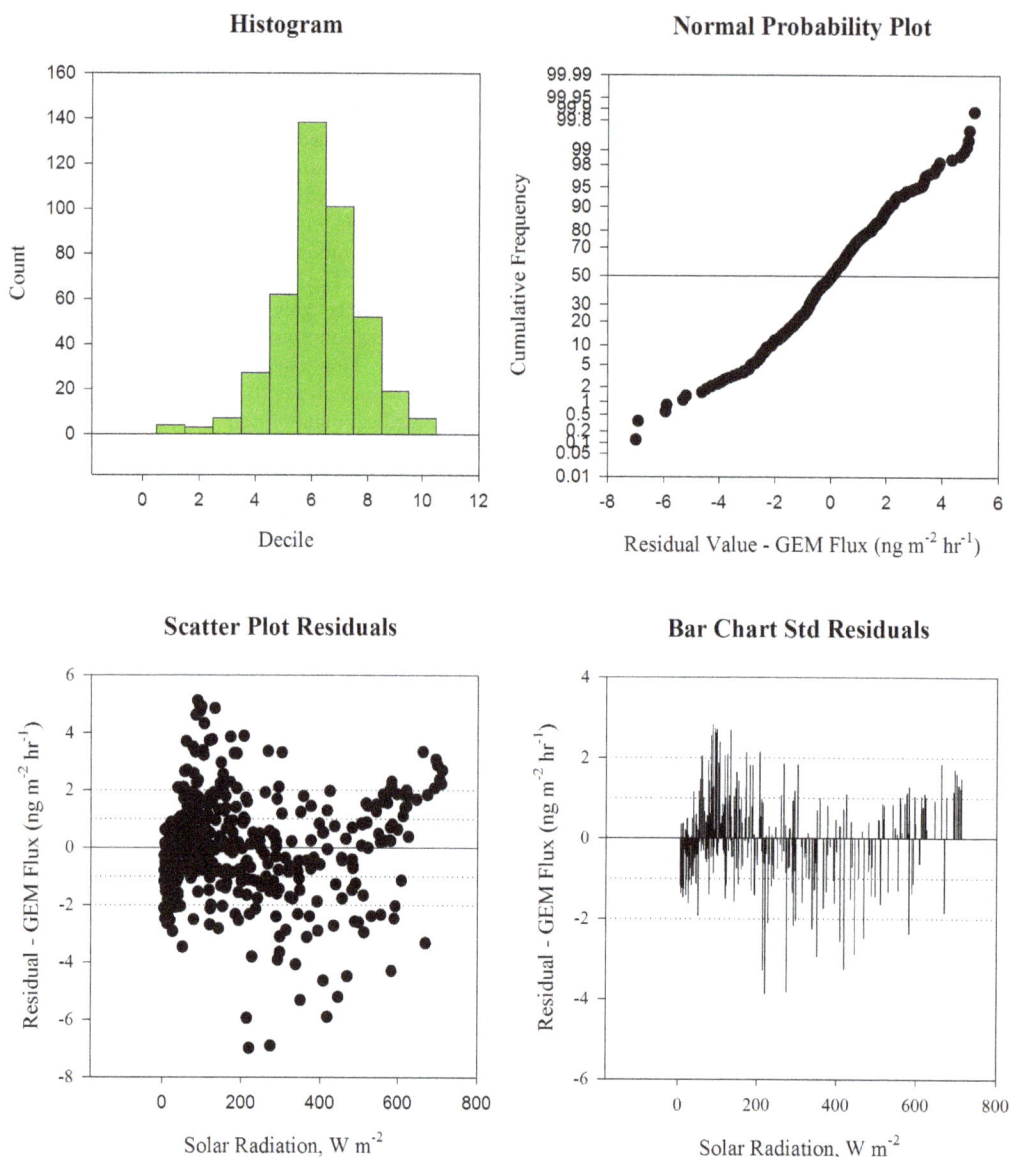

Figure 2. Histograms and Residual Plots of Daytime GEM Flux and Solar Radiation During Winter 2011.

Table 1. Measured Daytime And Nighttime GEM Flux Over 5 Winter 2011 Sampling Periods.

Date	Diurnal Period	GEM Flux (ng m^{-2} hr^{-1})					
		Mean	Std. Dev.	Range	Max	Min	Median
21–24 Jan	Daytime	1.13	1.37	5.60	4.44	−1.16	0.94
	Nighttime	−0.44	0.29	1.47	0.14	−1.33	−0.43
26–31 Jan	Daytime	2.65	1.92	8.20	6.57	−1.63	2.69
	Nighttime	−0.21	0.42	1.95	0.69	−1.26	−0.22
15–18 Feb	Daytime	1.88	2.96	12.56	8.09	−4.47	1.30
	Nighttime	−0.57	0.45	2.10	0.39	−1.71	−0.50
23–24 Feb	Daytime	3.03	2.60	9.38	8.13	−1.25	2.46
	Nighttime	−0.29	0.23	1.00	0.25	−0.75	−0.30
08–09 Mar	Daytime	3.50	3.08	9.97	9.89	−0.08	2.29
	Nighttime	−0.08	0.22	0.87	0.27	−0.60	−0.10
Overall	Daytime	2.37	2.48	14.36	9.89	−4.47	1.89
	Nighttime	−0.35	0.41	2.41	0.69	−1.71	−0.33

Table 2. Pearson Product-Moment Correlation Coefficients and P-Values For Correlations Between GEM Flux and Temperature, Relative Humidity, and Solar Radiation.

Date	Diurnal Period	Temperature (°C)		Relative Humidity (%)		Solar Radiation (W m^{-2})	
		Coefficient	P Value	Coefficient	P Value	Coefficient	P Value
21–24 Jan	Daytime	0.562	0.000	−0.494	0.000	0.546	0.000
	Nighttime	−0.004	0.959	−0.192	0.026	−	−
26–31 Jan	Daytime	0.189	0.022	−0.046	0.585	0.304	0.000
	Nighttime	−0.183	0.009	−0.319	0.000	−	−
15–18 Feb	Daytime	−0.518	0.000	0.673	0.000	0.820	0.000
	Nighttime	−0.553	0.000	−0.600	0.000	−	−
23–24 Feb	Daytime	0.300	0.027	−0.629	0.000	0.875	0.000
	Nighttime	0.000	0.997	−0.053	0.745	−	−
08–09 Mar	Daytime	0.446	0.001	−0.787	0.000	0.942	0.000
	Nighttime	−0.518	0.000	0.251	0.129	−	−
Overall	Daytime	0.103	0.035	−0.385	0.000	0.684	0.000
	Nighttime	−0.222	0.000	−0.132	0.002	−	−

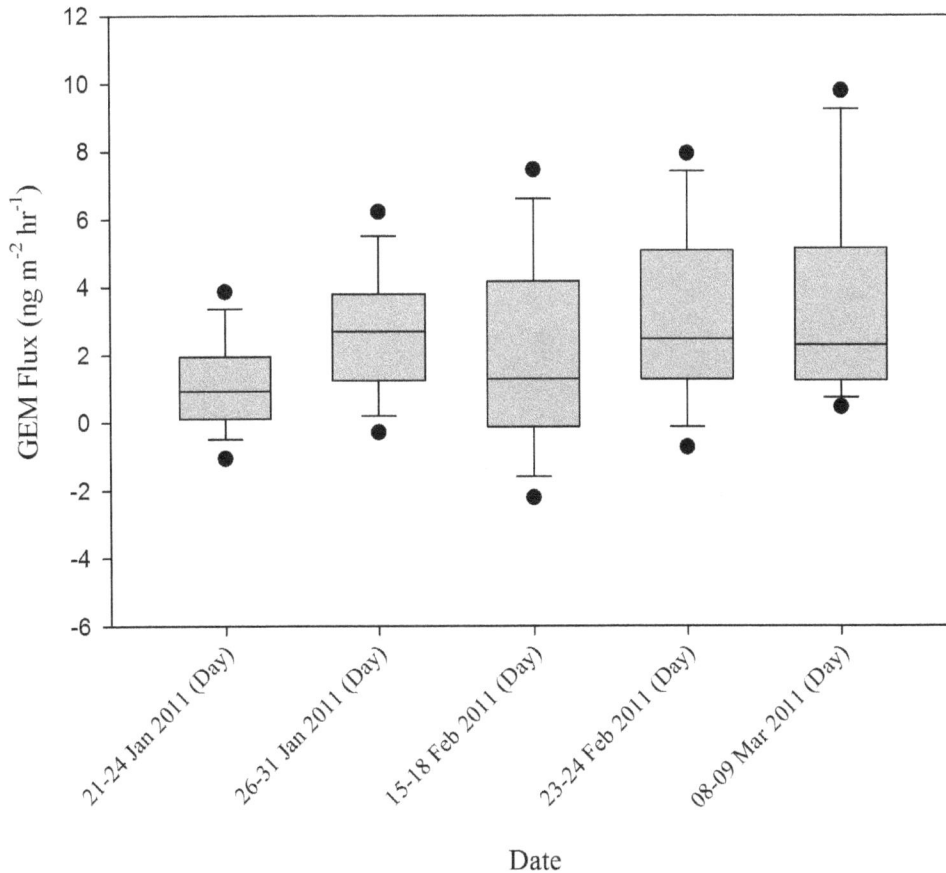

Figure 3. Average Daytime GEM Flux Measurements Made For Each Sampling Conducted During Winter 2011.

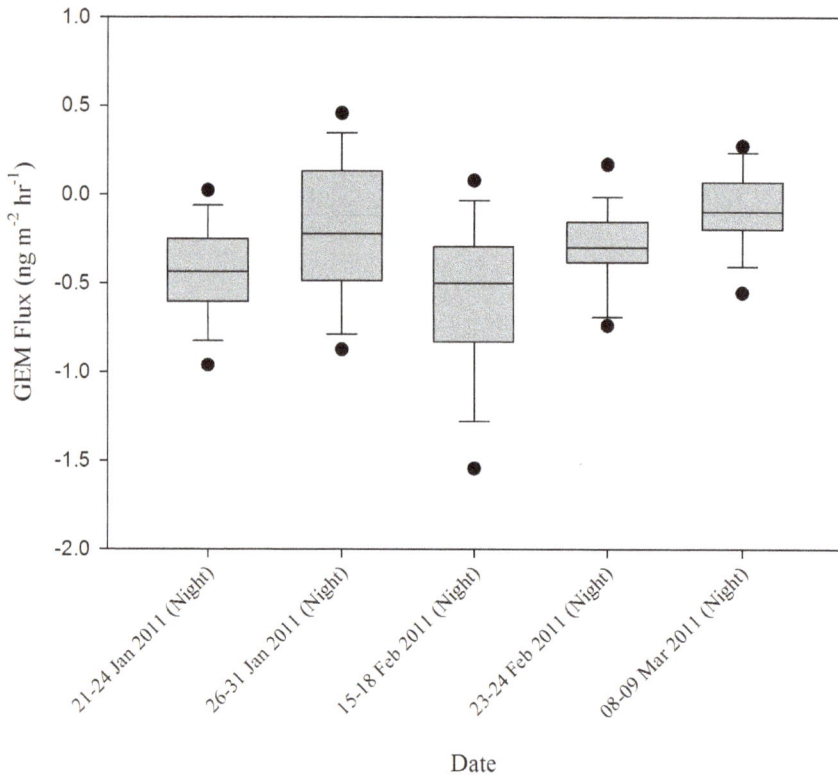

Figure 4. Average Nighttime GEM Flux Measurements Made For Each Sampling Conducted During Winter 2011.

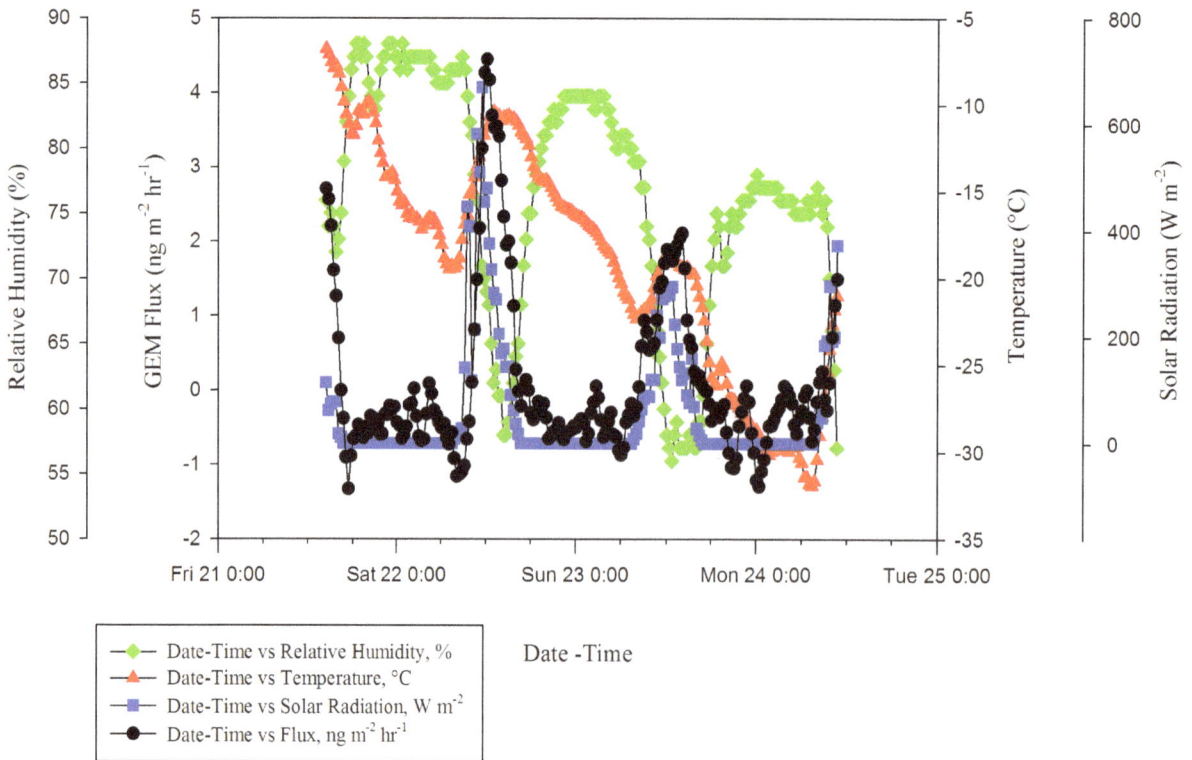

Figure 5. Diurnal Pattern Of GEM Flux For 21–24 January 2011 Sampling With Temperature, Relative Humidity, And Solar Radiation.

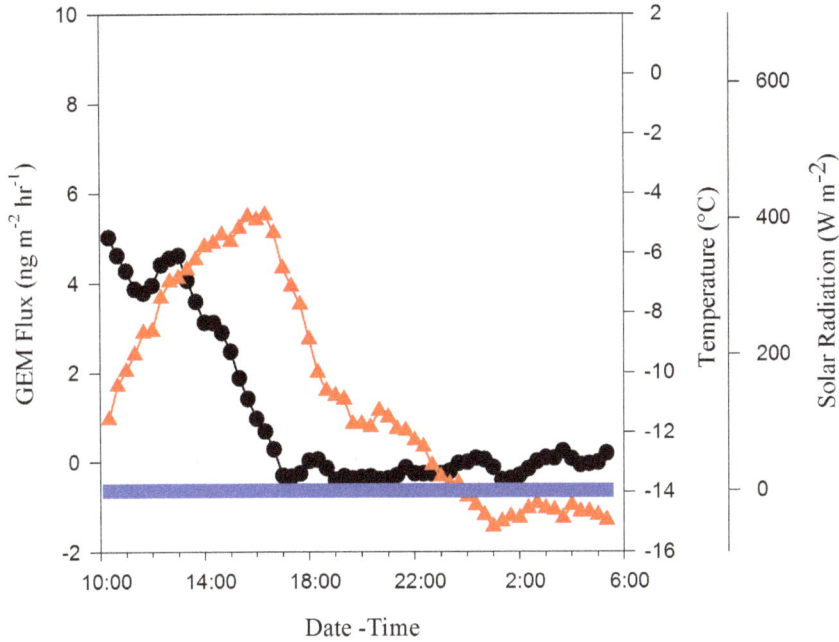

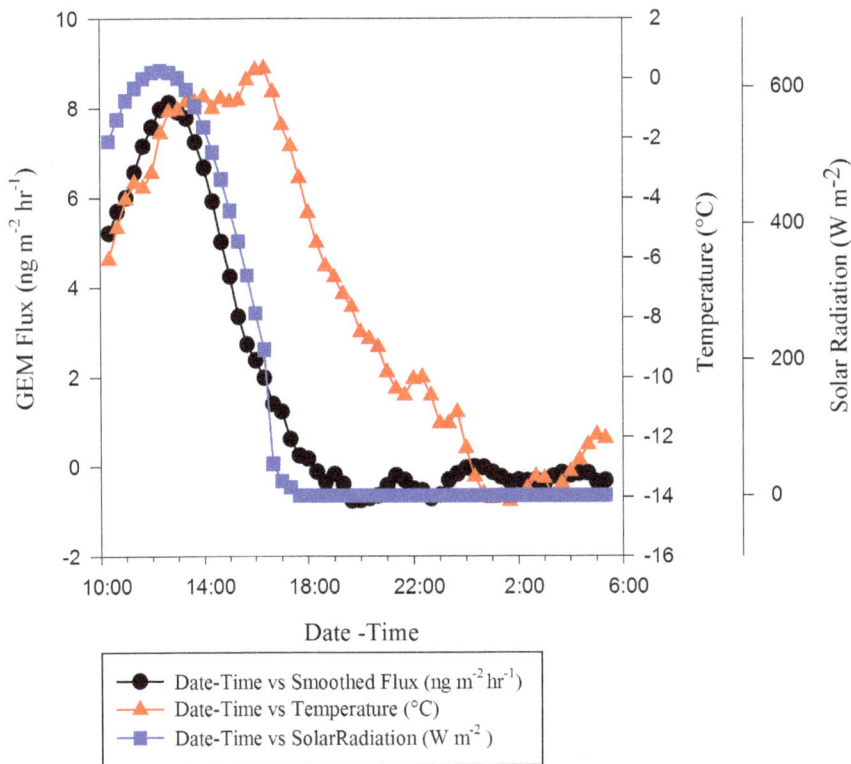

Figure 6. GEM Fluxes Measured Using Covered And Uncovered Chambers To Determine The Impact Of Solar Radiation On GEM Flux.

Table 3. PPMCs For Covered And Uncovered Chamber Tests For Impact Of Solar Radiation On GEM Flux.

Date	Diurnal Period	Temperature (°C)		Relative Humidity (%)		Solar Radiation (W m^{-2})	
		Coefficient	P Value	Coefficient	P Value	Coefficient	P Value
22–23 Feb (Covered)	Daytime	−0.624	0.002	0.489	0.021	–	–
	Nighttime	−0.374	0.025	0.141	0.412	–	–
23–24 Feb (Uncovered)	Daytime	0.300	0.027	−0.629	0.000	0.875	0.000
	Nighttime	0.000	0.997	−0.053	0.745	–	–

were used to determine the correlation coefficients (PPMC) between the variables [24].

Results and Discussion

Flux Measurements

During the 2011 winter sampling season, the flux was measured over five sampling periods, each lasting from one to six days. The measured flux ranged from a minimum −4.47 to a maximum 9.89 ng GEM m^{-2} h^{-1} (Table 1). The average daytime flux was 2.37 ± 2.48 ng GEM m^{-2} h^{-1} and the average nighttime flux was -0.35 ± 0.41 ng GEM m^{-2} h^{-1}. Measured nighttime Hg emission fluxes from other snowpack studies have been ≈ 0 ng GEM m^{-2} h^{-1} [8], while daytime fluxes have been shown to be much higher, ≈ 30–50 ng GEM m^{-2} h^{-1} [8]. Daytime fluxes were strongly correlated with solar radiation (PPMC value $= 0.684$, p-value $= 0.000 < 0.050$) and to a lesser extent temperature and relative humidity (PPMC value $= 0.103$, p-value $= 0.035 < 0.050$ and PPMC value $= -0.385$, p-value $= 0.000 < 0.050$ respectively) (Table 2). This strong correlation with solar radiation suggests that the daytime Hg emissions from the snow surface are a result of the photoreduction of GOM associated with the snow to GEM. Similar results have been reported in Ferrari, et al. (2005), where it was also reported that GEM emissions from the snowpack were negligible in comparison to emissions caused by solar irradiation at the surface. Nighttime fluxes were only weakly correlated with both temperature (PPMC value $= -0.222$, p-value $= 0.000 < 0.050$) and relative humidity (PPMC value $= -0.132$, p-value $= 0.002 < 0.000$) (Table 2) and showed a statistically significant difference from zero.

Overall, peak fluxes tended to increase later in the sampling season (Figure 3 and Figure 4). Emissions were highest during the last sampling period, 08–09 March, corresponding with the highest solar radiation peak (Max: 712 W m^{-2}). Fluxes also tended to follow a diurnal pattern (Figure 5) with peaks occurring during the day following increased exposure to solar radiation, and deposition occurring at night, similar to patterns reported in other literature [15,16].

Impact of Solar Radiation

To test the impact of solar radiation on GEM fluxes, the chamber was covered with aluminum foil to simulate zero UV conditions. The uncovered measurements were made on 23–24 February 2011, while the covered measurements were made on 22–23 February 2011 (Figure 6). During the uncovered and covered tests, the average GEM fluxes were 1.76 ± 3.06 and 0.99 ± 1.81 ng GEM m^{-2} h^{-1} respectively. The covered DFC daytime measurements were negatively correlated with temperature (PPMC coefficient $= -0.624$, p-value $= 0.000 < 0.050$) (Table 3). The slow decline in GEM flux after covering the

chamber is likely a result of diffusion of GEM from the interstitial air in the snowpack into the DFC. The uncovered DFC daytime measurements were positively correlated with solar radiation, and to a lesser degree, temperature (PPMC coefficient $= 0.875$, p-value $= 0.000 < 0.050$ and PPMC coefficient $= 0.300$, p-value $= 0.027 < 0.050$) (Table 3), similar to what has been reported in other arctic studies [8]. Overall, solar radiation had the highest positive impact on GEM emissions, and though temperature and relative humidity were correlated to GEM flux, their correlation with solar radiation (PPMC coefficient $= 0.711$ & -0.686 respectively, p-value $= 0.000 < 0.050$) indicate that their influence was likely a result of their codependence on solar radiation.

Modeling

In the past, empirical models have been developed using meteorological data in order to estimate surface GEM flux from soils in temperate regions of eastern North America [17,21,25]. However, no model exists to estimate GEM flux from snow in the temperate climate of northern New York. Previous models for this region [17] excluded winter fluxes from snow surfaces. In order to better model GEM flux throughout the winter season, two multiple linear regression models were developed based on aggregated seasonal flux data:

Winter 2011 (Daytime): ($R^2 = 0.481$)

$$F = 0.722 + 0.0358(T) + 0.00906(SR)$$

Winter 2011 (Nighttime): ($R^2 = 0.0616$)

$$F = -0.167 - 0.00939(T) - 0.00344(RH)$$

where F is GEM flux in ng m^{-2}hr^{-1}, T is ambient temperature in °C, RH is relative humidity in %, and SR is solar radiation in W m^{-2}. Fluxes predicted by this model for the 22–24 January sampling period are shown in Figure 7.

Several nonlinear polynomial and power equation fits and variable transformations were conducted using SigmaPlot, ver. 12 in order to develop a more precise correlative model structure. However, the dynamic fits showed little improvement.

Using the multiple linear regression models in conjunction with 5-year winter (December-March, 2005–2010) EPA Clean Air Status and Trends Network (CASTNET) meteorological data from the National Atmospheric Deposition Program (NADP) site, NY20, it is estimated that the average snow surface emissions from the open Huntington Wildlife Forest (HWF) site range from -0.10 ± 0.07 ng m^{-2} hr^{-1} (nighttime) to 1.53 ± 1.69 ng m^{-2} hr^{-1} (daytime) or ~ 17.22 ng m^{-2} year^{-1}. During the same time period Mercury Deposition Network (MDN) data from the same site yield

21-24 Jan 2011 Daytime GEM Flux Model Comparison

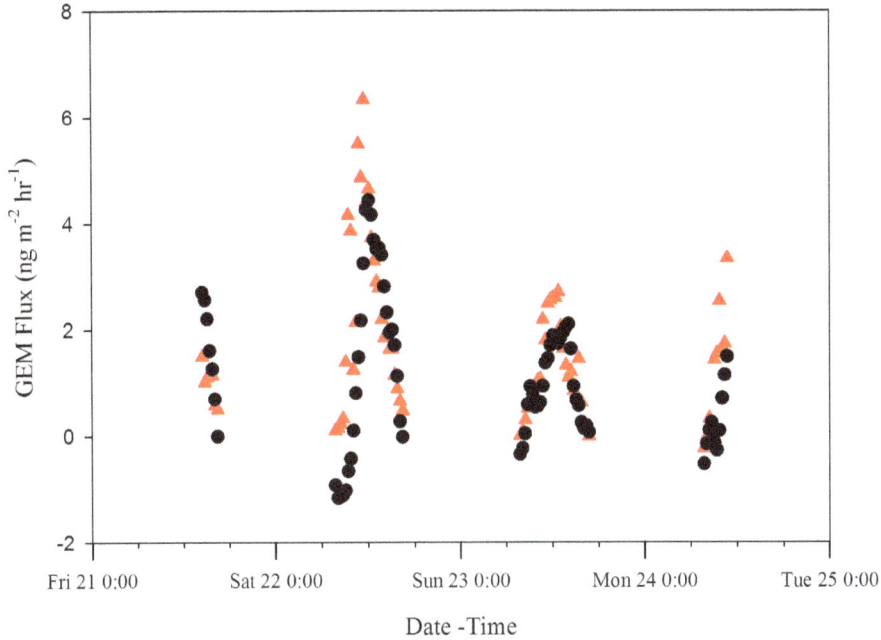

21-24 Jan 2011 Nighttime GEM Flux Model Comparison

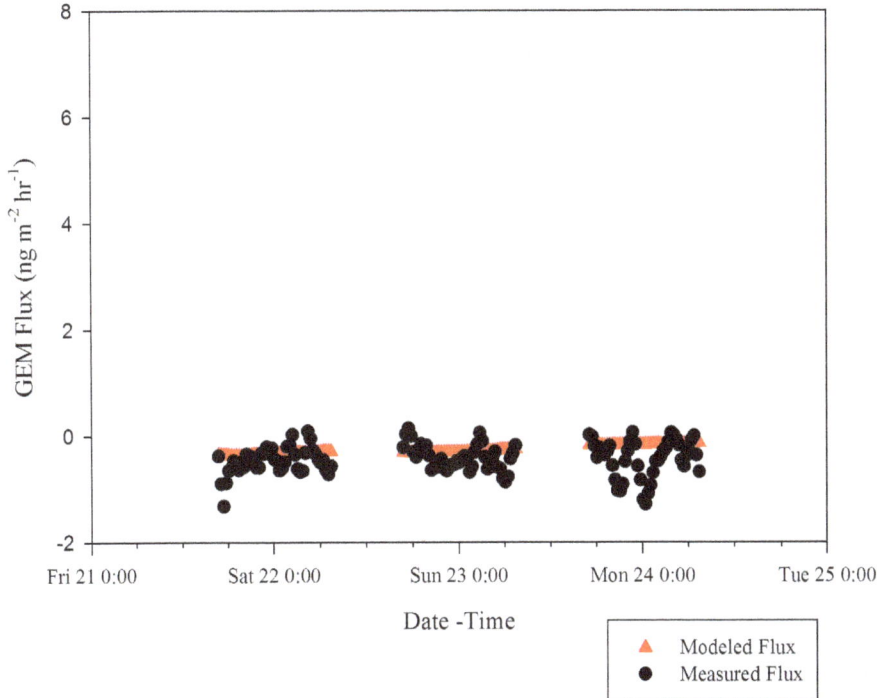

Figure 7. Daytime And Nighttime GEM Flux Model Comparison For 21–24 January Sampling Period.

a similar value with an average deposition flux of 0.48 ± 0.41 ng m^{-2} hr^{-1} or 11.52 ± 9.84 ng m^{-2} $year^{-1}$. The reason for the slightly higher modeled flux compared to the measured flux is likely due to the fact that some of the measurements used to make

the empirical model were made after fresh snowfall when GEM fluxes would be at their maximum values.

Overall, these models suggest that most if not all the Hg deposited to snow surfaces is promptly recycled. Similar reemission

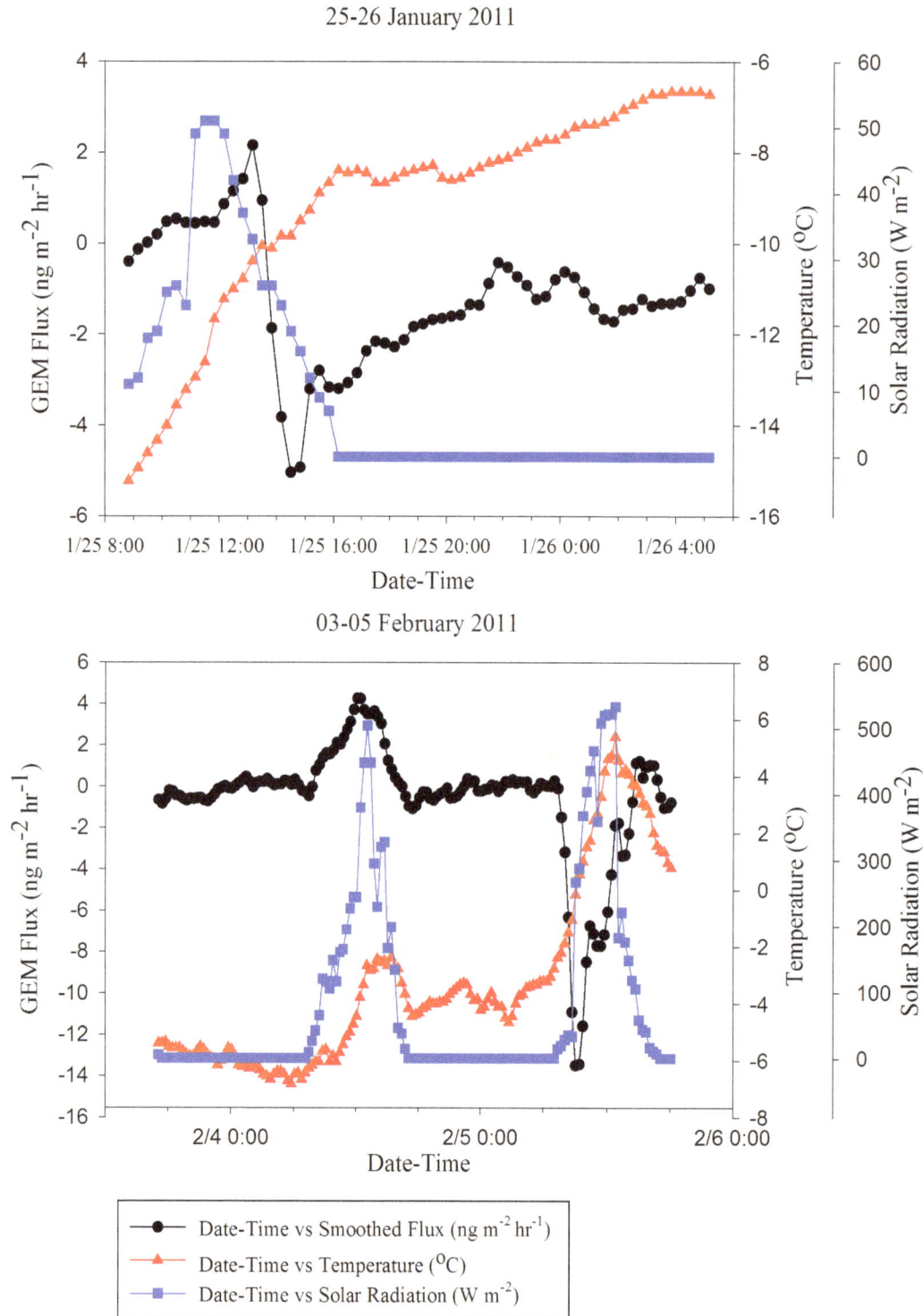

Figure 8. GEM Deposition Events For 25–26 January And 03–05 February 2011 Sampling Period.

phenomena have been reported by other research groups [8,12] with mean emission fluxes of 2–5 ng m^{-2} hr^{-1}, zero change in surface snow Hg concentration after deposition events, and up to 54% GEM reemission during the first 24 hours after a snowfall.

Deposition Events

Two unique deposition events with fluxes as high as −14 ng m^{-2} hr^{-1} occurred during separate sampling periods, one on 25 January and one on 05 February. Both of these event followed snowfalls ≥3 cm (Table 4) and melting also occurred during the 03–05 February sampling (Figure 8). During both of these events,

Table 4. Field Observations Made During Various Measurement Periods Throughout the 2011 Winter Sampling Season.

Date	Observation
21–24 Jan.	No recent snow
26–27 Jan.	Fresh snow (dusting, <2.5 cm) prior to sampling
27–30 Jan.	Fresh snow (dusting, <2.5 cm) and intermittent light snowing throughout sampling (no enough to cover inlets of chamber)
30–31 Jan.	No recent snow
15–18 Feb.	No recent snow
23–24 Feb.	No recent snow
08–09 Mar.	Fresh snow (≈7.5 cm); during end of season melt

fluxes were negatively correlated with temperature (PPMC coefficient $= -0.421$ and -0.439 respectively, p-value $= 0.000 < 0.050$) and displayed patterns opposite to the diurnal patterns typically seen. This sudden deposition event is similar to atmospheric Hg depletion events (AMDEs) witnessed in arctic regions during polar sunrise [9–11].

During an AMDE, rapid oxidation of GEM forms GOM that is subsequently deposited to the snow surface. Arctic AMDEs are springtime phenomenon that occur as a result of reactions with ozone and other halogen compounds, especially bromine oxides [26]. Though the cause of the two deposition events seen in Potsdam is unclear, they could coincide with sudden increases in atmospheric oxidant concentrations including free halogens.

Conclusions

Snow surface-to-air exchange of gaseous elemental Hg (GEM) was measured using a modified Teflon fluorinated ethylene propylene (FEP) dynamic flux chamber (DFC) in a remote, open site in Potsdam, New York during the winter months of 2011. The surface GEM flux ranged from -4.47 ng m^{-2} hr^{-1} to 9.89 ng m^{-2} hr^{-1}. For most sample periods, the daytime GEM flux was strongly correlated with solar radiation. The average nighttime GEM flux was slightly negative and was weakly correlated with all of the measured meteorological variables. Overall, preliminary models indicate that most if not all the Hg being deposited to snow surfaces is being reemitted back into the atmosphere. Two unique deposition events with fluxes as high as -14 ng m^{-2} hr^{-1} occurred during separate sampling periods following snowfalls ≥ 3 cm. During both of these events, fluxes were negatively correlated (PPMC coefficient $= -0.421$ and -0.439 respectively, p-value $= 0.000 < 0.050$) with temperature and displayed patterns opposite to the diurnal patterns typically seen.

Acknowledgments

A special thank you to Dr. Andrea Ferro, Dr. Stephan Grimberg, and Dr. Jiaoyan Huang for their help in reviewing this work, and James Laing for assistance with sampling.

Author Contributions

Conceived and designed the experiments: JAM TMH. Performed the experiments: JAM. Analyzed the data: JAM SM. Contributed reagents/materials/analysis tools: JAM TMH SM. Wrote the paper: JAM.

References

1. EPA (2013) Laws and Regulations | Mercury | US EPA. Available: http://www.epa.gov/hg/regs.htm. Accessed 14 December 2012.

2. EU (2011) DIRECTIVE 2011/65/EU of the European Parliament and of the Council of 8 June 2011 on the restriction of the use of certain hazardous substances in electrical and electronic equipment (recast): 88–110. Available: http://eur-lex.europa.eu/LexUriServ/LexUriServ.do?uri = OJ:L:2011:174:0088:0110:EN:PDF.

3. Mergler D, Anderson HA, Chan LHM, Mahaffey KR, Murray M, et al. (2007) Methylmercury Exposure and Health Effects in Humans: A Worldwide Concern. AMBIO: A Journal of the Human Environment 36: 3–11. Available: http://dx.doi.org/10.1579/0044-7447(2007)36[3:MEAHEI]2.0.CO;2. Accessed 14 December 2012.

4. Lin C-J, Pehkonen SO (1999) The chemistry of atmospheric mercury: a review. Atmospheric Environment 33: 2067–2079. Available: http://dx.doi.org/10.1016/S1352-2310(98)00387-2. Accessed 14 December 2012.

5. Schroeder WH, Munthe J (1998) Atmospheric mercury–An overview. Atmospheric Environment 32: 809–822. Available: http://dx.doi.org/10.1016/S1352-2310(97)00293-8. Accessed 14 December 2012.

6. Lindqvist O, Rodhe H (1985) Atmospheric mercury-a review. Tellus B 37B: 136–159. Available: http://www.tellusb.net/index.php/tellusb/article/view/15010. Accessed 14 December 2012.

7. Selin NE, Jacob DJ, Yantosca RM, Strode S, Jaegle L, et al. (2008) Global 3-D Land-Ocean-Atmosphere Model for Mercury: Present- Day Versus Preindustrial Cycles and Anthropogenic Enrichment Factors for Deposition. Global Biogeochemical Cycles 22: 13.

8. Ferrari C, Gauchard P, Aspmo K, Dommergue A, Magand O, et al. (2005) Snow-to-air exchanges of mercury in an Arctic seasonal snow pack in Ny-Ålesund, Svalbard. Atmospheric Environment 39: 7633–7645. Available: http://dx.doi.org/10.1016/j.atmosenv.2005.06.058. Accessed 14 December 2012.

9. Ariya PA, Dastoor AP, Amyot M, Schroeder WH, Barrie L, et al. (2004) The Arctic: a sink for mercury. Tellus B 56: 397–403. Available: http://www.tellusb.net/index.php/tellusb/article/view/16458. Accessed 14 December 2012.

10. Skov H, Christensen JH, Goodsite ME, Heidam NZ, Jensen B, et al. (2004) Fate of Elemental Mercury in the Arctic during Atmospheric Mercury Depletion Episodes and the Load of Atmospheric Mercury to the Arctic. Environmental Science & Technology 38: 2373–2382. Available: http://dx.doi.org/10.1021/es030080h. Accessed 14 December 2012.

11. Lindberg SE, Brooks S, Lin C-J, Scott KJ, Landis MS, et al. (2002) Dynamic Oxidation of Gaseous Mercury in the Arctic Troposphere at Polar Sunrise. Environmental Science & Technology 36: 1245–1256. Available: http://dx.doi.org/10.1021/es0111941. Accessed 14 December 2012.

12. Lalonde JD, Poulain AJ, Amyot M (2002) The Role of Mercury Redox Reactions in Snow on Snow-to-Air Mercury Transfer. Environmental Science & Technology 36: 174–178. Available: http://dx.doi.org/10.1021/es010786g. Accessed 14 December 2012.

13. Lalonde JD, Amyot M, Doyon M-R, Auclair J-C (2003) Photo-induced Hg(II) reduction in snow from the remote and temperate Experimental Lakes Area (Ontario, Canada). Journal of Geophysical Research: Atmospheres (1984–2012) 108: 4200.

14. Nelson SJ, Fernandez IJ, Kahl JS (2010) A review of mercury concentration and deposition in snow in eastern temperate North America. Hydrological Processes 24: 1971–1980.

15. Faïn X, Grangeon S, Bahlmann E, Fritsche J, Obrist D, et al. (2007) Diurnal production of gaseous mercury in the alpine snowpack before snowmelt. Journal of Geophysical Research 112: D21311. Available: http://www.agu.org/pubs/crossref/2007/2007JD008520.shtml. Accessed 14 December 2012.

16. Dommergue A, Ferrari CP, Poissant L, Gauchard P-A, Boutron CF (2003) Diurnal Cycles of Gaseous Mercury within the Snowpack at Kuujjuarapik/Whapmagoostui, Québec, Canada. Environmental Science & Technology 37:

3289–3297. Available: http://dx.doi.org/10.1021/es026242b. Accessed 14 December 2012.

17. Choi H-D, Holsen TM (2009) Gaseous mercury fluxes from the forest floor of the Adirondacks. Environmental pollution (Barking, Essex□: 1987) 157: 592–600. Available: http://dx.doi.org/10.1016/j.envpol.2008.08.020. Accessed 14 December 2012.

18. Zhang H, Lindberg SE, Barnett MO, Vette AF, Gustin MS (2002) Dynamic flux chamber measurement of gaseous mercury emission fluxes over soils. Part 1: simulation of gaseous mercury emissions from soils using a two-resistance exchange interface model. Atmospheric Environment 36: 835–846. Available: http://dx.doi.org/10.1016/S1352-2310(01)00501-5. Accessed 14 December 2012.

19. Lindberg SE, Zhang H, Vette AF, Gustin MS, Barnett MO, et al. (2002) Dynamic flux chamber measurement of gaseous mercury emission fluxes over soils: Part 2–effect of flushing flow rate and verification of a two-resistance exchange interface simulation model. Atmospheric Environment 36: 847–859. Available: http://dx.doi.org/10.1016/S1352-2310(01)00502-7. Accessed 14 December 2012.

20. Eckley CS, Gustin M, Lin C-J, Li X, Miller MB (2010) The influence of dynamic chamber design and operating parameters on calculated surface-to-air mercury fluxes. Atmospheric Environment 44: 194–203. Available: http://dx.doi.org/10.1016/j.atmosenv.2009.10.013. Accessed 14 December 2012.

21. Carpi A, Frei A, Cocris D, McCloskey R, Contreras E, et al. (2007) Analytical artifacts produced by a polycarbonate chamber compared to a Teflon chamber for measuring surface mercury fluxes. Analytical and bioanalytical chemistry 388: 361–365. Available: http://www.ncbi.nlm.nih.gov/pubmed/17260134. Accessed 14 December 2012.

22. Savitzky A, Golay MJE (1964) Smoothing and Differentiation of Data by Simplified Least Squares Procedures. Analytical Chemistry 36: 1627–1639. Available: http://dx.doi.org/10.1021/ac60214a047. Accessed 31 October 2012.

23. Shapiro SS, Wilk MB (1965) An Analysis of Variance Test for Normality (Complete Samples). Biometrika 52: 591–611. Available: http://www.jstor.org/stable/10.2307/2333709.

24. Pearson K (1895) No Title. Royal Society Proceedings. 241.

25. Gbor P, Wen D, Meng F, Yang F, Zhang B, et al. (2006) Improved model for mercury emission, transport and deposition. Atmospheric Environment 40: 973–983. Available: http://dx.doi.org/10.1016/j.atmosenv.2005.10.040. Accessed 14 December 2012.

26. Steffen A, Douglas T, Amyot M, Ariya P, Aspmo K, et al. (2008) A synthesis of atmospheric mercury depletion event chemistry in the atmosphere and snow. Atmospheric Chemistry and Physics 8.

16

The Ozone-Iodine-Chlorate Clock Reaction

Rafaela T. P. Sant'Anna, Emily V. Monteiro, Juliano R. T. Pereira, Roberto B. Faria*

Instituto de Química, Universidade Federal do Rio de Janeiro, Rio de Janeiro, RJ, Brazil

Abstract

This work presents a new clock reaction based on ozone, iodine, and chlorate that differs from the known chlorate-iodine clock reaction because it does not require UV light. The induction period for this new clock reaction depends inversely on the initial concentrations of ozone, chlorate, and perchloric acid but is independent of the initial iodine concentration. The proposed mechanism considers the reaction of ozone and iodide to form HOI, which is a key species for producing non-linear autocatalytic behavior. The novelty of this system lies in the presence of ozone, whose participation has never been observed in complex systems such as clock or oscillating reactions. Thus, the autocatalysis demonstrated in this new clock reaction should open the possibility for a new family of oscillating reactions.

Editor: Spencer J. Williams, University of Melbourne, Australia

Funding: This work was supported by Conselho Nacional de Desenvolvimento Científico e Tecnológico, grant 303988/2009-6 (www.cnpq.br); Fundação Carlos Chagas Filho de Amparo à Pesquisa do Estado do Rio de Janeiro, E-26/110-107/2007 (www.faperj.br); and Coordenação de Aperfeiçoamento de Pessoal de Nível Superior (www.capes.gov.br). The funders had no role in study design, data collection and analysis, decision to publish, or preparation of the manuscript.

Competing Interests: The authors have declared that no competing interests exist.

* E-mail: faria@iq.ufrj.br

Introduction

A clock reaction is a special chemical phenomenon of which the Landolt clock reaction is among the best examples [1,2]. Some of these reactions have captured the attention of a many students and others due to their sudden color changes that occur after a lag time or induction period. These clock reactions not only are beautiful but also belong to a class of nonlinear chemical phenomena including propagation of wave fronts, oscillating reactions and Turing structures, which are related to several natural processes and rhythms [3–6].

The definition of a clock reaction is a matter of dispute that remains unresolved. Some authors [7] consider the term 'clock reaction' appropriate only for reactions that present a stoichiometric ratio such that the end of the induction period is marked by the exhaustion of one reagent. However, the paradigmatic Landolt reaction between sulfite and iodate has been shown [8] to depend not only on the depletion of sulfite but also on the autocatalytic production of iodide and H^+ to show the abrupt formation of iodine. Thus, we may consider any reaction to have clock-type behavior if it presents an induction period followed by an abrupt change in the concentration of some species. For simplicity, our use of the term 'clock reaction' indicates a reaction that exhibits clock-type behavior.

The chlorate-iodine clock reaction [10] is among the last clock reactions to be discovered (the last one is based on molybdenum blue [9]). This reaction presents a rapid consumption of iodine after an induction period that is inversely proportional to the initial chlorate and H^+ concentrations but is independent of the initial iodine concentration. Galajda *et al.* [11] demonstrated that this reaction only occurs in the presence of the UV light from the deuterium lamp of a diode-array spectrophotometer, which breaks apart the iodine molecules, producing I• radicals that react with the chlorate.

In this work we show that addition of ozone to the chlorate-iodine system produces a clock reaction without requiring UV light.

Materials and Methods

The reaction was followed at the iodine absorbance band maximum ($\lambda = 460$ nm, $\varepsilon = 740$ L mol^{-1} cm^{-1}) [12] using a 10-mm optical path quartz cuvette and an Agilent 8453 UV-Vis spectrophotometer with only the tungsten lamp turned on (the deuterium lamp was turned off) to avoid exposing the sample to UV light. Some experiments were repeated following the reaction using a conventional double-beam Cary 1E spectrophotometer with identical results. Each experimental curve represents a set of several curves obtained for each set of initial reagent concentrations as indicated in the legends.

The ozone solutions were prepared by feeding high-purity oxygen into a commercial ozone generator. The ozone was bubbled in 0.6 mol L^{-1} perchloric acid to reduce the rate of ozone decomposition and obtain a concentration of approximately 3.0×10^{-5} mol L^{-1}. The ozone concentration was measured using its absorbance at 260 nm ($\varepsilon = 3000$ L mol^{-1} cm^{-1}) [13].

All reagents were used as received. The iodine was resublimed for some experiments, but the results were identical. All solutions were prepared using conductivity water (minimum 18 MΩ), and all experiments were conducted at $25 \pm 0.1°$C.

A simulation of the proposed mechanism was developed via the numerical integration of the model presented in Table 1 using a semi-implicit Runge-Kutta method [14] codified in Turbo Pascal.

Results and Discussion

While repeating the Galajda *et al.* [11] experiments, we attempted to place a quartz cuvette containing an aqueous iodine solution into a diode array spectrophotometer to illuminate it with

Table 1. Mechanism for the ozone-iodine-chlorate clock reaction.

Number[a]	Reaction	Rate law[b]
1*	$I_2 + H_2O \leftrightarrow HOI + I^- + H^+$	1.98×10^{-3} $[I_2]/[H^+]$ - 3.67×10^9 $[HOI][I^-]$
2*	$I_2 + H_2O \leftrightarrow H_2IO^+ + I^-$	5.52×10^{-2} $[I_2]$ - 3.48×10^9 $[H_2IO^+][I^-]$
3*	$HClO_2 + I^- + H^+ \rightarrow HOI + HOCl$	7.8 $[HClO_2][I^-]$
4**	$HClO_2 + HOI \rightarrow HIO_2 + HOCl$	6.9×10^7 $[HClO_2][HOI][H^+]$
5*	$HClO_2 + HIO_2 \rightarrow IO_3^- + HOCl + H^+$	1.0×10^6 $[HClO_2][HIO_2]$
6*	$HOCl + I^- \rightarrow HOI + Cl^-$	4.3×10^8 $[HOCl][I^-]$
7*	$HOCl + HIO_2 \rightarrow IO_3^- + Cl^- + 2 H^+$	1.5×10^3 $[HOCl][HIO_2]$
8*	$HIO_2 + I^- + H^+ \leftrightarrow 2 HOI$	1×10^9 $[HIO_2][I^-][H^+]$ - 22 $[HOI]^2$
9*	$2 HIO_2 \rightarrow IO_3^- + HOI + H+$	25 $[HIO_2]^2$
10*	$HIO_2 + H_2IO^+ \rightarrow IO_3^- + I^- + 3 H^+$	110 $[HIO_2][H_2IO^+]$
11*	$HOCl + Cl^- + H^+ \leftrightarrow Cl_2 + H_2O$	2.2×10^4 $[Cl_2][H^+]$ - 22 $[HOCl][Cl^-]$
12*	$Cl_2 + I_2 + 2 H_2O \rightarrow 2 HOI + 2 Cl^- + 2 H^+$	1.5×10^5 $[Cl_2][I_2]$
13*	$Cl_2 + HOI + H_2O \rightarrow HIO_2 + 2 Cl^- + 2 H^+$	1.0×10^6 $[Cl_2][HOI]$
14*	$HClO_2 \leftrightarrow ClO_2^- + H^+$	2×10^8 $[HClO_2]$ - 1×10^{10} $[ClO_2^-][H^+]$
15*	$HOI + H^+ \leftrightarrow H_2OI^+$	1×10^{10} $[HOI][H^+]$ - 3.4×10^8 $[H_2IO^+]$
16*	$I_2 + I^- \leftrightarrow I_3^-$	5.52×10^9 $[I_2][I^-]$ - 7.5×10^6 $[I_3^-]$
17	$O_3 + I^- + H^+ \rightarrow HOI + O_2$	1.2×10^9 $[O_3][I^-][H^+]$
18	$ClO_3^- + HIO_2 \rightarrow IO_3^- + HClO_2$	20 $[ClO_3^-][HIO_2][H^+]$
19	$ClO_3^- + H_2IO^+ + H^+ \rightarrow HIO_2 + HClO_2 + H^+$	1.4 $[ClO_3^-][H_2IO^+][H^+]$

a) * Reactions taken from Lengyel et al. [16]. ** Modified from the Lengyel et al. [16] model by including the $[H^+]$ effect in the rate law.
b) Rate constant units are s^{-1}, $L\,mol^{-1}\,s^{-1}$, $L^2\,mol^{-2}\,s^{-1}$ for the first-, second- and third-order processes, respectively.

the UV light from the deuterium lamp. After a given irradiation time, we removed some of this solution from the diode array spectrophotometer and mixed it with the other reagent solutions in another cuvette inside a conventional double-beam spectrophotometer. Surprisingly, the system clocked. Additional experiments indicated that the observed clock time was inversely proportional to the time that the iodine solution was exposed to UV light as depicted in Figure 1.

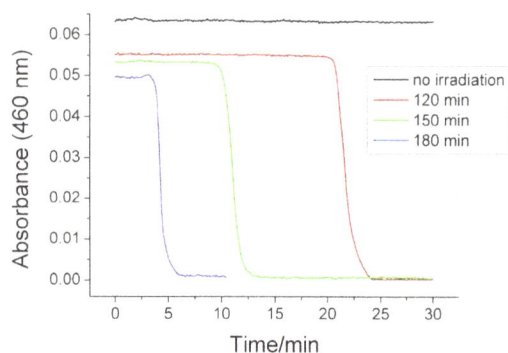

Figure 1. Effect of the iodine solution irradiation time on the clock time of the chlorate-iodine clock reaction. A 2.0×10^{-4} mol L^{-1} iodine solution was irradiated using the deuterium lamp of a diode array spectrophotometer during the indicated time. The reaction was followed at 460 nm using a conventional double-beam spectrophotometer. The initial concentrations after mixing the reactants in the cuvette were as follows: $[NaClO_3] = 0.0251$ mol L^{-1}, $[HClO_4] = 0.948$ mol L^{-1}, and $[I_2] = 7.3 \times 10^{-5}$ mol L^{-1}. The reactants were added to the solutions in the following order: $NaClO_3$, irradiated iodine solution, water, and $HClO_4$.

Considering that the lifetime of the I• is small based on its recombination rate (2 I• (aq) $\rightarrow I_2$ (aq); $k = 8 \times 10^9$ L mol^{-1} s^{-1} [15]), these radicals could not survive long enough (at least 10 s) to remove the cuvette containing the iodine solution from the diode array spectrophotometer, pipette this solution and mix it with the other solutions. In another experiment, we placed pure water into a quartz cuvette in the diode array spectrophotometer and irradiated it for a given time (90 min, 120 min or 150 min). Then, we mixed this irradiated water with the iodine, chlorate and perchloric acid solutions and observed that this system also clocked. The clock time was also inversely proportional to the irradiation time (figure not shown).

Because UV light is not expected to have any effect on water, one can naturally suppose that UV light split the dissolved oxygen ($O_2 + h\nu \rightarrow 2$ O) and formed ozone ($O_2 + O \rightarrow O_3$) that could then participate in a sequence of reactions to produce the clock behavior. This interpretation was confirmed by preparing an ozone solution, mixing it with the other reagents and observing the clock behavior.

The addition of a hydrogen peroxide solution did not produce clock behavior, which suggests that we cannot assume any oxidizing species will produce this type of nonlinear behavior. The effect of adding HOCl or chlorite (which becomes $HClO_2$) as observed by Galajda et al. [11] should be due to the participation of these species in the autocatalytic sequence of reactions as shown by the proposed mechanism (Table 1).

The effect of the initial ozone, chlorate, acid, and iodine concentrations on the induction period is demonstrated in Figures 2 to 5. All curves begin at 15 s, which is the time required to mix the reagents in the cuvette and put it in the cuvette holder inside the spectrophotometer. The reagents were added using a digital pipette in the following order: water, $NaClO_3$, iodine,

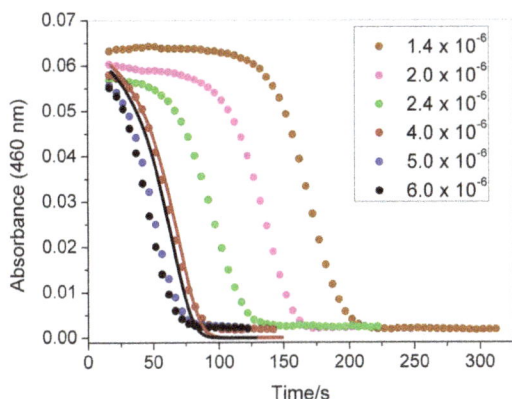

Figure 2. Effect of the ozone concentration on the clock time. The initial concentrations after mixing the reactants in the cuvette were as follows: $[O_3]$ as indicated in the figure (mol L^{-1}), $[NaClO_3] = 0.0251$ mol L^{-1}, $[HClO_4] = 0.474$ mol L^{-1}, and $[I_2] = 8.8 \times 10^{-5}$ mol L^{-1}. Experimental data (symbols); modeling results (continuous line).

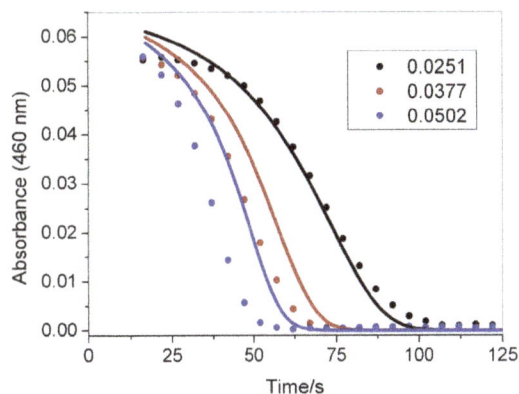

Figure 4. Effect of the acid concentration on the clock time. The initial concentrations after mixing the reactants in the cuvette were as follows: $[HClO_4]$ as indicated in the figures (mol L^{-1}), $[O_3] = 2.0 \times 10^{-6}$ mol L^{-1}, $[NaClO_3] = 0.0251$ mol L^{-1}, $[I_2] = 8.8 \times 10^{-5}$ mol L^{-1}. (a) experimental data (symbols); (b) modeling results (continuous line).

$HClO_4$ and ozone. As indicated in these figures, increasing the initial concentrations of these species reduces the induction period, except iodine which does not change the clock time but increases the initial absorbance.

The observed clock behavior can be understood by considering that ozone reacts with iodide to form HOI, which is a key species in producing the nonlinear autocatalytic behavior as previously demonstrated [16]. The full mechanism presented in Table 1 contains a core of reactions (Reactions 1 to 16) proposed by Lengyel *et al.* [16] between chlorine and iodine with species that provide the autocatalytic path necessary for the clock behavior. Although this mechanism core only qualitatively reproduces the chlorite-iodide clock behavior [17], it remains the best available model for this system. To this core group of reactions we added Reaction (17) between ozone and iodide [13], and added Reactions (18) and (19) to introduce the chlorate, as we made in our first model of the chlorate-iodine clock reaction [10], but with slightly different rate constants. To obtain qualitative agreement as demonstrated in Figure 4 for the acid concentration effect, we included one additional H^+ in the rate laws for Reactions (4) and

(19). In Reaction (4), this additional H^+ accounts for the shifts in the $[ClO_2^-]/[HClO_2]$ and $[HOI]/[H_2OI^+]$ ratios that should be affected by changes in $[H^+]$ as indicated by Jowsa *et al.* [17]. The same can be said for Reaction (19), which can be affected by the protonation of chlorate or some reaction intermediate. Other modifications of the manner in which H^+ participates in other reactions could improve the agreement in the acid concentration effect (Figure 4), but we decided to minimize these modifications,

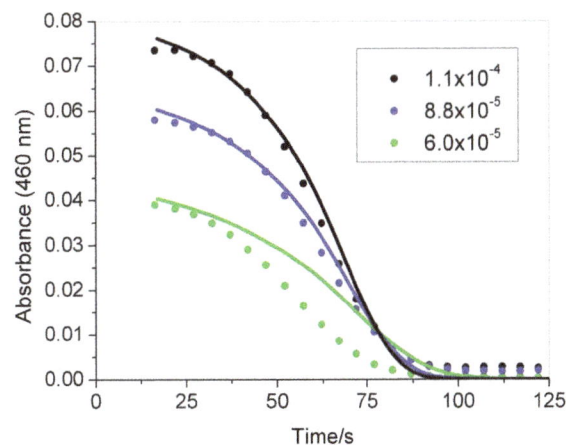

Figure 3. Effect of the chlorate concentration on the clock time. The initial concentrations after mixing the reactants in the cuvette were as follows: $[NaClO_3]$ as indicated in the figure (mol L^{-1}), $[O_3] = 3.0 \times 10^{-6}$ mol L^{-1}, $[HClO_4] = 0,474$ mol L^{-1}, $[I_2] = 8.8 \times 10^{-5}$ mol L^{-1}. Experimental data (symbols); modeling results (continuous line).

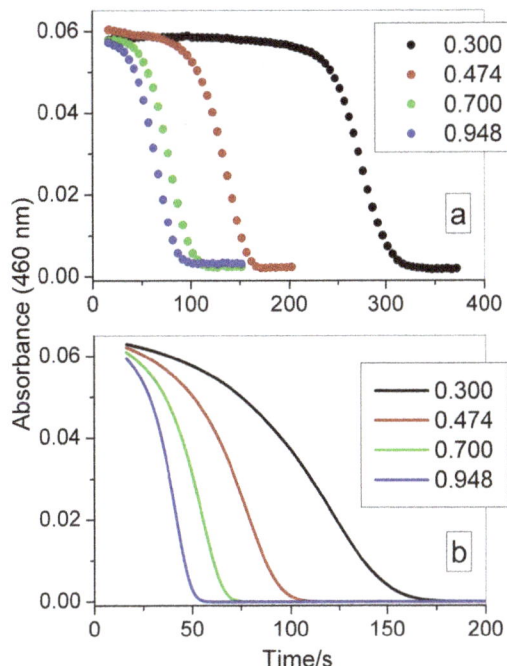

Figure 5. Effect of the iodine concentration on the clock time. Initial concentrations after mixing the reactants in the cuvette: $[I_2]$ as indicated in the figure (mol L^{-1}), $[NaClO_3] = 0.0251$ mol L^{-1}, $[HClO_4] = 0.474$ mol L^{-1}, $[O_3] = 4.0 \times 10^{-6}$ mol L^{-1}. Experimental data (symbols); modeling results (continuous line).

following only the suggestions made by Jowsa *et al.* [17]. As indicated in Figures 2 to 5, the proposed mechanism can reproduce the behavior observed in the experiments, demonstrating good agreement with the chlorate and iodine effect. For the ozone effect (Figure 2), the model produces good agreement for $[O_3]_0 = 4 \times 10^{-6}$ mol L^{-1}, but for other ozone concentrations, the agreement is only qualitative. For the acid concentration effect (Figure 4), the model produces a shorter clock time than the experimental value, but is still in good qualitative agreement.

Additional experiments have indicated that mixing ozone and iodine solutions decreased only the I_3^- bands at 350 and 288 nm, indicating the oxidation of iodide (Reaction 17) followed by a shift in Reaction (16) to the left, allowing one to conclude that the ozone does not react directly with the iodine. Because the reaction between ozone and iodide has a high rate constant, we did not include the reactions between ozone and other low-concentration species in the model.

In conclusion, without UV light, the addition of ozone to the chlorate-iodine system initiates an autocatalytic sequence of reactions that results in the rapid consumption of iodine after an induction period. However, if UV light is present, its effect over

the iodine molecules cannot be ignored, especially in systems containing oxychlorine species, as observed in the photoinduced reactions between $ClO_2\bullet$ and I_2 [18] and between ClO_3^- and I_2 [10,11].

To our knowledge, this is the first time that the participation of ozone in a system to produce complex nonlinear behavior has been demonstrated. This research can serve as the first step to discover oscillating reactions or other nonlinear dynamic behavior involving ozone.

Acknowledgments

R.T.P.S. thanks Coordenação de Aperfeiçoamento de Pessoal de Nível Superior-CAPES for scholarship. E.V.M. and J.R.T.P. thank PIBIC-CNPq for scholarship. We thank Dr. Istvan Lengyel for the use of the Turbo Pascal code employed in the simulations.

Author Contributions

Conceived and designed the experiments: RTPS EVM JRTP RBF. Performed the experiments: RTPS EVM JRTP RBF. Analyzed the data: RTPS EVM JRTP RBF. Wrote the paper: RTPS EVM JRTP RBF.

References

1. Landolt H (1886) Ueber die Zeitdauer der Reaction zwischen Jodsäure und schwefliger Säure. Ber. Dtsch. Chem. Ges. 19: 1317–1365.
2. Shakhashiri BZ (1985) Clock Reactions in Chemical Demonstrations. A Handbook for Teachers of Chemistry. Vol. 2, Madison: The University of Wisconsin Press, pp. 3–89.
3. Epstein IR, Pojman JA (1998) An Introduction to Nonlinear Chemical Dynamics: Oscillations, Waves, Patterns, and Chaos. New York: Oxford University Press.
4. Chance B, Pye EK, Ghosh AK, Hess B (1973) Biological and Biochemical Oscillators. New York: Academic Press.
5. Goldbeter A (1997) Biochemical Oscillations and Cellular Rhythms. Cambridge: University Press.
6. Danino T, Mondragón-Palomino O, Tsimring L, Hasty J (2010) A synchronized quorum of genetic clocks. Nature 463: 326–330.
7. Lente G, Bazsa G, Fábián I (2007) What is and what isn't a clock reaction? New J. Chem. 31: 1707.
8. Edblom EC, Orbán M, Epstein IR (1986) A New Iodate Oscillator: The Landolt Reaction with Ferrocyanide in a CSTR. J. Am. Chem. Soc. 108: 2826–2830.
9. Neuenschwander U, Negron A, Jensen KF (2013) A Clock Reaction Based on Molybdenum Blue. J. Phys. Chem. A 117: 4343–4351.
10. Oliveira AP, Faria RB (2005) The Chlorate-Iodine Clock Reaction. J. Am. Chem. Soc. 127: 18022–18023.

11. Galajda M, Lente G, Fábián I (2007) Photochemically Induced Autocatalysis in the Chlorate Ion-Iodine System. J. Am. Chem. Soc. 129: 7738–7739.
12. Rábai G, Beck MT (1987) Kinetics and Mechanism of the Autocatalytic Reaction between Iodine and Chlorite Ion. Inorg. Chem. 26: 1195–1199.
13. Liu Q, Schurter LM, Muller CE, Aloisio S, Francisco JS, et al. (2001) Kinetics and Mechanisms of Aqueous Ozone Reactions with Bromide, Sulfite, Hydrogen Sulfite, Iodide, and Nitrite Ions. Inorg. Chem. 40: 4436–4442.
14. Kaps P, Rentrop P (1979) Generalized Runge-Kutta methods of order four with stepsize control for stiff ordinary differential equations. Numer. Math. 33: 55–68.
15. de Violet PF, Bonneau R, Joussot-Dubien J (1973) Étude en spectroscopie par impulsion laser de l'iode dissous dans des solvants polaires, mise en évidence des complexes par transfert de charge du type atome d'iode-solvant. J. Chim. Phys. 70: 1404–1409.
16. Lengyel I, Li J, Kustin K, Epstein IR (1996) Rate Constants for Reactions between Iodine- and Chlorine-Containing Species: A Detailed Mechanism of the Chlorine Dioxide/Chlorite-Iodide Reaction. J. Am. Chem. Soc. 118: 3708–3719.
17. Jowza M, Sattar S, Olsen RJ (2005) Modeling Chlorite-Iodide Reaction Dynamics Using a Chlorine Dioxide-Iodide Reaction Mechanism. J. Phys. Chem. A 109: 1873–1878.
18. Rábai G, Kovács KM (2001) Photoinduced Reaction between Chlorine Dioxide and Iodine in Acidic Aqueous Solution. J. Phys. Chem. A 105: 6167–6170.

17

Investigating the Role of State Permitting and Agriculture Agencies in Addressing Public Health Concerns Related to Industrial Food Animal Production

Jillian P. Fry[1,2]*, **Linnea I. Laestadius**[1,3], **Clare Grechis**[1,4], **Keeve E. Nachman**[1,2,4], **Roni A. Neff**[1,2,4]

1 Center for a Livable Future, Johns Hopkins University, Baltimore, Maryland, United States of America, 2 Department of Environmental Health Sciences, Johns Hopkins Bloomberg School of Public Health, Baltimore, Maryland, United States of America, 3 Joseph J. Zilber School of Public Health, University of Wisconsin-Milwaukee, Milwaukee, Wisconsin, United States of America, 4 Department of Health Policy and Management, Johns Hopkins Bloomberg School of Public Health, Baltimore, Maryland, United States of America

Abstract

Objectives: Industrial food animal production (IFAP) operations adversely impact environmental public health through air, water, and soil contamination. We sought to determine how state permitting and agriculture agencies respond to these public health concerns.

Methods: We conducted semi-structured qualitative interviews with staff at 12 state agencies in seven states, which were chosen based on high numbers or rapid increase of IFAP operations. The interviews served to gather information regarding agency involvement in regulating IFAP operations, the frequency and type of contacts received about public health concerns, how the agency responds to such contacts, and barriers to additional involvement.

Results: Permitting and agriculture agencies' responses to health-based IFAP concerns are constrained by significant barriers including narrow regulations, a lack of public health expertise within the agencies, and limited resources.

Conclusions: State agencies with jurisdiction over IFAP operations are unable to adequately address relevant public health concerns due to multiple factors. Combining these results with previously published findings on barriers facing local and state health departments in the same states reveals significant gaps between these agencies regarding public health and IFAP. There is a clear need for regulations to protect public health and for public health professionals to provide complementary expertise to agencies responsible for regulating IFAP operations.

Editor: Suminori Akiba, Kagoshima University Graduate School of Medical and Dental Sciences, Japan

Funding: Salary support for all authors was provided by a gift agreement from the GRACE Communications Foundation (www.gracelinks.org). The funders had no role in study design, data collection and analysis, decision to publish, or preparation of the manuscript.

Competing Interests: The authors have declared that no competing interests exist.

* E-mail: jfry@jhsph.edu

Introduction

A dramatic series of changes in the landscape of animal agriculture have taken place over the past seventy years, accompanied by multiple public health concerns [1]. Small farms raising a diversity of crops and food animals have increasingly and steadily given way to a model of industrial food animal production (IFAP) that raises large numbers of animals in concentrated quarters, often on farms whose only crops are animal feed grown on spray fields. Hog production, for example, has shifted significantly since the 1980s. As Figure 1 illustrates, from 1987 to 2007 the number of hog operations in the US decreased from over 320,000 to about 75,000 [2]. Mid-aggregate enterprise size is a measurement of farm production that shows the point where half of a product comes from larger farms, and half from smaller farms. In 1987, the mid-aggregate enterprise size of US hog farms was 1,200 hogs, according to the USDA. By 2007, this measurement increased to 30,000, reflecting a 2,400% increase [3], as shown in Figure 1. The mid-aggregate enterprise size of dairy and broiler

(chicken meat) production operations increased by 613% and 127%, respectively, during the same interval [3]. There has also been geographic concentration in production. In 2007, 75% of all U.S. hogs were raised in only 220 counties, down from 508 counties in 1987 [3]. These changes in the model, methods, and system within which animals are produced for food pose both a regulatory challenge and clear concerns for environmental health and public health more generally.

Research linking IFAP to public health concerns and impacts continues to increase. In addition to posing respiratory health risks to those residing near operations [4]–[8] due to emissions that include hydrogen sulfide [9], particulate matter [9], endotoxins [10], ammonia [11], allergens [12], and volatile organic compounds [13], [14], odor generated by IFAP operations and spray fields has been associated with a broad range of health problems. Public access to information regarding hazardous airborne releases from IFAP operations is hindered due to exemptions in federal laws that require disclosure of such releases [15], despite research linking chronic exposure to odors from IFAP

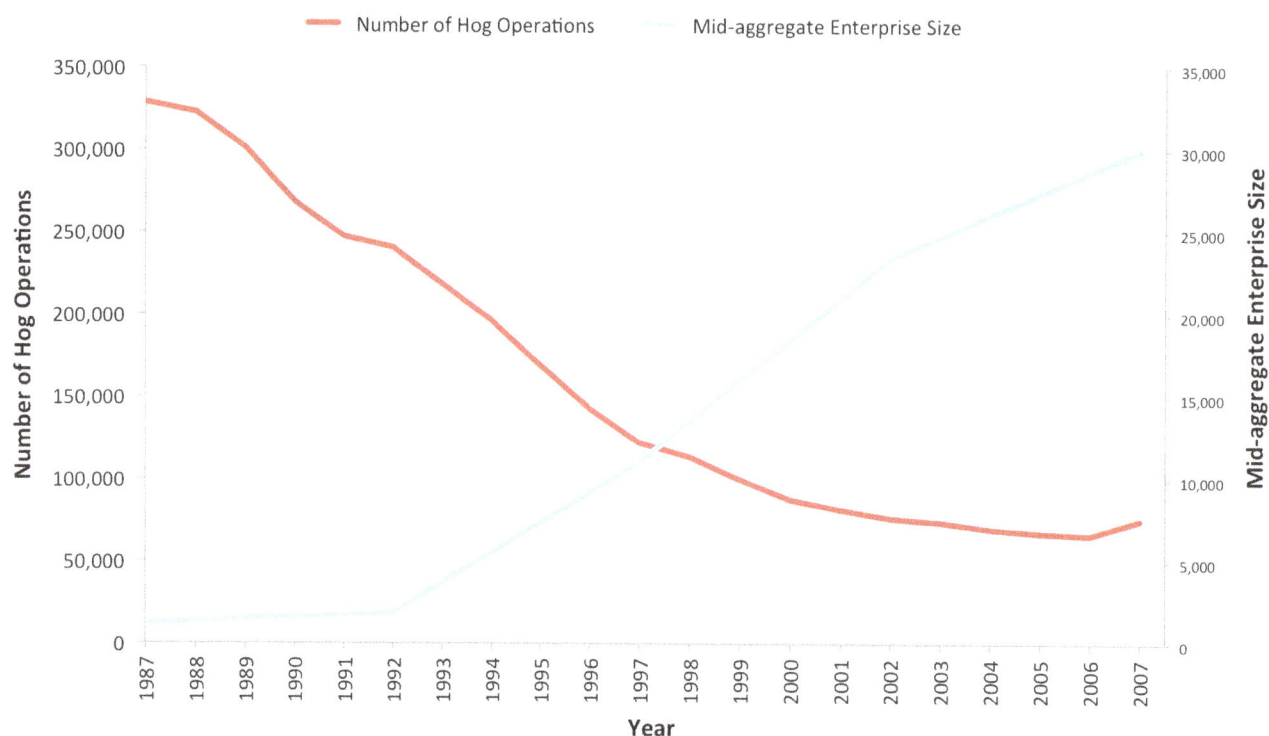

Figure 1. Hog production in the US: Number of operations and mid-aggregate enterprise size; 1987–2007. Data from the United States Department of Agriculture (USDA) shows the decline in number of hog operations and growth in operation size from 1987 to 2007. Sources: The Changing Organization of U.S. Farming, USDA Economic Research Service (http://www.ers.usda.gov/amber-waves/2011-december/changing-farming-practices.aspx#.UgqVWOu9xEo) and USDA Census of Agriculture QuickStats, various years (http://quickstats.nass.usda.gov/?source_desc = CENSUS).

to headaches, nausea, upset stomach, mood disorders, high blood pressure, and sleep problems [16]–[20]. Additionally, there is growing evidence that livestock can transmit methicillin-resistant *Staphylococcus aureus* (MRSA) to humans [21]–[23]. Rural water supplies are also at risk, as IFAP-generated animal waste contaminants, including nitrates, pathogens, pharmaceuticals, metals, and hormones, can leach into ground water [24], [25]. All of these concerns may be compounded by the fact that IFAP operations are disproportionately located in low-income communities with high-percentages of minority populations [26]–[28], which are more likely to experience limited political power [29] and barriers to healthcare access [30].

Despite health risks posed by these operations, regulation of IFAP is limited and characterized by a patchwork of different regulatory approaches from state to state. Under the Clean Water Act (CWA), IFAP operations that are designated as Concentrated Animal Feeding Operations (CAFOs) are required to obtain National Pollution Discharge Elimination System (NPDES) permits in order to discharge into U.S. waterways. Designation as a CAFO is determined by size and potential to pollute the surrounding environment. States are granted authority to determine some requirements, issue NPDES permits, monitor compliance, and impose penalties by the U.S. Environmental Protection Agency if the state adopts federal requirements into law. As of 2008, 44 states had permitting authority for CAFOs [31]. Animal operations that fail to meet the technical definition of a CAFO may still cause numerous concerns, thus the more inclusive term IFAP is utilized here.

The state agency responsible for permitting IFAP operations varies among states, with almost all delegating the responsibility to

Departments of Environmental Protection/Management, Natural Resources, or Agriculture [31]. The organization and jurisdiction of state agencies responsible for regulations pertaining to the environment and natural resources varies significantly by state [32]; so defining the broader roles and responsibilities of these agencies is difficult. We focused on environmental agencies that have been given authority by the EPA to implement the NPDES permitting program for CAFOs, and we refer to these agencies as "permitting agencies". Responsibilities of Departments of Agriculture vary less by state, and in general they include promotion and regulation of agriculture in the state, as well as other common responsibilities such as conservation and farmland protection [33]. Regulatory authority is shared among agencies in some states, and the transfer of regulatory authority to Departments of Agriculture has been favored by industry [34]. Some states have developed additional regulations for smaller IFAP operations that do not meet the size threshold to be considered a CAFO. Additionally, states vary regarding resources available to monitor and enforce regulations, and potential penalties for violations [31], [35]. Common across most states, however, is delegating the permitting to an agency without a primary mandate to address public health [31], raising concerns that public health issues may not be adequately monitored or addressed by the agencies tasked with regulating IFAP operations.

A 2008 report by the National Conference on State Legislatures provided an overview of agencies in each state responsible for regulating IFAP [31], but the report did not address how state agencies respond to community health concerns arising from IFAP operations. We conducted interviews with state permitting agency and agriculture department staff members to determine the ability of these agencies to: 1) monitor IFAP operations and enforce current regulations, and 2) respond to citizen health concerns arising from IFAP operations. This manuscript represents the second part of a study examining how state and local agencies address health concerns associated with IFAP. Our previously published research [36] found that despite the evidence linking IFAP to public health concerns, state and local health departments play a limited role in addressing health issues linked to IFAP operations. Health departments report a lack of formal jurisdiction over IFAP operations as the primary barrier to regulatory response. As a result, we found that many health departments refer IFAP concerns, including community member concerns related to health, to departments of agriculture and permitting agencies.

Assessing the roles of the main agencies with jurisdiction over issues relevant to public health is essential to a full understanding of how health concerns, including those related to IFAP, are addressed. In fact, defining inter-organizational relationships, evaluating capacity as financial priorities shift, and identifying gaps in services or jurisdiction are key priorities in public health systems research [37], [38]. We aimed not only to characterize the current role, capacity, and barriers of the primary agencies potentially involved with IFAP and public health, but also to highlight gaps in agency responsibilities that would not be identified if only one type of agency were studied.

Methods

Given the absence of research on the role of state agriculture and permitting agencies in responding to health concerns stemming from IFAP, an exploratory qualitative approach was chosen for data collection and analysis. See Fry (2013) for further details on the initial study, which examined state and county health department responses [36]. An inductive qualitative research approach was used, with a focus on flexibility rather than allegiance to any specific theoretical approach [39], [40]. Important aspects of study design, data collection, and data analysis are described below.

We used a purposive sampling strategy to select the states included in this study. First, all U.S. counties were ranked by USDA Census of Agriculture hog inventory data in two ways: 1) 2007 county hog inventory of operations with 1,000 or more hogs, and 2) increase in hog inventory of 1,000+ head operations between 2002 and 2007. It was anticipated that many hogs produced in intensive settings or large increases in hog numbers could lead community members or others to request that agencies take action to address health concerns. We also sought to examine settings where hog operations were near residential areas by ranking the top sixty counties from each list by population density using 2000 Census data. We then selected the top fifteen counties by population density from the two lists. These counties were located in eight states. Although states were chosen based on hog production and census data, the interviews asked about IFAP in general, and did not focus solely on hog operations.

We contacted state agriculture and permitting agency employees in the eight states to perform semi-structured interviews. Both types of agencies were included in order to understand their roles and to identify potential gaps in agency responses to health concerns. Permitting agencies and some agriculture departments were identified through the National Conference on State Legislatures' Survey of State Policies on CAFOs [31]. Through web-based searching and investigation, we confirmed the information from the report and collected contact information for additional agriculture departments. Most commonly, permitting

responsibilities fell under the authority of departments of environmental protection or natural resources. We contacted staff members who worked on livestock permitting and/or livestock production, identified using agency websites. The specific roles and titles of interviewees varied across states and included, among others, livestock permitting and water quality specialists, state veterinarians, and community-relations liaisons. In some cases, staff we contacted referred us to others in their agencies who could better answer our questions.

All interviews were conducted over the telephone by JF or LL, with a note-taker present on each call to document responses. Interviews were not audio recorded. The questionnaires included mostly open-ended and some closed-ended questions (See Appendix S1 for survey instruments), and some follow-up or clarifying questions were asked based on participants' responses. We also specifically queried permitting agency staff members about their states' permitting regulations. Interview questions were developed based on our prior familiarity with the topic, which is drawn from extensive background reading, original research, and interactions with community members impacted by IFAP operations. We did not use the terms, IFAP or CAFO, in the interviews. Before beginning each interview, we read participants a confidentiality statement. The Johns Hopkins Bloomberg School of Public Health Institutional Review Board (IRB) determined the study was exempt from IRB oversight and did not require an informed consent process.

After data collection was complete, notes from three interviews were double coded through an inductive coding process using *HyperRESEARCH 3.0.3* (ResearchWare, Randolph, MA). Codes were jointly discussed to develop a uniform codebook, which was then applied to the remaining interviews. As this is a qualitative study with a small, purposive sample, we provide limited numeric information to avoid implying that the findings are generalizable to a larger population [41]. Instead, we describe important themes identified in the interviews that allow better understanding of the situation in states with significant industrial hog production.

Results

We conducted telephone interviews with staff members from permitting agencies in seven states and departments of agriculture in five states. In one state, both the permitting agency and department of agriculture declined to participate, and agriculture department staff in two other states also declined. In total, twelve interviews in seven states were conducted between November 2010 and October 2011. Most of the interviews lasted between 35 and 45 minutes.

Regulatory Oversight

In nearly all instances, state-level environment or natural resource departments managed NPDES permits. One state delegated permitting authority to counties. In two states, departments of agriculture shared regulatory authority with other departments, and in another state the permitting agency held only narrow authority over NPDES permitting with many activities transferred to the agriculture department in recent years.

As specified in the federal Clean Water Act, NPDES permitting requirements apply only to certain IFAP operations that match the defined criteria for CAFOs. States also reported additional non-NPDES regulatory measures. For example, some states require permits for IFAP operations that do not match the federal CAFO definition, although requiring less information for approval than for CAFOs. Some permitting agency staff members also said they were trying to develop regulations that would strengthen reporting requirements for small/medium size operations, thereby bringing them in line with CAFO permitting requirements. One agriculture department staff member noted that, "Issues are more frequently from smaller farms, but people are quick to blame bigger farms".

We asked interviewees if their agencies had any setback/zoning policies related to human health concerns. While several counties and states require that IFAP operations be a certain distance away from property lines, wells, waterways, homes, churches, schools, and/or parks, many interviewees said the requirements are not necessarily based on health standards. One staff member said,

"There are setbacks, but those are more for odor rather than health. This can be from residences or populated areas, it's just a form of odor management. There are some setbacks for wells/private water wells too, that have more direct relevance to public health issues."

Setbacks can reduce public health impacts of odor and poor air quality, but interviewees indicated that most setbacks are not based on evidence-based health standards specifically designed to limit exposure to gases and particulate matter emitted from IFAP operations.

One permitting agency staff member said two counties in the state had established more stringent requirements for setbacks through health ordinances. Two states reported having air quality standards that apply to IFAP operations, but in one state the standard only applies to new operations.

Inspections

States varied significantly in the frequency of inspections of permitted CAFOs for compliance with regulations. One permitting agency staff member said that they "can only afford to [inspect] on a complaint basis" because they "don't have staff or money." By contrast, a staff member from another state's agriculture department indicated that they inspected CAFOs every six months. The staff member also stated, "before reforms a few years ago, some facilities had gone 25 years without inspection."

Contacts and Concerns Reported

All interviewees noted that their agencies had been contacted by people concerned about issues associated with living or spending time near animal production operations and/or manure (i.e., manure storage or spray fields). Estimates from permitting agency staff ranged from 25 to 120 contacts per year, and agriculture agency staff estimates ranged from fewer than 5 times per year to an average of 5 times per week. Some interviewees said that calls were generally more frequent when new or expanding operations were proposed and that they believed that calls had decreased due to the implementation of additional regulations. The absence of regulations pertaining to common community concerns with IFAP also appeared to reduce contacts as people began to learn what agencies were able to do in response to concerns. One permitting agency staff member said there "used to be more calls about odor, but there are no odor regulations, so there is nothing we can do about it; the public learned there's no point in calling about odor complaints."

We read to interviewees a list of types of people who may have contacted their agency about public health concerns (Appendix S1). All participants said they had been contacted by individuals describing their own concerns, and most also said they had been contacted by members of an organized campaign regarding

CAFOs/IFAP. Few agencies were contacted by health care providers.

We also questioned staff members about topics related to animal agriculture that their agencies may have ever been contacted about. A list of topics was provided, and all interviewees answered affirmatively regarding odor. Table 1 presents a ranking based on the number of interviewees reporting their agency had been contacted about the issue as it relates to animal agriculture. Rankings varied little between permitting and agriculture agency staff.

Response to Citizen Concerns

Permitting agency staff reported a variety of responses depending on the concern; they often mentioned gathering additional information and checking the validity of a concern by contacting the person reporting the issue or by investigating the issue. Inspection activities were frequently delegated to regional field staff or county conservation district staff. Permitting agencies also made referrals to other agencies including departments of agriculture and health, as well as state Farm Bureaus (agriculture trade group), when issues were beyond their jurisdictions. Staff indicated that at times they would contact the person who had raised the concern to let them know that they could not address it. One interviewee said: "If the problem is not covered under the agency, it might be a phone call or email to let people know why we can't address their concerns. Water issues are our primary jurisdiction. There are no state/federal regulations over air emissions." The person was referring to a lack of air quality regulations in their state. As mentioned above, only two states reported that they were able to address some air quality concerns under their current regulatory authority. A staff member in a state with an air quality standard that applies to all IFAP operations (i.e., sites that existed when the regulation was adopted and new sites) noted that if there were a "major complaint" they would use their equipment to take air samples.

Some agency staff reported reaching out to producers directly to try to resolve issues. One interviewee described an agreement with the state Farm Bureau in which the agriculture group would send someone to talk to the producer "farmer to farmer" in the event of a reported concern. They thought that producers are more willing to speak to another farmer than to an agency staff member, and that this process results in issues being resolved faster. Another staff member described talking to producers initially to see if rules are being followed. Then they explain the concern and determine if it can be resolved without further agency action. Some concerns could be related to a regulatory violation, but few agency personnel described a process for enforcement in response to community member concerns. Based on interviewees' responses,

Table 1. Topics of concern ranked in descending order according to number of agency staff stating people have ever contacted the agency about the issue.

Odor
Respiratory health
Ground water quality/contaminated well
Violations of regulations
Waste getting on property
General health
Traffic

this phenomenon appeared to reflect factors including limited jurisdiction over common concerns and preference for voluntary and informal solutions to violations of regulations. All permitting agency interviewees stated that records were maintained of reported concerns, but it was not clear if those records always included concerns that were not formally investigated.

Agriculture department staff members' responses to concerns varied due to the state-by-state differences in agriculture agencies' jurisdiction over IFAP. In general, staff said that they investigate and/or refer to relevant agencies in response to concerns. Staff members from several states said they would refer concerns to the state permitting agency, either right away or after learning more about the situation. Most department of agriculture interviewees indicated that they had limited authority over many IFAP concerns and said the state permitting agencies would be more involved. Issues related to manure or dead animals, however, were more likely to be dealt with by agriculture departments. A small number of interviewees said they might contact the state health department in response to a call from a concerned citizen, although one said specifically that they would not. As with permitting agencies, several department of agriculture staff members said they try to work directly with producers to resolve issues. No interviewees in agriculture departments talked about enforcement or penalties that would be imposed by their department after investigating a reported concern. Most staff members said no records were kept of concerns; one said they kept records and another said records are kept only if they investigate the concern.

Potential Role of Health Departments

The majority of permitting agency staff members and all agriculture department personnel interviewed thought that health departments should play a role in resolving citizen concerns related to IFAP. Interviewees indicated that they believed health departments should respond to health related concerns, work on air quality issues, provide technical assistance, collect and disseminate relevant data, and/or participate and provide input when regulations are under consideration. A small number of permitting agency and department of agriculture interviewees were not sure what the role of health departments should be, since health departments did not have regulatory mandates to enforce. Some staff stressed the importance of health department involvement due to permitting and agriculture agencies' lack of health expertise. One permitting agency staff member stated, "If anyone's going to [address health issues], it would have to be the health departments. From our perspective, we don't really have expertise in that area," and another said, "Obviously the health department has more expertise in the health area than us." Some agriculture agency staff members thought that involving health departments would provide additional assurance to the public that departments of agriculture are unable to provide on health issues.

Additional Engagement

Education. We asked interviewees if they perform any health education activities related to "animal production farms," and the majority of permitting agency staff members said that they present information to producer/farmer groups in order to keep them informed on current regulations and requirements. A few talked about presenting to community groups when asked to do so, and that such presentations also addressed current regulatory requirements. Agriculture agency staff members also said that they provided information to farmers/producers and agriculture groups on current regulations, and at times presented information to community groups. No staff member, in permitting or agriculture

agencies, said that they provided information regarding potential health issues related to IFAP.

Data collection. In response to a query about collection of environmental monitoring or health data, most permitting agency participants said that their agencies did little or no routine data collection. Data collection was usually performed in response to a concern, and generally involved surface water and well testing. One interviewee said, "If we have a well water complaint we'll do sampling case by case. We need to be sure there is good evidence the well is tainted or everyone would want their well water tested at our expense." One staff member said that some permits they issue require water monitoring and they do some monitoring at inspections, and that the data they collect is used for internal purposes only. One permitting staff member said they collected data on air quality in the past because they were directed to do so by the state legislature and that that monitoring led to a rulemaking process for farm emissions. The legislature, however, ultimately prevented the rule from being finalized and implemented. No agriculture department staff member reported that they collected environment or health data, but one interviewee said their department was working with a university to look into air quality issues.

Contact with community/environment groups. All permitting agency personnel reported some contact with organizations or groups of citizens that work to address local animal production farm issues. A few interviewees said their contact with community groups primarily consisted of receiving comments when new operations were proposed or receiving repeated requests for increased monitoring and stricter regulations. In one state, citizens petitioned the U.S. EPA to have the permitting agency's delegated authority over CAFOs revoked because they believed the agency was not doing enough to regulate CAFOs in the state. At the time of the interview, the action was still pending and six new staff had been hired in response to the petition. Another interviewee said that environment groups serve as "watch dogs" and are "good partners from that standpoint" because they look for problems, conduct their own monitoring, and quickly contact the agency about problems so they can investigate. One staff member spoke about serving on an advisory committee with community representatives and another said a group working to revise their state's manure manual, a guidance document for producers, includes citizen groups.

Most agriculture department staff said community/environment groups contact them, and some said that they have worked with them to address issues of concern. One interviewee said there are fewer groups now due to improved regulations in their state, and another said contacts from these groups increase in frequency when new operations are proposed.

Barriers and Needs

While most interviewees appeared to consider their agency's current level of engagement with IFAP to be appropriate, many staff members did describe a number of barriers preventing them from implementing their current oversight efforts more fully and from expanding their efforts to monitor IFAP operations and respond to concerns.

Barriers. Permitting agency staff described the main barriers affecting their oversight of IFAP as limited budgets, staff size, and political factors. Some also mentioned that recent budget and staff reductions due to the economic downturn have affected their departments' ability to meet their regulatory obligations. Interviewees spoke about the effects of budget cuts on inspection frequency and farm visits, resources available for environmental monitoring, and ability of staff to conduct or attend in-person meetings and provide technical assistance. One staff member said "face to face meetings are much more effective when working with agriculture, and staff cutbacks are not good for that," and another said that they "would like to spend more time working with producers so they know the regulations and what causes water problems." Some interviewees said political factors have prevented them from more effectively regulating IFAP. For example, one permitting agency staff member explained that they had attempted to begin surface water monitoring at farms in response to concerns from environmental groups, but that financial and political issues had prevented this from moving forward. Only one agriculture agency staff member responded to an open-ended question regarding barriers with concerns about the agency's budget, but when agriculture departments were presented with a list of potential needs (Table 2), all staff members responded affirmatively regarding the need for increased funding.

The threat of legal action was an additional difficulty and barrier brought up by some permitting agency staff members and one agriculture department representative. The agriculture department interviewee said they "feel uncomfortable responding when a group brings in a lawyer" because they are not prepared to respond to legal action against the agency. The permitting agency participants who mentioned lawsuits said specifically that federal regulations are unclear about who needs a permit, and that they worry about lawsuits stemming from the lack of clarity.

Needs. We asked staff an open-ended question about items that could increase their agencies' effectiveness when addressing IFAP-related issues and further prompted them with a list of potential items (Appendix S1). The items most frequently identified as needs by interviewees are listed in Table 2. Some wording was adapted slightly in order to be more appropriate for and relevant to each agency. The most commonly expressed

needs/resource improvements identified by permitting agency staff were: increased funding, more staff for environmental health, and educational materials for distribution to producers or the general public.

Even though most agriculture agency staff members did not mention funding as a barrier in an open-ended question about barriers/needs, it was the only item on the list that all agriculture interviewees said would be helpful to their department. Other needs identified were: updated information from researchers regarding the health effects of concern and more staff.

Discussion

Our study reveals that sampled state permitting and agriculture agencies have taken limited actions to prevent and/or respond to public health concerns arising from IFAP operations. The main barriers identified that prevent further engagement include narrow or inadequate regulations, a lack of public health expertise within the agencies, and limited resources. There was widespread agreement among permitting and agriculture agency interviewees that health departments (HDs) should play a role in regulating IFAP operations, partly due to their own agencies' limited mandates and available expertise in public health. Yet previously published findings show limited involvement by local and state HDs due to political barriers and a lack of jurisdiction, expertise, and resources [36].

These results indicate a fragmented system to protect public health where no agency has ownership of monitoring or addressing the impact of IFAP on people's health. In short, HDs generally lack jurisdiction over IFAP operations [36] and permitting and agriculture agencies generally lack jurisdiction over and the capacity to address public health concerns. A

Table 2. Needs indicated by interviewees from list (Descending order).

Permitting Agency Staff

More staff dedicated to environmental health

Increased funding for environmental health

Educational materials for distribution

Funding specifically for animal production activities

Training for staff on public health issues relevant to animal production farms

Different political climate

Information on health effects of concern

Environmental quality tracking tools

Clearer federal regulations/guidelines (item added during study in response to data collected; not asked of all interviewees)

Connections to experts (i.e., university researchers)

Agriculture Department Staff

Increased funding

Updated information from researchers on health effects of concern

More staff

Funding specifically for animal production activities

Educational materials for distribution

Training for staff on issues relevant to animal production farms

Different political climate

Connections to experts

Environmental quality tracking tools

growing divide between environmental and public health agencies was identified in the 1990's as a trend that threatens public health protections [42]. Research has found that the main foci of environment agencies have shifted to permitting, enforcement, record keeping, and standard setting, and away from public health evaluations [43]. Our findings are consistent with these trends.

Ideally, a new role for HDs in responding to community concerns over IFAP operations would be defined by legislation aimed at remedying the lack of explicitly health-focused protections in current IFAP regulations. Unfortunately, this seems unlikely at present due to economic and political factors at the local, state, and federal levels [44]. In the absence of needed legislative and regulatory reforms, we suggest that partnerships between HDs and agriculture and permitting agencies could begin to improve the situation by including public health considerations in local and state-level decisions regarding the permitting, monitoring, and enforcement of regulations pertaining to IFAP operations. For example, greater agriculture and permitting agency collaboration with HDs could result in more comprehensive setback requirements for IFAP operations or the implementation of air quality standards for IFAP emissions linked to health concerns. Given the aforementioned support that several study participants voiced for greater involvement of HDs in IFAP regulation, agencies appear to be receptive to these partnerships. The drawback of this approach, as compared to new regulations, is that these relationships would have to be created one-by-one and maintained over time in the absence of regulations providing HDs with a defined role. We also suggest a potential approach whereby panels of experts on IFAP and public health could develop training programs and provide technical support to agencies interested in such partnerships in order to help foster their development and improve effectiveness. Future work should seek to examine evidence-based strategies that can inform and serve as models for these types of collaborations.

We did not seek to fully characterize how IFAP regulations are implemented by sampled agencies, but we are concerned that there is significant variability in inspection frequency among states. The inconsistency reflects varying requirements by state with no federal requirement for inspection frequency [31]. Further investigation is also needed of the practice of state agencies contacting farmer and agriculture trade groups for assistance when called about a concern. It is possible that this is an effective way to deal with minor issues; however, if problems that could result in fines and other regulatory action are routinely handled in an informal manner, the deterrent function of penalties for poor practices may be weakened and/or lost [45]. In addition, concerns about vulnerability to legal action stemming from a lack of clarity in federal regulations could cause agencies to be less aggressive in their enforcement of regulations. Additional research is needed to determine if this more informal approach to enforcement puts the environment and public health at greater risk than a formal approach to the enforcement of IFAP regulations.

Strengths and Limitations

To our knowledge, this is the most in-depth study to date investigating how state agencies with jurisdiction over IFAP operations respond to health concerns. The value of the findings is amplified when combined with previously published results on the engagement of HDs with this issue. Including multiple agencies that could be involved with IFAP and public health–health, permitting, and agriculture agencies–allowed us to compile a more comprehensive profile of a system to protect public health that has substantial gaps. This approach was especially critical since

previously published results showed that HDs routinely refer concerned citizens to permitting and agriculture agencies, with unknown outcomes for those referrals [36]. These findings may hold relevance for other issues characterized by problematic public health system gaps.

This study provides new information on a relatively unexplored topic; but the sample size was small, and the results cannot be interpreted as representing all agriculture and permitting agencies in the US. Also, we interviewed only one staff member per agency, and other staff members at the sampled agencies might have provided different responses to our questions. Finally, as with all studies of this nature, our sample reflects only those agencies that agreed to participate in the study.

Conclusion

In light of steadily increasing evidence regarding the multifaceted impact of IFAP on the public's health, it is crucial to examine how regulatory agencies respond to concerns and to understand what factors encourage and restrict agency staff from addressing health issues. A fragmented regulatory approach, narrow regulations, a lack of health expertise among agency staff, and limited resources are barriers preventing effective responses by sampled state permitting and agriculture agencies, despite the fact that jurisdiction over IFAP lies with these agencies. These findings are particularly troubling in light of prior research indicating that HDs also face multiple barriers to engagement with IFAP. The human health implications of IFAP operations have thus largely fallen by the wayside from a regulatory perspective, with rural communities suffering the consequences.

Given the near absence of explicit public health protections in current IFAP regulations, and the findings about responses from government agencies when issues are brought to their attention, there is a clear need for a more comprehensive public health response to IFAP. New regulations giving HDs a formal role in regulating IFAP and/or requiring public health experts on staff at regulatory agencies would go a long way toward addressing current gaps in the system. Short of new regulations, many actions could be taken to encourage capacity building and partnering among agencies. Future efforts should determine best practices for establishing these types of partnerships and evaluate their effectiveness in addressing the current disconnect between environmental regulations, public health, and IFAP. That said, voluntary efforts should not be seen as a replacement for much needed regulatory reforms. The public health implications of IFAP are increasingly clear, and regulations should ensure proper monitoring, oversight, and response by government agencies to protect public health.

Acknowledgments

We thank the permitting and agriculture agency staff members who participated in the study.

Author Contributions

Conceived and designed the experiments: JF LL CG KN RN. Performed the experiments: JF LL CG KN RN. Analyzed the data: JF LL. Wrote the paper: JF LL KN RN.

References

1. The Pew Commission on Industrial Farm Animal Production (2008) *Putting Meat on the Table: Industrial Farm Animal Production in America*. Pew Charitable Trusts and Johns Hopkins Bloomberg School of Public Health. Available: http://www.ncifap.org/_images/PCIFAPFin.pdf. Accessed 2013 April 12.

2. US Department of Agriculture. Census of Agriculture Quick Stats. Available: http://quickstats.nass.usda.gov/?source_desc = CENSUS. Accessed 2013 April 12.

3. O'Donoghue EJ, Hoppe RA, Banker DE, Ebel R, Fuglie K, et al. (2011) The Changing Organization of U.S. Farming. EIB-88. U.S. Department of Agriculture, Economic Research Service. Accessed 2013 April 12.

4. Schinasi L, Horton RA, Guidry VT, Wing S, Marshall SW, et al. (2011) Air Pollution, Lung Function, and Physical Symptoms in Communities Near Concentrated Swine Feeding Operations. Epidemiology 22: 208–215.

5. Thu K, Donham K, Ziegenhorn R, Reynolds S, Thorne P, et al. (1997) A control study of the physical and mental health of residents living near a large-scale swine operation. J Agric Saf Health 3: 13–26.

6. Merchant JA, Naleway AL, Svendsen ER, Kelly KM, Burmeister LF, et al. (2005) Asthma and farm exposures in a cohort of rural Iowa children. Environ Health Perspect 113: 350–356.

7. Radon K, Schulze A, Ehrenstein V, van Strien RT, Praml G, et al. (2007) Environmental exposure to confined animal feeding operations and respiratory health of neighboring residents. Epidemiology 18: 300–308.

8. Pavilonis BT, Sanderson WT, Merchant JA (2013) Relative exposure to swine animal feeding operations and childhood asthma prevalence in an agricultural cohort. Environ Research 122: 74–80.

9. Schiffman SS, Bennett JL, Raymer JH (2001) Quantification of odors and odorants from swine operations in North Carolina. Agr Forest Meteorol 108: 213–240.

10. Schenker MB, Christiani D, Cormier Y, Dimich-Ward H, Doekes G, et al. (1998) American Thoracic Society: Respiratory Health Hazards in Agriculture. Am J Resp Crit Care Med 158: S1–S76.

11. Wilson SM, Serre ML (2007) Examination of atmospheric ammonia levels near hog CAFOs, homes, and schools in Eastern North Carolina. Atmos Environ 41: 4977–4987.

12. Williams DA, Breysse P, McCormack M, Diette G, McKenzie S, et al. (2011) Airborne cow allergen, ammonia and particulate matter at homes vary with distance to industrial scale dairy operations: an exposure assessment. Environ Health 10: 72. doi:10.1186/1476-069X-10-72.

13. Schinasi L, Horton RA, Guidry VT, Wing S, Marshall SW, et al. (2011) Air Pollution, Lung Function, and Physical Symptoms in Communities Near Concentrated Swine Feeding Operations. Epidemiology 22: 208–215.

14. Trabue S, Scoggin K, Li H, Burns R, Xin H, et al. (2010) Speciation of volatile organic compounds from poultry production. Atmos Environ 44: 3538–3546.

15. Smith TJS, Rubenstein LS, Nachman KE (2013) Availability of Information about Airborne Hazardous Releases from Animal Feeding Operations. PLoS ONE 8(12): e85342. doi:10.1371/journal.pone.0085342.

16. Schiffman SS, Miller EA, Suggs MS, Graham BG (1995) The effect of environmental odors emanating from commercial swine operations on the mood of nearby residents. Brain Res Bull 37: 369–375.

17. Avery RC, Wing S, Marshall SW, Schiffman SS (2004) Odor from industrial hog farming operations and mucosal immune function in neighbors. Arch Environ Health 59: 101–108.

18. Nimmermark S (2004) Odour influence on well-being and health with specific focus on animal production emissions. Ann Agr Environ Med 11: 163–173.

19. Horton RA, Wing S, Marshall SW, Brownley KA (2009) Malodor as a trigger of stress and negative mood in neighbors of industrial hog operations. Am J Public Health 99: S610–615.

20. Wing S, Horton RA, Rose KM (2013) Air Pollution from Industrial Swine Operations and Blood Pressure of Neighboring Residents. Environ Health Perspect 121(1): 92–96.

21. Voss A, Loeffen F, Bakker J, Klaassen C, Wulf M (2005) Methicillin-resistant *Staphylococcus aureus* in Pig Farming. Emerg Infect Dis 11(12): 1965–1966.

22. Rinsky JL, Nadimpalli M, Wing S, Hall D, Baron D, et al. (2013) Livestock-Associated Methicillin and Multidrug Resistant *Staphylococcus aureus* Is Present among Industrial, Not Antibiotic-Free Livestock Operation Workers in North Carolina. PLoS ONE 8(7): e67641. doi:10.1371/journal.pone.0067641.

23. Casey JA, Curriero FC, Cosgrove SE, Nachman KE, Schwartz BS (2013) High-Density Livestock Operations, Crop Field Application of Manure, and Risk of Community-Associated Methicillin-Resistant *Staphylococcus aureus* Infection in Pennsylvania. JAMA Intern Med, 173(21): 1980–90. doi:10.1001/jamainternmed.2013.10408.

24. Osterberg D, Wallinga D (2004) Addressing externalities from swine production to reduce public health and environmental impacts. Am J Public Health 94: 1703–1708.

25. Burkholder J, Libra B, Weyer P, Heathcote S, Kolpin D, et al. (2007) Impacts of waste from concentrated animal feeding operations on water quality. Environ Health Perspect 115: 308–312.

26. Wing S, Cole D, Grant G (2000) Environmental injustice in North Carolina's hog industry. Environ Health Perspect 108: 225–231.

27. Wilson SM, Howell F, Wing S, Sobsey M (2002) Environmental injustice and the Mississippi hog industry. Environ Health Perspect 110: 195–201.

28. Mirabelli MC, Wing S, Marshall SW, Wilcosky TC (2006) Race, poverty, and potential exposure of middle-school students to air emissions from confined swine feeding operations. Environ Health Perspect 114: 591–596.

29. Freudenberg N, Pastor M, Israel B (2011) Strengthening Community Capacity to Participate in Making Decisions to Reduce Disproportionate Environmental Exposures. Am J Public Health 101(S1): S123–S130.

30. US Census Bureau (2010) Income, Poverty, and Health Insurance Coverage in the United States: 2009. Available: http://www.census.gov/prod/2010pubs/p60-238.pdf. Accessed 2013 July 29.

31. Hendrick S, Farquhar D (2008) Concentrated Animal Feeding Operations A Survey of State Policies. National Conference of State Legislatures. Available: http://www.ncifap.org/bin/k/w/CAFO_State_Survey_Introduction.pdf. Accessed 2013 April 12.

32. Environmental Council of the States. Organization of Environmental Agencies. Available: http://www.ecos.org/section/states/natural_resources_org. Accessed 2014 Jan 6.

33. National Association of State Departments of Agriculture. NASDA & Members: A General Overview. Available: http://www.nasda.org/File.aspx?id = 4120. Accessed 2014 Jan 6.

34. Centner TJ, Newton GL (2011) Reducing concentrated animal feeding operations permitting requirements. J Anim Sci 89(12): 4364–4369.

35. Koski C (2007) Examining State Environmental Regulatory Policy Design. J Environ Plann Man 50(4): 483–502.

36. Fry JP, Laestadius LI, Grechis C, Nachman KE, Neff RA (2013) Investigating the Role of State and Local Health Departments in Addressing Public Health Concerns Related to Industrial Food Animal Production Sites. PLoS ONE 8(1): e54720. doi:10.1371/journal.pone.0054720.

37. Lenaway D, Halverson P, Sotnikov S, Tilson H, Corso L, et al. (2006) Public Health Systems Research: Setting a National Agenda. Am J Public Health 96: 410–413.

38. Gerding J, Sarisky J (2011) Environmental Public Health Systems and Services Research. J Environ Health 73: 24–25.

39. Smith J, Bekker H, Cheater F (2011) Theoretical versus pragmatic design in qualitative research. Nurse Researcher 18(2): 39–51.

40. Avis M (2003) Do we need methodological theory to do qualitative research? Qual Health Res 13(7): 995–1004.

41. Maxwell JA (2010) Using Numbers in Qualitative Research. Qual Inq 16: 475–482.

42. Burke TA, Shalauta NM, Tran NL (1995) Strengthening the Role of Public Health in Environmental Policy. Policy Stud J 23(1): 76–84. doi: 10.1111/j.1541-0072.1995.tb00507.x.

43. Burke TA, Tran NL, Shalauta NM (1995) Who's in Charge? 50-State Profile of Environmental Health and Protection Services. Available: http://www.asph.org/uploads/whos2.pdf. Accessed 2014 Jan 6.

44. Johns Hopkins Center for a Livable Future (2013) Industrial Food Animal Production in America: Examining the Impact of the Pew Commission's Priority Recommendations. Available: http://www.jhsph.edu/research/centers-and-institutes/johns-hopkins-center-for-a-livable-future/_pdf/research/clf_reports/CLF-PEW-for%20Web.pdf. Accessed 2014 Jan6.

45. Zinn MD (2002) Policing Environmental Regulatory Enforcement: Cooperation, Capture, and Citizen Suits. Stan Envtl L J 21: 81.

18

Leaching Behavior of Heavy Metals and Transformation of Their Speciation in Polluted Soil Receiving Simulated Acid Rain

Shun-an Zheng[1,2,3]*, Xiangqun Zheng[1,2,3], Chun Chen[1,2,3]

1 Agro-Environmental Protection Institute, Ministry of Agriculture, Tianjin, People's Republic of China, 2 Key Laboratory of Production Environment and Agro-Product Safety, Ministry of Agriculture, Tianjin, People's Republic of China, 3 Tianjin Key Laboratory of Agro-Environment and Agro-Product Safety, Tianjin, People's Republic of China

Abstract

Heavy metals that leach from contaminated soils under acid rain are of increasing concern. In this study, simulated acid rain (SAR) was pumped through columns of artificially contaminated purple soil. Column leaching tests and sequential extraction were conducted for the heavy metals Cu, Pb, Cd, and Zn to determine the extent of their leaching as well as to examine the transformation of their speciation in the artificially contaminated soil columns. Results showed that the maximum leachate concentrations of Cu, Pb, Cd, and Zn were less than those specified in the Chinese Quality Standards for Groundwater (Grade IV), thereby suggesting that the heavy metals that leached from the polluted purple soil receiving acid rain may not pose as risks to water quality. Most of the Pb and Cd leachate concentrations were below their detection limits. By contrast, higher Cu and Zn leachate concentrations were found because they were released by the soil in larger amounts as compared with those of Pb and Cd. The differences in the Cu and Zn leachate concentrations between the controls (SAR at pH 5.6) and the treatments (SAR at pH 3.0 and 4.5) were significant. Similar trends were observed in the total leached amounts of Cu and Zn. The proportions of Cu, Pb, Cd, and Zn in the EXC and OX fractions were generally increased after the leaching experiment at three pH levels, whereas those of the RES, OM, and CAR fractions were slightly decreased. Acid rain favors the leaching of heavy metals from the contaminated purple soil and makes the heavy metal fractions become more labile. Moreover, a pH decrease from 5.6 to 3.0 significantly enhanced such effects.

Editor: Stephen J. Johnson, University of Kansas, United States of America

Funding: This work was financially supported by Central Public Research Institutes Basic Funds for Research and Development (Agro-Environmental Protection Institute, Ministry of Agriculture). The funders had no role in study design, data collection and analysis, decision to publish, or preparation of the manuscript.

Competing Interests: The authors have declared that no competing interests exist.

* E-mail: zhengshunan@gmail.com

Introduction

Acid rain has been a well-known environmental problem for decades and can lead to acidification of surface waters and soils. Acid deposition is formed from SO_2 and NO_x emitted to the atmosphere, largely because of fossil-fuel combustion. The most important sources are energy production, especially coal- and oil-fired power plants, and transportation sources, such as vehicles and ships. The air pollutants are transformed in the atmosphere to H_2SO_4 and HNO_3, transported across distances potentially as far as hundreds of kilometers, and deposited as precipitation (wet deposition) and as gas and particles (dry deposition) [1,2]. Acid deposition is also an environmental problem of increasing concern in China, where acid rain is mainly distributed in the areas of Yangtze River to the south, Qinghai-Tibet Plateau to the east, and in the Sichuan Basin. About 40% of the total territory of China is affected by the acid rain [3]. Sichuan Basin is one of the most severely hit-area of acid rain in southern China, where the rapid industrialization for last few decades has caused fast growth in sulfur emissions. Based on the monitoring data for 21 cities in Sichuan Province within the State-Controlled-Network of China, the number of cities with the annual average pH value of acid rain lower than 5.6 was 19. According to the environmental protection

and monitoring agencies in Sichuan Province, the direct economic loss due to acid rain is estimated to be U. S. $ 3 billion for one year [4].

Sichuan basin is also known as the "Red Basin" because it is mainly covered by red or purple rock series of the Trias–Cretaceous system, from which the purple soils, one of the most important soils for agricultural production in subtropical areas of China, are developed and formed. Purple soil, classified as Eutric Regosols in FAO Taxonomy or Pup-Calric-Entisol in the Chinese soil taxonomy, is typically characterized by thin soil horizons and inherited many of the characteristics of parent materials or rocks, such as its color ranging from purple to red. But other changes in purple soil properties have taken place as a result of land use changes, agricultural practices, or eco-environment disturbances [5,6]. Due to the soil background and human activities, the soil has been severely contaminated by heavy metals (commonly including Cu, Pb, Cd and Zn) in many areas, resulting in potential risk to local human health and environment [7,8].

There is accumulating evidence that acid rain is able to enhance metal mobilization in soil ecosystems. Previous studies demonstrated that the H^+ ion in the acidic water displaces the cations from their binding sites, reduces the cation exchange capacity

(CEC), and increases the concentrations of these cations in the soil-water system. The negatively charged sulfate and nitrate ions in the acid rain can act as "counter-ions", which allow cations to be leached from the soil. Through a series of chemical reactions, cations such as K^+, Na^+, Ca^{2+} and Mg^{2+} are leached out and become unavailable to plants as nutrients [9,10]. Likewise, toxic ions, such as Cu, Pb and Cd, usually bound to the negatively charged surface of soil particles can be displaced by H^+ ion too [11].

Leaching is the process by which contaminants are transferred from a stabilized matrix to liquid medium, such as water or other solutions. However, the influence of acid rain on the leaching behavior of heavy metals and transformation of their speciation in polluted purple soil is not investigated in detail. The objectives of this study were to evaluate the leaching of heavy metals Cu, Pb, Cd and Zn in a contaminated purple soil affected by simulated acid rain (SAR) over a range of pH, and to identify how simulated acid rain influences the chemical speciation of these metals in purple soil. Column experiments were used in this study to provide information about element release, transport in soil, and chemistry of soil and leachates.

Materials and Methods

Soil Sample Collection and Analysis

The analyzed soil sample was collected from a 0 cm to 20 cm layer of agricultural purple soil in the Pengzhou Agro-ecological Station of the Chinese Academy of Agricultural Sciences in Sichuan Province, China. The study area has a subtropical humid monsoon climate with an average annual precipitation of 850 mm to 1000 mm and an average annual temperature of 15°C to 16°C. The soil samples were air-dried, ground, and passed through a 2 mm sieve. Selected soil characteristics determined by standard methods [12] were 6.27 for pH, 0.64 $g \cdot kg^{-1}$ for $CaCO_3$, 26.51 $g \cdot kg^{-1}$ for organic matter, 20.81 $g \cdot kg^{-1}$ for CEC, 265.41 $g \cdot kg^{-1}$ for clay (<0.002 mm), 51.84 $mg \cdot kg^{-1}$ for total copper, 43.92 $mg \cdot kg^{-1}$ for total lead, 0.36 $mg \cdot kg^{-1}$ for total cadmium, 121.23 $mg \cdot kg^{-1}$ for total zinc.

Contaminated Soil Preparation

Artificially contaminated soils were composed of the collected purple soil. The load quantities of Cu [$Cu(NO_3)_2 \cdot 3H_2O$], Pb [$Pb(NO_3)_2$], Cd [$Cd(NO_3)_2 \cdot 4H_2O$], and Zn [$Zn(NO_3)_2 \cdot 6H_2O$] were as follows: Cu+Pb+Cd+Zn = 400 $mg \cdot kg^{-1}$+500 $mg \cdot kg^{-1}$+1 $mg \cdot kg^{-1}$+500 $mg \cdot kg^{-1}$. Based on the levels of polluted soil defined by the Chinese Environmental Protection Agency, the selected concentrations in this study represented moderately contaminated soils (Grade III of the National Soil Heavy Metals Standards GB15618-1995). The thoroughly mixed soil samples were stored and incubated at a 75% water holding capacity for 12 mon. The chemical speciation of aged soil was then determined by the sequential extraction procedure of Tessier et al. [13]. The chemical reagents, extraction conditions, and their corresponding fractions are defined as follows:

1. Exchangeable fraction (EXC): 2 g of the soil sample (oven-dry weight), 16 mL 1.0 $mol \cdot L^{-1}$ $MgCl_2$, pH 7, shake vigorously in a reciprocating shaker for 1 h, 20°C.

2. Carbonate-bound fraction (CAR): 16 mL of pH 5, 1.0 $mol \cdot L^{-1}$ sodium acetate, shake vigorously in the reciprocating shaker for 5 h, 20°C.

3. The Fe/Mn oxide-bound fraction (OX): 40 mL of 0.04 $mol \cdot L^{-1}$ $NH_4OH \cdot HCl$ in 25% (v/v) acetic acid at pH 3 for 5 h at 96°C with occasional agitation.

4. The OM-bound fraction: 6 mL of 0.02 $mol \cdot L^{-1}$ HNO_3 and 10 mL of 30% H_2O_2 (pH adjusted to 2 with HNO_3), water bath, 85°C for 5 h with occasional agitation. 10 mL of 3.2 $mol \cdot L^{-1}$ NH_4OAc in 20% (v/v) HNO_3, shake vigorously in the reciprocating shaker for 30 min.

5. Residual fraction (RES): Dried in a force-air oven at 40°C, 24 h. Subsamples after sieving with 0.149 mm openings were used for determining Cu, Zn, Pb and Cd contents.

Extractions were performed in 100 mL polypropylene centrifuge tubes. Between each successive extraction, the supernatant was centrifuged at 1500×g for 30 min and then filtered using a membrane filter (0.45-μm nominal pore size).

Simulated Acid Rain Preparation

Simulated acid rain was designed according to the main ion composition and pH of the local rain water (the pH of rainfall varied from 3.0 to 5.6) [4]. Synthetic acid rain with pH values of 3.0, 4.5, and 5.6 was prepared from a stock H_2SO_4–HNO_3 solution (4:1, v/v). The concentrations of Ca^{2+}, NH_4^+, Mg^{2+}, SO_4^{2-}, CO_3^{2+}, Cl^-, and K^+ in SAR were 1.5, 2.62, 1.00, 10.00, 2.61, 11.17, and 1.78 $mg \cdot L^{-1}$, respectively.

Leaching Experiment

The column (Fig. 1A) was oriented vertically and slowly saturated from the bottom with deionized water until it reached the field-holding capacity. The soil column was allowed to stabilize for 24 h. The feed solutions were composed of SAR at pH 3.0, 4.5, and 5.6, with the last group as the control. The feed solutions were then introduced into the system using a peristaltic pump to percolate through the packed soil columns at a flow rate of (60±5) $mL \cdot h^{-1}$, which corresponds to the field infiltration velocity of 3.0 $cm \cdot h^{-1}$ [14] (Fig. 1B). The redox potential was measured in one-third of the columns to check the aeration of columns and to avoid waterlogging conditions. Each column was flushed with 3000 mL (1530 mm) of the incoming solution, which corresponds to 1.5 yr of precipitation (rain) in the study area. Experiments were conducted in triplicate at each pH treatment.

The leachate from the soil column was filtered through a 0.45 μm membrane filter, and 200 mL of the filtered leachate (equivalent to 100 mm of added SAR) was sampled using glass collectors to determine the heavy metal concentrations, pH, and electrical conductivity. After the tests, the columns were separated and extruded. All of the soils from the entire depth range were thoroughly mixed, dried, and ground to analyze the chemical speciation of the heavy metal residues using the sequential extraction procedure [13].

Metal Determination and Quality Control

The Cu, Zn, Pb, and Cd concentrations in soil were determined by digesting 0.5 g of the soil samples (oven-dry weight) with a HNO_3–HF–$HClO_4$ mixture followed by elemental analysis. The concentrations of these metals in all the solutions were analyzed by graphite furnace atomic absorption spectrometry (AA220Z; Varian, USA). The detection limits for Cu, Zn, Pb, and Cd were 2, 2, 1, and 0.5 $\mu g \cdot L^{-1}$, respectively. All the reagents used for analysis were of analytical grade or higher. All the containers were soaked in 10% HCl, rinsed thoroughly in deionized water, and dried before use. The standard substances such as the geochemical standard reference sample soil in China (GSS-15) were used to examine the precision and accuracy of determination. The relative errors (REs) between the sum of the metal concentration in individual fractions and the measured total metal concentration in

A

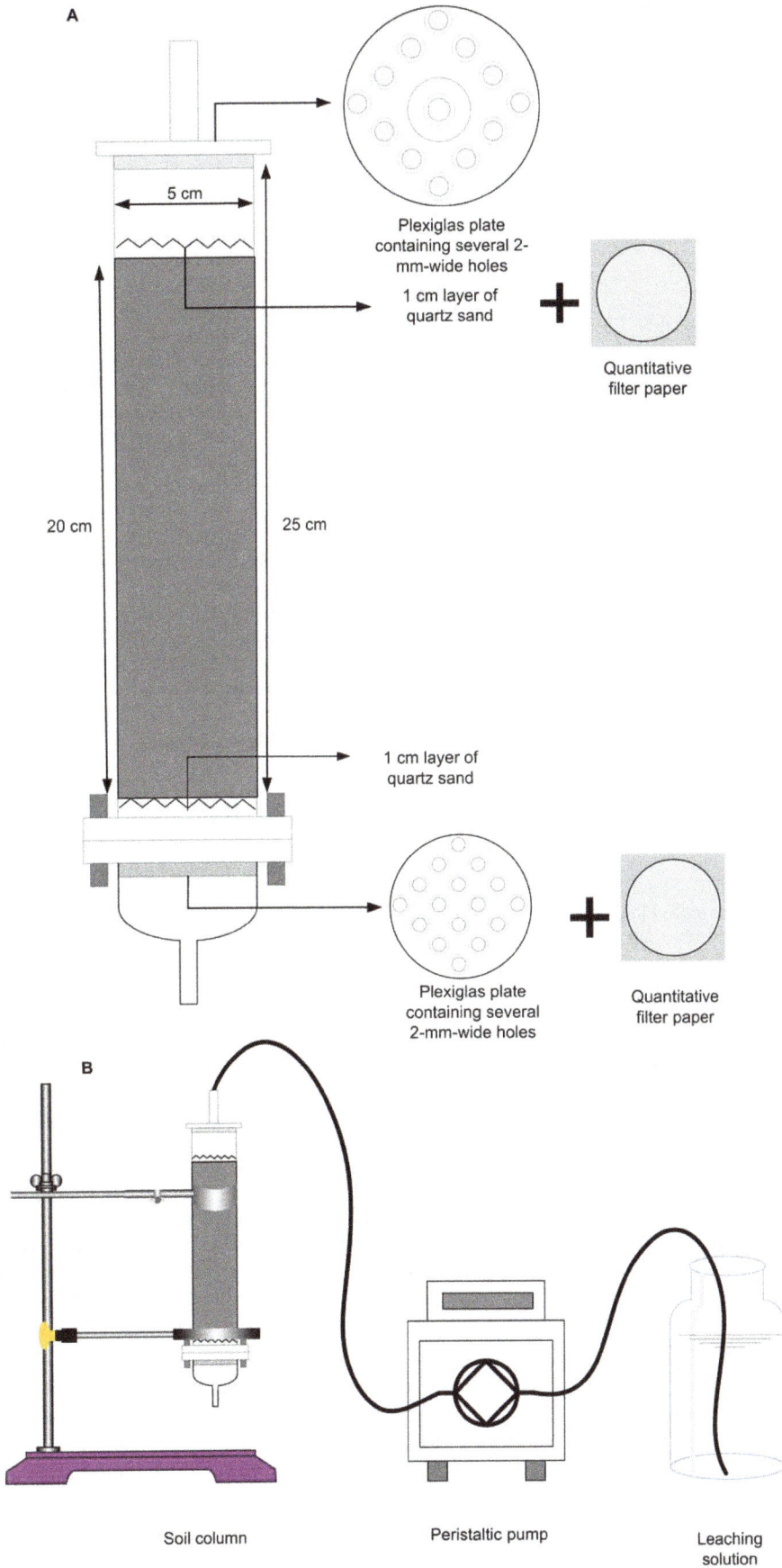

Plexiglas plate
containing several 2-
mm-wide holes

1 cm layer of
quartz sand

Quantitative
filter paper

5 cm

20 cm   25 cm

1 cm layer of
quartz sand

Plexiglas plate
containing several
2-mm-wide holes

Quantitative
filter paper

B

Soil column   Peristaltic pump   Leaching
solution

Figure 1. Leaching experimental design. (A) Schematic diagram of soil column. (B) Schematic analytical setup for the measurement of metal concentrations in the leaching experiment.

the soil samples, which ranged from –12.36% to 9.44%, were calculated to check the reliability of the sequential extraction procedure.

Statistical Analysis

Data were analyzed using the Origin 8.5.1 for Windows software at the 5% and 1% significance levels.

Results and Discussion

Heavy Metal Concentrations in the Leachate

The Pb and Cd concentrations in the leachate from SAR treatments at pH 3.0, 4.5, and 5.6 were generally very low throughout the leaching period, with most of them below their detection limits (data not shown). The maximum Pb and Cd concentrations in the leachate were only (17.0 ± 2.16) and (2.8 ± 0.42) $\mu g \cdot L^{-1}$, respectively, which are both less than the limits of the Chinese drinking water quality standards (50 and 10 $\mu g \cdot L^{-1}$, respectively). No significant differences were observed among the treatments of SAR at different pH levels. This result suggested that the polluted purple soil, which received SAR at a given pH in our study, may not cause groundwater contamination by Pb and Cd leaching. Low leachate concentrations of Pb and Cd can probably be attributed to the low mobility of Pb in soil as well as the comparatively low levels of Cd in artificially contaminated purple soil [15,16].

The Different results were observed with Cu and Zn. The leaching concentrations of Cu and Zn under SAR conditions are shown in Fig. 2. Higher Cu and Zn leachate concentrations were observed throughout the entire leaching period (ranging from 0.021 $mg \cdot L^{-1}$ to 1.49 $mg \cdot L^{-1}$ for Cu; 0.019 $mg \cdot L^{-1}$ to 2.34 $mg \cdot L^{-1}$ for Zn) as compared with those of Pb and Cd. However, the maximum Cu and Zn concentrations were still below the Chinese Quality Standards for Groundwater (Grade IV), thereby suggesting that leaching from the polluted purple soil under acid rain at the pH used in this study is unlikely to cause Cu and Zn contamination in water systems.

Fig. 2A shows the Cu concentrations in the leachates when SAR was used as the leaching reagent. Similar trends were observed for the changes in the Cu concentration of each treatment. The application of SAR produced a pulse of Cu in the leachates during the first stage of the experiment, with peaks of 1.49, 1.11, and 0.88 $mg \cdot L^{-1}$ at a pH of 3.0, 4.5, and 5.6, respectively. The concentrations decreased thereafter and reached <0.2 $mg \cdot L^{-1}$ at the end of the experiment. A small amount of Cu (<0.18 $mg \cdot L^{-1}$) was leached from the soil column in the first 200 mm of rainfall, which may be attributed to the required time for Cu to move from the upper surface layer to the bottom of the soil column and for the Cu-bound compounds to functionalize with the soil components. The SAR treatments at pH 3.0 and 4.5 yielded higher Cu concentrations than those in the control (SAR at pH 5.6), particularly before 1000 mm of SAR was added. The Cu leachate concentration increased as the pH values of SAR decreased. However, the Cu leachate concentrations for the treatments at pH 3.0 and 4.5 were approximately equal to those in the control as the leaching amounts were increased, thereby suggesting that the leachable Cu in all of the treatments was decreased. After 1200 mm of SAR was added, no significant differences were observed between the control and the treatments at pH 3.0 and 4.5.

The changes in the Zn leachate concentrations against the leaching amounts in the SAR treatments at different pH levels are shown in Fig. 2B. The variation curves of Zn with the three treatments were generally similar. Two peak values of the Zn concentration were found during the leaching experiment. The first peak appeared after approximately 300 mm to 500 mm of continuous leaching, and the second one appeared after 1200 mm of SAR was added. The peak values of the Zn concentration in the leachates were 2.34, 1.92, and 1.46 $mg \cdot L^{-1}$ for the treatments at pH 3.0, 4.5, and 5.6, respectively. These results suggested that the release of Zn from the contaminated purple soil can be observed in two stages. In the first leaching stage, leached Zn mainly existed as water soluble and exchangeable fractions. After being leached out, these mobile Zn began to adsorb and desorb in the soil column with the downward movement of leachate, and after 300–500 mm rainfall Zn concentration in the leachate reached the first peak value. During the second stage, Zn was released from different fractions such as the CAR-, OX-, and OM-bound Cu. A comparable amount of Zn was shifted from those fractions, which revealed the second peak values after 1200 mm of rainfall was added. Many studies have demonstrated that Zn in EXC and the water soluble fraction usually account for only 0.5% to 7% of the total Zn concentration [17–19], which could explain why the second peak values were higher than the first ones.

Total Leaching Amounts of Heavy Metals

An analysis of the total leaching amounts of heavy metals can directly reflect the leaching strength of the said heavy metals. The total leaching amounts of Pb and Cd were not calculated because most of their concentrations in the leachate were below the detection limits. The total leaching amounts of Cu and Zn in each treatment are shown in Table 1. Two-way ANOVA showed that the total amounts of the leached metals from the soil columns were significantly affected by the pH level and metal species as well as the interaction between these two factors (Table 2). Zn leached more easily from the soil column than Cu at each pH level (Tables 1 and 2). Similarly, Guo et al. [20] revealed that more Zn flowed out of the soil column than Cu in acidic soils under SAR treatments. The amount of leached metals increased at pH 3.0 and 4.5 as compared with that in the controls (pH 5.6), whereas the magnitude of this increase depended on the metal. Metallic elements become more soluble (leachable) under acidic conditions [21]. pH may likewise control the nature of the interactions between metals and soil surfaces [22]. Thus, higher Cu and Zn concentrations were induced by SAR and dissolved in columns at lower pH (Table 1). The H+ ion in acid rain displaces the cations from their binding sites, which causes the increased amount of heavy metal desorption, as shown in the following reactions [Eq. (1–2)]:

$$SOH \cdots X^{2+} + H^+ \leftrightarrow SOH_2 + X^{2+} \tag{1}$$

$$SOX^{2+} + H^+ \leftrightarrow SOH + X^{2+} \tag{2}$$

The different affinities of the binding sites to metals may cause the different rates and amounts of the desorbed metal. For Zn, greater amounts were released (Table 1). H+ ions can only replace

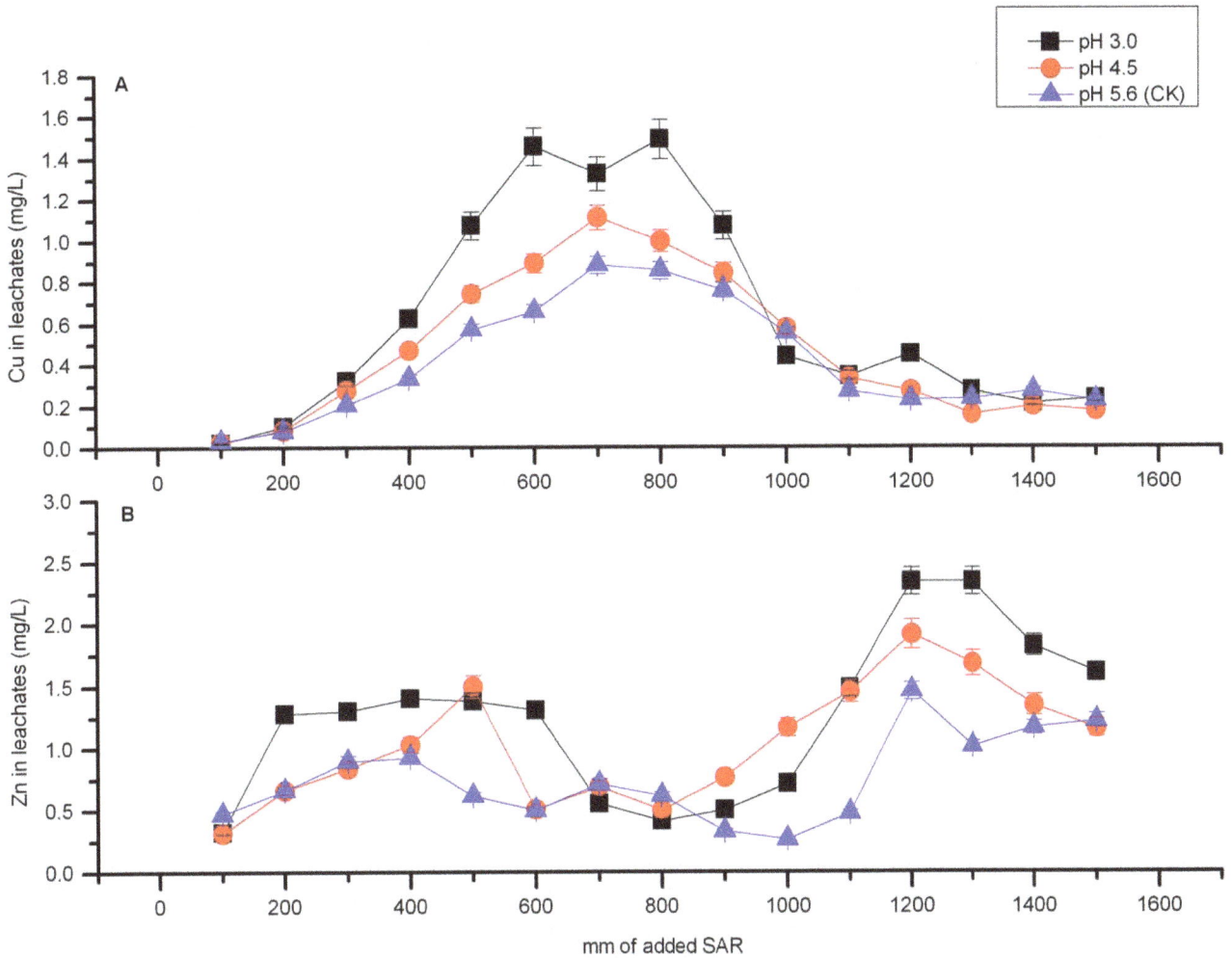

Figure 2. Metal concentrations in leachates as functions of addition of SAR. The concentrations of Cu and Zn in the leachates as functions of addition of SAR at different pHs. (A) Copper. (B) Zinc.

some metal cations, whereas the surface OH^- groups attached to the Fe or Al atoms can accept protons from the solution, thereby increasing the positive surface charges via the protonation of mineral surfaces [23,24]. Protonation is a very rapid reaction during the initial stage and depends on the amounts and properties of ferric oxides.

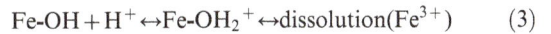

$$Fe\text{-}OH + H^+ \leftrightarrow Fe\text{-}OH_2{}^+ \leftrightarrow dissolution(Fe^{3+}) \qquad (3)$$

The surface protonation in soil promotes the dissolution of ferric oxides in acidic solutions [Eq. (3)], which exhibit high adsorptive capacities for heavy metals and consequently increase the

Table 1. Total amounts of the metals leached from the soil columns in mg of metal per kg of soil.

Treatment	Total leaching amounts (mg·kg⁻¹)	
	Cu	Zn
pH 3.0	(3.93±0.44) a	(7.81±0.61) a
pH 4.5	(2.98±0.36) b	(6.47±0.47) b
pH 5.6	(2.27±0.39) c	(4.72±0.33) c

Note: The values are means ± standard deviation. Different lower case letters show significant differences in the same treatment (ANOVA/LSD, $P<0.05$).

Table 2. Two-way ANOVA of pH levels and metal species effects for total amounts of the metals leached from the soil columns.

Source of variance	SS	MS	F
pH	48.22	48.22	245.75**
Metal species	16.93	8.46	43.14**
pH × metal species	1.64	0.82	4.18*
Error	2.35	0.20	
Total	69.14		

Note: **significant at 99% probability level, *significant at 95% probability level.

solubilization of these metals in their oxide fractions.

Chemical Speciation of Heavy Metals in Purple Soil Affected by SAR

The chemical fractions of Cu, Pb, Cd, and Zn in soils before and after the column leaching tests are shown in Fig. 3. For the original contaminated purple soil, Cu and Cd were dominantly associated with RES (37% to 41%), followed by OM (31% to 33%). Generally, EXC- and CAR-bound Cu and Cd accounted for <9% of the total amount of the respective metals. A significant fraction of Pb in the soils was bound in RES (66.31%), and four fractions accounted for <35% of the total Pb. Some studies have shown [15,25] that Pb is mostly present in the RES of soil at the surface or profile scale and is widely considered to exhibit very low geochemical mobility. The amount of Zn in soils was mainly associated with RES (42.69%), then with the OX (25.96%), followed by the CAR (14.44%), and OM (12.23%). The percent of exchangeable Zn in the soils was relatively low (4.69%).

Some variations in the proportions of the four metals in five fractions after the tests are shown by Fig. 3. The relative concentrations of the elements in the EXC and OX fractions were generally increased after the leaching experiment was conducted for the three treatments, whereas those of RES, OM,

and CAR fractions were slightly decreased. The Pb and Cd concentrations in EXC ranged from 3.28% to 4.39% and 4.47% to 6.79%, respectively. By contrast, these metals increased by more than 3% in OX. The increase in the Cu concentrations in EXC was greater than that in the OX (EXC 3% to 7% vs. OX <2%). The relative concentrations of Zn in OM and OX significantly increased, particularly in the EXC fraction (average 4.32%).

The EXC fraction was usually the first to be brought into the solution and is readily available for plant uptake. The four trace elements in EXC, with high bioavailability and mobility, were increased from ~1.2% to 5.03%, which indicated the increased direct risk of Cu, Pb, Cd, and Zn contamination in the soil/groundwater system as caused by acid rain. The amounts of the non-residual fractions represent the amounts of potentially active trace elements [25]. Generally, the high proportion of trace elements in the non-residual fractions of soils may suggest the large contributions of anthropogenic elements. The non-residual fractions of Cu, Pb, Cd, and Zn in the soil zones after the leaching tests averaged at 62.55%, 37.43%, 65.95%, and 61.61%, respectively. The non-residual fractions of Cu, Pb, Cd, and Zn in the original soil samples averaged at 59.75%, 33.69%, 62.04%, and 57.31%, respectively. The increased non-residual fractions represented the increased potential risk of Cu, Pb, Cd, and Zn contamination in the soil/groundwater system. Simple correlation

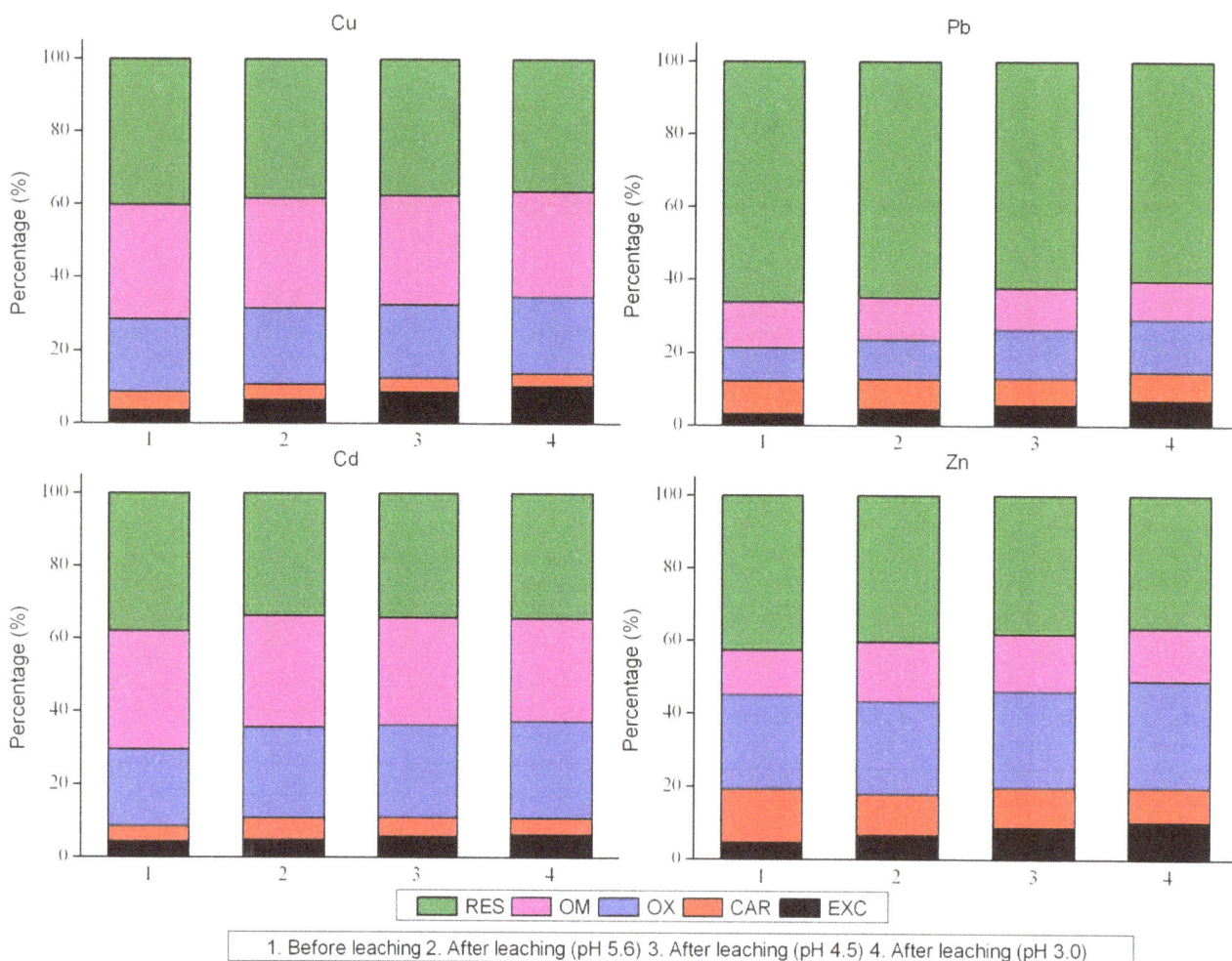

1. Before leaching 2. After leaching (pH 5.6) 3. After leaching (pH 4.5) 4. After leaching (pH 3.0)

Figure 3. Variation of the percentage of metals in different fractions. Species distribution of Cu, Pb, Cd and Zn extracted with Tessier scheme before and after the column tests.

analysis indicated that the pH was significantly correlated with the exchangeable heavy metal content ($r = 0.974$, $P < 0.01$). Likewise, the pH was significantly correlated with the non-residual heavy metal content ($r = 0.968$, $P < 0.01$). These data suggested that acid rain favors the leaching of heavy metals from the contaminated purple soil. Consequently, acid rain can affect the mobilization potential of heavy metals. The decrease in the pH from 5.6 to 3.0 significantly enhanced such effects of acid rain, which are likely to cause more serious harmful effects on soil-vegetation systems.

Conclusions

The maximum leachate concentrations of Cu, Pb, Cd, and Zn from the soil that received SAR were less than those in the Chinese Quality Standard for Groundwater (Grade IV). Moreover, most of the Pb and Cd leachate concentrations were below their detection limits. By contrast, higher Cu and Zn leachate concentrations were observed because these metals were released in higher amounts as compared with Pb and Cd. The differences in the Cu and Zn leachate concentrations between the control (SAR at pH 5.6) and the treatments (SAR at pH 3.0 and 4.5) were significant. Similar trends were observed in the total leaching amounts of Cu and Zn. The proportions of Cu, Pb, Cd, and Zn in the EXC and OX

fractions were generally increased after the leaching experiment at three pH levels, whereas those of the RES, OM, and CAR fractions were slightly decreased. Acid rain favors the leaching of heavy metals from the contaminated purple soil and causes the heavy metal fractions to become more labile. A decrease in the pH from 5.6 to 3.0 significantly enhanced such effects. However, the experiment is only an indoor simulation that used artificially contaminated purple soil. Thus, long-term field experiments on soils contaminated with acid rain should be conducted to study the cycling effect of heavy metal residues in the soil or those that leach into the groundwater.

Acknowledgments

The authors are especially grateful to Dr. Yirong Guo for critical review and perceptive comments on the manuscript. The comments and suggestions of the editor and two anonymous referees are appreciated.

Author Contributions

Conceived and designed the experiments: SZ. Performed the experiments: SZ CC. Analyzed the data: SZ XZ. Contributed reagents/materials/analysis tools: XZ. Wrote the paper: SZ.

References

1. Larssen T, Lydersen E, Tang D, He Y, Gao J, et al. (2006) Acid rain in China. Environ Sci Technol. 40: 418–425.
2. Singh A, Agrawal M (2008) Acid rain and its ecological consequences. J Environ. Biol. 29: 15–24.
3. Xie Z, Du Y, Zeng Y, Li Y, Yan M, et al. (2009) Effects of precipitation variation on severe acid rain in southern China. J Geogr Sci. 19: 489–501.
4. Ma L, Wang B, Yang J (2008) Spatial- Temporal Distribution of Acid Rain in Sichuan Province. Envion Sci Manage. 33: 26–29.
5. Li ZB, Li P, Han JG, Li M (2009) Sediment flow behavior in agro-watersheds of the purple soil region in China under different storm types and spatial scales. Soil Till Res. 105: 285–291.
6. Gao Y, Zhu B, Zhou P, Tang JL, Wang T, et al. (2009) Effects of vegetation cover on phosphorus loss from a hillslope cropland of purple soil under simulated rainfall: a case study in China. Nutr Cycl Agroecosys. 85: 263–273.
7. Wu F, Yang W, Zhang J, Zhou L (2010) Cadmium accumulation and growth responses of a poplar (Populus deltoids×Populus nigra) in cadmium contaminated purple soil and alluvial soil. J Hazard Mater. 177: 268–273.
8. Li Q, Wei C, Huang Y, Wang L, Wang D (2010) A study on the characteristics of heavy metals in orange ecosystem. Chin J Geochem. 29: 100–106.
9. Zhang JE, Ouyang Y, Ling DJ (2007) Impacts of simulated acid rain on cation leaching from the Latosol in south China. Chemosphere 67: 2131–2137.
10. Ling DJ, Huang QC, Ouyang Y (2010) Impacts of simulated acid rain on soil enzyme activities in a latosol. Ecotoxicol Environ Saf. 73: 1914–1918.
11. Wang M, Gu B, Ge Y, Liu Z, Jiang D, et al. (2009) Different responses of two Mosla species to potassium limitation in relation to acid rain deposition. J Zhejiang Univ-Sci B. 10: 563–571.
12. Lu RK (2000) Methods of soil and agrochemical analysis (in Chinese). Beijing: China Agricultural Science and Technology Press.

13. Tessier A, Campbell P, Bisson M (1979) Sequential extraction procedure for the speciation of particulate trace metals. Anal Chem. 51: 844–851.
14. Grolimund D, Borkovec M, Barmettler K, Sticher H (1996) Colloid-facilitated transport of strongly sorbing contaminants in natural porous media: a laboratory column study. Environ Sci Technol. 30: 3118–3123.
15. Martinez CE, Motto HL (2000) Solubility of lead, zinc and copper added to mineral soils. Environ Pollut. 107: 153–158.
16. Lee SZ, Chang L, Yang HH, Chen CM, Liu MC (1998) Adsorption characteristics of lead onto soils. J Hazard Mater. 63: 37–49.
17. Kabala C, Singh BR (2001) Fractionation and mobility of copper, lead, and zinc in soil profiles in the vicinity of a copper smelter. J Environ Qual. 30: 485–492.
18. Ma LQ, Rao GN (1997) Chemical fractionation of cadmium, copper, nickel, and zinc in contaminated soils. J Environ Qual. 26: 259–264.
19. Xiang HF, Tang HA, Ying QH (1995) Transformation and distribution of forms of zinc in acid, neutral and calcareous soils of China. Geoderma 66: 121–135.
20. Guo ZH, Liao BH, Huang CY (2005) Mobility and speciation of Cd, Cu, and Zn in two acidic soils affected by simulated acid rain. J Environ Sci. 17: 332–334.
21. Wilson MJ, Bell N (1996) Acid deposition and heavy metal mobilization. Appl Geochem. 11: 133–137.
22. Zhang H, Davison W, Tye AM, Crout NMJ, Young SD (2006) Kinetics of zinc and cadmium release in freshly contaminated soils. Environ Toxicol Chem. 25: 664–670.
23. Zhu MX, Jiang X, Ji GL (2005) Investigation of time-dependent reactions of H^+ ions with variable and constant charge soils: a comparative study. Appl Geochem. 20: 169–178.
24. Zhang FS, Zhang XN, Yu TR (1991) Reactions of hydrogen ions with variable charge soils: I. Mechanisms of reaction. Soil Sci. 151: 436–442.
25. Tack F, Verloo MG (1995) Chemical speciation and fractionation in soil and sediment heavy metal analysis: a review. Int J Environ An Ch. 59: 225–238.

19

A Combined Proteomic and Transcriptomic Analysis on Sulfur Metabolism Pathways of *Arabidopsis thaliana* under Simulated Acid Rain

Tingwu Liu[1,2ꝰ], **Juan A. Chen**[1ꝰ], **Wenhua Wang**[1,3], **Martin Simon**[1], **Feihua Wu**[1], **Wenjun Hu**[1], **Juan B. Chen**[1], **Hailei Zheng**[1,4]*

1 Key Laboratory of the Ministry of Education for Coastal and Wetland Ecosystems, College of the Environment and Ecology, Xiamen University, Xiamen, Fujian, P. R. China, **2** Department of Biology, Huaiyin Normal University, Huaian, Jiangsu, P. R. China, **3** Department of Biology, Duke University, Durham, North Carolina, United States of America, **4** State Key Laboratory of Marine Environmental Science, Xiamen University, Xiamen, Fujian, P. R. China

Abstract

With rapid economic development, most regions in southern China have suffered acid rain (AR) pollution. In our study, we analyzed the changes in sulfur metabolism in *Arabidopsis* under simulated AR stress which provide one of the first case studies, in which the systematic responses in sulfur metabolism were characterized by high-throughput methods at different levels including proteomic, genomic and physiological approaches. Generally, we found that all of the processes related to sulfur metabolism responded to AR stress, including sulfur uptake, activation and also synthesis of sulfur-containing amino acid and other secondary metabolites. Finally, we provided a catalogue of the detected sulfur metabolic changes and reconstructed the coordinating network of their mutual influences. This study can help us to understand the mechanisms of plants to adapt to AR stress.

Editor: Keqiang Wu, National Taiwan University, Taiwan

Funding: This study was financially supported by the Natural Science Foundation of China (NSFC) (30930076, 31260057, 30770192 and 30670317), the Scholarship Award for Excellent Doctoral Student granted by Ministry of Education, the Foundation of the Chinese Ministry of Education (20070384033, 209084), the Program for New Century Excellent Talents in Xiamen University (NCETXMU X07115) and a Changjiang Scholarship (X09111). The funders had no role in study design, data collection and analysis, decision to publish, or preparation of the manuscript.

Competing Interests: The authors have declared that no competing interests exist.

* E-mail: zhenghl@xmu.edu.cn

ꝰ These authors contributed equally to this work.

Introduction

Acid rain (AR), as a worldwide environmental issue, has been a serious global problem for several decades, especially in southern China [1]. As for plants, it has caused a series of damages, such as necrosis, thin crown, premature abscission, branch dieback, and has been treated as a new abiotic stress factor [2–4]. Acid rain is formed from SO_2 and nitrous oxides (NOx) emitted to the atmosphere, largely due to fossil-fuel combustion [5]. Different from other regions in the world, AR in China contains a lot of sulfate [1] due to the aggravated combustion of ubiquitous sulfur-containing coal [6]. As a result of significant emissions and subsequent deposition of sulfur (S), widespread AR is observed in southern and southwestern China [1]. However, studies are rarely focused on the plant's response in S metabolism to AR, and molecular details of this process are poorly understood [7].

S is an essential mineral element that is required in large amount in plants, animals, and microorganisms [8]. It is uptaken as sulfate and is then assimilated into organic compounds. S is found in two amino acids including cysteine (Cys) and methionine (Met), in oligopeptides including glutathione (GSH) and phytochelatins, in some vitamins and cofactors including biotin, molybdenum cofactor, thiamine and coenzyme A, in phytosulfokin hormones and in a variety of secondary products, all of which

are essential in plant nutrition [9]. Finally, S is integrated into some S-containing proteins. S also plays a critical role in catalytic and electrochemical functions in these biomolecules. Disulfide bonds between polypeptides, mediated by Cys, are of great importance in protein assembly and structure [10]. The regulation of sulfate uptake and assimilation has been dissected in great detail [11–13], and dynamic adaptations of the integrative gene-metabolite network in response to S deficiency have been deciphered [12,14,15].

Proteomic, transcriptomic, and metabolomic approaches can provide the comprehensive profiles of large numbers of gene expression products [16]. The use of these approaches to obtain comprehensive data sets increased rapidly in recent years, especially with respect to the mechanisms underlying plant growth and plant responses to stress [14,17]. The new high-throughput tools have provided the potential to systematically analyze biological systems and monitor their responses. By conceiving the network architecture and thus the interrelation and regulation of its components, it can be envisioned that it will be possible to comprehend the whole system.

In the present study, we explored whole-cellular processes of S metabolism at the levels of transcriptome and proteome in *Arabidopsis* under AR stress by applying a DNA array and a combination of proteomic and transcrpimic analysis. We depicted a whole picture for the changes of plant S metabolism under AR by combining an amount of multidimensional data. These data can provide novel indications as to reveal the response of the processes related to S metabolism to AR at the levels of the transcriptome and proteome.

Materials and Methods

Plant Materials and Growth Conditions

Seeds of *Arabidopsis thaliana*, ecotype Columbia-0 (Col-0) were planted in the mixed matrix with vermiculite and cover soil (2:1) after vernalization. Then, plants were grown in controlled growth chamber with a light/dark regime of 16/8 hr, temperature of 23/20°C and a light intensity of 150 μmol m^{-2} s^{-1} photosynthetically active radiation (PAR). After 3 weeks, the seedlings were sprayed by simulated acid rain (AR, pH 3.0) at 5 ml per seedling, meanwhile, the seedlings were sprayed with control solution (CK, pH 5.6) which had the same ion composition as AR. The AR solution was prepared from H_2SO_4 and HNO_3 in the ratio of 5 to 1 by chemical equivalents, which represents the average ion composition of rainfall in South China [18]. The final concentrations of H_2SO_4 and HNO_3 in the spray solution were 0.45 and 0.09 mM, respectively. The leaves were collected after AR treatment for 3 days and they were immediately frozen in liquid nitrogen (N_2) and stored at 70°C for subsequent protein/RNA extraction and enzyme for protein and RNA extraction assays. The phenotype of the treated and control groups were shown in Figure S1 after AR treatment. Each experiment was repeated at least three times.

Microarray Analysis

For Affymetrix GeneChip analysis, the materials were treated the same as described above. 20 mg of total RNA from leaves of *Arabidopsis* with or without AR treatment was extracted using the RNeasy plant mini kit (Qiagen), and the product was used to make biotin-labeled cRNA targets. The Affymetrix *Arabidopsis* ATH1 genome array GeneChip, which contains >22,500 probe sets representing-24,000 genes, was used. Hybridization, washing, and staining were performed according to the manufacturer's instructions. Image processing was performed using Affymetrix Gene-Chip Operating System (GCOS). Normalization and expression estimate computation were calculated from the. CEL output files from the Affymetrix GCOS 1.1 software using RMA implemented in R language using standard settings. Statistical testing for differential expression was performed with logic-t analysis. All microarray expression data are available at the Gene Expression Omnibus under the series entry GSE52487. Functional categories were assigned to genes using the AGI number to search the MIPS database (http://mips.gsf.de/cgi-bin/proj/thal/) and the *Arabidopsis* Information Resource website, TAIR (http://www.arabidopsis.org/).

Total Protein Extraction and Two-dimensional Electrophoresis

Proteins were extracted under denaturing conditions, according to the phenol procedure [19]. Briefly, one gram of frozen lyophilized tissue powder was re-suspended in 3 mL ice-cold extraction buffer (100 mM PBS, pH 7.5) containing 100 mM EDTA, 1% PVPP w/v, 1% Triton X-100 v/v, 2% b-mercaptoethanol v/v. After centrifugation at 4°C, 15,000 g for 15 min, the upper phase was transferred to a new centrifuge tube. Two volumes of Tris-saturated phenol (pH 8.0) were added and then the mixture was further vortexed for 10 min. Proteins were precipitated by adding five volumes of ammonium sulfate

saturated-methanol, and incubating at −20°C for at least 4 h. After centrifugation as described above, the protein pellet was resuspended and rinsed with ice-cold methanol followed by ice-cold acetone twice, and spun down at 15,000 g and 4°C for 5 min after each washing. Finally, the washed pellets were air-dried and recovered with lysis buffer containing 7 M urea, 2 M thiourea, 2% CHAPS, 13 mM DTT and 1% IPG buffer. The sample containing 800 μg of total proteins was subsequently loaded onto an IPG strip holder with length 17 cm, pH 4–7 linear gradient IPG strips (GE Healthcare, Sweden), and rehydrated for 24 h at room temperature. Strips were covered with mineral oil to prevent evaporation. Then IEF was performed as the following: 300 V for 1 h, 600 V for 1 h, 1000 V for 1 h, a gradient to 8000 V for 2 h, and kept at 8000 V for 64,000 V·h. After focusing, the strips were equilibrated with equilibration solution (50 mM Tris, pH 8.8, 6 M urea, 30% glycerol, 2% SDS) containing 1% DTT, and subsequently 4% iodoacetamide for 15 min for each equilibration solution. The separation of proteins in the second dimension was performed with SDS polyacrylamide gels (12%) on an Ettan DALT System (GE Healthcare, Sweden) and sealed in with 0.5% agarose, and run at 10 mA for electrophoresis. Each separation was repeated 3 times to ensure the protein pattern reproducibility.

Protein Staining and Image Analysis

The SDS-PAGE gels were stained by the CBB R250. 2-DE gels were scanned at 600 dots per inch (dpi) resolution with a scanner (Uniscan M3600). 2-D gel analysis was performed by PDQuest software (Bio-Rad). For each gel, a set of three images was generated, corresponding to the original 2-D scan, the filtered image, and the Gaussian image. The Gaussian image containing the three-dimensional Gaussian spots was used for the quantification analysis. The intensity of each protein spot was normalized relative to the total abundance of all valid spots. After normalization and background subtraction, a matchset was created by comparing the control gels. All spots were then submitted to further analysis to test whether or not their expression levels were affected by AR treatment and those that increased or decreased significantly more than 2-fold change were then identified by MALDI TOF/MS. The apparent *Mr* of each protein in gel was determined with protein markers.

Protein Identification

Excised gel spots were washed several times with destaining solutions (25 mM NH_4HCO_3 for 15 min and then with 50% v/v ACN containing 25 mM NH_4HCO_3 for 15 min). Gel pieces were dehydrated with 100% ACN and dried, then incubated with a reducing solution (25 mM NH_4HCO_3 containing 10 mM DTT) for 1 h at 37°C, and subsequently with an alkylating solution (25 mM NH_4HCO_3 containing 55 mM iodoacetamide) for 30 min at 37°C. After reduction and alkylation, gels were washed several times with the destaining solutions and finally with pure water for 15 min, before dehydration with 100% ACN. Depending on protein amount, 2–3 μL of 0.1 mg μL^{-1} modified trypsin (Promega, sequencing grade) in 25 mM NH_4HCO_3 was added to the dehydrated gel spots. After 30 min incubation, 7 μL of 25 mM NH_4HCO_3 were added to submerge the gel spots at 37°C overnight.

After digestion, the protein peptides were collected and vacuum-dried. 0.5 μL peptide mixture was mixed with 0.5 μL matrix solution (HCCA at half saturation in 60% ACN/0.1% TFA v/v). A total of 1 μL of reconstituted in-gel digest sample was spotted initially on Anchorchip target plate. The dried sample on the target plate was washed twice with 1 μL of 0.1% TFA, left for 30 s before solvent removal. MALDI TOF MS analysis (Re-

FlexTMIII, Bruker) was used to acquire the peptide mass fingerprint (PMF). The spectra were analyzed with the flexAnalysis software (Bruker-Daltonics). All spectra were smoothed, and internally calibrated with trypsin autolysis peaks. Then, the measured tryptic peptide masses were transferred through MS BioTool program (Bruker-Daltonics) as inputs to search against the taxonomy of *Arabidopsis thaliana* (thale cress) in NCBI (NCBInr) database. The PMF searched parameters were 100 ppm tolerance as the maximum mass error, MH$^+$ monoisotopic mass values, allowance of oxidation (M) modifications, allowed for one missed cleavage, and fixed modification of cysteine by carboxymethyl (Carbamidomethylation, C). The match was considered in terms of a higher Mascot score, the putative functions, and differential expression patterns on 2-DE gels. Good matches were classified as those having a Mascot score higher than 60 (threshold). The identification was considered only with a higher MASCOT score, maximum peptide coverage and additional experimental confirmation of the protein spots on the 2-DE gels. The identified proteins were searched within the UniProt and TAIR database to find out if their function was known, then they were further classified using Functional Catalogue software (http://mips.gsf.de/projects/funcat).

Real-time Quantitative PCR

Verification of differential gene expression was performed by real-time quantitative PCR (qRT-PCR) in the Rotor-GeneTM 6000 real-time analyzer (Corbett Research, Mortlake, Australia) using the FastStart Universal SYBR Green Master (ROX, Roche Ltd., Mannheim, Germany) according to the manufacturer's instructions. Reaction conditions (10 µL volumes) were optimized by changing the primer concentration and annealing temperature to minimize primer-dimer formation and to increase PCR efficiency. The following PCR profile was used: 95°C for 5 min, 40 cycles of 95°C for 30 s, the appropriate annealing temperature for 30 s and 72°C for 30 s, a melting curve was then performed to verify the specificity of the amplification. Each run included standard dilutions and negative reaction controls. Successive dilutions of one sample were used as a standard curve. All the results presented were standardized using the housekeeping gene *Actin2*. The results of the mRNA expression level of genes were expressed as the normalized ratio using the $_{\Delta\Delta}$Ct method according to Livak and Schmittgen [20]. Ct values of each target gene were calculated by Rotor-Gene 6000 Application Software, and the $_{\Delta}$Ct value of the *Actin2* rRNA gene was treated as an arbitrary constant for analyzing the $_{\Delta\Delta}$Ct value of samples. Three independent pools for each target gene were averaged, and the standard error of the mean value was recorded. The primer sequences used for the gene amplification are described in Table S1.

Physiological Index

Glutathione (GSH) Content. Glutathione (GSH) Content was estimated fluorimetrically according to Karni et al [21]. Half a gram plant material was frozen in liquid nitrogen and ground in 0.5 mL of 25% H_3PO_3 and 1.5 mL of 0.1 M sodium phosphate-EDTA buffer (pH 8.0). The homogenate was centrifuged at 10,000 g for 20 min to obtain supernatant for the estimation of GSH. The supernatant was diluted four times with phosphate-EDTA buffer (pH 8.0). The assay mixture for GSH estimation contained 100 mL of the diluted supernatant, 0.9 mL of phosphate-EDTA buffer and 100 mL of *O*-phthalaldehyde solution (1 mg : 1 mL). After thorough mixing and incubation at room temperature for 15 min, the solution was transferred to a quartz cuvette and the fluorescence at 420 nm was measured after excitation at 350 nm.

Ser Acetyltransferase Activity. Ser acetyltransferase (SAT) activity was measured according to the method described by Youssefian et al [22]. The incubation mixture with final volume of 240 µL contained 12 µM KPO$_3$, 16 µM Ser, 30 µg BSA, 0.5 µM acetyl CoA, 1 µM Na$_2$S, and an appropriate amount of extracts. The reaction was started by addition of the extracts and continued for 20 min at 25°C and was terminated by addition of 400 µL 4 M HCl. The tubes were centrifuged at 15,000 g for 3 min and to an aliquot of 200 µL supernatant, 200 µL modified ninhydrine reagent was added. The mixture was heated at 100°C for 10 min and cooled rapidly on ice, then 400 µL 98% ethanol was added and the absorbance was determined at 560 nm. The calibration curve was established by adding known amounts of L-Cys to the assay mixture and measuring these without incubation.

Amino Acid Content. The samples of plant material (0.5 g) were mixed with 1 ml of extraction solution (60% methanol, 25% chloroform, and 15% water) at 42°C for 10 min. After brief centrifugation, the supernatant was collected and the residue was extracted with the same mixture solution again, then both supernatants were combined. After adding the chloroform (1 mL) and water (1 mL), the resulting mixture was centrifuged again and the upper water-methanol phase was collected. Then the supernatants were dried in a vacuum desiccator, and then dissolved in 200 µL of water. The concentration of free amino acids was determined using *O*-phthalaldehyde reagent, followed by measuring the 335/447 nm fluorescence. Amino acid analyses were performed by the ion-exchange chromatography technique with a Hitachi model L-8800 amino acid analyzer (Hitachi Co. Ltd., Tokyo, Japan) with a column packed with Hitachi custom ion-exchange resin.

Statistical Analysis

Each experiment was repeated at least three times. Values in figures and tables were expressed as means ± SE. The statistical significance of the data was analyzed using univariate analysis of variance ($p < 0.05$) (one-way ANOVA; SPSS for Windows, version 11.0).

Results and Discussion

Integrative Proteomic and Transcriptomic Analysis on S Metabolism

In order to investigate the expression changes of proteins related to S metabolism under AR treatment, we analyzed the expression patterns of AR responsive proteins using a proteomic approach. The proteins were separated by 2-DE. On CBB-stained 2-DE gels, over 1500 highly reproducible protein spots in the pI range of 4–7 were revealed. 2-DE maps of the leaf proteome are shown in Figure 1A. Close-up views of several protein spots are shown in Figure 1B. Sixteen proteins related to S metabolism were identified and thereafter the functional categories were assigned to proteins using the AGI number to search the MIPS database (Figure 2A). Detailed information including the description of proteins, the MOWSE scores, theoretical pI values, molecular weights (*Mr*) and peptides matched of those 16 proteins which are related to S assimilation and primary/secondary metabolism are shown in Table 1 and Table S2.

To further examine the responses of *Arabidopsis* to AR, we applied transcript profiling employing the Affymetrix AH1 chips covering 24,000 genes to analyze the changes in gene expression patterns. In total, 13 genes which dramatically changed their expression were found related to S metabolism (Table 2). A list of

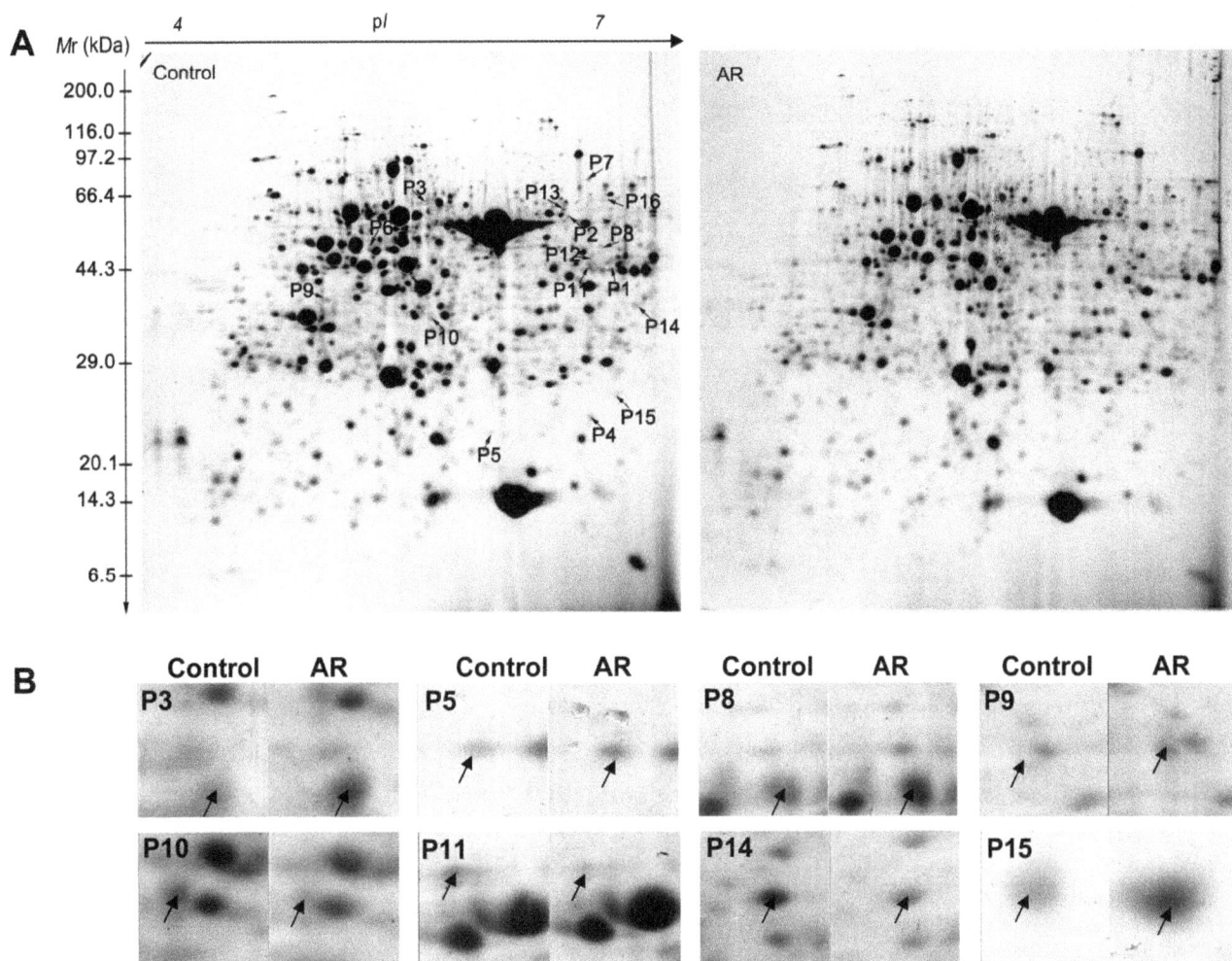

Figure 1. Protein expressions of *Arabidopsis thaliana* leaves after simulated acid rain (AR) treatment for 3 days (A). Molecular weight (*Mr*) in kilodaltons and p*I* of proteins are indicated on the left and top of the representative gel, respectively. Sixteen spots related to sulfur metabolism with at least a 2-fold change under AR stress are indicated. Close-up view of some differentially expressed protein spots (B).

the 13 S metabolism related genes significantly regulated at the transcript level, having been re-annotated and classified into functional classes as defined by MIPS database, is provided in Table 2.

From our results, the differentially expressed proteins and genes under AR covered each step of S metabolism pathways according to their functional categories, including S uptake, transportation, reduction, assimilation and S-containing amino acids and other derivates synthesis metabolisms (Figure 2A and B). A Venn diagram of regulated cytosolic mRNA versus regulated proteins shows an overlap of 4 genes (Figure 2C), indicating that a large number of genes are solely regulated either at mRNA or protein level. Similar results were also found in earlier studies [14,17,23,24]. Here are some reasons that may clarify the results. Firstly, proteomic studies suffer from inherent technical short-comings associated with, for example, protein insolubility, fractionation losses, extreme p*I*, etc [17]. Secondly, despite recent improvements, proteomic technique remains poorly suitable to separate highly hydrophobic, basic or low-abundant proteins [23]. Thus, subcellular membrane proteome, and especially their integral protein moieties, remain poorly accessible [25,26]. On the other hand, mRNA degradation, alternative splicing, and post-

transcriptional regulation of gene expression could also lead to the lack of strong correlations with protein expression status [27].

In order to further confirm and extend the results obtained from proteomic and transcriptomic analysis, we performed quantitative real-time PCR (qRT-PCR) analysis on 12 genes, all of which are very crucial in S metabolism pathways, including S uptake (Sulfate transporter1;2 gene, *SULTR1;2*), reduction (ATP sulfurylase gene, *APS*; APS reductase gene, *APR*) as well as on the genes related to S-containing amino acids synthesis (*O*-acetylserine(thiol)lyase gene, *OASA*; Cysteine synthase gene, *OASB*; Glutathione synthetase gene, *GSH2*) and other S derivates synthesis metabolisms (Glutathione S-transferase gene, *GST3*; Glutathione peroxidase gene, *GPX6*; Cytosolic thioredoxin gene, *TRX5*; Myrosinase gene, *TGG2*; S-adenosylmethionine synthetase gene, *MTO3*; S-adeno-sylmethionine decarboxylase gene, *SAMDC*). qRT-PCR analysis showed that transcript expression level of genes related to primary sulfur assimilation, such as *APR*, *APS1*, *OASA1 GSH2* and *GST3*, were up-regulated (Figure 3). However, the synthesis genes of some S-containing amino acids and derivatives (*MTO3* and *SAMDC*) were down-regulated (Figure 3). The results were highly correlated with those of the array data, thus confirming the results from proteomic and transcriptomic studies. However, the change

A   Protein

B   mRNA

- ■ Sulfur assimilation
- ■ Glutathione metabolism
- ■ Methionine metabolism
- ■ Glutamate metabolism
- ■ Glycine, serine and threonine metabolism
- ■ Sencondary metabolism

C

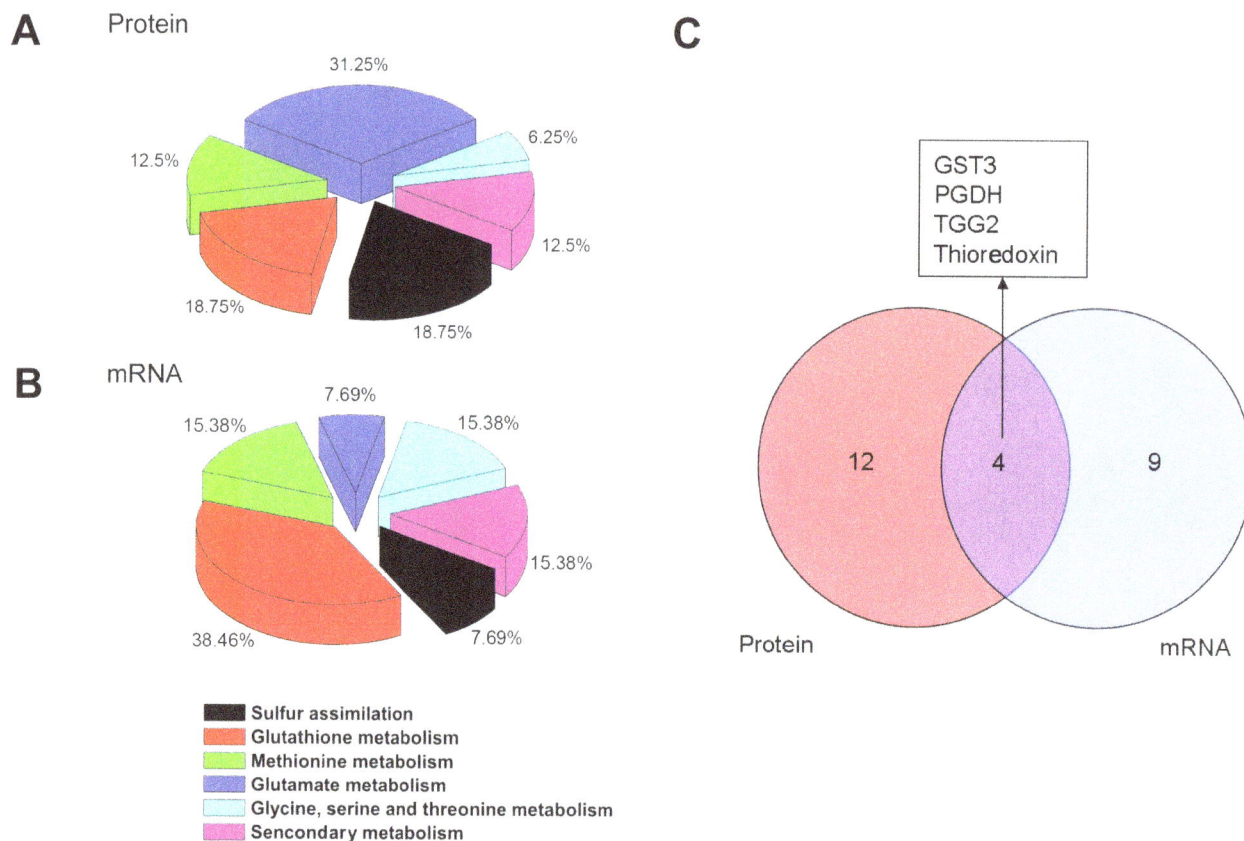

Figure 2. Functional classification of the significant differential expression proteins (A) and genes (B) after simulated acid rain (AR) treatments in *Arabidopsis thaliana*. Venn diagram shows the number of overlapped genes or proteins between gene and protein expression profiles after AR treatment (C).

level of differential expression of a single gene was a little different with microarray as described previously [14]

Primary S Assimilation was Activated under AR

The combined proteomic and transcriptomic analysis on our experimental data sets provided a superior view of the complex physiology of *Arabidopsis* in response to AR compared to either proteomic or transcriptomic approach alone. As shown in Figure 4, the proteins/genes data were obtained from the proteomic and genomic microarray expreiments, which revealed a possible systematic AR-responsive mechacnism of S assimilation and related pathways in *Arabidopsis* under AR treatment.

Sulfate (SO_4^{2-}) is the most oxidized and thus a stable form of S presented in the soil. Uptake of S into roots from the soil is almost exclusively via sulfate uptake [8]. In our experiment, we found the expression of sulfate transporter gene (*SULTR 1;2*), which has an important function in S uptake, was reduced under AR treatment. A number of genes encoding the sulfate transporter have been reported in *Arabidopsis* [28–30]. They are classified into five subfamilies, named SULTR1 to 5, according to their deduced amino acid sequences. The members in SULTR1 are high-affinity transporters for sulfate. SULTR1;1 and SULTR1;2 of *Arabidopsis* are inducible by sulfate depletion, responsible for initial uptake of sulfate from outside of the plant cell [29]. The transporter is well known to show a strong repression in expression in the presence of an adequate S supply. Transport activity, mRNA pool size and

protein expression all decrease under conditions of excess S supply [31,32]. In our study, AR treatment increased soil sulfate, hence it is not surprising that sulfate transporter gene expression was down-regulated".

For assimilation, sulfate must be activated by APS, in which sulfate is linked by an anhydride bond to a phosphate residue by consumption of ATP and concomitant release of pyrophosphate [33]. This reaction is catalyzed by APS and is the sole entry step for S metabolism. It is reported that APS mediates the reduction reaction of sulfate to sulfite by APS reductase (APR) in plants, which is subsequently reduced to sulfide by sulfite reductase [32]. Many studies have found that APR is another key enzyme in sulfate assimilation in plants [9,34]. In our experiment, we found that the increase of APR mRNA accumulation contributes to the higher sulfate assimilation from outside into plant under AR treatment (Figure 4).

The final step in the assimilation of reduced sulfate is the incorporation of S into thiol-containing amino-acid, Cys [11]. Two enzymes, Ser acetyltransferase (SAT) and *O*-acetylserine (thiol) lyase (OASTL), are committed for this step. SAT catalyzes the formation of *O*-acetylserine (OAS) from Ser and acetyl-CoA. Many reports have found that SAT plays an important role in regulating Cys biosynthesis [10,35,36]. While the plants were exposed to AR, the expression in gene level of SAT was up-regulated, however, the OASTL and OASTL isoform oasB (OASB) were down-regulated. All of the expression changes lead

Table 1. Identification of protein spots with a significant 2-fold changes in AR compared with control treatment for *Arabidopsis thaliana* leaves.

Spot	Accession number	Protein identity	Theo. M_r (kDa)/pI	Expt. M_r (kDa)/pI	SC (%)	Mascot score	Fold change	SP/TP	p-value
Sulfur metabolism									
P1	AT4G14880	O-acetylserine (thiol) lyase (OASA1)	34/5.9	42/7.0	32%	88	−3.6±0.21	5/13	1.1×10^{-3}
P2	AT2G43750	(OAS-TL) isoform oasB (OASB)	42/8.2	42/7.0	21%	89	−4.0±0.30	6/15	9.8×10^{-6}
P3	AT3G22890	ATP sulfurylase (APS)	51/6.8	51/6.3	23%	91	2.4±0.12	7/13	4.8×10^{-16}
Glutathione metabolism									
P4	AT5G41670	6-phosphogluconate dehydrogenase family protein (6-PDG)	53/5.5	54/5.6	19%	71	3.5±0.22	6/16	1.6×10^{-4}
P5	AT5G16710	glutathione dependent dehydroascorbate reductase (DHAR3)	29/7.9	28/7.0	41%	68	4.4±0.31	11/31	6.0×10^{-13}
P6	AT1G02920	glutathione S-transferase (GST3)	24/6.6	24/6.1	50%	67	2.9±0.16	8/29	1.2×10^{-9}
Methionine metabolism									
P7	AT3G17390	S-adenosylmethionine synthetase (MTO3)	43/5.6	43/5.5	46%	95	2.3±0.14	15/32	7.6×10^{-9}
P8	AT5G17920	Methionine synthase (MetS)	84/6.5	85/6.1	31%	139	2.3±0.11	17/24	1.5×10^{-4}
Glutamate metabolism									
P9	AT1G23310	glutamate-glyoxylate aminotransferase (GGT1)	53/6.9	49/6.6	24%	87	2.1±0.05	21/36	1.2×10^{-3}
P10	AT1G66200	glutamine synthase clone R2 (GSR2)	39/4.9	39/5.1	28%	95	−2.4±0.14	10/17	1.5×10^{-2}
P11	AT3G17820	glutamate-ammonia ligase (GLD)	39/5.9	39/5.7	47%	92	−3.0±0.18	14/22	6.1×10^{-7}
P12	AT5G63570	glutamate-1-semialdehyde 2,1-aminomutase (GSA)	50/6.9	51/6.4	27%	99	−3.9±0.21	11/27	1.5×10^{-13}
P13	AT3G17240	lipoamide dehydrogenase (LPD)	54/7.0	54/6.6	32%	86	2.9±0.22	10/23	5.6×10^{-3}
Glycine, serine and threonine metabolism									
P14	AT4G34200	phosphoglycerate dehydrogenase (EDA9)	63/6.5	64/6.3	23%	106	−3.3±0.20	15/27	6.5×10^{-2}
Others									
P15	AT4G03520	thiol-disulfide exchange intermediate (TRX5)	20/9.6	21/6.9	34%	90	4.0±0.18	5/9	2.4×10^{-9}
P16	AT5g25980	thioglucoside glucohydrolase (TGG2)	63/7.5	62/7.1	33%	180	3.6±0.14	14/21	2.1×10^{-4}

Table 2. Differentially expressed transcripts induced by AR for *Arabidopsis thaliana* leaves.

Probe ID	NO.	Gene name	Accession number	Description	Fold change	p-Value
	Sulfur metabolism					
260602_at	G1	SAT	At1g55920	Serine acetyltransferase	2.3±0.03	4.3×10^{-4}
	Glutathione metabolism					
266746_s_at	G2	GSTF3	At2g02930	Glutathione S-transferase	2.3±0.06	3.7×10^{-3}
264383_at	G3	GPX1	At2g25080	Glutathione peroxidase	2.7±0.02	6.5×10^{-5}
254890_at	G4	GPX6	At4g11600	Ggutathione peroxidase	2.5±0.01	6.2×10^{-5}
246785_at	G5	GSH2	At5g27380	Glutathione synthetase	2.3±0.04	1.7×10^{-3}
262932_at	G6	Micro-GST	At1g65820	Microsomal glutathione S-transferase	2.0±0.05	1.7×10^{-3}
	Glutamate metabolism					
249581_at	G7	GSR1	At5g37600	Cytosolic glutamine synthetase	−2.4±0.03	3.2×10^{-4}
260309_at	G8	AOAT2	At1g70580	Glutamate-glyoxylate transaminase	−2.6±0.07	1.1×10^{-2}
	Methionine metabolism					
246490_at	G9	SAMDC	At5g15950	S-adenosylmethionine decarboxylase	−3.5±0.07	1.3×10^{-3}
	Glycine, serine and threonine metabolism					
253162_at	G10	PSAT	At4g35630	Phosphoserine amintransferase	−2.5±0.02	2.4×10^{-4}
259403_at	G11	PGDH	At1g17745	Phosphoglycerate dehydrogenase	−2.5±0.01	1.3×10^{-4}
	Other metabolism					
260943_at	G12	TRX5	At1g45145	Cytosolic thioredoxin	3.3±0.06	1.9×10^{-2}
265058_s_at	G13	TGG2	At1g52040	Thioglucoside glucohydrolase	5.4±0.09	2.1×10^{-2}

to the increased assimilation of inorganic S into Cys. Consequently, the activity of SAT was also up-regulated under AR treatment due to high concentration of sulfate (Figure 5A). Meanwhile, higher level of Cys content was observed in our study (Figure 5B).

Cys is the pivotal sulfur-containing compound regarded as the terminal metabolite in S assimilation and the starting point for biosynthesis of Met, GSH, and a variety of other S-containing metabolites [32]. Therefore, the increased content of Cys eventually led to the increase in Met and GSH contents (Figure 5C and D), as well as the increase in expression of several Met and GSH biosynthesis related genes at transcriptional and protein levels (Figure 4) under AR.

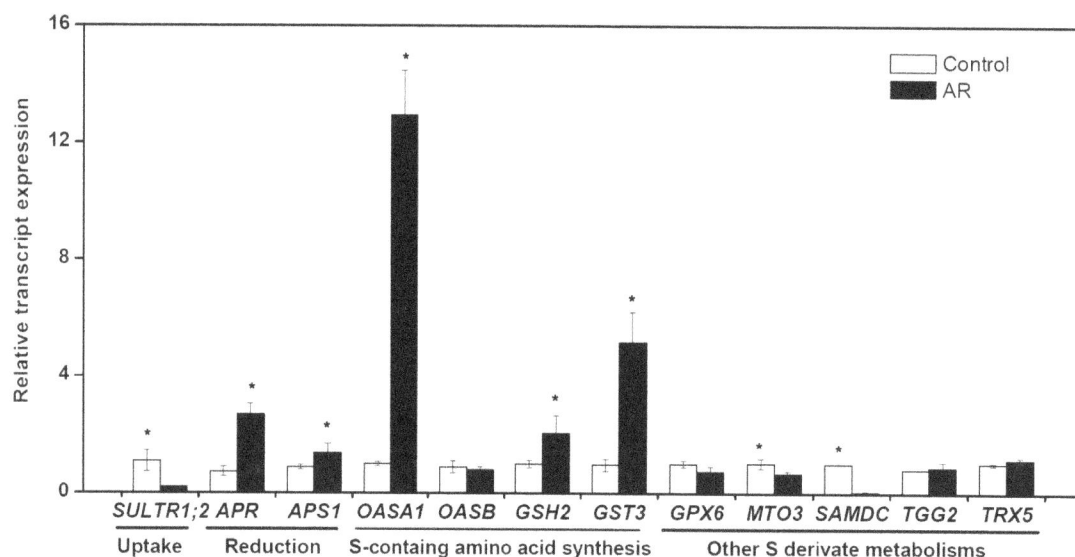

Figure 3. Relative changes in transcript level quantified by qRT-PCR for twelve genes related to sulfur metabolism in *Arabidopsis thaliana* leaves under simulated acid rain (AR). The fold-change values were derived from the average of three replicate measurements. The asterisk indicates significance at $p<0.05$.

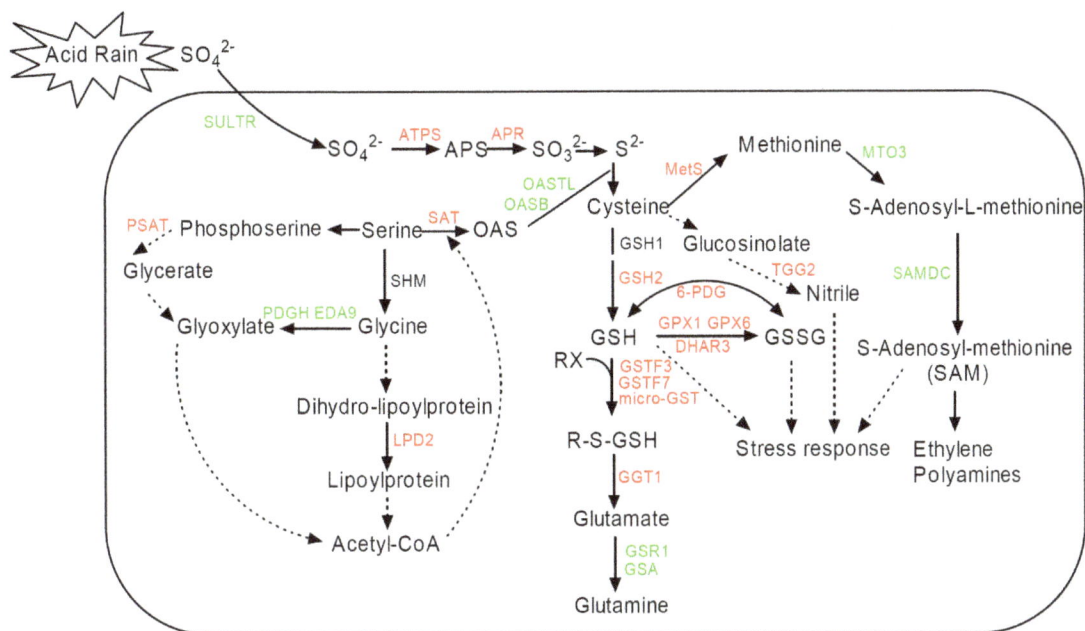

Figure 4. Schematic representation of a possible systematic response mechanism related to sulfur metabolism in *Arabidopsis thaliana* **under simulated acid rain (AR) stress.** The up- and down- regulated proteins or genes are labeled in red and green. The direct and indirect interaction of proteins or genes are indicated as solid or dotted line, respectively. Combined with our results, this figure was developed from the reviews of Hawkesford *et al.* [56] and Saito [11].

Downstream Genes and Proteins in S Metabolism Pathway Were Depressed by AR

Different from the primary S assimilation, where more Cys, Met and GSH synthesis process was induced under AR (Figure 4 and 5), the metabolism of some other amino acids and derivatives was depressed under AR. In our study, two Glu synthetase (GS) genes (glutamine synthase clone R2 gene, *GSR1*; glutamate-ammonia ligase gene, *GLD*) were down-regulated. GS is a key enzyme in this nitrogen assimilatory process, as it catalyzes the first step in the conversion of inorganic nitrogen (ammonium) into its organic form [37]. Consistently, we detected that the Glu content was greatly decreased under AR treatments (Figure 5E). These results suggested that nitrogen (N) assimilation was inhibited under AR treatments. On the other hand, glutamate-1-semialdehyde 2,1-aminomutase (GSA) is the first committed precursor of porphyrin synthesis in organelles and organisms that use the carbon (C) 5 pathway [38]. As we know, lots of important organic molecules such as chlorophyll are closely related with porphyrin, indicating the C fixation is influenced by AR through GSA mediated porphyrin synthesis pathway. Earlier studies have shown that AR could inhibit respiration and photosynthesis, and further inhibit plant growth [2,39], which could be an indirect proof of our data. Generally, these results indicated that N and C metabolism were disordered under AR treatment.

In Met cycle, Met is converted to S—Adenosyl Methionine (SAM), which is a methyl donor for numerous reactions. SAM is also a substrate for ethylene, polyamine and phytosiderophore synthesis [40]. The expression of SAMS3, which is a key enzyme in SAM synthesis, was greatly inhibited by AR, suggesting that the Met cycle was influenced by AR treatment. Besides, S-adenosylmethionine decarboxylase gene (*SAMDC*) expression was also depressed under AR in our study (Figure 4). Recently, *Arabidopsis* mutant analysis has indicated that *SAMDC* is essential for plant polyamine biosynthesis pathway and play an important role in plant growth and development [41]. In plants, polyamines are not only important

for both stress responses and developmental processes but also essential for plant survival [42]. The disorder of polyamines metabolism by AR would lead to more serious plant damage.

GSH Plays a Crucial Role in Reactive Oxygen Species (ROS) Scavenging under AR

Cys availability has been shown to be the main factor limiting GSH production, both in normal plants and in those that overexpress genes for GSH biosynthesis [43]. Cys is incorporated into GSH that is one of the major redox controllers that plays significant roles in scavenging ROS through the GSH-ascorbate cycle [25] in which the dehydroascorbate reductase (DHAR) reduces dehydroascorbate to ascorbate, while oxidizing GSH to glutathione disulfide [27]. Many enzymes involved in this process were up-regulated in our study which led to the synthesis of more GSH in plant cells (Figure 4). Although some reports indicated that Cys and GSH are negative regulators of gene expression responding to S assimilation [44,45], there were some other reports that the level of Cys increased in SO_2 fumigated beech leaves [24]. In spruce trees, exposure to SO_2 increased the accumulation of GSH and the activation of several scavenging enzymes [46]. An additional effect of SO_2 fumigation was an increased level of sulfate suggesting that increased content of thiols in response to excessive S deposition is a common phenomenon.

As we know, ROS as a typical secondary stress triggered by AR, can cause severe damage to plants including growth and photosynthesis reduction, and premature senescence as well [47–49]. To prevent damage to membranes, chlorophylls and proteins, ROS have to be detoxified by scavenging systems that are consisting of low-molecular weight antioxidants and antioxidative enzymes in the apoplast and the symplast of plant cells [50]. Accumulating evidence further suggests that these adaptive responses of plants to increased ROS levels are mediated by changes in cellular GSH concentrations or the redox status of GSH pool [43].

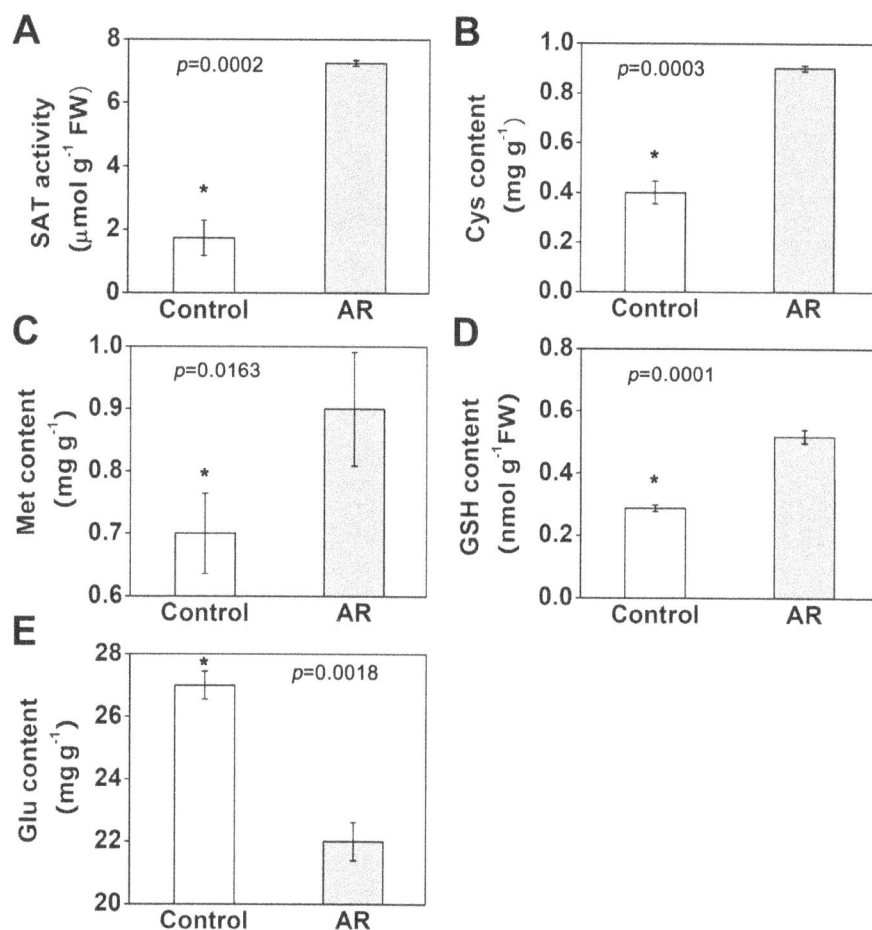

Figure 5. Effects of simulated acid rain (AR) on the reduced glutathione (GSH) content (A), Ser acetyltransferase (SAT) activity (B) and three S-containing amino acid contents (C–E) in *Arabidopsis thaliana.* The values were derived from the average of three replicate measurements. The asterisk indicates significance at $p < 0.05$.

Direct evidence showing ROS is a signaling component in plants is not yet available. However, several genes involved in S assimilation and synthesis of S-containing amino acids were induced by exposure to O_2^- [48]. In this study, we identified a set of genes and proteins related to GSH metabolism pathways which were greatly up-regulated. For example, the expression of two genes *(GPX1, GPX6)* encoding glutathione peroxidases (GPX) was induced under AR in our study. As we know, GPX is the general name of the enzyme family with peroxidase activity whose main biological role is to protect the organism from oxidative damage. Evidence suggested that GPX activity also plays a role in stress-related signal transduction [51]. The plant glutathione transfer-ases, formerly known as glutathione S-transferases (GSTs) are a large and diverse group of enzymes that catalyze the conjugation of electrophilic xenobiotic substrates with GSH [52]. Besides, GSTs are also important components of the cellular defense against oxidative stress [38]. We found GST genes or proteins were induced in both proteomic and transcriptomic experiments. Consistently, numerous studies have revealed that members of the GST super-family are expressed in response to microbial infection, cell division and environmental stresses [53,54] as well as AR treatment in our pervious study [55].

Surely, acid rain not only disordered S metablism, but also effected many other pathways. Lots of publications have demon-strated that AR causes a series of damages to plants, which includes destruction of the cell membrane, inhibition of respiration and photosynthesis, as well as disorders in metabolism of glucose, lipids and amino acids. Our transcriptomic and proteomic analysis also revealed that the expression of a set of genes and proteins related to primary metabolism, photosynthesis, metabolism of ROS, cellular transport, and signal transduction, were influenced by AR treatment. Due to the complexity of the emerging pattern, further work is required to delineate and confirm the precise effects of AR on metabolism and physiology.

Conclusion

Using proteomic and transcriptomic methods, we studied the responses of S uptake and metabolic pathways in *Arabidopsis* seedlings exposed to simulated AR. By summarizing the informa-tion on the coordination between different metabolic changes, a network of mutual cross-influences in the AR-stress response could be assembled. Apparently, the entire network of S metabolism was coordinately regulated under AR stress. First of all, sulfate uptake and acquisition, that totally control the input of sulfate into S metabolic pathways, have been identified to be in positive correlation with AR. Furthermore, the activation of sulfate also increased as AR was imposed. Thirdly, the biosynthesis from sulfate to S-containing amino acid, for example, Cys and Met and other secondary metabolites were up-regulated under AR stress. Finally, we depicted the coordinating network of S metabolism including S uptake, activation, S-containing amino acid biosynthesis and other S-containing metabolites synthesis under AR stress. This study can help us to understand the mechanisms by which plants adapt to AR environment by alteration of the S metabolism.

Supporting Information

Figure S1 Injury phenotype of Arabidopsis leaves under simualted acid rain treatment.

Table S1 Primer pairs used in qRT-PCR analysis for 12 sulfur metabolism related genes. Actin 2 was used as a standard to normalize the content of cDNA.

Table S2 The details of identified acid rain stress-responsive proteins in Arabidopsis.

Acknowledgments

We are grateful to Chen Lei for assistance in experiments, and Mr. Sieh Kargbo for editing the manuscript.

Author Contributions

Conceived and designed the experiments: TWL JAC FHW HLZ. Performed the experiments: TWL JBC WJH WHW. Analyzed the data: TWL JBC FHW HLZ. Contributed reagents/materials/analysis tools: TWL FHW. Wrote the paper: TWL JAC HLZ MS.

References

1. Larssen T, Lydersen E, Tang DG, He Y, Gao JX, et al. (2006) Acid rain in china. Environmental Science & Technology 40: 418–425.
2. Likens GE, Driscoll CT, Buso DC (1996) Long-term effects of acid rain: response and recovery of a forest ecosystem. Science 272: 244–246.
3. Rogasik J, Schroetter S, Schnug E (2002) Impact of air pollutants on agriculture. Phyton-Annales Rei Botanicae 42: 171–182.
4. Karnosky DF (2001) Impacts of air pollution on forest ecosystems - Preface. Environmental Pollution 115: 317–317.
5. Likens GE, Weathers KC, Butler TJ, Buso DC (1998) Solving the acid rain problem. Science 282: 1991–1992.
6. Brychkova G, Xia ZL, Yang GH, Yesbergenova Z, Zhang ZL, et al. (2007) Sulfite oxidase protects plants against sulfur dioxide toxicity. Plant Journal 50: 696–709.
7. Lee Y, Park J, Im K, Kim K, Lee J, et al. (2006) Arabidopsis leaf necrosis caused by simulated acid rain is related to the salicylic acid signaling pathway. Plant Physiology and Biochemistry 44: 38–42.
8. Bick JA, Leustek T (1998) Plant sulfur metabolism - the reduction of sulfate to sulfite. Current Opinion in Plant Biology 1: 240–244.
9. Hell R (1997) Molecular physiology of plant sulfur metabolism. Planta 202: 138–148.
10. Rausch T, Wachter A (2005) Sulfur metabolism: a versatile platform for launching defence operations. Trends in Plant Science 10: 503–509.
11. Saito K (2004) Sulfur assimilatory metabolism. The long and smelling road. Plant Physiology 136: 2443–2450.
12. Kopriva S, Rennenberg H (2004) Control of sulphate assimilation and glutathione synthesis: interaction with N and C metabolism. Journal of Experimental Botany 55: 1831–1842.
13. Zhu GH, Zhuang CX, Wang YQ, Jiang LR, Peng XX (2006) Differential expression of rice genes under different nitrogen forms and their relationship with sulfur metabolism. Journal of Integrative Plant Biology 48: 1177–1184.
14. Gallardo K, Firnhaber C, Zuber H, Hericher D, Belghazi M, et al. (2007) A combined proteome and transcriptome analysis of developing Medicago truncatula seeds. Molecular & Cellular Proteomics 6: 2165–2179.
15. Klikocka H, Haneklaus S, Bloem E, Schnug E (2005) Influence of sulfur fertilization on infection of potato tubers with Rhizoctonia solani and Streptomyces scabies. Journal of Plant Nutrition 28: 819–833.
16. Hesse H, Kreft O, Maimann S, Zeh M, Willmitzer L, et al. (2001) Approaches towards understanding methionine biosynthesis in higher plants. Amino Acids 20: 281–289.
17. Li LY, Li QB, Rohlin L, Kim U, Salmon K, et al. (2007) Quantitative proteomic and microarray analysis of the archaeon Methanosarcina acetivorans grown with acetate versus methanol. Journal of Proteome Research 6: 759–771.
18. Fan HB, Wang YH (2000) Effects of simulated acid rain on germination, foliar damage, chlorophyll contents and seedling growth of five hardwood species growing in China. Forest Ecology and Management 132: 285–285.
19. Carpentier SC, Witters E, Laukens K, Deckers P, Swennen R, et al. (2005) Preparation of protein extracts from recalcitrant plant tissues: An evaluation of different methods for two-dimensional gel electrophoresis analysis. Proteomics 5: 2497-2507.
20. Livak KJ, Schmittgen TD (2001) Analysis of relative gene expression data using real-time quantitative PCR and the 2(T)(-Delta Delta C) method. Methods 25: 402–408.
21. Karni L, Moss SJ, Telor E (1984) Glutathione-reductase activity in heterocysts and vegetative cells of the Cyanobacterium Nostoc-Muscorum. Archives of Microbiology 140: 215–217.
22. Youssefian S, Nakamura M, Orudgev E, Kondo N (2001) Increased cysteine biosynthesis capacity of transgenic tobacco overexpressing an O-acetylserine(thiol) lyase modifies plant responses to oxidative stress. Plant Physiology 126: 1001–1011.
23. Resch A, Leicht S, Saric M, Pasztor L, Jakob A, et al. (2006) Comparative proteome analysis of Staphylococcus aureus biofilm and planktonic cells and correlation with transcriptome profiling. Proteomics 6: 1867–1877.
24. Marmagne A, Brabant P, Thiellement H, Alix K (2010) Analysis of gene expression in resynthesized Brassica napus allotetraploids: transcriptional changes do not explain differential protein regulation. New Phytologist 186: 216–227.
25. Cordwell SJ, Thingholm TE (2010) Technologies for plasma membrane proteomics. Proteomics 10: 611–627.
26. Gorg A, Drews O, Luck C, Weiland F, Weiss W (2009) 2-DE with IPGs. Electrophoresis 30: S122–S132.
27. Mitchell P (2010) Proteomics retrenches. Nature Biotechnology 28: 665–670.
28. Shibagaki N, Rose A, McDermott JP, Fujiwara T, Hayashi H, et al. (2002) Selenate-resistant mutants of Arabidopsis thaliana identify Sultr1;2, a sulfate transporter required for efficient transport of sulfate into roots. Plant Journal 29: 475–486.
29. Yoshimoto N, Takahashi H, Smith FW, Yamaya T, Saito K (2002) Two distinct high-affinity sulfate transporters with different inducibilities mediate uptake of sulfate in Arabidopsis roots. Plant Journal 29: 465–473.
30. Hawkesford MJ (2003) Transporter gene families in plants: the sulphate transporter gene family - redundancy or specialization? Physiologia Plantarum 117: 155–163.
31. Nikiforova V, Freitag J, Kempa S, Adamik M, Hesse H, et al. (2003) Transcriptome analysis of sulfur depletion in Arabidopsis thaliana: interlacing of biosynthetic pathways provides response specificity. Plant Journal 33: 633–650.
32. Maruyama-Nakashita A, Inoue E, Watanabe-Takahashi A, Yarnaya T, Takahashi H (2003) Transcriptome profiling of sulfur-responsive genes in Arabidopsis reveals global effects of sulfur nutrition on multiple metabolic pathways. Plant Physiology 132: 597–605.
33. Koralewska A, Buchner P, Stuiver CEE, Posthumus FS, Kopriva S, et al. (2009) Expression and activity of sulfate transporters and APS reductase in curly kale in response to sulfate deprivation and re-supply. Journal of Plant Physiology 166: 168–179.
34. Loudet O, Saliba-Colombani V, Camilleri C, Calenge F, Gaudon V, et al. (2007) Natural variation for sulfate content in Arabidopsis thaliana is highly controlled by APR2. Nature Genetics 39: 896–900.
35. Harms K, von Ballmoos P, Brunold C, Hofgen R, Hesse H (2000) Expression of a bacterial serine acetyltransferase in transgenic potato plants leads to increased levels of cysteine and glutathione. Plant Journal 22: 335–343.
36. Toda K, Takano H, Nozaki J, Kuroiwa T (2001) The second serine acetyltransferase, bacterial-type O-acetylserine (thiol) lyase and eukaryotic-type O-acetylserine (thiol) lyase from the primitive red alga Cyanidioschyzon merolae. Journal of Plant Research 114: 291–300.
37. Peterman TK, Goodman HM (1991) The Glutamine-synthetase gene family of Arabidopsis thaliana - light regulation and differential expression in leaves, roots and seeds. Molecular & General Genetics 230: 145–154.
38. Oliveira IC, Coruzzi GM (1999) Carbon and amino acids reciprocally modulate the expression of glutamine synthetase in arabidopsis. Plant Physiology 121: 301–309.
39. Tomlinson GH (2003) Acidic deposition, nutrient leaching and forest growth. Biogeochemistry 65: 51–81.
40. Goto DB, Ogi M, Kijima F, Kumagai T, van Werven F, et al. (2002) A single-nucleotide mutation in a gene encoding S-adenosylmethionine synthetase is associated with methionine over-accumulation phenotype in Arabidopsis thaliana. Genes & Genetic Systems 77: 89–95.
41. Ge CM, Cui X, Wang YH, Hu YX, Fu ZM, et al. (2006) BUD2, encoding an S-adenosylmethionine decarboxylase, is required for Arabidopsis growth and development. Cell Research 16: 446–456.
42. Walters DR (2003) Polyamines and plant disease. Phytochemistry 64: 97–107.
43. Noctor G, Arisi ACM, Jouanin L, Foyer CH (1998) Manipulation of glutathione and amino acid biosynthesis in the chloroplast. Plant Physiology 118: 471–482.

44. Hirai MY, Klein M, Fujikawa Y, Yano M, Goodenowe DB, et al. (2005) Elucidation of gene-to-gene and metabolite-to-gene networks in Arabidopsis by integration of metabolomics and transcriptomics. Journal of Biological Chemistry 280: 25590–25595.

45. Fabio F, Clarissa L, Barbara G, Gian AS (2007) Sulfur metabolism and cadmium stress in higher plants. Global Science Books.

46. Rennenberg H, Herschbach C, Haberer K, Kopriva S (2007) Sulfur metabolism in plants: Are trees different? Plant Biology 9: 620–637.

47. Bandurska H, Borowiak K, Miara M (2009) Effect of two different ambient ozone concentrations on antioxidative enzymes in leaves of two tobacco cultivars with contrasting ozone sensitivity. Acta Biologica Cracoviensia Series Botanica 51: 37–44.

48. Langebartels C, Wohlgemuth H, Kschieschan S, Grun S, Sandermann H (2002) Oxidative burst and cell death in ozone-exposed plants. Plant Physiology and Biochemistry 40: 567–575.

49. Yano A, Suzuki K, Shinshi H (1999) A signaling pathway, independent of the oxidative burst, that leads to hypersensitive cell death in cultured tobacco cells includes a serine protease. Plant Journal 18: 105–109.

50. Asada K (2004) Functions of the water-water cycle in chloroplasts. Plant and Cell Physiology 45: S11–S11.

51. Winfield MO, Lu CG, Wilson ID, Coghill JA, Edwards KJ (2010) Plant responses to cold: transcriptome analysis of wheat. Plant Biotechnology Journal 8: 749–771.

52. Dixon DP, Skipsey M, Edwards R (2010) Roles for glutathione transferases in plant secondary metabolism. Phytochemistry 71: 338–350.

53. Hatzios KK (1999) Functions and regulation of plant glutathione S-transferases. Abstracts of Papers of the American Chemical Society 218: U117–U117.

54. Moons A (2005) Regulatory and functional interactions of plant growth regulators and plant glutathione S-transferases (GSTS). Plant Hormones 72: 155–202.

55. Liu TW, Fu B, Niu L, Chen J, Wang WH, et al. (2011) Comparative proteomic analysis of proteins in response to simulated acid rain in Arabidopsis. Journal of Proteome Research 10: 2579–2589.

56. Hawkesford MJ, De Kok IJ (2006) Managing sulphur metabolism in plants. Plant, Cell and Environment 29: 382–395.

20

Predicting Greenhouse Gas Emissions and Soil Carbon from Changing Pasture to an Energy Crop

Benjamin D. Duval[1,2¤a], Kristina J. Anderson-Teixeira[1¤b], Sarah C. Davis[1¤c], Cindy Keogh[3], Stephen P. Long[1,2,4], William J. Parton[3], Evan H. DeLucia[1,2,4]*

1 Energy Biosciences Institute, University of Illinois at Urbana-Champaign, Urbana, Illinois, United States of America, **2** Global Change Solutions, Urbana, Illinois, United States of America, **3** Natural Resource Ecology Laboratory, Fort Collins, Colorado, United States of America, **4** Department of Plant Biology, University of Illinois at Urbana-Champaign, Urbana, Illinois, United States of America

Abstract

Bioenergy related land use change would likely alter biogeochemical cycles and global greenhouse gas budgets. Energy cane (*Saccharum officinarum* L.) is a sugarcane variety and an emerging biofuel feedstock for cellulosic bio-ethanol production. It has potential for high yields and can be grown on marginal land, which minimizes competition with grain and vegetable production. The DayCent biogeochemical model was parameterized to infer potential yields of energy cane and how changing land from grazed pasture to energy cane would affect greenhouse gas (CO_2, CH_4 and N_2O) fluxes and soil C pools. The model was used to simulate energy cane production on two soil types in central Florida, nutrient poor Spodosols and organic Histosols. Energy cane was productive on both soil types (yielding 46–76 Mg dry mass·ha^{-1}). Yields were maintained through three annual cropping cycles on Histosols but declined with each harvest on Spodosols. Overall, converting pasture to energy cane created a sink for GHGs on Spodosols and reduced the size of the GHG source on Histosols. This change was driven on both soil types by eliminating CH_4 emissions from cattle and by the large increase in C uptake by greater biomass production in energy cane relative to pasture. However, the change from pasture to energy cane caused Histosols to lose 4493 g CO_2 eq·m^{-2} over 15 years of energy cane production. Cultivation of energy cane on former pasture on Spodosol soils in the southeast US has the potential for high biomass yield and the mitigation of GHG emissions.

Editor: Chenyu Du, University of Nottingham, United Kingdom

Funding: This research was funded by the Energy Bioscience Institute. The funders had no role in study design, data collection and analysis, decision to publish, or preparation of the manuscript.

Competing Interests: Please note that Duval, Long and DeLucia are affiliated with a company, Global Change Solutions LLC.

* E-mail: delucia@illinois.edu

¤a Current address: Dairy Forage Research Center, United States Department of Agriculture, Agricultural Research Service, Madison, Wisconsin, United States of America
¤b Current address: Conservation Ecology Center, Smithsonian Conservation Biology Institute, National Zoological Park, Front Royal, Virginia, United States of America
¤c Current address: Voinovich School for Leadership and Public Affairs, Ohio University, Athens, Ohio, United States of America

Introduction

Land use has a pervasive influence on atmospheric greenhouse gas (GHG) concentrations and thereby on climate [1,2,3]. Carbon emissions from land use change, often to make way for agriculture, have contributed substantially to anthropogenic increases in the atmospheric CO_2 concentration [2]. For example, C emissions from tropical deforestation have been estimated at 10.6 ± 1.8 Pg CO_2 per year between 1990 and 2007, equal to ~40% of global fossil fuel emissions [3]. Likewise, it is estimated that 40–52 Pg CO_2 have been released by plowing high-C native prairie soils [4]. Agricultural practices are important to global GHG budgets, with agroecosystems contributing ~14% of global anthropogenic GHG emissions [1]. Agricultural practices can also reduce GHG emissions and enhance soil carbon, and have the potential to mitigate climate change [4,5].

Land use and land management changes associated with the emerging bioenergy industry are likely to have substantial impacts on global GHG budgets [6,7,8]. A change from fossil fuels to an energy economy more reliant on plant-derived biofuels has the potential to reduce GHG emissions [9]. The prospect of lowering emissions is one factor leading to the United States' mandate to produce 136 billion liters of renewable fuel by 2022 [10]. However, meeting this mandate will require substantial land area [11,12], which implies potentially major changes to regional biogeochemical cycling [12,13].

Corn grain (*Zea mays*) is the dominant crop used for ethanol production in the US [14]. However, the ability of corn ethanol to reduce GHG emissions is questionable [15,16], and corn production exacerbates nitrogen pollution and other environmental problems [17–19]. Of particular concern is the possibility that diversion of corn for ethanol production will increase global grain prices and trigger agricultural expansion and deforestation elsewhere in the world [7]. The emerging commercial technology to convert ligno-cellulose to ethanol could redress the reliance on corn grain as an ethanol feedstock [20]. This could be particularly beneficial if cellulosic biofuel crops are grown on land that is not important for food production, while having lower GHG emissions than traditional row-crop agriculture [21,22]. Therefore, considerable research has focused on understanding the soil C and greenhouse gas consequences of replacing traditional agriculture used for bioenergy with perennial grasses like switchgrass (*Panicum virgatum* L.), Miscanthus (*Miscanthus x giganteus* J. M. Greef & Deuter ex Hodk. & Renvoize), or restored prairie cropping systems in the Midwestern United States [17,23–27].

The Southeastern United States holds particular potential for cultivation of second-generation biofuel crops [28,29]. In comparison with the corn-soy and wheat belts of the Midwestern US, this region's longer growing season, high precipitation and relatively lower land costs make it attractive for biofuel crop production. However, far less is known about the biogeochemical consequences of land-use change to biofuel crop production in this region.

Energy cane, a promising crop for ligno-cellulosic fuel production, is a variety of sugarcane (*Saccharum officinarum*) that is higher yielding, more cold tolerant and has lower sucrose content than commercially produced sugarcane [28]. Because of its lower sugar concentration, it has not been widely cultivated, but has been of interest commercially as a genetic stock for improving cold tolerance in higher sucrose sugarcane strains [28]. With the development of ligno-cellulosic ethanol conversion technologies, sucrose concentration is less important for ethanol production, and energy cane could become an important biofuel feedstock as yields are high, ranging from 25–74 Mg·ha^{-1}·yr^{-1} dry mass (Table 1).

Florida is the largest sugarcane producing state in the US and is therefore a likely location for large-scale energy cane production [30]. Currently, 466,000 hectares of land in Florida are used for low-intensity grazing, and converting some portion of this land could provide an option for growing energy cane [31,32]. However, it is unknown if converting pasture to cultivated land will affect GHG exchange with the atmosphere and soil carbon storage. More frequent soil disturbance and the presence of larger quantities of litter from growing energy cane could increase CO_2 efflux to the atmosphere [33,34], while removing cattle from the landscape will displace methane (CH_4) efflux [35]. If fields are fertilized, nitrous oxide (N_2O) emissions may increase because of greater substrate availability for denitrifying microbes [36], and

indeed, high rates of N_2O efflux have been measured from sugarcane grown on highly fertilized soils in Australia [37]. However, considering the entire suite of greenhouse gasses, there may be an overall reduction in GHG flux due to the offset provided by greater atmospheric carbon uptake into the crop.

The region of Florida where energy cane is likely to be grown has two distinct soil types. The most common soils are Spodosols, which are low nutrient and low organic matter sands requiring significant fertilizer to maintain agricultural productivity [38]. Substantial sugarcane production in Florida also occurs on Histosols, which are high organic matter "mucks" that are not typically fertilized, as production on these soils can be maintained by N mineralization from organic matter [39]. The cultivation of Histosols began by draining swamplands, where organic matter had accumulated under anaerobic conditions. Drainage accelerates decomposition and further cultivation of these organic soils is associated with rapid oxidation of organic matter, resulting in significant soil C loss and emissions of CO_2 and N_2O to the atmosphere [9,40,41].

Theoretical [13,25] and empirical research [42,43] indicate that the conversion of land in the rain-fed Midwest currently used to produce corn for ethanol to perennial biofuel feedstocks such as switchgrass or Miscanthus (a close relative of sugarcane) would greatly reduce or reverse the emission of GHG to the atmosphere and rebuild depleted carbon stocks in the soil. Prior studies with Miscanthus in Europe have measured substantial decreases in nitrogen use, and large increases in soil biomass and organic matter relative to other agricultural land uses [44,45]. There have been no experimental studies that address how changing a landscape to cultivate energy cane will impact GHG emissions and soil C stocks. This is addressed here by using the process-based biogeochemical model DayCent to run *in silico* experiments to ask how land use change from pasture to energy cane production changes ecosystem GHG flux and soil C storage. We test the hypotheses that converting pastures to energy cane will lead to reductions in GHG flux to the atmosphere and increase soil C stocks, and that soil type is an important modulator of that change.

Methods

Plant and Soil Analyses

To parameterize the DayCent model, plants and soils were collected on private land in Highlands County, Florida (27° 21′ 49″ N, 81° 14′ 56″ W) in May 2011. Paired 4-m^2 plots ($n = 3$) were randomly located in energy cane fields that had been recently (<2 months) converted from pasture and in adjacent non-cultivated pasture on both Spodosols (hyperthermic Arenic Alaquods) and Histosols (hyperthermic Histic Glossaqualfs). We harvested all aboveground biomass from each plot. Soil samples were taken from the pastures in areas not yet under energy cane cultivation. Three soil cores to a depth of 1 m were extracted from each plot with a 1.75-cm diameter wet sampling tube (JMC product # PN010, Newton, IA). Soil cores were separated by depth (0–30 cm, >30 cm). Plant material and soils were oven dried at 65°C (plant material) and 105°C (soils) until they reached constant mass. Dried soils were coarse ground with a mill (model F-4, Quaker City, Phoenixville, PA), and then fine ground with a coffee grinder (Sunbeam Products Inc., Boca Raton, FL). Total C and N content may have been slightly underestimated from the dried Histosols due to volatilization, but the values we measured (Table 1) fall well within the range reported by NRCS Web Soil Survey [46]. Plant material was ground to pass a 425-μm mesh (Wiley mill, Thomas Scientific, Swedesboro, NJ, USA). Plant and soil subsamples

Table 1. Input parameters (mean and one standard error of the mean; SEM) for carbon and nitrogen concentration of energy cane and soils collected from the Highlands Ethanol farm, Highlands County, Florida.

		%C		%N	
		Mean	SEM	Mean	SEM
Energy cane	Live leaves	43.68	0.22	1.80	0.18
	Dead leaves	39.77	0.22	0.52	0.03
	Stalks	41.18	0.33	0.87	0.10
Soils	Soil Depth				
Histosols	0–60 cm	7.77	2.48	0.50	0.20
	60–100 cm	7.77	2.48	0.50	0.20
Spodosols	0–30 cm	0.77	0.17	0.04	<0.01
	30–60 cm	0.36	0.03	0.02	0.01
	60–100 cm	0.36	0.03	0.02	0.01

When site-specific data were not available, plant information was used from reference [65], and soil data were collected from the NRCS Web Soil Survey (http://websoilsurvey.nrcs.usda.gov/).

Table 2. Site information for studies used in DAYCENT model validation.

Site	Lit. Yield	Model Yield	Max. Temp.	Min. Temp.	Precipitation	Latitude	Longitude	Reference
Auburn, AL	26.1	25.4	24.2	9.8	1160	32.67	−85.44	Woodard and Prine, 1993
Belle Glade, FL	25.0	28.3	27.8	16.4	1378	26.68	−80.67	Korndorfer, 2009
EREC, FL	51.3	43.5	29.1	17.7	1181	26.65	−80.63	Gilbert et al., 2006
Gainesville, FL	35.6	27.3	27.0	13.7	1123	29.68	−82.27	Woodard and Prine, 1993
Hendry, FL	39.2	55.9	28.4	18.3	1362	27.78	−82.15	USDA, 2011
Hillsboro, FL	60.7	62.5	28.5	18.3	1547	27.90	−82.49	Gilbert et al., 2006
Houma, LA (1st ratoon)	36.6	38.2	25.2	14.8	500	29.57	−90.65	Legendre and Burner, 1995
Houma, LA (2nd ratoon)	34.9	37.6	25.2	14.8	500	29.57	−90.65	Legendre and Burner, 1995
Hundley, FL	73.5	62.0	28.5	18.1	1457	26.30	−80.16	Gilbert et al., 2006
Jay, FL (plant cane)	35.8	33.6	26.9	16.4	1321	28.65	−80.82	Woodard and Prine, 1993
Jay, FL (1st ratoon)	27.8	32.8	26.9	16.4	1321	28.65	−80.82	Woodard and Prine, 1993
Lakeview, FL	71.3	62.5	28.5	17.4	1275	26.30	−80.15	Gilbert et al., 2006
Hidalgo, TX	34.6	42.5	28.7	18.1	576	26.17	−97.93	Weidenfeld, 1995
Ona, FL (1st ratoon)	40.5	38.1	28.6	16.0	1160	27.48	−81.92	Woodard and Prine, 1993
Ona, FL (2nd ratoon)	30.2	31.6	28.6	16.0	1160	27.48	−81.92	Woodard and Prine, 1993
Pahokee, FL	60.5	65.5	28.4	17.5	1269	26.82	−80.66	Glaz and Ulloa, 1993
Palm Beach, FL	32.3	35.4	29.0	16.6	851	26.67	−80.15	USDA, 2011
Quincy, FL (1st ratoon)	26.3	25.1	25.8	12.9	1445	30.59	−84.58	Woodard and Prine, 1993
Quincy, FL (2nd ratoon)	27.8	21.7	25.8	12.9	1445	30.59	−84.58	Woodard and Prine, 1993
Shorter, AL	26.4	25.4	24.9	10.9	1119	32.40	−85.94	Sladden et al., 1991
Sundance, FL	42.1	43.5	28.6	17.5	1303	26.60	−80.87	Gilbert et al., 2006

Yield values from the literature and modeled yields for energy cane and sugarcane represent total aboveground biomass expressed as Mg ha^{-1} on a dry mass basis. Climate variables include mean annual maximum and minimum temperature (°C) and mean annual precipitation (mm).

within each plot were combined, and C and N concentrations were measured for depth-stratified soil samples (Table 1) and total above ground biomass with a flash combustion chromatographic separation elemental analyzer (Costech 4010 CHNSO Analyzer, Costech Analytical Technologies Inc. Valencia, CA). The instrument was calibrated with acetanilide obtained from Costech Analytical Technologies, Inc. Other physical soil attributes, including texture, bulk density and water holding capacity were obtained from the NRCS Web Soil Survey [46] for Highlands County, Florida.

The DayCent Model

The DayCent model [47,48] was developed to simulate ecosystem dynamics for agricultural, forest, grassland and savanna ecosystems [49–51]. The model is a daily time step version of the Century model [52,53], using the same soil carbon and nutrient cycling submodels to simulate soil organic matter dynamics (C and N) and nitrogen mineralization. DayCent uses more mechanistic submodels than Century to simulate daily plant production, plant nutrient uptake, trace gas fluxes (N_2O, CH_4), NO_3 leaching, and soil water and temperature [48,54–58].

The DayCent soil organic matter model is widely used to simulate the impacts of management practices on soil carbon dynamics and nutrient cycling. Specifically, the soil organic matter submodel has been used to simulate the impacts of soil tillage practices; no-tillage, minimum tillage and conventional tillage [59,60], crop rotations [59], and biofuel crops; woody biomass, switchgrass (*Panicum virgatum*), Miscanthus (*Miscanthus* X *giganteus*), and sugarcane [13,61] on soil carbon dynamics for agricultural systems. These studies test model performance against observed

data and demonstrate general success in simulating changes in soil carbon levels associated with management practices.

The soil trace gas submodel has been extensively tested using observed soil CH_4 and N_2O data sets from agricultural and natural ecosystems, and once parameterized with plant production data, provides accurate predictions of trace gas fluxes. Specifically, DayCent has successfully simulated the observed impacts of N fertilizer additions and cropping systems [50,58,59] on soil N_2O and CH_4 fluxes. The model results and observed data sets demonstrate that increasing N fertilizer levels increases soil N_2O fluxes and that soil N_2O fluxes are much lower for perennial crops as compared to annual crops.

The DayCent model has been used extensively to simulate grassland and crop yields [50,58,59,62], and to evaluate the environmental impacts of growing crops. Adler et al. [63] used the DayCent model to simulate net greenhouse gas fluxes (soil C status and soil CH_4 and N_2O fluxes) associated with the use of corn, soybeans, alfalfa, hybrid popular, reed canary grass and switchgrass for biofuel energy production in Pennsylvania. Davis et al. [25] used the DayCent model to simulate the environmental impacts of growing switchgrass and Miscanthus in Illinois and compared simulated plant production for switchgrass and Miscanthus with observed yield data. The authors also compared the net soil greenhouse gas fluxes (soil C changes and soil CH_4 and N_2O fluxes) associated with growing switchgrass and Miscanthus and growing corn and soybeans. Davis et al. [13] recently used the DayCent model to simulate the environmental impact of replacing the corn currently grown for ethanol production in the Corn Belt with perennial grasses (Miscanthus and switchgrass) for second-generation biofuel production. The authors found that the

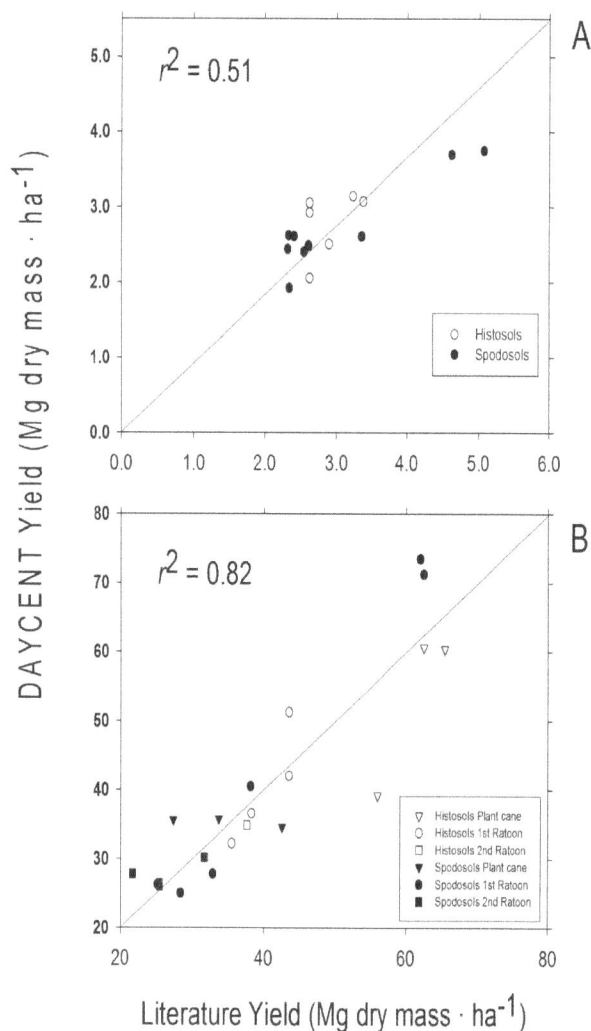

Figure 1. Regression analysis used in DayCent model valida-tion. Model output for dry mass yield was compared to literature values for A) pasture yield values from the USDA-NASS database, B) sugarcane and energy cane dry mass yield. Data points are compared to a 1:1 line.

DayCent model successfully predicted corn, Miscanthus, and switchgrass biomass production for U.S. sites with multiple N fertilizer levels. They also showed that the DayCent model successfully simulated observed annual soil N_2O fluxes from corn and switchgrass grown with multiple N fertilizer levels and showed that soil N_2O fluxes are much lower for fertilized switchgrass than for corn.

Furthermore, the basis for the DayCent model, Century, has been used to simulate sugarcane production in Brazil [61,64] and Australia [65]. These authors show that the Century/DayCent soil organic matter sub-model can correctly simulate the impacts of fertilizer, and organic matter additions on soil carbon levels and surface litter decay.

Model Parameterization

Energy cane is a variety of sugarcane, and thus parameterizing DayCent for this crop required only minor changes to the previously published input data used for sugarcane [61,65]. Energy cane differs from sugarcane in that it has increased cold

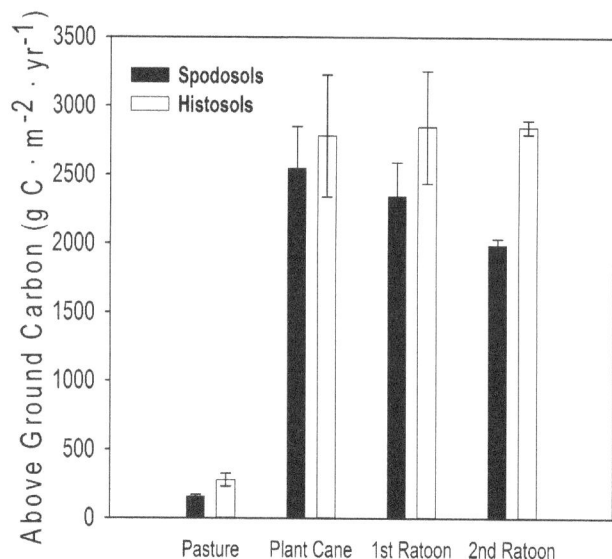

Figure 2. Modeled above ground production of grazed pasture and energy cane in Highlands County, Florida. Values are mean above ground carbon (g $C·m^{-2}·yr^{-1}$, ± SD) for 15 years in pasture, and for 5 × 3-year ratoon cycles in energy cane (each bar represents the average of 5 values, one for each year for each stage in the planting cycle).

tolerance, decreased sucrose content, and higher cellulose content. We adjusted parameters based on direct measurement of energy cane tissue traits described above. The principal changes from sugarcane to energy cane were reducing the minimum C:N ratio of leaves from 28.6 to 22.1 and changing the C:N of stems from 160 to 30.5. Because of this change in C:N, the parameter for C allocation to stems in DayCent also was modified (from 60% to 40%), to reflect the lower C content of stems relative to N for energy cane versus the previously modeled sugarcane parameters [65]. Bahiagrass (*Paspalum notatum* Flueggé) pasture was simulated using the existing DayCent model parameters for warm season grasses [66,67].

Histosols are challenging to model as organic matter and C content typically are uniform throughout the soil profile [68], but they also are known to subside because of oxidation of the highly labile organic C pools characteristic of these soils [69]. This subsidence was calculated from the modeled rate of organic matter loss and bulk density. DayCent simulates soil C flux to a depth of 30 cm [47], so as soil was lost with subsidence new soil and organic matter became part of this upper 30 cm column from below. This assumes that loss only occurred in the upper 30 cm, which is reasonable since this is the disturbed and aerated part of the soil. The C and N added from low in the soil profile was calculated from the rate of subsidence and the measured elemental contents and bulk density of the soil that was below 30 cm, when sampled, which is at time zero in our model. However, model output calculates GHG and soil C to soil depths to 30 cm.

Model Validation

Literature values of aboveground production (dry mass) for grazed pasture, sugarcane and energy cane (Table 2) were used to validate DayCent. Validation focused on aboveground biomass production because this variable has been measured widely across a range of sites. While there were insufficient data on trace gas flux or changes in soil C in sugarcane or energy cane for validation of

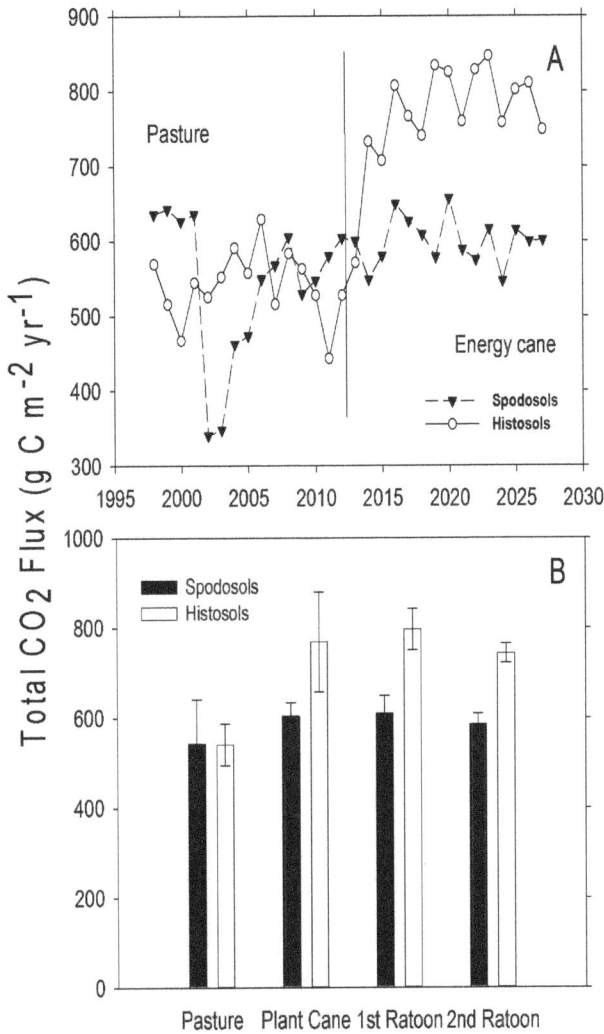

Figure 3. Modeled total soil CO_2 flux from pasture and land converted to energy cane in Highlands County, Florida. A) Total annual soil CO_2 flux (expressed as g C·m^{-2}). Dashed line represents year of land use conversion from pasture to energy cane. B) Mean total soil CO_2 flux (g C·m^{-2}·yr^{-1}, ± SD) for 15 years in pasture, and for 5, 3-year ratoon cycles in energy cane (each bar represents the average of 5 values, one for each year for each stage in the planting cycle).

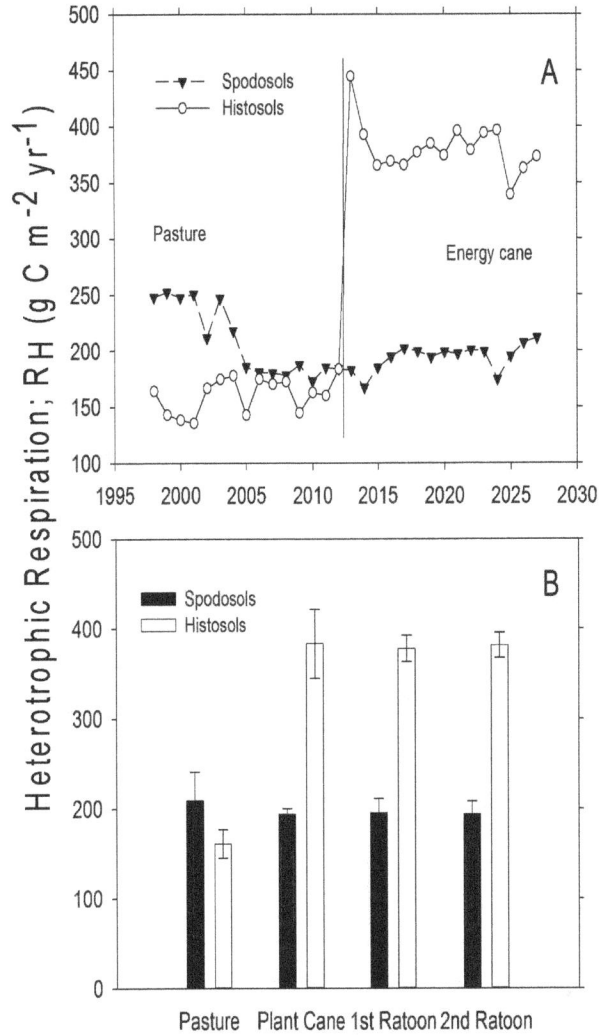

Figure 4. Modeled heterotrophic respiration (R_H) from pasture and land converted to energy cane in Highlands County, Florida. A) Total annual heterotrophic respiration (g C·m^{-2}). Dashed line represents year of land use conversion from pasture to energy cane. B) Mean heterotrophic respiration (g C·m^{-2}·yr^{-1}, ± SD) for 15 years in pasture, and for 5, 3-year ratoon cycles in energy cane (each bar represents the average of 5 values, one for each year for each stage in the planting cycle).

these variables, validation based on productivity for other crops reliably predicts trace gas flux [59,60,63,70–72].

We compiled a literature database of 17 sites that had reliable data on sugarcane and energy cane yield. There were also pasture productivity data for 15 of those sites [73]. In some instances we were able to contact researchers directly to access unpublished data (Table 2). The geographic range of sites represents the breadth of sugarcane production in the continental United States, and the potential range of energy cane production on currently grazed pastures. For all sites, daily weather data inputs (minimum and maximum temperature, daily precipitation) from 1980 to 2002 were obtained from the DayMet database [74]. The model was run using the DayCent growing degree-day subroutine to determine plant emergence, senescence and death, based on plant phenological characters and daily weather data. Soil data for the validation sites were obtained from the NRCS Web Soil Survey [75]. Using the same schedule of management events used for the

in silico experiments (described below), DayCent was run with site-specific soil and weather data for each sites. The fit of modeled to measured above ground dry mass production (Mg dry matter ha^{-1}) of our simulations of grazed pasture and energy cane were separately tested via linear regression, using the linear model function in R [76].

Initial Simulation Conditions

A "spin-up" period in DayCent based on historical land use and vegetation type was used to set initial soil conditions. The dominant, historic vegetation type for this area of south-central Florida was savanna, with a mixture of grasses and several species of scrub-oak, or sawgrass for the swamp areas [77]. A mix of perennial C_3 grasses species and symbiotic N_2 fixing plants, were used as initial conditions for the savanna simulation (initial vegetation type "savanna" in DayCent). A period of 2000 years was simulated to obtain an initial soil C and N conditions prior to

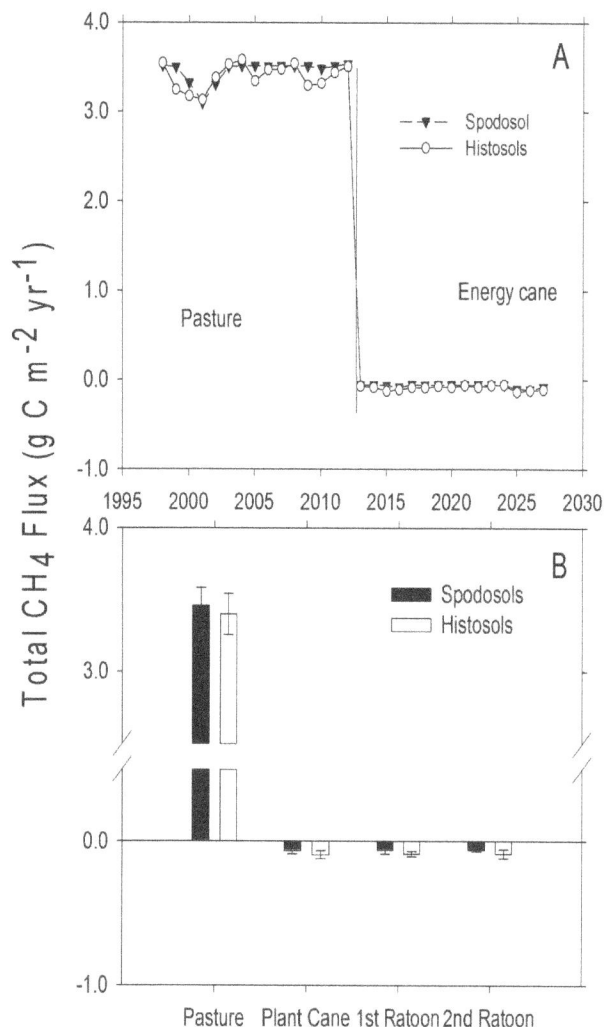

Figure 5. Modeled CH₄ flux from pasture and land converted to energy cane in Highlands County, Florida. A) Total annual CH₄ flux (g C·m⁻²). The solid vertical line represents year of land use conversion from pasture to energy cane, positive values indicate CH₄ efflux and negative values indicate CH₄ uptake. B) Mean CH₄ flux (g C·m⁻²·yr⁻¹, ± SD) for 15 years in pasture, and for 5, 3-year ratoon cycles in energy cane (each bar represents the average of 5 values, one for each year for each stage in the planting cycle).

Figure 6. Changes in total soil organic C from pasture and land converted to energy cane in Highlands County, Florida. A) Total annual SOC flux (g C·m⁻²). The solid vertical line represents year of land use conversion from pasture to energy cane. B) Mean SOC flux (g C·m⁻²·yr⁻¹, ± SEM) for 15 years in pasture, and for 5, 3-year ratoon cycles in energy cane (each bar represents the average of 5 values, one for each year for each stage in the planting cycle).

our *in silico* experiments. The model was run for spin ups and all subsequent experiments using the growing degree-day sub-routine.

In silico Experiments

Model simulations were then run to determine the GHG soil-atmosphere exchange and change in soil C predicted for conversion of pasture to energy cane on the two dominant soil types, Spodosols and Histosols. We used daily weather data inputs (minimum and maximum temperature, daily precipitation) from 1951 to 2002, which was the longest time period available for Highlands County, Florida obtained from the DayMet database [74]. This weather file is used by DayCent to create a mean and standard deviation of weather parameters, thus the more weather data available for a given site, the more accurately the variability of a site will be captured by the model.

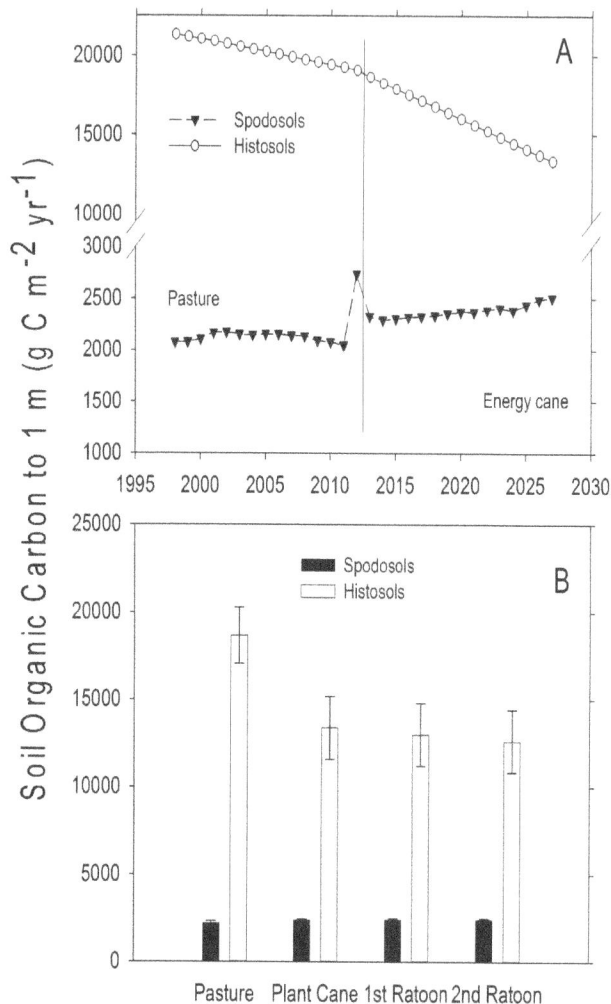

To initiate the experimental simulations, in 1998 we converted the savanna by removing all above ground biomass and plowing to a depth of 30 cm. A landscape conversion to a grazed Bahiagrass (*Paspalum notatum* Flueggé) ecosystem was then simulated. Bahiagrass is a common forage grass for this part of Florida that would be considered "improved pasture", although usually not fertilized or irrigated [78]. We simulated grazing in our modeling experiment by annually removing 10% of live shoot and 1.0% of standing dead shoots. Prior to planting energy cane, another plow event to 30 cm was initiated to remove the pasture vegetation and simulate the physical land use change.

The simulated cycle of energy cane planting and harvest was based on the sugarcane literature [65,79,80] and discussions with University of Florida and USDA sugarcane agronomists [81,82]. In the simulations, energy cane was planted in January of the first year (2013), followed by a two-year ratoon (crop regenerated from remaining biomass) from which 80% of the above ground biomass was harvested in December. At the end of the second ratoon, the crop was removed and the land plowed before planting a new

Table 3. Modeled ecosystem carbon, nitrogen and greenhouse gas fluxes after converting pasture to energy cane on nutrient poor Spodosols and organic matter rich Histosols.

	Spodosols			Histosols		
	Pasture	**Energy cane**	**Δ**	**Pasture**	**Energy cane**	**Δ**
SOC (g C·m^{-2})	2736	2513	−224	16087	10373	−5715
Nitrogen Mineralization (g N·m^{-2})	134	203	69	216	293	77
Heterotrophic Respiration (g C·m^{-2})	3130	2913	−218	2413	5715	3302
Total Soil CO_2 Efflux (g C·m^{-2})	8148	8993	845	8111	11540	3429
CH_4 (g CO_2eq·m^{-2})	2980	−33	−3013	2958	−46	−3004
N_2O (g CO_2eq·m^{-2})	214	649	435	6713	1742	−4970
Total System C Flux (g CO_2eq·m^{-2})	−1159	−2812	−1653	−1367	924	2291
Total Greenhouse Gas Flux (g CO_2eq·m^{-2})	2035	−2196	−4231	8304	2620	−5684

Greenhouse gas and N mineralization values are the sum of values from pasture 15 years prior to conversion to energy cane and the sum values for 15 years following the conversion to energy cane. Positive values indicate a flux to the atmosphere and negative values indicate uptake from the atmosphere by the ecosystem. Soil organic matter values are the differences between the last year of energy cane production and the last year of pasture. Total GHG values are the sums of CH_4, N_2O and total system C flux (calculated in DayCent as the difference between all C uptake and storage versus efflux from respiration) expressed as CO_2e. Differences (Δ) represent the values for energy cane minus pasture.

plant crop. This cycle of ratooning and planting was repeated in the simulation for fifteen years following conversion from pasture; i.e. five cycles of three years each. This three-year planting cycle is typical for sugarcane production in Florida [83,84].

Irrigation events were scheduled every month throughout the dry season, and every two months during the rainy season to maintain soil water at field capacity. Fertilizer ($NH_4^+ - NO_3^-$) was applied in mid February and mid June of each year of the simulation, at a rate of 102 kg N·ha^{-1} per fertilization event for Spodosols. No fertilizer was added to the organic rich Histosols. This fertilization schedule was based on studies that suggest that a split fertilization regime at this rate maximizes sugarcane yield, and that fertilizing above this level does not increase yield but increases N_2O efflux [38,65,85]. The input files used to drive DayCent (e.g. schedule files, plant input parameters, and soil input files) are available online [86].

Calculations and Statistical Analyses

We summed daily GHG and soil C fluxes from DayCent to calculate yearly fluxes and report those in g C or N·m^{-2} yr^{-1}, with the exception of total GHG values which are reported as CO_2 equivalents [87] and factored by warming potential ($CO_2 = 1$, $CH_4 = 23$, $N_2O = 296$; ref. 85). Total ecosystem C flux was calculated as the annual change in total ecosystem C storage between the beginning and end of a year and represents the net ecosystem carbon balance expressed in CO_2eq [88,89].

Because the model experiments were performed using the same site with the same weather data, but controlled for soil type, the simulations had the structure of a paired design where each year was a replicate [90]. We therefore used paired t-tests to determine differences between soil types within a plant type ($n = 15$) and between plant types within a soil type ($n = 15$). The variation reported with mean annual values represents inter-annual variation in the predicted variables. Heteroscedasticity was examined with the Fligner-Killeen test, and output data distributions, which did not meet variance assumptions, were compared with the Wilcoxon rank-sum test. The routines t.test (paired = TRUE) and wilcox.test were performed using R [76,90].

Because of the large number of pair-wise comparisons of our model results, the False Discovery Rate (FDR) test was used to account for multiple comparisons. The FDR test is less conser-

vative than a P-value adjustment such as the Bonferroni correction, and determines the probability of a Type I error. We calculated a FDR of 0.024 for our matrix of tests, and therefore justified the use of multiple paired t-tests without P-value adjustment [91].

Results

Predicted harvested yields for both pasture and energy cane in our validation sites agreed well with measured values from the literature (Pasture: $r^2 = 0.52$, Energy cane: $r^2 = 0.82$, Figure 1A & 1B), indicating that our modeled predictions provided a good representation of the productivity that drives the biogeochemical dynamics of DayCent.

For our modeled site, DayCent estimated a large increase in aboveground plant biomass production after conversion of pasture to energy cane (Figure 2); annual aboveground biomass production increased by a factor of 14 on Spodosols and by a factor of 10 on Histosols, relative to pasture. Energy cane production ranged from 1911–3153 g C m^{-2} yr^{-1} (46–76 Mg dry biomass·ha^{-1}). Predicted energy cane production remained high through the three harvests on Histosols, but declined through the modeled ratoon cycle on Spodosols (Figure 2).

There was considerable temporal variation in predicted soil CO_2 efflux from pasture in the 15 years simulated prior to the conversion to energy cane (Figure 3a). This variation was particularly evident for pasture on Spodosols and was driven primarily by variation in precipitation. Total soil CO_2 efflux was similar for pasture on both soil types, but significantly increased when averaged over 15 years after conversion to energy cane on the Histosols (Figure 3a; $t = 10.65$, d.f. $= 14$, $P<0.001$). Land use conversion did not increase CO_2 efflux on Spodosols ($t = 0.58$, d.f. $= 14$, $P = 0.57$). Following conversion to energy cane CO_2 efflux from Histosols was significantly higher than energy cane on Spodosols (Figure 3b; $t = 9.56$, d.f. $= 14$, $P<0.001$).

The conversion of land from pasture to energy cane had no significant effect on the predicted heterotrophic component of soil respiration (R_H) on Spodosols (Figure 4a), but caused a large increase in R_H from the Histosols (Figure 4a; $t = 31.86$, d.f. $= 14$, $P<0.001$) and resulted in higher R_H on Histosols than Spodosols following the conversion to energy cane (Figure 4b; $t = 23.68$, d.f.

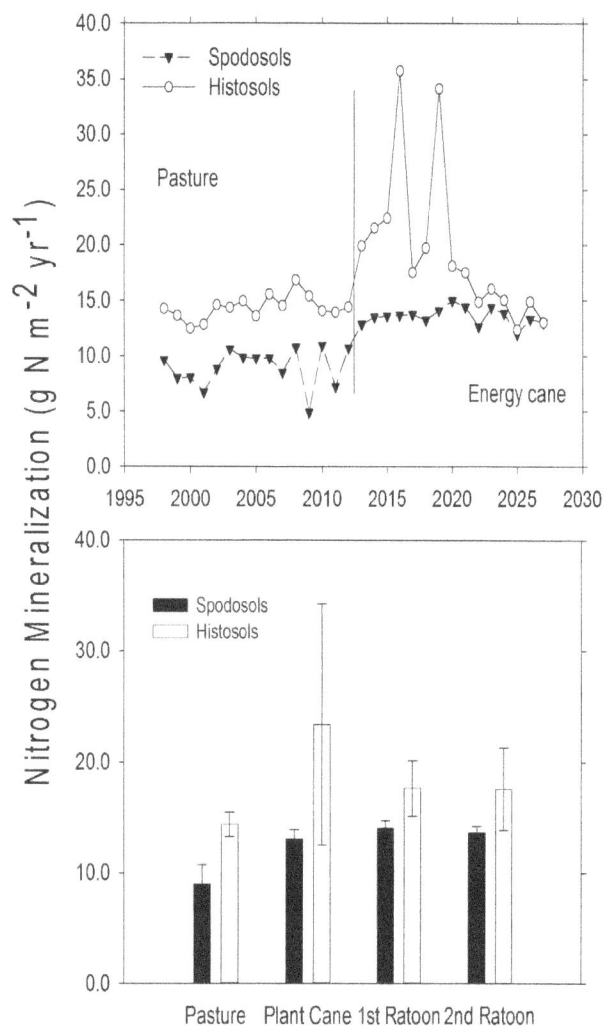

Figure 7. Modeled nitrogen mineralization rates from pasture and land converted to energy cane in Highlands County, Florida. A) Total annual N mineralization rate (g $N \cdot m^{-2}$). The solid vertical line represents year of land use conversion from pasture to energy cane. B) Mean N mineralization rate (g $N \cdot m^{-2} \cdot yr^{-1}$, ± SD) for 15 years in pasture, and for 5, 3-year ratoon cycles in energy cane (each represents the average of 5 values, one for each year for each stage in the planting cycle).

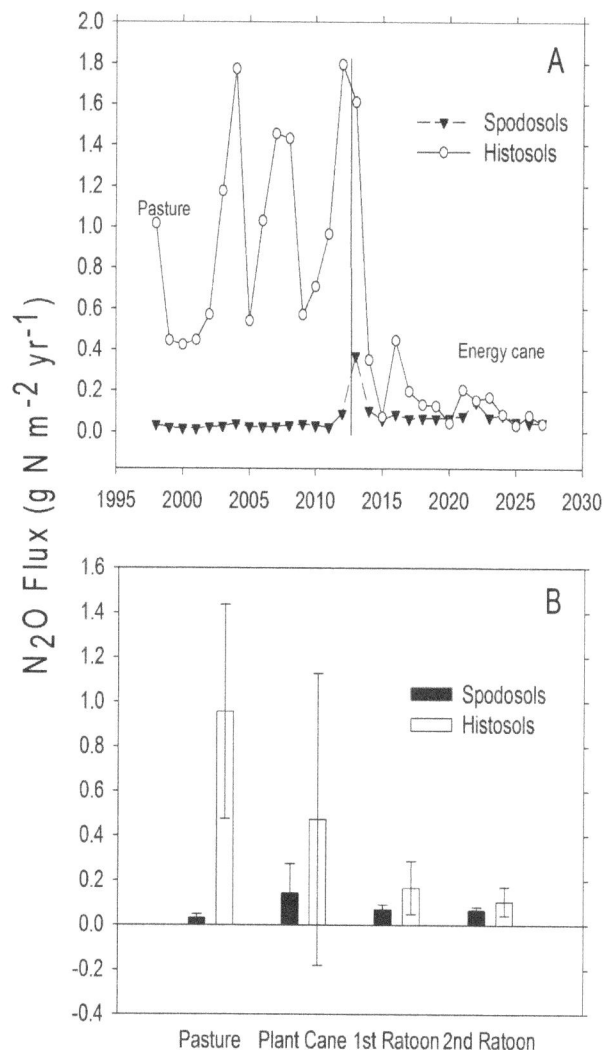

Figure 8. Modeled N_2O flux from pasture and land converted to energy cane in Highlands County, Florida. A) Total annual N_2O flux (g $N \cdot m^{-2}$). The solid vertical line represents year of land use conversion from pasture to energy cane, positive values indicate N_2O efflux and negative values indicate N_2O uptake. B) Mean N_2O flux (g $N \cdot m^{-2} \cdot yr^{-1}$, ± SD) for 15 years in pasture, and for 5, 3-year ratoon cycles in energy cane (each bar represents the average of 5 values, one for each year for each stage in the planting cycle).

$= 14$, $P < 0.001$). Prior to the conversion to energy cane, modeled (R_H) was slightly higher in pasture on Spododols than on Histosols (Figure 4; $t = 31.86$, d.f. $= 14$, $P < 0.001$).

On both soil types, the removal of cattle associated with the conversion of pasture to energy cane caused a substantial change in predicted CH_4 flux ($t = 185$, d.f. $= 14$, $P < 0.001$ on Spodosols; $t = 167$, d.f. $= 14$, $P < 0.001$ on Histosols; Figure 5a). Without cattle, pastures were a small CH_4 sink (0.16–0.60 g $C \cdot m^{-2} \cdot yr^{-1}$ uptake in Spodosols, 15 year sum $= 112$ g $CO_2eq \cdot m^{-2}$, 0.12–0.57 g$C \cdot m^{-2} \cdot yr^{-1}$ uptake for Histosols, 15 year sum $= 135$ g $CO_2eq \cdot m^{-2}$). Introducing cattle at stocking rates and grazing intensity typical for this region (1 head cattle·ha^{-1}: ref. 31), caused pasture on both soil types to be a substantial source of CH_4 to the atmosphere (Figure 5).

Changes in vegetation and management practices altered soil organic carbon (SOC), and these changes were particularly evident on the Histosols (Table 2; Figure 6). Histosols had a

larger pool of active C (weekly to monthly turnover) than Spodosols under both pasture and energy cane (pasture, $t = 19.25$, d.f. $= 14$, $P < 0.001$; energy cane, $t = 14.21$, d.f. $= 14$, $P < 0.001$). Comparing the remaining total SOC pools between the end of pasture and the last year of the energy cane simulation, Histosols lost a large amount of soil organic C; 5714 g $C \cdot m^{-2}$ to 1 m depth (Figure 6; $t = 296$, d.f. $= 14$, $P < 0.001$), compared to the SOC loss from Spodosols of 224 g $C \cdot m^{-2}$ to 1 m (Table 3).

Nitrogen mineralization increased after pasture was converted to energy cane on both the fertilized Spodosols ($t = 9.02$, d.f. $= 14$, $P < 0.001$) and the non-fertilized Histosols ($t = 2.72$, d.f. $= 14$, $P = 0.02$). After conversion to energy cane, Histosols had higher rates of N mineralization than Spodosols (Figure 7; $t = 3.43$, d.f. $= 14$, $P = 0.004$), and this increase in available N likely accounted for the continued high yields on Histosols.

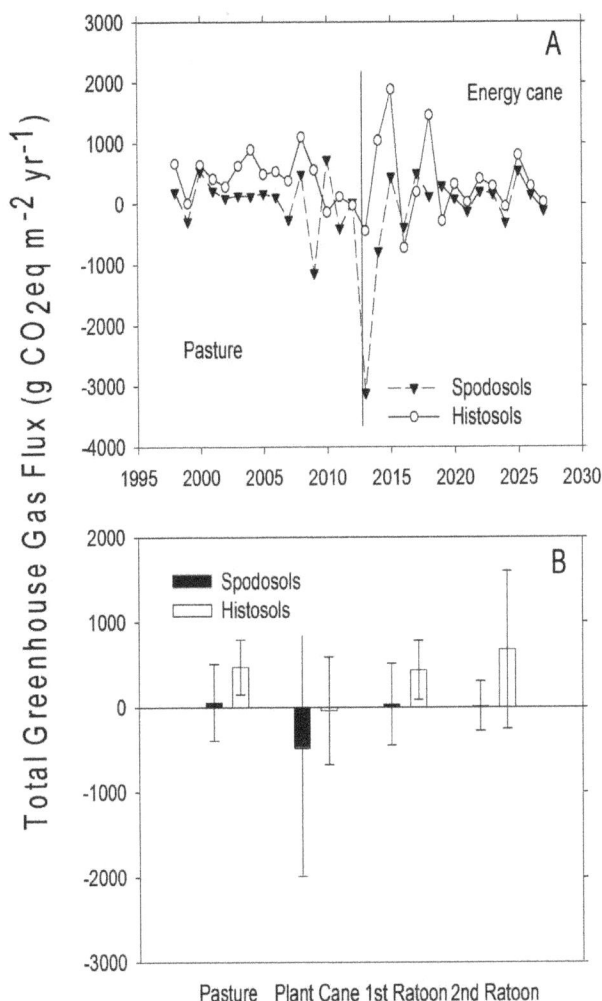

Figure 9. Changes in total greenhouse gas (GHG) from pasture and land converted to energy cane in Highlands County, Florida. Positive values indicate GHG efflux and negative values indicate GHG uptake. A) Total annual GHG flux, reported as CO_2 equivalents converted to account for differences in warming potential (g $CO_2e \cdot m^{-2}$). The solid vertical line represents year of land use conversion from pasture to energy cane. B) Mean greenhouse gas flux in CO_2 equivalents converted to account for differences in warming potential (g $CO_2e \cdot m^{-2} \cdot yr^{-1}$, ± SD) for 15 years in pasture, and for 5, 3-year ratoon cycles in energy cane (each bar represents the average of 5 values, one for each year for each stage in the planting cycle).

Prior to conversion to energy cane, N_2O efflux was higher in pastures on Histosols compared to Spodosols (Figure 8; Wilcoxon rank sum, W = 225, P<0.001). After conversion to energy cane, Histosols remained greater sources of N_2O than Spodosols (t = 12.15, d.f. = 14, P<0.001). Conversion of pasture to energy cane decreased N_2O efflux on Histosols (Figure 8a; t = 4.30, d.f. = 14, P<0.001), but increased N_2O efflux on Spodosols (Figure 8b; t = 2.87, d.f. = 14, P = 0.01). It is likely that N_2O emission from Histosols decreased following conversion because the increase in productivity resulted in a higher uptake of nitrate that would otherwise be available for denitrification.

Total GHG exchange (global warming potential) was calculated by converting the fluxes of CH_4 and N_2O to CO_2 equivalents based on their warming potential relative to CO_2 [87] and summing these with total system C flux (Table 2). Variation in weather caused substantial inter-annual variation in total GHG flux, with both pasture and energy cane varying between net GHG sinks and sources (Figure 9); no significant differences in annual GHG flux were resolved on either soil type (Spodosols: t = 1.15, d.f. = 14, P = 0.27; Histosols: t = 0.13, d.f. = 14, P = 0.90). When the cumulative GHG emission were calculated for the fifteen years prior to conversion, pasture was a net source to GHGs to the atmosphere on both soil types, and pasture was a stronger source on Histosols (8304 $gCO_2eq \cdot m^{-2}$) than on Spodosols (2035 $gCO_2eq \cdot m^2$; Table 3). Conversion of pasture to energy cane caused the Spodosols to transition from a source to a sink for GHGs and reduced the flux of GHGs to the atmosphere on Histosols. On both soil types, the reduction GHG emission to the atmosphere was associated with a large decrease in CH_4 emissions caused by the elimination of cattle grazing. On the Histosols, the reduction in N_2O emissions to the atmosphere also contributed to reduced emission of GHGs. This analysis of GHG emissions and their corresponding global warming potentials did not account for the displacement of fossil fuel emissions by the biofuel product.

Discussion

Parameterization of the DayCent model for energy cane, an emerging bioenergy crop, successfully simulated biomass production across the southeast United States (Figure 1). Our simulations suggested high yields for energy cane on former pastureland in a subtropical climate when Spodosols are highly fertilized (200 kg $N \cdot ha^{-1} \cdot yr^{-1}$), and when microbial activity in Histosols leads to high rates of N mineralization (rates were 44% higher on Histosols). When integrated over 15 years (Table 3), conversion of pasture to energy cane on Spodosols converted a net source of GHG (due to cattle CH_4 emissions) to a sink driven by the removal of cattle and the increase in C uptake by energy cane. While Histosols were a net GHG source under both pasture and energy cane, the source was reduced by the land use conversion (Table 3). The GHG improvement resulting from this conversion from pasture to energy cane would be even greater if fossil fuel displacement by cellulosic ethanol had been included.

The range of our simulated energy cane yields was 46–76 Mg ha^{-1} dry mass per year on fertilized Spodosols and unfertilized Histosols. Using published values for the conversion efficiency for the production of cellulusoic ethanol [22], a hypothetical energy cane farm of 10,000 ha could therefore produce between 142–236 million liters of ethanol [92]. In comparison, equal areas of land devoted to corn grain and Miscanthus in the Midwest would yield between 25 and 73% this amount of ethanol, respectively, assuming the maximum yields reported by other authors [11,22].

Typically, sugarcane yield declines with ratooning, the repeated harvests of aboveground material generated by vegetative growth [93]. The model reproduced the yield decline for energy cane on Spodosols but not on Histosols, but the model in its current configuration probably failed to capture the mechanisms that would normally cause a decline in yield. Various factors ranging from increases in nematode populations and ratoon stunting disease, to mechanical compaction of the soil have been implicated in ratoon decline [94,95], and these were not accounted for in the model. Although sugarcane in Florida typically is grown for three years and three annual harvests before re-planting, if it were grown for more years between re-planting, we would expect a continuing yield decline on the Spodosols. In contrast, continued mineralization of organic matter on Histosols may sustain high yields beyond the 3-year period simulated in the model. On both soils the GHG benefits would be improved with longer ratoon cycles because of less soil disturbance due to decreased frequency of soil disturbance for replanting.

Organic matter (OM) content of soils is important for sustaining high yields of sugarcane, in part because OM mineralization provides the labile N necessary to sustain plant growth. Spodosols had much less OM than Histosols (Appendix I; Figure 6; [65,96]). The high CO_2 efflux rates from Histosols (Figures 3–4) and the patterns of SOC loss following land use change (Figure 6) correspond to higher rates of OM mineralization. The associated higher rates of N mineralization (Table 3; Figure 7) on Histosols provided additional N to energy cane and improved crop yield (Figure 2). Although energy cane on Spodosols was fertilized to offset the low N content of these soils, rates of nitrification (the process by which NH_4^+ is converted into the highly mobile NO_3^- anion) were higher on these soils. The fertilizer applied to energy cane crops on Spodosols in the simulations was NH_4^+ - NO_3^-, a labile substrate for nitrification [66]. Spodosols had consistently higher nitrification rates than Histosols, and therefore higher NO_3^- content because of fertilization, and it is possible that some fraction of fertilizer was lost before plant uptake [97]. We hypothesize that a combination of NO_3^- leaching from fertilizer before plant uptake, lower initial N content, and lower mineralization rates may have created a stronger N limitation to yield on Spodosols but not on Histosols.

Before land use change, pasture on both soil types was a net source of GHGs to the atmosphere (Table 3). This is consistent with both direct measurements [98] and modeling efforts [99] that have found grazed pastures to be net sources of GHGs, but this is also a function of grass species present and animal stocking density [100]. The model estimated that pastures were sinks for CO_2, with total C uptake of 1159 g CO_2 m^{-2} and 1367 g CO_2 m^{-2} over 15 years on Spodosols and Histosols, respectively (Table 3). In the absence of cattle, both soil types were CH_4 sinks (112 and 135 g CO_2eq, respectively), but including reasonable estimates of CH_4 efflux from cattle (Figure 5) and N_2O efflux from soils (Figure 8) resulted in net GHG emission to the atmosphere on both pasture soils (Table 3). Following conversion to energy cane, the production of N_2O on Spodosols increased (Figure 8) within the range of N_2O flux rates previously reported for Australian sugarcane fertilized at similar rates to this study [37]. The increase in N_2O was offset by uptake of CO_2 and the change from a source to a sink for CH_4 (Figure 5), with the net effect that Spodosols became a net GHG sink (Table 3). Indeed, over 15 years energy cane on Spodosol was a GHG sink of >40 Mg CO_2eq per hectare (Table 3). On Histosols, eliminating grazing following the conversion of pasture to energy cane caused a similar decrease in CH_4 efflux to the atmosphere (Figure 5) and this land use change also reduced N_2O emissions (Figure 8; Table 3). However, following land conversion this system switched from a net CO_2 sink to a source, and this change in total system C prevented energy cane on Histosols from becoming a net sink for GHGs. The driver for GHG production on Histosols was higher R_H, and significant losses of soil organic matter [69] that resulted in total C efflux from these soils (Table 3).

The model successfully simulated energy cane biomass production across a range of sites across the southern United States (Figure 1). Previous studies have shown that DayCent reliably predicts soil biogeochemistry and GHG exchange when parameterized for net primary production [51], suggesting that the estimates of GHG flux and soil C dynamics were reasonable. Eddy-flux measurements of GHG exchange that are now being initiated at this site will provide an independent test of the predictions of GHG effects of conversion made here.

Indirect land use change (ILUC) – the stimulation of deforestation or increased agriculture in other parts of the world driven by diversion of current agricultural land to bioenergy production – potentially poses an environmental risk of bioenergy production [7,101]. Growing energy cane on land converted from low stocking density pasture would be unlikely to trigger significant increases in food price or ILUC in the way that large-scale shifts from corn or soy production in the Midwestern United States would motivate greater production of those crops elsewhere [13]. Indeed, the recommended stocking density for Bahiagrass pasture in this region is ~1 animal·ha^{-1} [31], and cattle and calf operations in Florida account for less than 6% of the state's annual agricultural revenue [102]. The loss in meat production could be redressed with minimal increases in current stocking rates, and would be unlikely to trigger the type of large-scale landscape changes that may occur through the diversion of midwestern agricultural land [7]. However, displacing cattle for energy cane production may potentially increase methane emissions elsewhere, which would negate the local benefit of reduced methane flux to the atmosphere.

The environmental impacts of changing land use from pasture to energy cane were highly dependent on the soil type. Whereas the cultivation of Histosols results in high CO_2 efflux and the reduction of soil carbon (Figures 3, 4, and 6), the model predicted that energy cane crops on Spodosols would act as a net C and GHG sink (Figure 6, Table 3). From both a biofuel and biogeochemical perspective, these results suggest that energy cane grown on nutrient poor soils, as opposed to organic soils, has the potential to be a high-yielding bio-ethanol feedstock that creates a GHG sink in the Southeastern United States.

Acknowledgments

We thank Michael Masters for lab analysis of plant and soil C and N. Lykes Brothers Inc. graciously provided access to energy cane plantations and pastures, which greatly helped in parameterizing our model. We also thank Dr. Robert Gilbert, Dr. Barry Glaz, and Mr. Pedro Korndorfer for sharing their knowledge on sugar and energy cane agronomy and data that aided our modeling effort.

Author Contributions

Conceived and designed the experiments: BDD KJAT WJP EHD. Performed the experiments: BDD SCD CK WJP. Analyzed the data: BDD SCD KJAT WJP SPL EHD. Contributed reagents/materials/analysis tools: CK WJP. Wrote the paper: BDD KJAT WJP SPL EHD.

References

1. IPCC (2007) Climate Change 2007: The Physical Science Basis. Contribution of Working Group I to the Fourth Assessment Report of the Intergovernmental Panel on Climate Change. Solomon S, Qin D, Manning M, Chen Z, Marquis M et al. Eds. Cambridge, UK, Cambridge, UK.

2. Le Quéré C, Raupach MR, Canadell JG, Marland G, Bopp L, et al. (2009) Trends in the sources and sinks of carbon dioxide. Nat Geosci 2: 831–836.

3. Pan Y, Birdsey RA, Fang J, Houghton R, Kauppi PE, et al. (2011) A large and persistent carbon sink in the world's forests. Science 333: 988–993.

4. Lal R (2004) Soil carbon sequestration impacts on global climate change and food security. Science 304: 1623–1627. doi: 10.1126/science.1097396.

5. Tilman D (1998) The greening of the green revolution. Nature 396: 211–212.

6. Fargione JE, Hill JD, Tilman D, Polasky S, Hawthorne P (2008) Land clearing and the biofuel carbon debt. Science 319: 1235–1238.

7. Searchinger T, Heimlich R, Houghton RA, Dong F, Elobeid A, et al. (2008) Use of U.S. croplands for biofuels increases greenhouse gases through emissions from land-use change. Science 319: 1238–1240.

8. Melillo JM, Reilly JM, Kicklighter DW, Gurgel AC, Cronin TW (2009) Indirect emissions from biofuels: How important? Science 326: 1397–1399.

9. Börjesson P (2009) Good or bad bioethanol from a greenhouse gas perspective – What determines this? Applied Energy 86: 589–594.

10. United States Congress (2007) The Energy Independence and Security Act of 2007 (H.R. 6). Available: http://frwebgate.access.gpo.gov/cgibin/getdoc.cgi?dbname = 110_cong_bills&docid = f: h6enr.txt.pdf. Accessed 2010 Dec 19.

11. Heaton EA, Dohleman FG, Long SP (2008) Meeting US biofuel goals with less land: the potential of Miscanthus. Glob Change Biol 14: 2000–2014.

12. Fargione JE, Plevin RJ, Hill JD (2010) The ecological impact of biofuels. Ann Rev Ecol Evol Syst 41: 351–377.

13. Davis SC, Parton WJ, Del Grosso SJ, Keough C, Marx E, et al. (2012) Impacts of second-generation biofuel agriculture on greenhouse gas emissions in the corn-growing regions of the US. Front Ecol Environ 10: 69–74. doi:10.1890/110003.

14. Dien BS, Bothast RJ, Nichols NN, Cotta MA (2002) The U.S. corn ethanol industry: an overview of current technology and future prospects. Int Sugar J 103: 204–211.

15. Davis SC, Anderson-Teixeira KJ, DeLucia EH (2009) Life-cycle analysis and the ecology of biofuels. Trends Plant Sci 14: 140–146.

16. O'Hare M, Plevin RJ, Martin JI, Jones AD, Kendall A, et al. (2009) Proper accounting for time increases crop-based biofuels' greenhouse gas deficit versus petroleum. Environ Res Lett 4: doi:10.1088/1748–9326/4/2/024001.

17. Donner SD, Kucharik CJ (2008) Corn-based ethanol production compromises goal of reducing nitrogen export by the Mississippi River. Proc Natl Acad Sci USA 105: 4513–4518.

18. Hill J, Polasky S, Nelson E, Tilman D, Huo H (2009) Climate change and health costs of air emissions from biofuels and gasoline. Proc Nat Acad Sci USA 106: 2077–2082.

19. Smeets EMW, Bouwman LF, Stehfest E, van Vuuren DP, Posthuma A (2009) Contribution of N2O to the greenhouse gas balance of first-generation biofuels. Glob Change Biol 15: 1–23.

20. Solomon BD, Barnes JR, Halvorsen KE (2007) Grain and cellulosic ethanol: history, economics, and energy policy. Biomass Bioenergy 31: 416–425.

21. Tilman D, Socolow R, Foley JA, Hill J, Larson E (2009) Beneficial biofuels-the food, energy and environment trilemma. Science 325: 270–271.

22. Somerville C, Youngs H, Taylor C, Davis SC, Long SP (2010) Feedstocks for lignocellulosic biofuels. Science 329: 790–792.

23. Tilman D, Hill J, Lehman C (2006) Carbon-negative biofuels from low-input high-diversity grassland biomass. Science 314: 1598–1600.

24. Anderson-Teixeira KJ, Davis SC, Masters MD, DeLucia EH (2009) Changes in soil organic carbon under biofuel crops. Glob Change Biol Bioenergy 1: 75–96.

25. Davis SC, Parton WJ, Dohleman FG, Smith CM, Del Grosso S, et al. (2010) Comparative biogeochemical cycles of bioenergy crops reveal nitrogen fixation and low greenhouse gas emissions in a Miscanthus x giganteus agro-ecosystem. Ecosystems 13: 144–156.

26. Robertson GP, Hamilton SK, Del Grosso SJ, Parton WP (2011) The biogeochemistry of bioenergy landscapes: carbon, nitrogen, and water considerations. Ecol App 21: 1055–1067. doi: 10.1890/09–0456.1.

27. Zeri M, Anderson-Teixeira KJ, Masters MD, Hickman G, DeLucia EH, et al. (2011) Carbon exchange by establishing biofuel crops in central Illinois. Agric Ecosyst Environ 144: 319–329.

28. Sladden SE, Bransby DI, Aiken GE, Prine GM (1991) Biomass yield and composition, and winter survival of tall grasses in Alabama. Biomass Bioenergy 1: 123–127.

29. Mark T, Darby P, Salassi M (2009) Energy cane usage for cellulosic ethanol: estimation of feedstock costs. Southern Agricultural Economics Association Annual Meeting, Atlanta, Georgia, January 31-February 3, 2009.

30. Baucum LE, Rice RW (2009) An overview of Florida sugarcane. University of Florida IFAS Extension document SS-AGR-232.

31. Hersom M (2005) Pasture stocking density and the relationship to animal performance. Animal Science Department, Florida Cooperative Extension Service, Institute of Food and Agricultural Sciences, University of Florida, Document number AN155.

32. Steiner J (2012) personal communication.

33. Bowden RD, Nadelhoffer KJ, Boone RD, Melillo JM, Garrison JB (1993) Contributions of aboveground litter, belowground litter, and root respiration to soil respiration in a temperate mixed hardwood forest. Can J For Res 23: 1402–1407.

34. Paustian K, Six J, Elliott ET, Hunt HW (2000) Management options for reducing CO2 emissions from agricultural soils. Biogeochemistry 48: 147–163.

35. DeRamus HA, Clement TC, Giampola DD, Dickison PC (2003) Methane emissions of beef cattle on forages: efficiency of grazing management systems. J Environ Qual 32: 269–277.

36. Mosier A, Kroeze C, Nevison C, Oenema O, Seitzinger S, et al. (1998) Closing the global N2O budget: nitrous oxide emissions through the agricultural nitrogen cycle. Nutr Cycl Agroecosys 52: 225–248.

37. Thorburn PJ, Biggs JS, Collins K, Probert ME (2010) Nitrous oxide emissions from Australian sugarcane production systems – are they greater than from other cropping systems? Agric Ecosyst Environ 136: 343–350.

38. Rice RW, Gilbert RA, Lentini RS (2002) Nutritional requirements for Florida sugarcane. Florida Cooperative Extension Service. UF/IFAS, Document SS-ARG-228. University of Florida Institute of Food Agricultural Science.

39. Morgan KT, McCray JM, Rice RW, Gilbert RA, Baucum LE (2009) Review of current sugarcane fertilizer recommendations: a report from the UF/IFAS sugarcane fertilizer standards task force. Document SL 295, Soil and Water

40. Morris DR, Gilbert RA, Reicosky DC, Gesch RW (2004) Oxidation potentials of soil organic matter in Histosols under different tillage methods. Soil Sci Soc Am J 68: 817–826.

41. Stehfest E, Bouwman LF (2006) N2O and NO emission from agricultural fields and soils under natural vegetation: summarizing available measurement data and modeling of global annual emissions. Nutr Cycl Agroecosys 74: 207–228.

42. Anderson-Teixeira KJ, Masters MD, Black CK, Zeri M, Hussain MZ, et al. (2012) Altered belowground carbon cycling following land use change to perennial bioenergy crops. Ecosystems, in press.

43. Smith CM, David MB, Mitchell CA, Masters MD, Anderson-Teixeira KJ, et al. (2013) Reduced nitrogen losses following conversion of row crop agriculture to perennial biofuel crops. J Env Qual 42: 219–228, doi: 10.2134/jeq2012.0210.

44. Beale CV, Long SP (1997) Seasonal dynamics of nutrient accumulation and partitioning in the perennial C-4-grasses Miscanthus x giganteus and Spartina cynosuroides. Biomass Bioenergy 12: 419–428.

45. Hansen EM, Christensen BT, Jensen LS, Kristensen K (2004) Carbon sequestration in soil beneath long-term Miscanthus plantations as determined by ¹³C abundance. Biomass Bioenergy 26: 97–105.

46. NRCS Web Soil Survey website. Available: http://websoilsurvey.nrcs.usda.gov/app/HomePage.htm. Accessed 2010 Oct 7.

47. Parton WJ. Hartman MD, Ojima DS, Schimel DS (1998) DAYCENT and its land surface submodel: description and testing. Glob Planet Change 19: 35–48.

48. Parton WJ, Holland EA, Del Grosso SJ, Hartmann MD, Martin RE, et al. (2001) Generalized model for NOx and N2O emissions from soils. J Geophys Res-Atmos 106: 17403–17420.

49. DayCent: Daily Century Model website. Available: http://www.nrel.colostate.edu/projects/daycent/. Accessed 2010 Jun 4.

50. Del Grosso SJ, Halvorson AD, Parton WJ (2008a) Testing DayCent model simulations of corn yields and nitrous oxide emissions in irrigated tillage systems in Colorado. J Environ Qual 37: 1383–1389, doi:10.2134/jeq2007.0292.

51. Parton WJ, Hanson PJ, Swanston C, Torn M, Trumbore SE, et al. (2010) ForCent model development and testing using the Enriched Background Isotope Study experiment. J Geophys Res 115: G04001.

52. Parton WJ, Schimel DS, Cole CV, Ojima DS (1987) Analysis of factors controlling soil organic levels of grasslands in the Great Plains. Soil Sci Soc Am J 51: 1173–1179.

53. Parton WJ, Ojima DS, Cole CV, Schimel DS (1994) A general model for soil organic matter dynamics: sensitivity to litter chemistry, texture and management, R.B. Bryant,R.W. Arnoldm, Editors, Quantitative Modeling of Soil Forming Processes, Soil Science Society of America, Madison, WI. Pp. 147–167.

54. Eitzinger J, Parton WJ, Hartman M (2000) Improvement and validation of a daily soil temperature submodel for freezing/thawing periods. Soil Sci 165: 525–534.

55. David MB, Del Grosso SJ, Hu X, McIsaac GF, Parton WJ, et al. (2009) Modeling denitrification in a tile-drained, corn and soybean agroecosystem of Illinois, USA. Biogeochemistry 93: 7–30.

56. Del Grosso SJ, Parton WJ, Mosier AR, Ojima DS, Hartmann MD (2000a) Interaction of soil carbon sequestration and N2O flux with different land use practices. In: van Ham J, Baede APM, Meyer LA, Ybema R (eds.), Non-CO2 Greenhouse Gases: Scientific Understanding, Control and Implementation. Kluwer Academic Publishers, The Netherlands. 303–311.

57. Del Grosso SJ, Parton WJ, Mosier AR, Ojima DS, Kulmala AE, et al. (2000b) General model for N2O and N2 gas emissions from soils due to denitrification. Global Biogeochem Cycles 14: 1045–1060.

58. Del Grosso SJ, Mosier AR, Parton WJ, Ojima DS (2005) DayCent model analysis of past and contemporary soil N2O and net greenhouse gas flux for major crops in the USA. Soil Tillage Res 83: 9–24.

59. Del Grosso SJ, Ojima DS, Parton WJ, Mosier AR, Peterson GA, et al. (2002) Simulated effects of dryland cropping intensification on soil organic matter and GHG exchanges using the DAYCENT ecosystem model. Environ Pollution 116, S75–S83.

60. Del Grosso SJ, Parton WJ, Ojima DS, Keough CA, Riley TH, et al. (2008b) DAYCENT simulated effects of land use and climate on county level N loss vectors in the USA. Pages 571–595 in: R.F. Follett, and J.L. Hatfield (eds.) Nitrogen in the Environment: Sources, Problems, and Management, 2nd ed. Elsevier Science Publishers, The Netherlands.

61. Galdos MV, Cerri CC, Cerri CEP, Paustian K, Van Antwerpen R (2010) Simulation of sugarcane residue decomposition and aboveground growth, Plant Soil 326: 243–259. DOI 10.1007/s11104–009–004–3.

62. Hartmann MD, Merchant EK, Parton WJ, Gutmann MP, Lutz SM, et al. (2011) Impact of historical land use changes in the U.S. Great Plains, 1883 to 2003. Ecol App 21: 1105–1119.

63. Adler PR, Del Grosso SJ, Parton WJ (2007) Life-Cycle assessment of net greenhouse-gas flux for bioenergy cropping systems. Ecol Appl 17: 675–691.

64. Galdos MV, Cerri CC, Cerri CEP, Paustian K, Van Antwerpen R (2009) Simulation of soil carbon dynamics under sugarcane with the CENTURY Model, Soil Sci Am J 73: 802–811.

65. Vallis I, Parton WJ, Keating BA, Wood AW (1996) Simulation of the effects of trash and N fertilizer management on soil organic matter levels and yields of sugarcane. Soil Tillage Res 38: 115–132.

66. Pepper DA, Del Grosso SJ, McMurtrie RE, Parton WJ (2005) Simulated carbon sink response of shortgrass steppe, tallgrass prairie and forest ecosystems to rising [CO_2], temperature and nitrogen input, Global Biogeochem Cycles 19: GB1004 doi:10.1029/2004GB002226.

67. Kelly RH, Parton WJ, Hartman MD, Stretch LK, Ojima DS, et al. (2000), Intra-annual and interannual variability of ecosystem processes in shortgrass steppe, J Geophys Res 105(D15) 20093–20100 doi:10.1029/2000JD900259.

68. Brady NC, Weil RR (2002) The nature and properties of soils. Prentice Hall, Upper Saddle River, New Jersey. 960 p.

69. Morris DR, Gilbert RA (2005) Inventory, crop use, and soil subsidence of Histosols in Florida. J Food Agr Environ 3: 190–193.

70. Newman Y, Vendramini J, Blount A (2010) Bahiagrass (*Paspalum notatum*): overview and management. University of Florida IFAS Extension. Publication #SS-AGR-332. http://edis.ifas.ufl.edu/ag342. Accessed 2011 Aug 7.

71. Valentine DW, Holland EA, Schimel DS (1994) Ecosystem and physiological controls over methane production in northern wetlands. J Geophys Res 99: 1563–1571.

72. Del Grosso SJ, Parton WJ, Mosier AR, Walsh MK, Ojima DS, et al. (2006) DayCent national scale simulations of N_2O emissions from cropped soils in the USA. J Environ Qual 35: 1451–1460.

73. US Department of Agriculture, National Agricultural Statistics Service website. Available: http://www.nass.usda.gov/Quick_Stats/. Accessed 2010 Feb 25.

74. DAYMET United States Data Center-A source for daily surface weather data and climatological summaries website. Available: www.daymet.org. Accessed 2010 Jun 1.

75. US Department of Agriculture Natural Resources Conservation Service, Web Soil Survey website. Available: http://websoilsurvey.sc.egov.usda.gov/app/HomePage.htm. Accessed 2010 Dec 12.

76. R Development Core Team (2007) R: A language and environment for statistical computing. R Foundation for Statistical Computing, Vienna, Austria.

77. Barbour MG, Billings WD (1988) North American terrestrial vegetation. Press Syndicate of the University of Cambridge. Melbourne, Australia.

78. Pitman WD, Portier KM, Chambliss CG, Kretschmer AE (1992) Performance of yearling steers grazing bahia grass pastures with summer annual legumes or nitrogen fertilizer in subtropical Florida. Trop Grasslands 26: 206–211.

79. Glaz B, Ulloa MF (1993) Sugarcane yields from plant and ratoon sources of seed cane. J Am Soc Sugar Cane Tech 13: 7–13.

80. Wiedenfeld RP, Enciso J (2008) Sugarcane responses to irrigation and nitrogen in semiarid south Texas. Agron J 100: 665–671.

81. Gilbert RA, Shine JM, Miller JD, Rice RW, Rainbolt CR (2006) The effect of genotype, environment and time of harvest on sugarcane yields in Florida, USA. Field Crop Res 95: 156–170.

82. Glaz B (2012) Personal communication.

83. Glaz B, Morris DR (2010) Sugarcane Responses to water-table depth and periodic flood. Agron J 102: 372–380.

84. US Environmental Protection Agency (2011) Florida Sugarcane Metadata. Environmental Protection Agency, Washington DC. Available: http://www.epa.gov/oppefed1/models/water/met_fl_sugarcane.htm. Accessed 2011 August 1.

85. Muchovej RM, Newman PR (2004) Nitrogen fertilization of sugarcane on a sandy soil: II soil and groundwater analysis. J Am Soc Sugar Cane Tech 24: 225–240.

86. University of Illinois, DeLucia Laboratory Public Data Archive website. Available: http://www.life.illinois.edu/delucia/Public%20Data%20Archive/. Accessed 2013 Jun 4.

87. Department of Energy and Climate Change (DECC) and the Department for Environment, Food and Rural Affairs website. Available: https://www.gov.uk/government/publications/2012-guidelines-to-defra-decc-s-ghg-conversion-factors-for-company-reporting-methodology-paper-for-emission-factors. Accessed: 2013 Jul 23.

88. Forster P, Ramaswamy V, Artaxo P, Berntsen T, Betts R, et al. (2007) Climate Change 2007: The Physical Science Basis. Contribution of Working Group I to the Fourth Assessment Report of the Intergovernmental Panel on Climate Change. Cambridge University Press, Cambridge, United Kingdom and New York, NY, USA.

89. Chapin FS, Woodwell G, Randerson J, Rastetter E, Lovett G, et al. (2006) Reconciling carbon-cycle concepts, terminology, and methods. Ecosystems 9: 1041–1050.

90. Crawley MJ (2007) The R Book. John Wiley and Sons, West Sussex, England. 942 p.

91. Storey JD (2003) The positive false discovery rate: a Bayesian interpretation and the q-value. Ann Stat 31: 2013–2035.

92. Graham-Rowe D (2011) Agriculture: beyond food versus fuel. Nature 474: S6–S8. doi: 10.1038/474S06a.

93. Ball-Coelho B, Sampaio EVSB, Tiessen H, Stewart JWB (1992) Root dynamics in plant and ratoon crops of sugar cane. Plant Soil 142: 297–305.

94. Hoy JW, Grisham MP, Damann KE (1999) Spread and increase of ratoon stunting disease of sugarcane and comparison of disease detection methods. Plant Disease 83: 1170–1175.

95. Stirling GR, Blair BL, Pattemore JA, Garside AL, Bell MJ (2001) Changes in nematode populations on sugarcane following fallow, fumigation and crop rotation, and implications for the role of nematodes in yield decline. Australas Plant Pathol 30: 323–335.

96. Yadav RL, Prasad SR (1992) Conserving the organic matter content of the soil to sustain sugarcane yield. Exp Ag 28: 57–62.

97. Chapin FS, Matson PA, Mooney HA (2002) Principals of Terrestrial Ecosystem Ecology. Springer Science, New York, New York, USA.

98. Rowlings D, Grace P, Kiese R, Scheer C (2010) Quantifying N_2O and CO_2 emissions from a subtropical pasture. 19th World Congress of Soil Science, Soil Solutions for a changing World. 1–6 August 2010, Brisbane, Australia.

99. Howden SM, White DH, Mckeon GM, Scanlan JC, Carter JO (1994) Methods for exploring management options to reduce greenhouse gas emissions from tropical grazing systems. Clim Change 27: 49–70.

100. Liebig MA, Gross JR, Kronberg SL, Phillips RL (2010) Grazing management contributions to net global warming potential: A long-term evaluation in the Northern Great Plains. J Environ Qual 39: 799–809.

101. Plevin RJ, O'Hare M, Jones AD, Torn MS, Gibbs HK (2010) Greenhouse gas emissions from biofuels: Indirect land use change are uncertain but may be much greater than previously estimated. Env Sci Tech 44: 8015–8021.

102. Florida Department of Agriculture and Consumer Services website. Available: http://www.fl-ag.com/agfacts.htm. Accessed 2011 Oct 1.

21

UVB Radiation as a Potential Selective Factor Favoring Microcystin Producing Bloom Forming Cyanobacteria

Yi Ding[1,2], Lirong Song[1], Bojan Sedmak[3]*

1 Key Laboratory of Algal Biology, Institute of Hydrobiology, Chinese Academy of Sciences, Wuhan, China, **2** University of the Chinese Academy of Sciences, Beijing, China, **3** Department of Genetic Toxicology and Cancer Biology, National Institute of Biology, Ljubljana, Slovenia

Abstract

Due to the stratospheric ozone depletion, several organisms will become exposed to increased biologically active UVB (280–320 nm) radiation, not only at polar but also at temperate and tropical latitudes. Bloom forming cyanobacteria are exposed to UVB radiation on a mass scale, particularly during the surface bloom and scum formation that can persist for long periods of time. All buoyant species of cyanobacteria are at least periodically exposed to higher irradiation during their vertical migration to the surface that usually occurs several times a day. The aim of this study is to assess the influence on cyanobacteria of UVB radiation at realistic environmental intensities. The effects of two UVB intensities of 0.5 and 0.99 W/m^2 in up to 0.5 cm water depth were studied *in vitro* on *Microcystis aeruginosa* strains, two microcystin producing and one non-producing. After UVB exposure their ability to proliferate was estimated by cell counting, while cell fitness and integrity were evaluated using light microscopy, autofluorescence and immunofluorescence. Gene damage was assessed by TUNEL assay and SYBR Green staining of the nucleoide area. We conclude that UVB exposure causes damage to the genetic material, cytoskeletal elements, higher sedimentation rates and consequent cell death. In contrast to microcystin producers (PCC7806 and FACHB905), the microcystin non-producing strain PCC7005 is more susceptible to the deleterious effects of radiation, with weak recovery ability. The ecological relevance of the results is discussed using data from eleven years' continuous UVB radiation measurements within the area of Ljubljana city (Slovenia, Central Europe). Our results suggest that increased solar radiation in temperate latitudes can have its strongest effect during cyanobacterial bloom formation in spring and early summer. UVB radiation in this period may significantly influence strain composition of cyanobacterial blooms in favor of microcystin producers.

Editor: Brett Neilan, University of New South Wales, Australia

Funding: The work was supported by Slovenian Research Agency (http://www.arrs.gov.si/en/dobrodoslica.asp), Research Program P1-0245ARRS: Ecotoxicology, Toxicogenomics and Carcinogenesis - (BS salary, reagents, materials, analysis tools), and Ministry of Defence Administration for Civil Protection and Dosaster Relief (Contract No URSZR 4300-1117/2009-1) "Cyanobacterial toxins in surface waters," which co-financed the Laminar Flow Cabinet. Chinese grants from the Natural Science Foundation of China-Yunnan Project (U0833604) (http://www.nsfc.gov.cn/e_nsfc/desktop/zn/0101.htm) and '973' Program (2008CB418000) funded the nine-month stay of Ding Yi at the National Institute of Biology in Ljubljana. The funders had no role in study design, data collection and analysis, decision to publish, or preparation of the manuscript.

Competing Interests: The authors have declared that no competing interests exist.

* E-mail: bojan.sedmak@nib.si

Introduction

Cyanobacteria evolved in an extreme environment under anoxic conditions, high temperature, large variations of available nutrients and strong solar radiation including UV. These are major adverse factors that influenced their early life on Earth [1]. For these reasons the current competitive strategies of cyanobacteria should also be considered in the framework of their morphophysiological adaptations from an evolutionary aspect. At the end of the last century depletion of the ozone layer that absorbs the harmful radiation from the sun increased the risk of exposure of aquatic organisms to biologically effective UVB radiation. Current best estimates suggest that slow recovery of the ozone layer may be expected only during the next half of the 21st century [2] and that it could take decades before the ozone layer recovers to pre−1980 values [3]. Freshwater ecosystems, especially those in mid to low latitudes, may be affected to a much higher degree by the persisting fluxes of UV radiation [4]. Thus, in relatively shallow freshwater bodies, climate change and stratospheric ozone

depletion may act synergistically to increase the exposure of organisms to UVB radiation.

The impacts of harmful UVB radiation in the water ecosystem are restricted to the upper part of the photic layer. The most vulnerable are buoyant organisms, with special emphasis on photoautotrophs that exploit light as their main source of energy. Bloom forming cyanobacteria are among the most numerous phytoplanktonic species frequently exposed on the water surface. Cyanobacterial species that possess an active buoyancy regulation mechanism appear at the water surface several times a day during their vertical migration through the water column and can even persist at the very surface for several hours a day, forming surface blooms and scums. Bloom forming cyanobacteria lacking this type of regulation are in this respect totally dependent for their existence on water mixing activity. Eutrophication, as a consequence of human activities, increasingly accelerates the incidence of harmful cyanobacterial blooms under stable climate conditions [5]. Given that the rising temperatures favor cyanobacterial dominance, Paerl and Huisman [6] claim that there is a link between global warming and the worldwide increasing incidence of harmful cyanobacterial blooms in both their abundance and frequency. Cyanobacterial blooms are a global problem for a great number of reasons, with microcystin (MC) production being the main cause of human fatalities [7]. Ozone depletion, with consequent increased UVB radiation at ground level, is an additional environmental factor that could alter the phytoplankton composition in surface waterbody ecosystems.

Several studies have demonstrated that UVB radiation induces severe damages to cyanobacteria. Genetic material and the photosystem are the main targets of radiation products. The two most mutagenic and cytotoxic DNA lesions, cyclobutane pyrimidine dimers and pyrimidine pyrimidone photoproducts [8] together with reactive oxygen species (ROS) [9] are such examples. Recent experimental data indicate that microcystin plays an important role in producing cyanobacteria under stress conditions [10]. In their attempt to clarify the biological role of microcystin, Zilliges and coworkers (2011) propose this cyanopeptide as protein-modulating metabolite and protectant against oxidative stress. They demonstrated that microcystin binding to several enzymes of the Calvin cycle, phycobiliproteins and two NADPH-dependent reductases is strongly enhanced under high light and stress conditions [11].

Our efforts are directed to understanding how UVB radiation influences cyanobacteria and cyanobacterial bloom formation and composition in an environment of increasing water eutrophication. The aim is to establish how, and to what extent if any, harmful UVB radiation can influence the global dominance of cyanobacteria.

Materials and Methods

Strains and Cultivation

Three unicellular *Microcystis aeruginosa* strains were used. Two axenic *M. aeruginosa* strains - the microcystin non-producing (MC non-producing) PCC7005, the microcystin producing (MC producing) PCC7806 from Institute Pasteur (Paris, France) - and the FACHB905 (MC producing) isolated from Dian Lake (Dianchi, Yunnan Province China) (Chinese Academy of Science Collection). The experiment was divided into two groups with appropriate controls. All strains were grown at 25°C and maintained under sterile conditions in 100 ml flasks in 50 ml BG-11 medium exposed to daylight. The first group (PAR-acclimated) was removed from the photosynthetically active radiation (PAR) at 10 a.m., exposed to UVB for six hours at lower intensity $0.5 \ W/m^2$, and then returned to the 24 h solar day/night cycle. The second group (Dark-acclimated) was dark acclimated for 24 h, exposed to UVB radiation and maintained in complete darkness for the following 48 h.

The experiment was repeated using only two PAR-acclimated cyanobacterial strains (MC non-producing PCC7005 and MC producing PCC7806) exposed to a higher $(0.99 \ W/m^2)$ UVB intensity for three hours and returned to the 24 solar day/night cycle.

The cell concentration used in the experiments was 5×10^6 cells ml^{-1}.

UVB Exposure and Cell Sampling

Both UVB intensities used in our experiments were chosen on the basis of environmental measurements. In late spring during their growth season cyanobacteria are frequently exposed to continuous UVB radiation reaching and exceeding intensities of $0.5 \ W/m^2$, while in summer the relative surface blooms can be exposed for several hours to intensities higher than $1 \ W/m^2$ (Figure 1).

Cell counts were determined with a hemocytometer (Blaubrand, Germany). Each of the three strains in covered plastic Petri dishes (90 mm diameter)was separately exposed in triplicate to UVB radiation in up to 0.5 cm BG-11 medium depth, at a concentration of 5×10^6 cells ml^{-1} in 20 ml BG11 medium. The plastic cover is an effective filter for UVC radiation [12]. A Q-Panel UV-B 313 fluorescent lamp (Cleveland, OH, USA) was used as the UVB source. The radiation was measured using a Radiometer (RM22, Dr. Gröbel Elektronik GmbH, Germany). Control groups were not exposed to UV radiation and were either maintained on the natural solar day/night cycle (PAR-acclimated) or kept in darkness (Dark-acclimated). Samples were collected immediately after UVB exposure (time 0), 24 and 48 h after irradiation. The cyanobacterial cells irradiated with the higher UVB intensity $(0.99 \ W/m^2)$ were under observation for the next twenty days exposed to PAR.

Terminal Deoxynucleotidyl Transferase Labeling (TUNEL) Assay

TUNEL assays were carried out with an In Situ Cell Death Detection Kit, Fluorescein (Roche Diagnostics, Cat.No.11 684 795 910, Mannheim, Germany). Control cells and cell samples at 0, 24 and 48 h after UVB irradiation $(0.51 \ W/m^2$ for six hours) were collected. They were fixed at room temperature for 1.5 h with 2% paraformaldehyde in PBS, washed with PBS and permeabilized for 20 min at 4°C in solutions containing 0.1% Triton X-100 and 0.1% sodium citrate. The labeling and signal conversions were carried out according to the manufacturer's instructions. Negative controls were treated as described above but labeled only with the label solution (without terminal transferase). Positive controls consisted of permeabilized cells pretreated with DNase I (Invitrogen, Cat.No.18047-019) for 20 min prior to labeling with the TUNEL reaction mixture. Labeled samples were analyzed under a fluorescence microscope (Olympus BX51, Japan) at an excitation wavelength in the range of 450–500 nm, and results were processed using the Image-Pro Express 6.0 software program. Representative images were taken after analysis of at least 500 cells per sample.

SYBR Green Staining

Control cells and cells collected after UVB irradiation were washed with BG-11, fixed in 3.7% paraformaldehyde in BG-11 for 60 min at room temperature, washed again three times with BG-11 medium and permeabilized in BG-11 medium containing 0.1% Triton X-100 (v/v) (Merck, Germany) for 20 min. The cells were then incubated with SYBR green (Sigma, USA) (1:500 in BG-11) for 15 min at room temperature in the dark. Images were recorded on an epifluorescence microscope Nikon Eclipse T300.

Immunostaining of the Cytoskeletal Framework

Cells were prepared and immunostained as reported [13]. The target in *M. aeruginosa* strains was the Unnamed protein product (accession number CAO86402) [14].

Phase Contrast and Epifluorescence Microscopy

Pigment autofluorescence was recorded using a G-2A filter with excitation from 541 to 551 nm (bandpass 546 CWL, dichromatic mirror cut-on 565 nm, barrier filter cut-on 590 nm). The cytoskeleton framework and the SYBR Green fluorescence were recorded using a B-2A filter with excitation in the blue light region from 450 to 490 nm (bandpass, 470 CWL, dichromatic mirror cut-on 500 nm, barrier filter cut-on 515 nm). The images were recorded on a Nikon Eclipse T300 microscope with Super high-

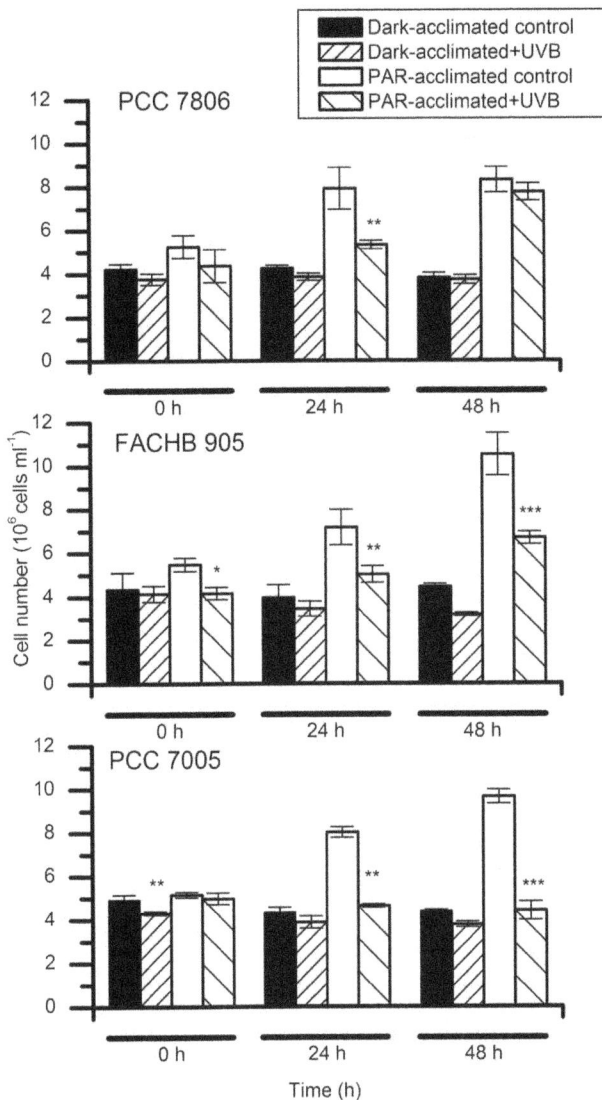

Figure 1. Data from continuous measurement of environmental UVB radiation in Ljubljana city (Central Europe). Panel A. Daily fluctuations in average UVB intensity for three selected months in the year with the lowest (1998) and in the year of the highest cumulative UVB radiation dose (2003). The dashed horizontal line indicates the average UVB intensity used for 6 hours exposure in *in vitro* experiments (Automatic Amp Station Ljubljana-Bežigrad). The city of Ljubljana is situated in Central Europe (latitude 46°3'5" N, longitude 14°30'20" E) at 291 m (954 ft) above sea level. Panel B. Average solar UVB hourly doses in spring (April), summer (July) and autumn (October) for the years 1998 and 2003. The dashed horizontal line indicates the dose (1800 J/m^2) used in *in vitro* experiments. Zero values derived from days without solar irradiation are omitted.

pressure mercury lamp power supply HB-10103AF using an oil immersion objective, magnification 1000X. The images were processed using NIS-Elements D (Nikon, Japan) software. Representative images of control cells and cells after UVB irradiation were taken after analysis of at least 500 cells per sample.

Sedimentation

The cyanobacterial strains were divided into two groups with related controls and irradiated for 6 hours at an intensity of

0.51 W/m^2 as described above. After UVB exposure the PAR-acclimated group was allowed to settle for 48 h in 100 ml flasks exposed to natural 24 h solar day/night cycle. The Dark-acclimated group was kept in the dark for 24 h before UVB exposure and returned to complete darkness for the next 48 h.

Assessment of Sedimentation and Proliferation

Each strain from each group was distributed equally into three culture flasks that were then kept undisturbed for 24 or 48 hours. The cell concentrations of the PAR- and Dark-acclimated groups were assessed by cell counting, first by taking samples from the middle of the water column of the undisturbed culture and then after thorough mixing of the culture. In this way the total cell concentrations for individual sedimentation periods and the average cell concentrations in the suspensions were both obtained. Sedimentation was quantified by subtracting the average cell concentration in suspension from the total cell concentration after mixing.

Measurement of Environmental UVB Radiation

The data for UVB radiation were obtained from the Slovenian Environmental Agency. Raw data for two years (1998 and 2003) from continuous 24 hours UVB radiation measurements (the automatic Amp Station Ljubljana-Bežigrad) were processed (Figure 1). The city of Ljubljana is situated in Central Europe (latitude 46°3'5" N, longitude 14°30'20" E) at 291 m (954 ft) above sea level and has an Oceanic climate (Köppen climate classification "Cfb") with continental characteristics. 1998 was chosen as the year with the lowest (cumulative UVB dose 2.5 10^6 J/m^2) and 2003 as the year with the highest UVB radiation (cumulative UVB dose 3.0 10^6 J/m^2) at ground level. The average daily UVB intensities for the months of April (spring), July (summer) and October (autumn) were calculated (Figure 1, Panel A), together with the UVB doses for hypothetical average individual days in the same months in both years. The highest UVB intensities were also determined. Data for the days without sun were omitted in the figures showing the hourly distribution of the UVB dose of the hypothetical day for single months (Figure 1, Panel B).

Data Analysis

All experiments were performed in triplicate. Data are presented as means ± standard deviations and analyzed using Microcal Origin Software (Version 8.0, Microcal Software Inc. Northampton, MA, USA). Significant differences between control and treated samples were determined by one-way ANOVA followed by a Tukey's test. Differences were considered to be significant when $P < 0.05$. UVB radiation data were presented using GraphPad Prism 5 (GraphPad Inc., USA).

Results

Influence of UVB Radiation on Cell Proliferation

After PAR exposure all three strains in controls were able to proliferate normally and approximately doubled their concentration in 48 h, while the Dark-acclimated group was unable to proliferate in the absence of light (Figure 2). After UVB exposure (0.51 W/m^2 for six hours) lower cell count was observed in all three PAR acclimated strains where both MC producers (PCC7806 and FACHB905) recovered from the stress induced and resumed proliferation. On the contrary the MC non-producing PCC7005 strain was unable to increase in cell number in the period of 48 hours after UVB exposure (Figure 2). The cell count for treated PCC7005 strain remained at the starting level throughout the experiment. There were no significant changes in

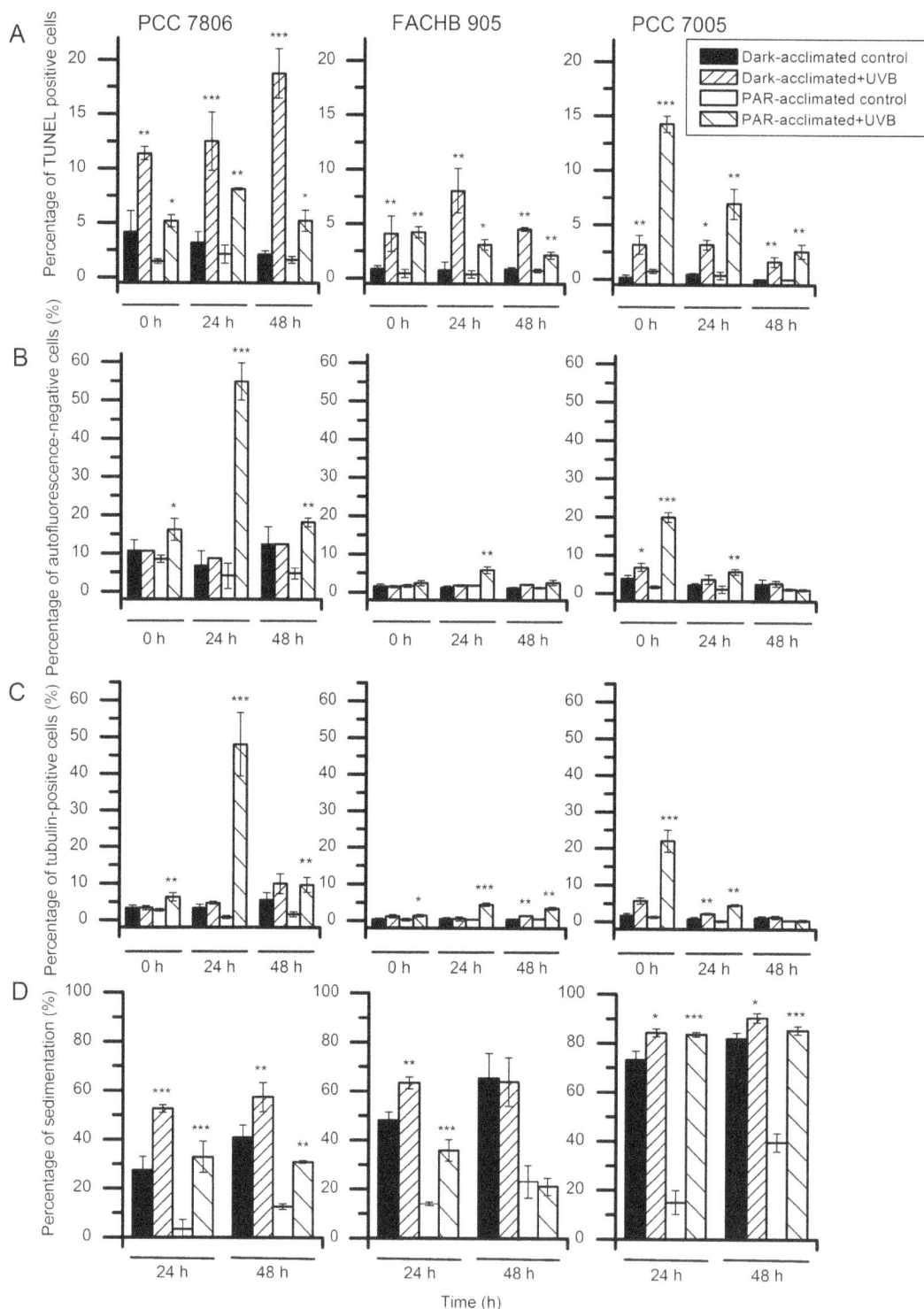

Figure 2. Proliferation of the three *M. aeruginosa* strains after a six-hour exposure to UVB radiation at intensity of 0.51 W/m². ($*p<0.05$, $**p<0.01$, $***p<0.001$).

cell number in the Dark acclimated MC producing strains after UVB exposure although FACHB905 strain showed a downward trend in cell concentration. In contrast Dark-acclimated MC non-producing PCC7005 strain exhibited a significant decrease in cell number immediately following UVB exposure (Figure 2).

Exposure to the higher UVB intensity (0.99 W/m²) induced cell granulation and massive cell lysis of both cyanobacterial strains (PCC7005 and PCC7806). Reliable cell counting was prevented by the loss of normal cell morphology, aggregation and large amounts of cellular debris (Figure 3, Panel A, Images d and g). Twentyfour hours after UVB irradiation individual cells of the

MC producing PCC7806 strain were still able to preserve some original morphology (arrow in Figure 3, Panel A, Image d) while the non-producing PCC7005 cells were all in the process of disintegration (Figure 3, Panel A, Image g). None of the irradiated strains were able to resume proliferation after 20 days of cultivation under PAR. Normal cell morphology is presented in Figure 3 (Panel A, Image a) showing PCC7005 strain as an example.

TUNEL Assay

DNA fragmentation was observed immediately after exposure to UVB radiation (intensity 0.51 W/m^2, six hours exposure) in both Dark-acclimated strains and in those exposed to PAR. DNA damage was followed in all three strains for the entire 48 hours of observation (Figure 4, Panel A).

In both MC-producing strains (FACHB905 and PCC7806), Dark-acclimated cells suffered more damage than cells maintained in the PAR environment. In contrast, in the MC non-producing PCC 7005 strain, the percentage of damaged Dark-acclimated cells remained low and stable while PAR-acclimated cells suffered major damage, with a trend towards reduced incidence with time. The highest percentage of damaged cells was observed in the Dark-acclimated PCC7806 MC-producing strain, with an increase

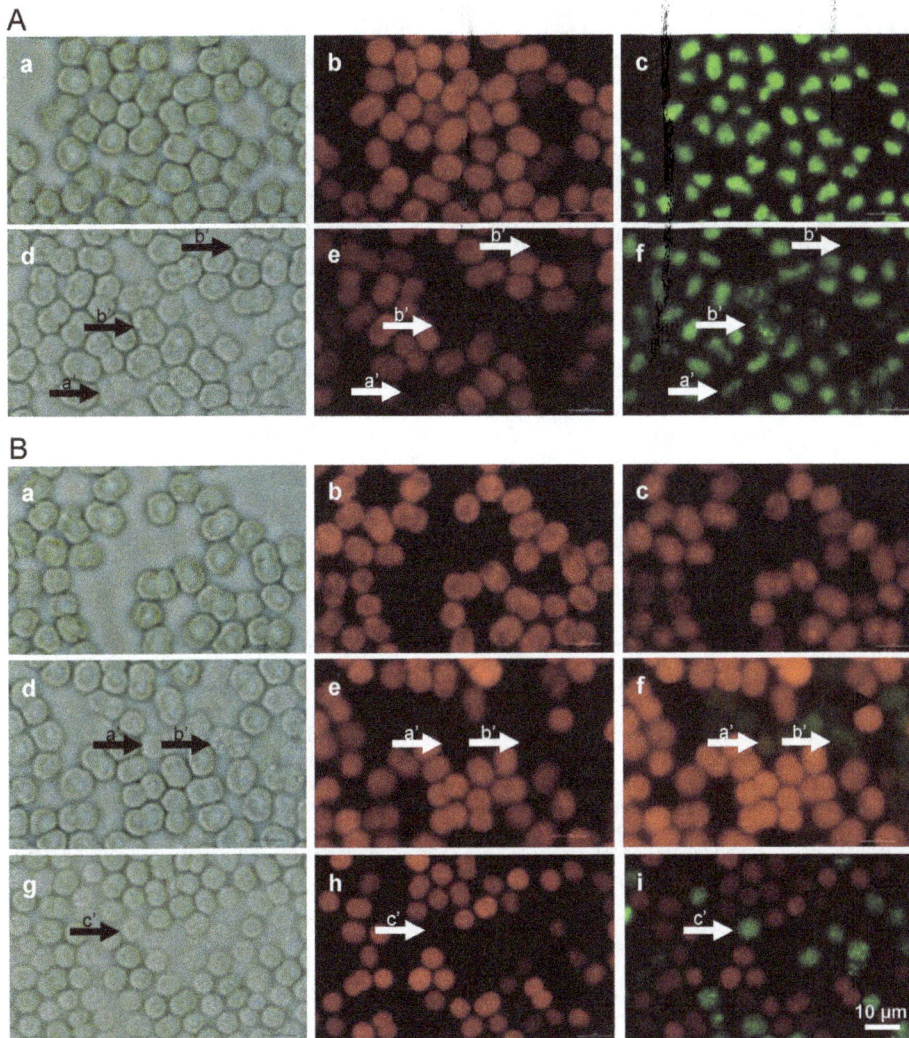

Figure 3. Study of both MC non-producing (PCC7005) and MC-producing (PCC7806) *M. aeruginosa* **strains exposed to UVB radiation for three hours at the intensity of 0.995 W/m^2.** Panel A. Upper row, control MC non-producing PAR acclimated PCC7005 strain, middle row - PCC7806 strain exposed to UVB radiation and lower row – PCC7005 strain exposed to UVB radiation. Microphotographs of the same area of *M. aeruginosa* cells taken under light microscopy (Images a, d and g), autofluorescence (Images b, e and h) and SYBR Green DNA staining (Images c, f and i). Control cells in the upper row show normal cell morphology (Image a) with homogeneous autofluorescence (Image b), and compact nucleoids (Image c). Middle row - Twentyfour hours after UVB exposure the majority of PCC7806 cells udergo the process of lysis (Image d), with reduced autofluorescence (Image e). The degree of autofluorescence was estimated by using exposure times (Image b 83 ms; Image e 83 ms; Inset Image e 833 ms and Image h 917 ms). Inset in Image e is a microphotograph of the same area of cells taken at an extended exposure time. SYBR Green staining reveals complete degradation of the genetic material 24 hours after exposure expressed also as a weak signal (Image c – control cells, exposure time 250 ms; Image f – PCC7806, 250 ms and Image I – PCC7005, 917 ms). Scale bar = 5 µm. Panel B. Macroscopic representation of the three *M. aeruginosa* cell strains exposed to UVB radiation for three hours at an intensity of 0.995 W/m^2. Cell cultures in Petri dishes immediately after UVB exposure (Image a), in cell culture flasks after being returned to the solar day/night cycle for 24 hours (Image b) and after a 20-day exposure to PAR (Image c) (control cells to the left, MC producing PCC7806 in the middle and MC non-producing PCC7005 to the right on all images).

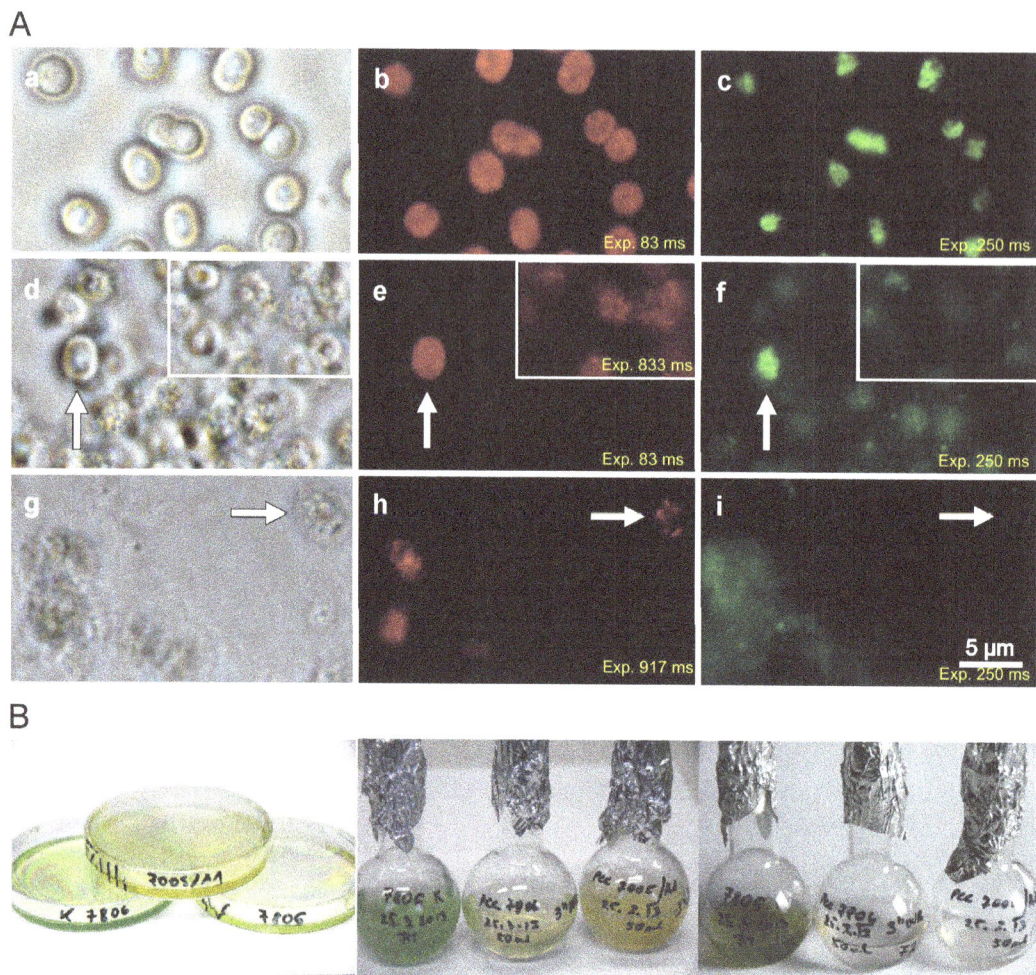

Figure 4. Detection of injuries induced in cyanobacteria after exposure to UVB radiation for six hours at an intensity of 0.51 W/m². Panel A. Percentage of TUNEL-positive *M. aeruginosa* cells after UVB radiation compared with control groups. (*$p<0.05$, **$p<0.01$, ***$p<0.001$). Panel B. Photobleaching as a result of exposure to UVB radiation. Only MC non-producing PCC7005 Dark-acclimated cells show an increase in the percentage of cells showing no autofluorescence after UVB exposure, while a subpopulation (20%) of PAR-acclimated cells (PCC7806) almost immediately loses the signal. PAR-acclimated cells with no autofluorescence are more numerous in both MC producing strains, with an increased proportion 24 hours after exposure. Immediately after irradiation, only MC non-producing PCC7005 strain cells exhibit photobleaching (*$p<0.05$, **$p<0.01$, ***$p<0.001$). Panel C. Increase in percentage of cytoskeleton positive *M. aeruginosa* cells after UVB exposure. The response of PAR-acclimated MC-non-producing strain PCC7005 to UVB irradiation is immediate, while both MC-producing strains FACHB905 and PCC7806 reach the maximum of cytoskeleton labeling in 24 hours subsequent to irradiation. Only a small percentage of Dark-acclimated cells show positive immunolabeling (*$p<0.05$, **$p<0.01$, ***$p<0.001$). Panel D. Sedimentation of *M. aeruginosa* cells is enhanced by UVB radiation exposure. Dark-acclimated cells show faster sedimentation than PAR- acclimated. (*$p<0.05$, **$p<0.01$, ***$p<0.001$).

in time reaching almost 20% at 48 hours of incubation (Figure 4, Panel A).

Due to the severe damages inflicted to cells exposed to the higher (0.99 W/m²) UVB intensity for three hours the assessment using TUNEL assay was not possible.

Autofluorescence and SYBR Green Labeling

The three cyanobacterial strains show similar responses to UVB radiation (0.51 W/m² for six hours). Cells in the control group are clearly demarcated and relatively homogeneous in their autofluorescence, with a compact nucleoid (Figure 5, Panel A, Images a, b, c). Phase contrast microscopy revealed some changes in cells exposed to UVB radiation. Accentuated granulation, with alterations at the cell wall/membrane level, was detected (Figure 5, Panel A, Image d). Their autofluorescence could be quenched to a null level (Figure 5, Panel A, Image e, arrow b'),

with a blurred nucleoid indicating loosening of the genetic material (Figure 5, Panel A, Image f). Degradation of the genetic material is evident in the heavily damaged cells that are less visible under light microscopy. The SYBR Green signal is very weak and the labeled area small or even absent (Figure 5, Panel A, Image f, arrows a' and b').

Regardless of whether the cells in the controls are Dark- or PAR-acclimated, only a small subpopulation with a maximum of 10% is devoid of autofluorescence (Figure 4, Panel B). The percentage of UV exposed cells with no autofluorescence is high in MC non-producing PCC7005 immediately following exposure while, in both MC-producing strains, they are evident only the next day, when they exceed 50% in the PCC7806 strain. The photobleaching is not detectable in Dark-acclimated MC non-producing cells.

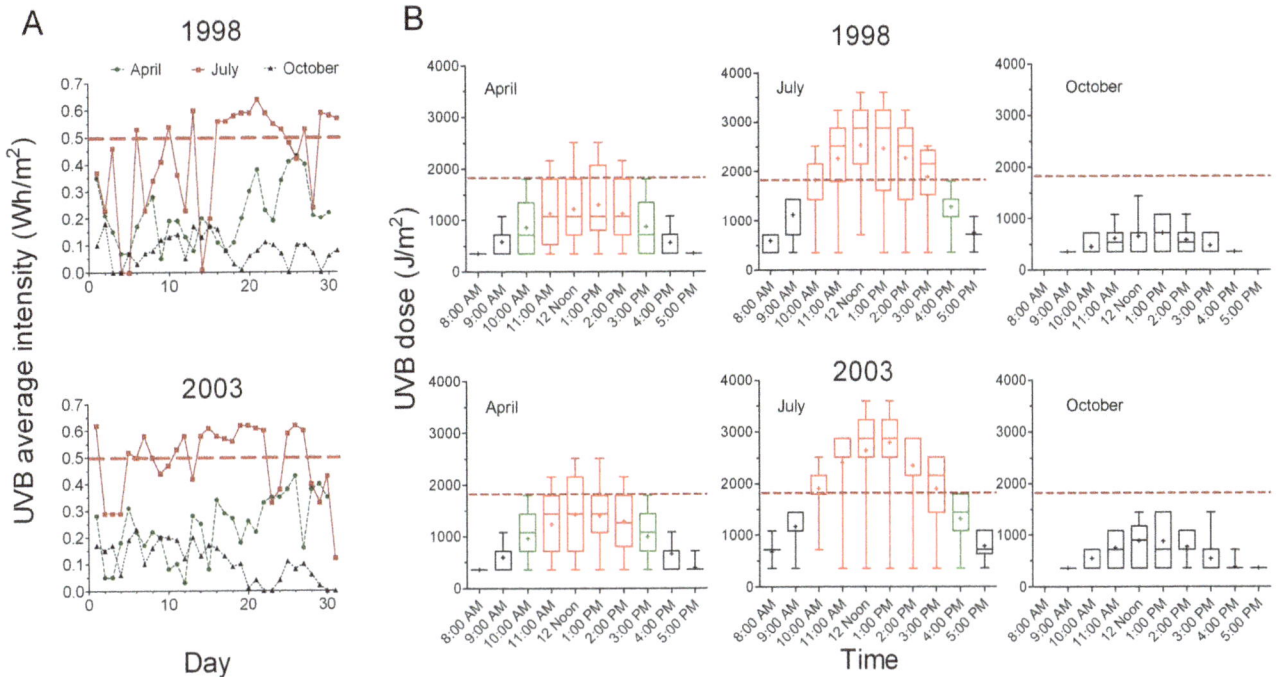

Figure 5. Study of *M. aeruginosa* **PCC7005 PAR-acclimated cells exposed to UVB radiation for six hours at an intensity of 0.51 W/m².** Panel A Cells in control under light microscopy -their morphology with clear demarcation (Image a), -their homogeneous autofluorescence (Image b), and compact nucleoids (Image c). After exposure to UVB radiation, some cells lose their contrast, as observed in light microscopy, consequently becoming poorly visible with accentuated granulation (Image d, arrows a' and b'), with reduced autofluorescence quenched to the null level (Image e, arrows a' and b'). SYBR Green DNA staining demonstrates that, following exposure to UVB, all cells suffer some damage, depicted as blurred nucleoids. The cells that suffered more damage additionally show degradation of the genetic material (Image f, arrows a' and b') expressed also as a weak signal (Image f, arrow a'). Scale bar = 5 μm. Panel B. Immunolabeling of cyanobacterial cytoskeletal elements with anti-bovine α-tubulin mouse monoclonal antibody following exposure to UVB radiation. Upper row, control PCC7005 PAR-acclimated cells, PCC7005 PAR-acclimated cells exposed to UVB radiation; middle row and PCC 7806 PAR -acclimated cells exposed to UVB radiation (24 h) – lower row. Microphotographs of the same area of *M. aeruginosa* cells taken under light microscopy (Panel B, Images a, d and g), autofluorescence (Images b, e and h) and immunofluorescence (Images c, f and i). Successful immunolabeling is achieved only on heavily damaged cells with very weak or no autofluorescence. These are the cells that are barely visible under light microscopy (Image d; arrow a'), frequently granulated (arrows b') and not detectable using autofluorescence (Image e, arrows a' and b'; Image h, arrow c'). It can be seen that PCC7806 in the middle row (Image f; arrows a' and b') is immunolabeled in a way that is different from MC non-producing PCC7005 shown in Image i, lower row; arrow c'. Scale bar = 5 μm.

Only PCC7806 strain exposed to the higher UVB intensity (0.99 W/m²) that retained some autofluorescence was perceived at longer exposure time with a very weak and dispersed SYBR Green signal (Figure 3, Panel A, Images e and f). It is possible to detect only individual cells of the MC non-producing PCC7005 strain that retain some feeble autofluorescence but with completely degraded genetic material (Figure 3, Panel A, Images h and i) compared to control cells with homogeneous autofluorescence and compact nucleoids (Figure 3, Panel A, Images b and c).

Cyanobacterial Cytoskeletal Framework Immunolabeling

In cyanobacterial cells that exhibit quenched autofluorescence the cytoskeletal framework becomes visible. Three pictures of the same area of control and of UVB radiation exposed cells (Figure 5, Panel B) were taken by phase contrast and by epifluorescence using G and B-2A filter sets. In the control, the cells are healthy and undamaged (Image a), their autofluorescence is homogeneous (Image b) and there is no anti-tubulin immunolabeling (Image c). On exposure to UVB radiation, some cells show damaged with immunostained cytoskeletal elements colored green-yellow, as is seen in Images f, I (Panel B). Different levels of damage can be discriminated: gradual loss of cell integrity which is reflected in their fusion with the background (Image g; arrow c') exhibiting

weak delimitation from the environment occasionally with accentuated granulation (Image d; arrow b'). The gradual loss of autofluorescence (Images e and h; arrows a', b' and c') is accompanied by the emergence of immunolabeling of cytoskeletal elements (Images f and i; arrows a', b' and c'), while cells in control show normal morphology (Image a), homogeneous autofluorescence (Image b) and no immunolabeling (Image c in Figure 5, Panel B).

Seriously damaged cyanobacterial cells can be detected using immunolabeling of cytoskeletal elements with anti-bovine α-tubulin mouse monoclonal antibody. Dark-acclimated cells show significantly less damage than PAR-acclimated cells. There is also an essential difference between MC producing and MC non-producing strains. The MC non-producing strain PCC7005 shows immediate immunolabeling following irradiation, in contrast to the two producing strains in which the labeling reaches the maximal percentage after 24 hours. Dark acclimated cells show no significant immunolabeling of cytoskeletal elements (Figure 4, Panel C). We were not able to successfully immunostain the cells exposed to the higher (0.99 W/m²) UVB intensity due to excessive damage that resulted in large amounts of immunolabeled cellular debris (not shown).

Sedimentation

The three strains in controls show similar sedimentation rates (5–15% of PAR-acclimated cells). Dark acclimatization significantly increases sedimentation of all three strains - the MC-non-producing strain (PCC7005) reaches the highest value of over 70%. Exposure to UVB radiation additionally decreases the floating capability of all strains examined. Regardless of the light conditions the most affected was the MC- non-producing PCC7005 strain in which almost the entire population (over 80%) lost their floating capability. MC-producing strains (FACHB905 and PCC7806) were also affected, although to lesser degree (Figure 4; Panel D). The sedimentation of cells exposed to higher (0.99 W/m^2) UVB intensity could not be evaluated because of the severe changes in cell morphology (aggregation and decay).

Environmental UVB Radiation

The calculated average all-day (from 8 AM to 5 PM) intensities presented in Figure 1 (Panel B) are lower than the average six hour intensities (the highest consecutive six hours - as the exposure in our *in vitro* experiments) to which the cyanobacterial blooms are exposed in nature. Average six hours measurements equal to or higher than 0.5 W/m^2 (the total dose of 1800 J/m^2 or higher on average per hour) were detected as early as mid April (spring) that persisted for even multiple (up to three) consecutive days at the end of the month (Figure 1, Panels A and B). In mid-autumn (October) the average UVB intensity no longer reached the average six hours intensity of 0.5 W/m^2. In summer (June/July) the values frequently doubled, reaching peak values of 1.1 W/m^2 that could be measured as early as in May (data not presented).

The hourly dose can on average, already at 10 AM in April of both chosen years (1998 with the lowest cumulative UVB dose and 2003 with the lowest cumulative UVB dose), reach (green box plots) and exceed (red box plots in Figure 1, Panel B) the biologically relevant UVB intensity of 0.5 W/m^2 and the dose of 1800 J/m^2 as used in *in vitro* experiments and can persist for six consecutive hours (Figure 1 Panel B). In summer the average duration of irradiance is prolonged by one to two hours of continuous UVB insolation with a two-fold higher hourly dose at midday. In mid-autumn (October) the total daily UVB irradiation falls to biologically negligible values at this altitude (Figure 1).

Discussion

The impact of UVB radiation on cyanobacterial bloom formation and, in this respect, the role of MC production has been assessed. The damage inflicted on subcellular level of cyanobacterial cells was evaluated by visualizing the nucleoid area and cytoskeletal elements. Intensities and UVB doses (0.51 W/m^2 for six hours = a cumulative hourly dose of 1836 J/m^2) to which the three unicellular *Microcystis aeruginosa* strains were exposed are common in temperate latitudes as early as springtime (Figure 1), when cyanobacteria start to proliferate. In summer average hourly intensities as high as 1.2 W/m^2 can persist for a few hours at midday (Figure 1, Panel B). With the increase in water temperature, dormant cyanobacteria are recruited from the sediment and the planktonic populations of bloom forming cyanobacterial species increase. The axenic *Microcystis* cyanobacterial species and strains cultivated *in vitro* are not directly comparable to the environmental ones since they are unicellular rather than having the natural colonial morphology, and therefore their buoyancy regulation is limited. On the other hand they display authentic damage without the interference of other organisms and allow robust statistical interpretation. As unicellular they may be more susceptible to UV radiation because of poorly expressed surface envelopes. Such unicellular morphology of cyanobacteria is frequently abundant in natural environment at the beginning of their growth season in spring.

Evaluation of UVB Cytotoxicity

The experiments have been performed with cyanobacterial axenic strains to overcome biotic interference and to be able to assess damage indisputably. The main difference between the three strains is in their capacity for MC synthesis. PC 7005 is not capable of MC-production while PCC7806 and FACHB905 are MC-producers. Using different techniques we were able to visualize injuries over a period of 48 hours after UVB exposure. A decrease in cell number of cyanobacteria can be observed after irradiation (Figure 2) as already reported on low bacterized uni-algal cultures [15]. Cell counting itself is not enough to determine whether the decrease in cell number is due to inhibition of proliferation or to cell lysis, and if the damages are restricted to the period of exposure or prolonged into the post-irradiation phase. We do not exclude the possible inhibition of proliferation on a subpopulation of cells. The relatively high cell concentration used in our experiments (5×10^6 cells/ml) allows mutual shading of the cells exposed to UV radiation and could consequently mitigate damages. By adopting the labeling of genetic material we have determined that the UVB exposure at an intensity of 0.5 W/m^2 provokes serious damage from which a subpopulation of cells cannot recover but undergo apoptotic-like programmed cell death (AL-PCD) [16] (Figure 4; Panel A and Figure 5; Panel A, Image f). This can be misinterpreted as inhibition of proliferation if we overlook the cell lysis that occurs. However the subpopulation of less damaged cells is still able to proliferate and together with those capable of repair can eventually re-enter the cell cycle. As shown in Figure 2, the proliferation of the still vital subpopulation following UVB exposure must actually be faster than in the control (as in the case of PCC7806 at 48 h after exposure) in order to establish almost the same concentration as in the control (Figure 2; upper row). In contrast the MC non-producing strain PCC7005 is not able to recover over a period of 48 hours of cultivation (Figure 2; lower row). To further explain and clarify the conclusion presented above we must follow the order of precedence derived from the following experiments. Immediately after UVB irradiation almost all cyanobacterial cells show morphological changes in the nucleoid area. The staining of the DNA double helix with SYBR Green reveals a blurred nucleoid, indicating the loosening and/or destabilization of the tertiary structure in UVB exposed cells (Figure 5, Panel A; Image f) compared to control (Figure 5, Panel A; Image c). Some of the cells lose their integrity (Figure 5; Panel A, Image d), autofluorescence (Figure 5; Panel A, Image e) and normal DNA nucleoid organization, emitting weak or no SYBR Green fluorescence (Figure 5; Panel A, Image f, arrows a' and b'). A more specific type of DNA damage is detected using the TUNEL assay (Figure 4; Panel A). This method has already been adopted successfully to pursue AL-PCD in *Microcystis* species of cyanobacteria [16]. PCD is a process that is triggered in cells unable to repair inflicted damage. The removal of such cells in bloom forming cyanobacteria results in increased fitness of the remaining population and an additional nutrient source in such heavily populated environments as cyanobacterial blooms are. Therefore still vital MC producing cyanobacteria scavenge on deteriorating ones.

The modes of AL-PCD observed in MC producing and MC non-producing strains are basically different. While a significant proportion (14%) of MC non-producing strain PCC7005 shows damage immediately after UVB irradiation, with a tendency to diminish in the following two day period, MC producing strains

(FACHB905 and PCC7806) show a gradual increase in the number of TUNEL positive cells with time (Figure 4; Panel A). All such labeled cells eventually die in the process of self-destruction. The overall low immunolabeling and photobleaching of FACHB905 strain (Figure 4, Panels B and C) can be attributed to the cell fragility that results in increased amounts of immunolabeled debris (not shown).

After exposure to UVB a subpopulation of cells shows a gradual loss of autofluorescence (Figure 4; Panel B and Figure 5; Panel B, Images e and h) with consecutive immunolabelling of cytoskeletal elements (Figure 4; Panel C and Figure 5; Panel B, Images f and i), providing evidence that cytoskeletal elements become visible only in cells with reduced or null autofluorescence. These are the same cells that show major damage of the nucleoid (Figure 5; Panel A, Images e and f). Photobleaching can result from the decoupling of phycobilisomes and photosystem II reaction centers with phycobilin breakdown [17].

It has already been demonstrated that cytoskeletal elements become visible in severely damaged cells and as immunolabeled cell debris, or in stressed cells with low autofluorescence [13]. This kind of visualization enables us to detect otherwise invisible cells or cyanobacterial cell "ghosts" [13]. These are frequently smaller, as the result of cytoskeleton collapse to the degraded interior described also as honeycomb skeletal structure in the SEM microscopy study by Gumbo and Cloete (2011) [18]. Some difference in cytoskeleton labeling between the two MC producing and the MC the non-producing strain are apparent. They can be attributed to differences in the distribution of thylakoids and/or to the effectiveness of cell permeabilization (Figure 5; Panel B, Images f and i). The harmful effects of UVB radiation on the cytoskeleton have already been corroborated by experiments on inhibition of cyanobacterial motility following UV exposure [19] as well as on the degradation of cytoskeletal proteins in phytoplanktons other than cyanobacteria [20].

Although resulting in similar cumulative doses (cca. 1.1×10^4 J/m^2), higher UVB intensity (0.99 W/m^2) in a shorter period of time produces significantly more lethal damages than the lower intensity (0.51 W/m^2) protracted for an adequately longer time span (compare Figure 5 with Figure 3, Panel A).

The main reason for the quick dissolution of cells exposed to high UVB intesity (0.99 W/m^2) is more likely due to acute physiological stress and chronic depression of key physiological processes that resulted in rapid necrosis rather than AI-PCD.

UVB Radiation and MC Production

A reasonable explanation of the apparent link between MC-production and higher resistance to UVB radiation could lie in factors that have influenced the past development of cyanobacteria. Their evolution started in an environment hostile to life where high UV radiation was one of the major threats to complex organic molecules [21]. In parallel to this, genetic studies have provided evidence that cyanobacteria were able to produce microcystins in their early history [22,23]. At a time when we are confronted with elevated UV irradiance due to ozone depletion, MC may regain their original role as protector against UV stress. Bloom forming cyanobacteria must cope with the negative effects of UV light by their obligatory light requirements for photosynthesis. Moderate UVB radiation results in evident oxidative stress [9]. There is scientific evidence that under such stress MC producers have a comparative advantage as MC acts as a protein-modulating metabolite and protectant, increasing the fitness of their host [11]. This protective role is possible thanks to the resistance of cyanobacterial phosphoprotein phosphatase (PPP) family to MC [24]. The exposure to high UVB radiation (0.99 W/

m^2) induces irreversible damages to cyanobacteria with more pronounced effects on MC non-producing strains (Figure 3, Panels A and B).

It is experimentally confirmed that light regulates MC synthesis on two basic levels; genetic and metabolic. Transcript levels of both MC synthetase genes; *mcy B* (peptide synthase) and *mcy D* (polyketide synthase) are increased under high light and decreased when moved into the dark [25]. Both MC production and its protective binding to proteins are stimulated by light [11,26]. However, timely delivery of unbound product to the compromised site is crucial for MC mediated protection. It can be deduced that not only the mere presence of MC, but also its production is important for MC-mediated protection. This is also supported by our experimental results showing that the Dark-acclimated MC-producing cyanobacteria, where their production is presumably stopped or slowed down, are more vulnerable to UVB exposure compared to PAR-acclimated MC-producers (Figure 4). The exact timing is still unknown but an efficacious protection is usually established before the injury takes place. It is not a coincidence that mechanisms of protection are trigerred by light since in natural environment high light intensities are usually accompanied by high intensities of biologically active UVB radiation.

Ecological Adaptations and UV Radiation

Cyanobacteria have two basic strategies for avoiding potential UV damage. The first is the production of UV protecting pigments [12,27] and the second, avoidance using the regulation of vertical migration. Gas vesicles enable cyanobacteria to adjust their vertical position in the water column [28]. During the vegetative period in spring, cyanobacteria are recruited from the bottom as unicellular entities or very small colonies. When cells appear on the surface their small dimensions hardly allow any vertical migration and may be exposed to irradiation for a long period of time during the day. In summer, as they grow older, the colonies become larger, capable of faster vertical migration and multiple reappearances on the water surface during the day. Not only protective pigments, the colony dimension itself can also confer UV screening due to self-shading of several layers of cells [29]. The formation of massive cell envelopes or slime layers embedding the colonies are frequently exopolysaccharide fibril structures associated with proteins [30]. Such a matrix surrounding mature colonies can also serve as an effective UV filter. Thus, microcystin production also gives in this respect an additional advantage to producing strains by promoting cell aggregation [31,32].

It is likely that various levels of PAR and UV affect the buoyancy of cyanobacteria in different ways, depending on their action on photosynthesis. In contrast to the results of UVB on *Atrhrospira* [33], the irradiation especially of MC-non-producing *Microcystis* strains strongly diminishes their floatation capability (Figure 4, Panel D), causing damaged cyanobacteria to be lost to the bottom of deeper lakes. It must be borne in mind that all these negative effects of UVB radiation are to a certain extent mitigated by the presence of the entire light spectrum that, by photoreactivation, can reduce and repair some of UVB inflicted damages [8].

There is additional evidence that corroborates the possible advantages of MC synthesis. In spite of the metabolic burden imposed by MC synthesis, producing cyanobacteria (PCC7806) show a competitive advantage under high light conditions over MC non-producing strains [34]. In a natural *Microcystis* population, MC producing colonies are always bigger and therefore probably older than non-producing ones, proving their endurance when exposed to natural irradiance [35]. The increase in the MC-producing cyanobacteria in the bloom as the consequence of UVB radiation can lower biodiversity [36] and consequently influence food web structure and energy transfer in the waterbody.

We can conclude stratospheric ozone depletion, climate change and increased water eutrophication can act synergistically to augment the frequency and hepatotoxicity of cyanobacterial blooms.

Conclusions

Exposure to environmental doses of UVB radiation in *in vitro* conditions provokes:

a. Photobleaching effects,

b. Damage to the nucleoid area and possible induction of apoptotic-like programmed cell death (AL-PCD),

c. Higher proliferation rates of the undamaged subpopulation probably because of the direct use of autologous molecular material leaking from damaged cells,

d. Cyanobacteria are particularly sensitive to UVB radiation in the light period of metabolic activity,

e. Environmental UVB intensities may significantly influence the development and composition of cyanobacterial blooms.

f. UVB irradiance may indirectly lower biodiversity by favoring MC-producing cyanobacteria.

Acknowledgments

We thank Professor Roger Pain for critical reading of the manuscript, Professor Alenka Gaberščik for helping with laboratory equipment, Andrej Meglič PhD for assistance with additional experiments, Tina Eleršek PhD for her help with box plot graphs and Karmen Stanič for technical assistance. We thank Slovenian Environmental Agency for the raw data from continuous 24 hours UVB radiation measurements.

Author Contributions

Conceived and designed the experiments: BS. Performed the experiments: YD. Analyzed the data: BS LS. Contributed reagents/materials/analysis tools: BS. Wrote the paper: BS.

References

1. Schopf JW (2000) The fossil record: tracing the roots of the cyanobacterial lineage. In: Whitton BA, Potts M, editors. The ecology of cyanobacteria. Kluwer Academic Publishers, Dordrecht, 13–35.
2. Madronich S, McKenzie RL, Björn LO, Caldwell MM (1998) Changes in biologically active ultraviolet radiation reaching the Earth's surface. Journal of Photochemistry and Photobiology B: Biology 46(1–3): 5–19. doi: 10.1016/S1011–1344(98)00182–1.
3. Weatherhead EC, Andersen SB (2006) The search for signs of recovery of the ozone layer. Nature 441: 39–45. doi: 10.1038/nature04746.
4. Morris DP, Zagarese H, Williamson CE, Balseiro EG, Hargreaves BR et al. (1995) The attenuation of solar UV radiation in lakes and the role of dissolved carbon. Limnology and Oceanography 40(8): 1381–1391.
5. Codd JA (2000) Cyanobacterial toxins, the perception of water quality, and the prioritisation of eutrophication control. Ecological Engineering 16(1): 51–60. doi: 10.1016/S0925–8574(00)00089–6.
6. Paerl HW, Huisman J (2008) Blooms like it hot. Science 320(5872): 57–58. doi: 10.1126/science.1155398.
7. Azevedo SMFO, Carmichael WW, Jochimsen EM, Rinehart KL, Lau S et al. (2002) Human intoxication by microcystins during renal dialysis treatment in Caruaru-Brazil. Toxicology 181–182: 441–446. doi: 10.1016/S0300–483X(02)00491–2.
8. Häder DP, Sinha RP (2005) Solar ultraviolet radiation-induced DNA damage in aquatic organisms: potential environmental impact. *Mutation Research* 571(1–2): 221–233. doi: 10.1016/j.mrfmmm.2004.11.017.
9. He YY, Häder DP (2002) Involvement of reactive oxygen species in the UV-B damage to the cyanobacterium *Anabaena* sp. Journal of Photochemistry and Photobiology B: Biology 66(1): 73–80. doi: 10.1016/S1011–1344(01)00278–0.
10. Dziallas C, Grossart HP (2011) Increasing oxygen radicals and water temperature select for toxic *Microcystis* sp. PloS ONE 6(9): e25569. doi: 10.1371/journal.pone.0025569.
11. Zilliges Y, Kehr JC, Meissner S, Ishida K, Mikkat S et al. (2011) The cyanobacterial hepatotoxin microcystin binds to proteins and increases the fitness of *Microcystis* under oxidative stress conditions. PloS ONE 6(3): e17615. doi: 10.1371/journal.pone.0017615.
12. Scherer S, Chen TW, Böger P (1998) A new UV-A/B protecting pigment in the terrestrial cyanobacterium *Nostoc commune*. Plant Physiology 88(4): 1055–1057. doi: 10.1104/pp.88.4.1055.
13. Sedmak B, Carmeli S, Pompe-Novak M, Tušek-Žnidarič M, Grach-Pogrebinsky O, et al., (2009) Cyanobacterial cytoskeleton immunostaining: the detection of cyanobacterial cell lysis induced by planktopeptin BL1125. Journal of Plankton Research 31(11): 1321–1330. doi: 10.1093/plankt/fbp076.
14. Frangeul L, Quillardet P, Castets AM, Humbert JF, Matthijs HCP et al. (2008) Highly plastic genome of *Microcystis aeruginosa* PCC7806, a ubiquitous toxic freshwater cyanobacterium. BMC Genomics 9: 274. doi: 10.1186/1471-2164-9-274.
15. Quesada A, Mouget J-L, Vincent WF (1995) Growth of Antarctic cyanobacteria under ultraviolet radiation: UVA counteracts UVB inhibition. Journal of Phycology 31(2): 242–248. doi: 10.1111/j.0022-3646.1995.00242.x.
16. Ding Y, Gan NQ, Li J, Sedmak B, Song LR 2012 Hydrogen peroxide induces apoptotic-like cell death in *Microcystis aeruginosa* (*Chroococcales, Cyanobacteria*) in a dose-dependent manner. Phycologia 51(5): 567–575. doi: 10.2216/11–107.1
17. Kulandaivelu G, Gheetha V, Periyanan S (1989) Inhibition of energy transfer reactions in cyanobacteria by different ultraviolet radiation. In: Singhal GS et

al., editors. Photosynthesis, Molecular Biology and Bioenergetics. Spriger-Verlag, New York 305–313.
18. Gumbo JR, Cloete TE (2011) Light and electron microscope assessment of the lytic activity of *Bacillus* on *Microcystis aeruginosa* African Journal of Biotechnology 10(41): 8054–8063. doi: 10.5897/AJB10.1311.
19. Donkor V, Häder DP (1991). Effects of solar ultraviolet radiation on motility, photomovement and pigmentation in filamentous, gliding cyanobacteria. FEMS Microbiology Letters 86(2): 159–168. doi: 10.1016/0378–1097(91)90661-S.
20. Ekelund NGA (1991) The effects of UV-B radiation on dinoflagellates. Journal of Plant Physiology 138(3): 274–278. doi: 10.1016/S0176–1617(11)80287–7.
21. Hessen D (2008) Solar radiation and the evolution of life. In: Bjertness E editor. Solar radiation and human health. The Norwegian Academy of Science and Letters, Oslo, 123–136.
22. Rantala A, Fewer DP, Hisbergues M, Rouhainen L, Vaitomaa J (2004) Phylogenetic evidence for the early evolution of microcystin synthesis Proceedings of the National Academy of Sciences of the United States of America 101(2): 568–573. doi: 10.1073/pnas.0304489101.
23. Jungblut AD, Neilan BA (2006) Molecular identification and evolution of the cyclic peptide hepatotoxins, microcystin and nodularin, synthetase genes in three orders of cyanobacteria. Archives of Microbiology 185(2): 107–114. Doi: 10.1007/s00203–005–0073–5.
24. Shi L, Carmicheal WW, Kennely PJ (1999) Cyanobacterial PPP family protein phosphatases possess multifunctional capabilities and are resistant to microcystin-LR. The Journal of Biological Chemistry 274(15): 10039–10046.
25. Kaebernick M, Neilan BA, Börner T, Dittmann E (2000) Light and transcriptional response of the microcystin biosynthesis gene cluster. Applied and Environmental Microbiology 66: 3387–3392. Doi: 10.1128/AEM.66.8.3387–3392.2000.
26. Utkilen H, Gjølme N (1992) Toxin production by *Microcystis aeruginosa* as a function of light in continuous cultures and its ecological significance. Applied and Environmental Microbiology 58(4): 1321–1325.
27. Sinha RP, Klisch M, Gröniger A, Häder DP (1998) Ultraviolet-absorbing/screening substances in cyanobacteria, phytoplankton and macroalgae. Journal of Photochemistry and Photobiology B: Biology 47(2–3): 83–94. doi: 10.1016/S1011–1344(98)00198–5.
28. Walsby A E (1987) Mechanism of buoyancy regulation by planktonic cyanobacteria with gas vesicles: In: Fay P and Van Baalen C editors. The cyanobacteria. Elsevier, Amsterdam, 377–414.
29. Pereira S, Zille A, Micheletti E, Moradas-Fereira P, De Philippis R et al. (2009) Complexity of cyanobacterial exopolysaccharides: composition, structures, inducing factors and putative genes involved in their biosynthesis and assebly. FEMS Microbiology Reviews 33(5): 917–941 doi: 10.111/j.1574-6976.2009.00183.x.
30. Harel M, Weiss G, Daniel E, Wilenz A, Hadas O et al. (2012) Casting a net: fibres produced by *Microcystis* sp. In field and laboratory populations Env. Microbiol. Rep. 4: 342–349. doi: 10.1111/j.1758-2229.2012.00339.x.
31. Sedmak B, Eleršek T (2005) Microcystins induce morphological and physiological changes in selected representative phytoplanktons. Microbial Ecology 50(2): 298–305. doi: 10.1007/s00248-004-0189-1.
32. Gan NQ, Xiao Y, Zhu L, Wu ZX, Liu J et al. (2012) The role of microcystins in maintaining colonies of bloom-forming *Microcystis* spp. Environmental Microbiology 14(3): 730–742. doi: 10.1111/j.1462-2920.2011.02624.x.

33. Ma Z, Gao K (2009) Photosynthetically active UV radiation act in an antagonistic way in regulating buoyancy of *Arthrospira (Spirulina) platensis* (cyanobacterium). Environmental and Experimental Botany 66(2): 265–269. doi: 10.1016/j.envexpbot.2009.02.006.

34. Phelan RR, Downing TG (2011) A growth advantage for microcystin production by *Microcystis* PCC7806 under high light. Journal of Phycology 47(6): 1241–1246. doi: 10.1111/j.1529–8817.2011.01056.x.

35. Kurmayer R, Christiansen G, Chorus I (2003) The abundance of microcystin-producing genotypes correlates positively with colony size in *Microcystis* sp. and determines its microcystin net production in Wannsee. Applied and Environmental. Microbiology 69(2): 787–795. doi: 10.1128/AEM.69.2.787–795.2003.

36. Sedmak B, Kosi G (1998) the role of microcystins in heavy cyanobacterial bloom formation. Journal of Plankton Research 29(4): 691–708. doi: 10.1093/plankt/20.4.691. Erratum 20(7): 1421. doi: 10.1093/plankt/20.7.1421.

22

Effects of Elevated CO_2 on Litter Chemistry and Subsequent Invertebrate Detritivore Feeding Responses

Matthew W. Dray[1]*, Thomas W. Crowther[1,2], Stephen M. Thomas[1], A. Donald A'Bear[1], Douglas L. Godbold[3], Steve J. Ormerod[1], Susan E. Hartley[4], T. Hefin Jones[1]

1 Cardiff School of Biosciences, Cardiff University, Cardiff, United Kingdom, 2 School of Forestry and Environmental Studies, Yale University, New Haven, Connecticut, United States of America, 3 Institute of Forest Ecology, University of Natural Resources and Life Sciences (BOKU), Vienna, Austria, 4 York Environmental Sustainability Institute, University of York, York, United Kingdom

Abstract

Elevated atmospheric CO_2 can change foliar tissue chemistry. This alters leaf litter palatability to macroinvertebrate detritivores with consequences for decomposition, nutrient turnover, and food-web structure. Currently there is no consensus on the link between CO_2 enrichment, litter chemistry, and macroinvertebrate-mediated leaf decomposition. To identify any unifying mechanisms, we presented eight invertebrate species from aquatic and terrestrial ecosystems with litter from *Alnus glutinosa* (common alder) or *Betula pendula* (silver birch) trees propagated under ambient (380 ppm) or elevated (ambient +200 ppm) CO_2 concentrations. Alder litter was largely unaffected by CO_2 enrichment, but birch litter from leaves grown under elevated CO_2 had reduced nitrogen concentrations and greater C/N ratios. Invertebrates were provided individually with either (i) two litter discs, one of each CO_2 treatment ('choice'), or (ii) one litter disc of each CO_2 treatment alone ('no-choice'). Consumption was recorded. Only *Odontocerum albicorne* showed a feeding preference in the choice test, consuming more ambient- than elevated-CO_2 birch litter. Species' responses to alder were highly idiosyncratic in the no-choice test: *Gammarus pulex* and *O. albicorne* consumed more elevated-CO_2 than ambient-CO_2 litter, indicating compensatory feeding, while *Oniscus asellus* consumed more of the ambient-CO_2 litter. No species responded to CO_2 treatment when fed birch litter. Overall, these results show how elevated atmospheric CO_2 can alter litter chemistry, affecting invertebrate feeding behaviour in species-specific ways. The data highlight the need for greater species-level information when predicting changes to detrital processing–a key ecosystem function–under atmospheric change.

Editor: Shuijin Hu, North Carolina State University, United States of America

Funding: Funding by Postgraduate Research Studentships to MWD (Cardiff University President's Research Scholarship) and ADA (NERC/I527861); a Yale Climate and Energy Institute Fellowship to TWC; and a Knowledge Economy Skills Scholarship (KESS) studentship to SMT. Bangor FACE facility development funded by the Science Research Investment Fund (SRIF), with Forestry Commission Wales and the Centre for Integrated Research in the Rural Environment (CIRRE) supporting running costs. The funders had no role in study design, data collection and analysis, decision to publish, or preparation of the manuscript. Websites for funding bodies as follows: Cardiff University President's Scholarship, http://www.cardiff.ac.uk/presidents/; NERC, http://www.nerc.ac.uk. Yale Climate and Energy Institute Fellowship, http://climate.yale.edu/grants-fellowships/postdoctoral-fellowships; KESS, http://www.higherskillswales.co.uk/kess/; SRIF, http://www.delni.gov.uk/index/further-and-higher-education/higher-education/role-structure-he-division/he-research-policy/research-capital-funding-srif.htm; Forestry Commission Wales, http://naturalresourceswales.gov.uk/splash?orig = /; and CIRRE, http://www.cirre.ac.uk/.

Competing Interests: The authors have declared that no competing interests exist.

* E-mail: draymw@cardiff.ac.uk

Introduction

Global concentrations of atmospheric carbon dioxide (CO_2) could more than double by 2100 [1]. Typically, CO_2 enrichment leads to increased plant photosynthesis, resulting in greater biomass and production [2]. Plant tissue chemistry is typically modified, with decreasing nitrogen concentrations and increasing carbon-nitrogen (C/N) ratios affecting herbivore life-history and feeding responses [3].

Approximately 90% of primary production in forest ecosystems escapes herbivory and forms detritus [4], providing a crucial energy pool that underpins the trophic structure of soils and adjacent freshwaters [5]. The effect of elevated CO_2 on the chemical composition of green foliar tissues reduces its palatability to detritivores when it falls as litter [6]. In particular, elevated CO_2 can reduce litter resource quality by decreasing litter nitrogen content [7,8], subsequently increasing C/N ratios [9,10]. Increases in structural [6,8,9] and defensive [10,11] compounds have also

been reported, along with both increases and decreases in phosphorus concentrations [12,13]. The potential for rising CO_2 concentrations to alter litter chemical composition is established, but the consequences for invertebrate-mediated decomposition – an important ecosystem function – remain unclear [14].

Detritivorous macroinvertebrates are functionally important in detritus-based ecosystems, as they are responsible for both comminution and consumption of litter, releasing nutrients for other organisms, such as saprophagous fungi [15,16]. To maintain optimal body nutrient concentrations, theoretical predictions and empirical evidence suggest that invertebrates can increase feeding rates of reduced-quality material (e.g. [17,18]), a process known as 'compensatory feeding' (as defined by [19]). Despite this, poor quality litter has also been shown to increase handling times [20], while reducing nutrient assimilation, slowing development rates, and increasing mortality [6,21]. These conflicting responses have resulted from studies focusing on a small number of species (e.g. [13,18]), which also fail to incorporate aquatic and terrestrial invertebrates, despite differences in detrital accumulation and energy flow between these habitats [22]. A broad-scale study incorporating a range of invertebrate species from different habitats is essential to identify the unifying mechanisms that govern invertebrate feeding responses to elevated-CO_2 litter.

We investigated the feeding preferences and consumption rates of eight detritivorous macroinvertebrate species presented with *Alnus glutinosa* (Linnaeus) Gaertner (common alder) and *Betula pendula* Roth (silver birch) leaf litter produced under ambient and elevated atmospheric CO_2. We tested the hypotheses that: (1) CO_2 enrichment will reduce leaf chemical quality and, given nitrogen-fixing ability in alder, responses will differ by tree species; (2) when presented with a choice between ambient and elevated CO_2 litter, invertebrates will prefer ambient material due to its higher quality; (3) when given litter of one CO_2 treatment only, consumption of elevated-CO_2 litter will be greater, to compensate for its reduced quality.

Methods

Leaf Litter Preparation

Alder and birch litters were produced at the BangorFACE facility, Bangor, UK [23] (Fig. 1). Trees were grown in eight identical plots (four ambient-CO_2 and four elevated-CO_2) to minimise infrastructure-induced artefacts. CO_2 enrichment was carried out using high velocity pure CO_2 injection, controlled using equipment and software modified from EuroFACE [24]. Elevated CO_2 concentrations, measured at 1 min intervals, were within 30% deviation from the pre-set target concentration of 580 ppm CO_2 (ambient +200 ppm) for 75–79% of the photosyn-thetically-active period (daylight hours from budburst until leaf abscission) of 2005–2008. Vertical profiles of CO_2 concentration measured at 50 cm intervals through the canopy showed a maximum difference of +7% from reference values obtained at the top of the canopy [23]. From the beginning of leaf senescence, fallen leaf litter was collected weekly until all leaves had abscised (October to December). Litter within each CO_2 treatment was homogenised and air-dried.

Initial chemical leaching and microbial colonisation of litter ('conditioning') are crucial steps in making litter palatable to detritivorous macroinvertebrates [25,26]. Prior to the start of the experiment, litter was conditioned in fine mesh bags (100 μm to permit microorganisms only) placed in plastic containers ($29\times29\times10$ cm; Fig. 1). For each tree species $\times CO_2$ treatment combination, one bag was placed in aerated stream water that was inoculated with stream-collected litter of mixed-species origin ('aquatic conditioning'); a second bag per tree species $\times CO_2$ treatment combination was inserted between field-collected soil and mixed deciduous leaf litter ('terrestrial conditioning'). Containers were maintained at $11\pm1°C$ with a 12:12 h light-dark cycle and terrestrial containers were sprayed with deionised water every three days to maintain humidity ($\sim$50%). These conditions were selected to represent natural conditioning processes in aquatic and terrestrial habitats in a controlled manner. After two weeks, leaf discs were cut using a 9 mm diameter cork-borer (avoiding the mid-vein), which were air-dried and weighed ($\pm$0.1 mg) prior to experimental use.

Litter samples allocated to chemical analyses (Fig. 1) were stored at –80°C before being oven-dried (50°C for 24 h) and ground into powder (120 s, 50 beats s^{-1}; Pulverisette 23 ball mill, Fritsch GmbH, Idar-Oberstein, Germany). Each sample was composed of litter from three separate leaves. For carbon, nitrogen and phosphorus analyses, five samples were processed per tree × CO_2 treatment × conditioning type combination; for lignin analysis, four samples were used. The percentage leaf dry mass (% leaf DM) of carbon and nitrogen, and the carbon-nitrogen (C/N) ratio, were determined by flash combustion and chromatographic separation of $\sim$1.5 mg leaf powder using an elemental analyser (Elemental Combustion System 4010 CHNS-O Analyzer, Costech Analytical Technologies, Inc., Milan, Italy), calibrated against a standard ($C_{26}H_{26}N_2O_2S$). Phosphorus concentrations (% leaf DM) were quantified using X-ray fluorescence (see [27] for detailed methodology). The percentage Acetyl-Bromide-Soluble Lignin (% ABSL) was determined following the acetyl bromide spectropho-tometric method [28]. Lignin-nitrogen (lignin/N) ratios were calculated for each tree species × CO_2 treatment × conditioning treatment combination.

Invertebrates

Eight macroinvertebrate species were selected for study (Table 1), representing a taxonomic range of litter consumers found in temperate forest habitats [29,30]. Aquatic species were collected from streams in the Brecon Beacons National Park, South Wales, UK (51°50′53″N, 3°22′16″W and 51°50′55″N, 3°33′43″W) and Roath Park, Cardiff, UK (51°30′00″N,

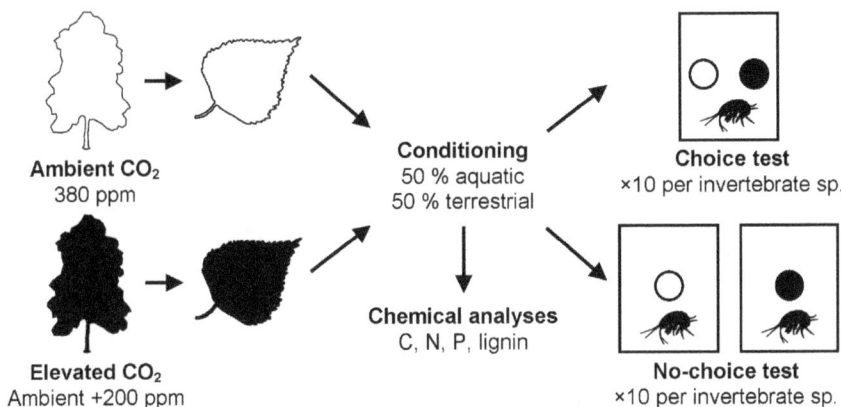

Figure 1. Overview of the experimental approach. Litter was produced under ambient- and elevated-CO_2 atmospheres at BangorFACE, UK. Half of the litter from each CO_2 treatment was conditioned aquatically and half terrestrially. Chemical analyses of the conditioned litter were undertaken, and litter discs were presented to aquatic and terrestrial invertebrates in choice and no-choice tests. Only one tree and one invertebrate species have been shown for clarity. Not to scale.

Within figure:

Ambient CO2
380 ppm

Elevated CO2
Ambient +200 ppm

Conditioning
50 % aquatic
50 % terrestrial

Chemical analyses
C, N, P, lignin

Choice test
×10 per invertebrate sp.

No-choice test
×10 per invertebrate sp.

$3°10'10''$W); terrestrial species were collected from soil-litter interfaces in Bute Park, Cardiff, UK ($51°48'49''$N, $3°18'24''$W). The National Park Authority granted general permission to access sites on common land in the Brecon Beacons National Park, South Wales, UK. Cardiff Council granted permission for access to sites in Cardiff, UK. No endangered or protected species were involved in collections from the field. All individuals were adults, apart from larval *Odontocerum albicorne* and *Sericostoma personatum* caddisflies. Individuals from within each species were selected for size similarity. Prior to experimental use, invertebrates were maintained for at least four weeks in single-species containers ($11\pm1°C$, 12:12 h light-dark cycle) and were fed *Fagus sylvatica* Linnaeus (common beech) litter conditioned as for experimental litter, preventing habituation to experimental alder and birch litter. Feeding was ceased two days prior to the experiments to allow for gut clearance.

Experimental Arenas

All experiments were conducted in $11\times16.5\times3.5$ cm lidded plastic arenas (Cater For You Ltd, High Wycombe, UK) lined with compacted sterilised aquarium gravel (Unipac, Northampton, UK) and were maintained at $11\pm1°C$ with a 12:12 h light-dark cycle. Aquatic microcosms were filled with 400 ml of filtered (100 μm mesh) stream water (circumneutral pH; collected from $51°50'53''$N, $3°22'16''$W) and aerated through a pipette tip (200 μl Greiner Bio-One) attached to an air-line. Terrestrial microcosms were sprayed with deionised water every three days to maintain moisture content and humidity ($\sim$50%). All arenas were uniquely labeled ('microcosm ID'). These standardised conditions were chosen to mimic natural habitats, while minimising the availability of supplementary organic material that could act as a confounding resource during the feeding trials.

For litter of each tree species, detritivores were presented with: (i) a choice between ambient- and elevated-CO_2 material, to provide a direct comparison of detritivore preferences, and (ii) a no-choice situation with each CO_2 treatment presented on its own, approximating litter consumption in current (ambient-CO_2) and future (elevated-CO_2) atmospheric conditions (Fig. 1). In each experiment, ten microcosms were set up for each invertebrate and tree species combination ($n = 160$). A single invertebrate was added to each arena and was placed in the end opposite the airline in aquatic arenas and equidistant to both discs in the choice test. In the choice test, one disc of each CO_2 treatment was pinned to the centre of the arena, 4 cm apart. Discs were replenished when at least 50% of the existing disc had been consumed. In the no-choice test, half of the microcosms contained one ambient-CO_2 disc and the other half one elevated-CO_2 disc, pinned to the centre of the

arena. Both experiments ended after 14 days, or when five (50%) of the individuals of a specific species consumed at least 50% of one disc (choice experiment only). For each invertebrate, the total mass of litter consumed was calculated (±0.1 mg). For choice experiment data, this value was divided by the number of days over which the test had taken place.

Additionally, control microcosms were set up to ensure that differences in mass loss between CO_2 treatments were due to invertebrate activity alone. For each experiment, ten microcosms were set up for each habitat type $\times$ tree species combination. Controls for the choice test each contained one disc of each CO_2 treatment; half of the no-choice control microcosms contained one ambient-CO_2 disc and the other half contained one elevated-CO_2 disc. Leaf discs were air-dried and weighed (±0.1 mg) after 14 days and their total mass loss calculated.

Data Analysis

Statistical analyses were performed separately for alder and birch litter using R version 3.0.1 [31]. Data available from http://dx.doi.org/10.6084/m9.figshare.791634. were checked for normality and homogeneity of variance following Crawley [32]; response variables were transformed using Box-Cox power transformations when assumptions were not met (*car* package [33]). Significance was set at $\alpha = 0.05$ for all analyses.

Two-way analysis of variance (ANOVA) was used to test the main and interactive effects of CO_2 treatment and microcosm type on each chemical variable (carbon, nitrogen, phosphorus and lignin concentrations, and C/N ratio). Planned contrasts (*lsmeans* package [34]) were used to compare the effects of CO_2 treatments for each conditioning treatment.

The main and interactive effects of CO_2 treatment and microcosm type were tested on the mass loss of control discs. Linear mixed-effects models were used to analyse choice control data (*nlme* package [35]), where non-independence of discs sharing the same microcosm was accounted for by including microcosm ID as a random term. The same fixed terms were used to analyse control data from the no-choice test using two-way ANOVA.

In the choice test, litter consumption per day was analysed using linear mixed-effects models (*nlme* package [35]) with the main and interactive effects of CO_2 treatment and invertebrate species as fixed effects and microcosm ID as a random effect. Planned contrasts were performed to compare consumption of ambient- and elevated-CO_2 discs within (i) each invertebrate species, and (ii) invertebrate species grouped by habitat of origin (*contrast* package [36]).

In the no-choice test, the main and interactive effects of CO_2 treatment and invertebrate species on litter consumption were

Table 1. Detritivorous macroinvertebrate species used in the study.

Habitat	Name	Authority	Order: Family
Aquatic	*Asellus aquaticus*	(Linnaeus 1758)	Isopoda: Asellidae
	Gammarus pulex	(Linnaeus 1758)	Amphipoda: Gammaridae
	Odontocerum albicorne	(Scopoli 1763)	Trichoptera: Odontoceridae
	Sericostoma personatum	(Kirby & Spence 1826)	Trichoptera: Sericostomatidae
Terrestrial	*Blaniulus guttulatus*	(Bosc 1792)	Julida: Blaniulidae
	Oniscus asellus	Linnaeus 1758	Isopoda: Oniscidae
	Porcellio scaber	Latreille 1804	Isopoda: Porcellionidae
	Tachypodoiulus niger	(Leach 1815)	Julida: Julidae

tested using two-way ANOVA. Planned contrasts were performed to test the effects of CO_2 treatment on disc consumption within (i) each invertebrate species (*lsmeans* package [34]) and (ii) invertebrate species grouped by habitat of origin (*gmodels* package [37]).

Results

Litter Chemistry

CO_2 enrichment altered leaf litter chemistry, but effects differed between tree species. For birch, CO_2-enriched litter contained lower nitrogen concentrations, and higher lignin concentrations and C/N ratios than ambient-CO_2 litter (Tables 2 and 3). Litter chemistry varied between conditioning types, with higher carbon concentrations in aquatically-conditioned litter and lower nitrogen concentrations in terrestrially-conditioned litter (Table 2). For both conditioning types, elevated-CO_2 litter contained lower nitrogen concentrations (aquatic, estimate = 0.76% DM, $P<0.001$; terrestrial, estimate = 1.17% DM, $P<0.001$; Table 3) and higher C/N ratios (aquatic, estimate = 8.31, $P<0.001$; terrestrial, estimate = 10.28, $P<0.001$; Table 3). For alder litter, the effect of CO_2 treatment was less predictable, with differential responses between conditioning types (Table 2). Elevated CO_2 increased alder nitrogen concentrations when conditioned terrestrially (estimate = 0.29% DM, $P=0.036$; Table 3), although there was no concurrent effect in aquatically-conditioned litter (estimate = 0.1% DM, $P=0.44$; Table 3). No treatment or species effects on litter phosphorus concentrations were observed (Tables 2 and 3).

Invertebrate Responses

For both tree species in the choice and no-choice control arenas, disc mass loss in the absence of invertebrates was unaffected by CO_2 treatment and conditioning type ($P>0.05$). Litter mass loss in the presence of invertebrates was therefore assumed to be a result of invertebrate feeding alone.

In the choice test, leaf palatability affected invertebrate feeding, but this was dependent on tree species. Birch litter consumption was higher for ambient- than elevated-CO_2 discs overall ($F_{1,72} = 10.48$, $P=0.002$); there was no effect of CO_2 on consumption of alder discs ($F_{1,72} = 187.21$, $P=0.34$). Consumption also varied between invertebrate species (alder, $F_{7,72} = 0.92$, $P<0.001$; birch, $F_{7,72} = 30.05$, $P<0.001$). The effect of CO_2 on birch consumption varied by invertebrate species ($F_{7,72} = 3.44$, $P=0.003$), where *O. albicorne* preferred ambient-CO_2 discs (estimate = 1.29 mg d^{-1}, $P<0.001$; Fig. 2B). The effect of CO_2 on litter preference did not vary between invertebrates feeding on

alder ($F_{1,72} = 0.5$, $P=0.83$; Fig. 2A). When grouped, aquatic species preferred ambient-CO_2 birch discs over those grown under elevated CO_2 (estimate = 1.09 mg d^{-1}, $P=0.008$), but no other preferences were exhibited (aquatic species fed alder, estimate = 0.02 mg d^{-1}, $P=0.585$; terrestrial species fed alder, estimate = 0.03 mg d^{-1}, $P=0.496$; terrestrial species fed birch, estimate = 0.06 mg d^{-1}, $P=0.061$).

In the no-choice test, consumption rates were higher when invertebrates fed on ambient- rather than elevated-CO_2 birch discs ($F_{1,64} = 6.39$, $P=0.014$). The trend was consistent across all invertebrate species, but no individual species showed a significant response (CO_2 treatment × invertebrate species: $F_{7,64} = 0.341$, $P=0.932$; Fig. 2D). This overall effect of CO_2 did not occur in alder leaves ($F_{1,64} = 3.6$, $P=0.062$), but the effect of CO_2 varied significantly between species ($F_{7,64} = 4.56$, $P<0.001$); more of the elevated-CO_2 discs were consumed by *G. pulex* (estimate = 2.89 mg, $P=0.002$) and *O. albicorne* (estimate = 3.22 mg, $P<0.001$), while *O. asellus* consumed more of the ambient-CO_2 discs (estimate = 2.86 mg, $P=0.0022$; Fig. 2C). When grouped by habitat, aquatic invertebrates ate more elevated-CO_2 than ambient-CO_2 alder (estimate = 1.965 mg, $P<0.001$) but there was no effect on birch (estimate = 0.1 mg, $P=0.073$). CO_2 treatment had no effect on consumption by terrestrial species fed either alder (estimate = 0.22 mg, $P=0.306$) or birch (estimate = 0.1 mg, $P=0.085$).

Discussion

Elevated atmospheric CO_2 and microbial conditioning type modified leaf litter chemistry, though effects differed between tree species (supporting Hypothesis 1). Individual invertebrate species varied in their responses, suggesting that caution has to be taken when extrapolating general trends from single-species studies.

Elevated atmospheric CO_2 reduced birch litter quality: the concentration of nitrogen decreased and the C/N ratio increased, regardless of conditioning type. Most species did not respond to this change; *O. albicorne* was the only species with behaviour that supported Hypothesis 2, showing a strong preference for ambient-CO_2 litter. Prior work supports this response: Ferreira *et al.* [13] showed that low C/N ratios reduced birch litter consumption by the caddisfly *Sericostoma vittatum* Rambur, while Cotrufo *et al.* [17] found that the woodlouse *P. scaber* preferred high quality (lower C/N ratio and lignin concentration) *Fraxinus excelsior* Linnaeus litter grown under ambient CO_2. Alder litter showed negligible chemical change as a result of elevated CO_2, perhaps due to symbiosis with nitrogen-fixing bacteria that help maintain nutrient supplies [38]. Unexpectedly, a slight increase in quality (increased

Table 2. ANOVA summary table of main and interactive effects of CO_2 treatment (CO_2) and conditioning type (CT) on litter chemistry.

Tree species	Variables	Carbon $F_{1,16}$	P	Nitrogen $F_{1,16}$	P	Phosphorus $F_{1,16}$	P	Lignin $F_{1,12}$	P	C/N $F_{1,16}$	P
Alder	CO_2	0.6	0.435	1.1	0.305	2.8	0.117	0.04	0.543	1.3	0.271
	CT	0.3	0.577	4.1	0.059	0.2	0.684	0.2	0.673	3.8	0.071
	CO_2 × CT	1.5	0.241	4.7	**0.045**	0.4	0.387	3.6	0.082	4	0.064
Birch	CO_2	0.1	0.712	791	**<0.001**	3.1	0.098	4.8	**0.048**	605.3	**<0.001**
	CT	12.1	**0.003**	95	**<0.001**	0.04	0.848	1	0.331	62.5	**<0.001**
	CO_2 × CT	3.6	0.077	36.4	**<0.001**	0.3	0.566	0.1	0.756	6.8	**0.019**

P values <0.05 are emboldened.

Table 3. Chemical composition of leaf litter (mean ± 1 SEM).

Tree species	CT	CO_2	Chemical composition				Chemical ratios	
			Carbon (% DM)	Nitrogen (% DM)	Phosphorus (% DM)	Lignin (% ABSL)	C/N	Lignin/N
Alder	Aquatic	Ambient	48.61 ± 0.37a	3.73 ± 0.16a	0.074 ± 0.009a	22.17 ± 2.64a	13.11 ± 0.16a	5.94
		Elevated	48.48 ± 0.25a	3.63 ± 0.091a	0.064 ± 0.009a	19.56 ± 2.74a	13.37 ± 0.36a	5.38
	Terrestrial	Ambient	48.04 ± 0.22a	3.35 ± 0.016a	0.084 ± 0.009a	19.16 ± 1.01a	14.33 ± 0.02a	5.71
		Elevated	48.68 ± 0.40a	3.65 ± 0.026b	0.062 ± 0.01a	24.34 ± 1.14a	13.35 ± 0.10a	6.68
Birch	Aquatic	Ambient	51.22 ± 0.13a	2.54 ± 0.018a	0.09 ± 0.008a	22.10 ± 3.28a	20.17 ± 0.11a	8.7
		Elevated	50.84 ± 0.13a	1.79 ± 0.004b	0.066 ± 0.01a	27.76 ± 1.69a	28.47 ± 0.08b	15.55
	Terrestrial	Ambient	49.86 ± 0.24a	3.08 ± 0.017a	0.082 ± 0.01a	25.09 ± 2.07a	16.19 ± 0.04a	8.15
		Elevated	50.44 ± 0.41a	1.91 ± 0.063b	0.07 ± 0.006a	29.32 ± 1.52a	26.47 ± 0.74b	15.33

Abbreviations: percent dry mass (% DM), percent acetyl-bromide-soluble lignin (% ABSL), conditioning type (CT).
Different lowercase letters indicate significant differences ($P<0.05$) between CO_2 treatments for each tree species × CT combination.

nitrogen concentration) under elevated CO_2 occurred when alder litter was conditioned terrestrially, but this did not result in any feeding preferences. Effects of conditioning type on litter chemistry may have occurred due to differences in chemical leaching and microorganism activity between aquatic and terrestrial environments [39]. Our data indicate that CO_2 enrichment will affect litter palatability to macroinvertebrate detritivores as a result of chemical change, though these effects will be plant and invertebrate species-specific.

In the no-choice test, invertebrates were expected to compensate for low-quality litter by increasing consumption relative to high-quality litter. In contrast to this expectation, compensatory feeding was not observed in either tree species. There was no clear pattern for alder; invertebrate responses were highly idiosyncratic, with *O. asellus* being the only species to consume more of the low-

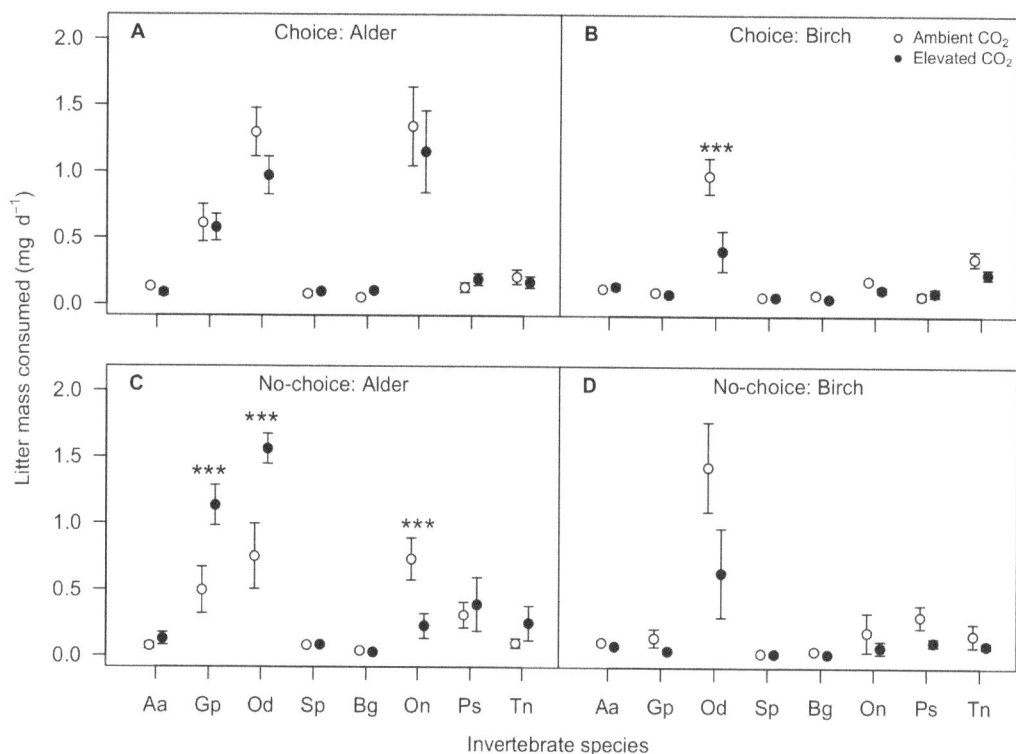

Figure 2. Effects of CO_2 treatment on feeding responses of each invertebrate species. The mean litter consumption (± 1 SE) of each invertebrate species is shown for (A) alder and (B) birch in the choice test, and (C) alder and (D) birch in the no-choice test. Asterisks indicate significant differences between CO_2 treatments within each invertebrate species (***$P<0.001$). Species are arranged by habitat of origin: aquatic species are *Asellus aquaticus* (Aa), *Gammarus pulex* (Gp), *Odontocerum albicorne* (Oa) and *Sericostoma personatum* (Sp); terrestrial species are *Blaniulus guttulatus* (Bg), *Oniscus asellus* (On), *Porcellio scaber* (Ps) and *Tachypodoiulus niger* (Tn).

quality resource (terrestrially-conditioned alder litter contained lower nitrogen when grown under ambient-CO_2). Hättenschwiler et al. [18] detected a similar compensatory response for O. asellus and another woodlouse, P. scaber: higher consumption rates were recorded on low-quality, CO_2-enriched F. sylvatica litter (low nitrogen concentration, high C/N ratio). The current study showed that G. pulex and O. albicorne consumed more elevated-CO_2 than ambient-CO_2 alder, despite no observed chemical differences. It is possible that elevated CO_2 reduced litter palatability by altering chemical constituents that were not quantified here, such as secondary metabolites. For example, phenolics and tannins have been shown to be affected by CO_2 levels [40]. Birch litter responses appeared less idiosyncratic, with no individual species increasing consumption of elevated-CO_2 litter. These results suggest that litter species identity determines the predictability of invertebrate feeding responses, but that compensatory feeding is not a unifying trend amongst detritivorous macroinvertebrates.

Feeding rates may have varied due to increased handling times associated with low quality birch litter (e.g. [20]), or because of differences in species' body chemistry and their ability to cope with elemental imbalances with CO_2-enriched resources [41,42]. Heterotrophs, such as the detritivores in our study, tend to maintain constant body elemental composition [43] and may alter feeding behaviour to achieve optimum chemical balance. Our results show that individual invertebrate species rarely demonstrated significant responses to CO_2 treatments in either test. This suggests that although individual species responses appear idiosyncratic, when considered as a whole, the invertebrate community generally shows consistent and predictable behavioural and functional responses to litter chemical changes induced by elevated CO_2.

Altered consumption of litter by macroinvertebrates will affect energy release from detritus, in turn affecting secondary production, and food-web structure and functioning [5]. Specifically, on the basis of invertebrate responses in our study, mineralisation of carbon and nutrients could slow down in forests dominated by birch or other tree species with similar chemistry. This is reinforced by our observations of high lignin/N and C/N ratios of elevated-CO_2 birch litter, which are predictors for slow decomposition rates [44]. Conversely, stands containing a lot of alder, or other species with lower C/N ratios, may show little response in terms of detrital processing and nutrient turnover. Differences between tree species make it difficult to predict overall decomposition rates, a task made more difficult by the prevalence of litter mixtures in temperate deciduous forests, which tend to exhibit non-additive decay [45].

Changes to litter quality as a result of elevated CO_2 may also affect invertebrate community composition, a potentially important determinant of decomposition rates [19]. This could be caused by changes to food selection [46] and increased patchiness of resource quality in litter mixtures on the forest floor [47]. Differential changes to feeding rates may alter competitive dynamics between invertebrate species, with advantages for species whose dietary breadth extends beyond leaf litter, such as G. pulex and S. personatum [48,49].

Our study provides, to date, the broadest assessment of detritivorous invertebrate species' feeding responses to CO_2-enriched litter, improving our mechanistic understanding of a key ecosystem process in temperate woodland ecosystems. Future elevations of atmospheric CO_2 are predicted to affect the breakdown of detritus indirectly by reducing leaf litter quality for macroinvertebrate detritivores. The study highlights that this process is highly tree species-specific, and there will be strong responses in some forest stands and minimal effects in others. Identifying the mechanisms governing such ecosystem variation in functional responses to climate change is essential if we are to predict the consequences of elevated CO_2 for forest carbon dynamics and nutrient cycling at regional and landscape-scales.

Acknowledgments

Stefan Reidinger, Deborah Coldwell and Simon McQueen-Mason at the University of York for assistance with chemical profiling. Adriana De Palma for comments on early versions of the manuscript. Jaimie Crowther for technical assistance.

Author Contributions

Conceived and designed the experiments: MWD TWC SMT ADA THJ. Performed the experiments: MWD SEH. Analyzed the data: MWD TWC SMT. Contributed reagents/materials/analysis tools: DLG SJO SEH. Wrote the paper: MWD. Wrote initial manuscript: MWD. Drafting and revising initial manuscript: MWD TWC SMT ADA DLG SJO SEH THJ.

References

1. IPCC (2007) Climate Change 2007: The physical science basis. In: Solomon SQD, Manning M, Chen Z, Marquis M, Averyt KB, et al., editors. Contribution of Working Group I to the Fourth Assessment Report of the Intergovernmental Report on Climate Change. Cambridge: Cambridge University Press. 749–845.

2. Curtis PS, Wang X (1998) A meta-analysis of elevated CO_2 effects on woody plant mass, form and physiology. Oecologia 113: 219–313.

3. Robinson EA, Ryan GD, Newman JA (2012) A meta-analytical review of the effects of elevated CO_2 on plant-arthropod interactions highlights the importance of interacting environmental and biological variables. New Phytol 194: 321–336.

4. Cebrian J (1999) Patterns in the fate of production in plant communities. Am Nat 154: 449–468.

5. Moore JC, Berlow EL, Coleman DC, de Ruiter PC, Dong Q, et al. (2004) Detritus, trophic dynamics and biodiversity. Ecol Letters 7: 584–600.

6. Tuchman NC, Wetzel RG, Rier ST, Wahtera KA, Teeri JA (2002) Elevated atmospheric CO_2 lowers leaf litter nutritional quality for stream ecosystem food webs. Glob Change Biol 8: 163–170.

7. Coûteaux M, Kurz C, Bottner P, Raschi A (1999) Influence of increased atmospheric CO_2 concentration on quality of plant material and litter decomposition. Tree Physiol 19: 301–311.

8. Norby RJ, Cotrufo MF, Ineson P, O'Neill EG, Canadell JG (2001) Elevated CO_2, litter chemistry, and decomposition: A synthesis. Oecologia 127: 153–165.

9. Cotrufo MF, Ineson P, Rowland AP (1994) Decomposition of tree leaf litters grown under elevated CO_2: Effect of litter quality. Plant Soil 163: 121–130.

10. Tuchman NC, Wahtera KA, Wetzel RG, Teeri JA (2003) Elevated atmospheric CO_2 alters leaf litter quality for stream ecosystems: An in situ leaf decomposition study. Hydrobiologia 495: 203–211.

11. Parsons WFJ, Lindroth RL, Bockheim JG (2004) Decomposition of Betula papyrifera leaf litter under the independent and interactive effects of elevated CO_2 and O_3. Glob Change Biol 10: 1666–1667.

12. Liu L, King JS, Giardina P (2007) Effects of elevated atmospheric CO_2 and tropospheric O_3 on nutrient dynamics: decomposition of leaf litter in trembling aspen and paper birch commuities. Plant Soil 299: 65–82.

13. Ferreira V, Gonçalves AL, Godbold DL, Canhoto C (2010) Effect of increased atmospheric CO_2 on the performance of an aquatic detritivore through changes in water temperature and litter quality. Glob Change Biol 16: 3284–3296.

14. Prather CM, Pelini SL, Laws A, Rivest E, Woltz M, et al. (2012) Invertebrates, ecosystem services and climate change. Biol Rev 88: 327–348.

15. Wallace JB, Webster JR (1996) The role of macroinvertebrates in stream ecosystem function. Annu Rev Entomol 41: 115–139.

16. Lavelle P, Decaëns T, Aubert M, Barot S, Blouin M, et al. (2006) Soil invertebrates and ecosystem services. Eur J Soil Biol 42: S3–S15.

17. Cotrufo MF, Briones MJI, Ineson P (1998b) Elevated CO_2 affects field decomposition rate and palatability of tree leaf litter: Importance of changes in substrate quality. Soil Biol Biochem 30: 1565–1571.

18. Hättenschwiler S, Bühler S, Körner C (1999) Quality, decomposition and isopod consumption of tree litter produced under elevated CO_2. Oikos 85: 271–281.

19. Gessner MO, Swan CM, Dang CK, McKie BG, Bardgett RD, et al. (2010) Diversity meets decomposition. Trends Ecol Evol 25: 372–380.

20. Ott D, Rall BC, Brose U (2012) Climate change effects on macrofaunal litter decomposition: the interplay of temperature, body masses and stoichiometry. Philos T Roy Soc B 367: 3025–3032.

21. Frost PC, Tuchman NC (2005) Nutrient release rates and ratios by two stream detritivores fed leaf litter grown under elevated atmospheric CO$_2$. Archiv Hydrobiol 163: 463–477.

22. Shurin JB, Gruner DS, Hillebrand H (2006) All wet or dried up? Real differences between aquatic and terrestrial food webs. Proc Roy Soc B 273: 1–9.

23. Smith AR, Lukac M, Hood R, Healy JR, Miglietta F, et al. (2013) Elevated CO$_2$ enrichment induces a differential biomass response in a mixed species temperate forest plantation. New Phytol 198: 156–168.

24. Miglietta F, Peressotti A, Vaccari FP, Zaldei A, deAngelis P, et al. (2001) Free-air CO$_2$ enrichment (FACE) of a poplar plantation: the POPFACE fumigation system. New Phytol 150: 465–476.

25. Daniel O, Schonholzer F, Ehlers S, Zeyer J (1997) Microbial conditioning of leaf litter and feeding by the wood-louse *Porcellio scaber*. Pedobiologia 41: 397–401.

26. Graça MAS, Cressa C, Gessner MO, Feio MJ, Callies KA, et al. (2001) Food quality, feeding preferences, survival and growth of shredders from temperate and tropical streams. Freshwater Biol 46: 947–957.

27. Reidinger S, Ramsey M, Hartley SE (2012) Rapid and accurate analyses of silicon and phosphorus in plants using a portable X-ray fluorescence spectrometer. New Phytol 195: 699–706.

28. Foster CE, Martin TM, Pauly M (2010) Comprehensive compositional analysis of plant cell walls (lignocellulosic biomass) part I: Lignin. J Vis Exp 37: e1745.

29. Moog O (2002) Fauna Aquatica Austriaca. Vienna: Federal Ministry of Agriculture, Forestry, Environment and Water Management.

30. Wurst S, De Deyn GB, Orwin K (2012) Soil biodiversity and functions. In: Wall DH, editor. Soil Ecology and Ecosystem Services. Oxford: Oxford University Press. 28–44.

31. R Development Core Team (2013) *R*: A language and environment for statistical computing. Vienna: *R* Foundation for Statistical Computing.

32. Crawley MJ (2007) The *R* Book. Chichester: John Wiley & Sons, Ltd. 942 p.

33. Fox J, Weisberg S (2011) *car*: Companion to applied regression. *R* package version 2.0–18.

34. Lenth RV (2013) *lsmeans*: Least-squares means. *R* package version 1.06–5.

35. Pinheiro J, Bates D, DebRoy S, Sarkar D, R Development Core Team (2013) *nlme*: Linear and nonlinear mixed effects models. *R* package version 3.1–109.

36. Kuhn M, Weston S, Wing J, Forester J (2011) *contrast*: A collection of contrast methods. *R* package version 0.1.

37. Warnes GR (2012) *gmodels*: Various *R* programming tools for model fitting. *R* package version 2.15.4.

38. Temperton VM, Grayston SJ, Jackson G, Barton CVM, Millard P, et al. (2003) Effects of elevated carbon dioxide concentration on growth and nitrogen fixation in *Alnus glutinosa* in a long-term field experiment. Tree Physiol 23: 1051–1059.

39. Taylor M, Zimmer M (2012) Drowned or dry: A cross-habitat comparison of detrital breakdown processes. Ecosystems 15: 477–491.

40. Lindroth RL (2012) Atmospheric change, plant secondary metabolites and ecological interactions. In: Iason GR, Dicke M, Hartley SE, editors. The Ecology of Plant Secondary Metabolites: From Genes to Global Processes. Cambridge: Cambridge University Press. 120–153.

41. Martinson HM, Schneider K, Gilbert J, Hines JE, Hambäck PA, et al. (2008) Detritivory: stoichiometry of a neglected trophic level. Ecol Res 23: 487–491.

42. Hladyz S, Gessner MO, Giller PS, Pozo J, Woodward G (2009) Resource quality and stoichiometric constraints on stream ecosystem functioning. Freshwater Biol 54: 957–970.

43. Sterner RW, Elser JJ (2002) Ecological Stoichiometry: The Biology of Elements from Molecules to the Biosphere. Princeton: Princeton University Press. 584 p.

44. Melillo JM, Aber JD, Muratore JF (1982) Nitrogen and lignin control of hardwood leaf litter decomposition dynamics. Ecology 63: 621–626.

45. Gartner TB, Cardon ZG (2004) Decomposition dynamics in mixed-species leaf litter. Oikos 104: 230–246.

46. Hättenschwiler S, Bretscher D (2001) Isopod effects on decomposition of litter produced under elevated CO$_2$, N deposition and different soil types. Glob Change Biol 7: 565–579.

47. Swan CM, Palmer MA (2006) Preferential feeding by an aquatic consumer mediates non-additive decomposition of speciose leaf litter. Oecologia 149: 107–114.

48. MacNeil C, Dick JTA, Elwood RW (1997) The trophic ecology of freshwater *Gammarus* spp. (Crustacea: Amphipoda): problems and perspectives concerning the functional feeding group concept. Biol Rev 72: 349–364.

49. Friberg N, Jacobsen DJ (1999) Variation in growth of the detritivore-shredder *Sericostoma personatum* (Trichoptera). Freshwater Biol 32: 133–142.

23

Personal Exposure to Household Particulate Matter, Household Activities and Heart Rate Variability among Housewives

Ya-Li Huang[1,2], Hua-Wei Chen[3], Bor-Cheng Han[2], Chien-Wei Liu[4], Hsiao-Chi Chuang[5,6], Lian-Yu Lin[7], Kai-Jen Chuang[1,2]*

1 Department of Public Health, School of Medicine, College of Medicine, Taipei Medical University, Taipei, Taiwan, **2** School of Public Health, College of Public Health and Nutrition, Taipei Medical University, Taipei, Taiwan, **3** Department of Cosmetic Application and Management, St. Mary's Junior College of Medicine, Nursing and Management, Yilan, Taiwan, **4** Department of Information Management, St. Mary's Junior College of Medicine, Nursing and Management, Yilan, Taiwan, **5** Division of Pulmonary Medicine, Department of Internal Medicine, Shuang Ho Hospital, Taipei Medical University, Taipei, Taiwan, **6** School of Respiratory Therapy, College of Medicine, Taipei Medical University, Taipei, Taiwan, **7** Department of Internal Medicine, Division of Cardiology, National Taiwan University Hospital, Taipei, Taiwan

Abstract

Background: The association between indoor air pollution and heart rate variability (HRV) has been well-documented. Little is known about effects of household activities on indoor air quality and HRV alteration. To investigate changes in HRV associated with changes in personal exposure to household particulate matter (PM) and household activities.

Methods: We performed 24-h continuous monitoring of electrocardiography and measured household PM exposure among 50 housewives. The outcome variables were $\log_{10}$-transformed standard deviation of normal-to-normal (NN) intervals (SDNN) and the square root of the mean of the sum of the squares of differences between adjacent NN intervals (r-MSSD). Household PM was measured as the mass concentration of PM with an aerodynamic diameter <2.5 μm ($PM_{2.5}$). We used mixed-effects models to examine the association between household $PM_{2.5}$ exposure and $\log_{10}$-transformed HRV indices.

Results: After controlling for potential confounders, an interquartile range change in household $PM_{2.5}$ with 1- to 4-h mean was associated with 1.25–4.31% decreases in SDNN and 0.12–3.71% decreases in r-MSSD. Stir-frying, cleaning with detergent and burning incense may increase household $PM_{2.5}$ concentrations and modify the effects of household $PM_{2.5}$ on HRV indices among housewives.

Conclusions: Indoor $PM_{2.5}$ exposures were associated with decreased SDNN and r-MSSD among housewives, especially during stir-frying, cleaning with detergent and burning incense.

Editor: Rudolf Kirchmair, Medical University Innsbruck, Austria

Funding: This study was supported by grants (NSC 101-2314-B-038-053-MY3 and TMU101-AE1-B08) from the National Science Council of Taiwan and Taipei Medical University. The funders had no role in study design, data collection and analysis, decision to publish, or preparation of the manuscript.

Competing Interests: The authors have declared that no competing interests exist.

* E-mail: kjc@tmu.edu.tw

Introduction

Air pollution exposure, particularly particulate matter (PM), has been associated with increased cardiovascular mortality and morbidity [1,2]. These associations have been partially supported by the association of PM with heart rate variability (HRV) changes, and previous panel studies have reported this association as a possible mechanism linking PM to increased risk for cardiovascular diseases [3]. Recently, several studies have reported that the association of cardiovascular endpoints with personal exposure to PM [4–6]. It is also known that people spend 87% of their time in enclosed buildings [7]. These findings imply that exposure to indoor PM may increase the cardiovascular effects of PM exposure. Moreover, the World Health Organization (WHO) considers indoor air pollution as the 3rd most important risk factor, responsible for 4.3% of the global burden of disease [8].

There are many household sources of PM, such as cooking, cleaning and tobacco smoke [7,9,10], which worsen indoor air quality and induce cardiovascular effects among housekeepers who spend most of their time in houses. However, studies of the adverse effect on HRV of personal exposure to household PM among housekeepers are lacking. The effect modification of household activities on the association of PM exposure with HRV changes remains unclear. The aim of this study was to investigate the association between personal exposure to household PM and HRV indices among housewives using four 24-h visits to monitor each participant's household PM exposure, HRV indices and household activity pattern in their private homes.

Materials and Methods

Ethics approval

The study was approved by the ethics committee of St. Mary's Medicine Nursing and Management College. Written informed consent was obtained from each participant before the study began.

Study participants and design

This panel study was designed to simultaneously and continuously monitor changes in household PM levels and HRV indices as well as household activities among housewives in their own homes. The selection criteria for volunteer participants were as follows: no history of smoking or drinking; no medication that might affect cardiac rhythm; and no cardiovascular diseases, such as coronary artery disease, arrhythmia, hypertension, diabetes mellitus and dyslipidemia. Ninety-two housewives responded to our recurring advertisement; 50 of them (54%) living in 50 homes in the Taipei metropolitan area met the criteria and were willing to participate in this study after our protocols had been explained.

The protocol included four home visits that entailed continuous 24-h monitoring of electrocardiography (ECG), household PM, noise, meteorological conditions and time-activity patterns at approximately one-season intervals in the years 2010 to 2012. Each of the 50 housewives had four home visits, for a total of 200 home visits. During their first visits, age, sex, and household characteristics were recorded using a questionnaire. Height and weight were measured and used to calculate body mass index (BMI). Information on indoor environmental measurements and time-activity patterns during the study periods were collected at all visits.

Household particulate matter, meteorological conditions, noise and time-activity patterns

We conducted 24-h continuous monitoring of household PM, meteorological conditions, noise and time-activity patterns during each visit for each housewife. Household PM less than 2.5 μm in diameter ($PM_{2.5}$), temperature and relative humidity were measured continuously using a personal dust monitor (DUSTcheck Portable Dust Monitor, model 1.108; temperature and humidity sensor, model 1.153FH; Grimm Labortechnik Ltd., Ainring, Germany), which measured and recorded 1-min mass concentrations of $PM_{2.5}$ as well as temperature and relative humidity. Noise level was measured using a portable noise dosimeter (Logging Noise Dose Meter Type 4443, Brücl & Kjær, Nærum, Denmark), which reported 1-min continuous equivalent sound levels (Leq) and the time-weighted-averages (TWAs) of noise doses. A range of 30–100 dBA was used to measure noise exposure with 1-min readings over 24 hours.

We asked each participant to carry the dust monitor personally from 0700 hr to 2200 hr to measure personal household $PM_{2.5}$ and noise exposure during the participants' household activities. Participants themselves recorded their time-activity patterns, such as indoor tobacco smoke exposure, indoor chemical dispersion, burning incense, stir-frying and cleaning with detergent every hour from 0700 hr to 2200 hr during the study periods. After sampling, the raw data for 1-min household $PM_{2.5}$, temperature, relative humidity and noise measurements were matched with the sampling time of HRV monitoring and then calculated as 1-, 2-, 3- and 4-h means if 75% of the data were present.

Heart rate variability monitoring

We performed continuous ambulatory ECG monitoring using a PACERCORDER 3-channel device (model 461A; Del Mar Medical Systems LLC, Irvine, CA, USA) with a sampling rate of 250 Hz (4 msec) from 0700 hr to 2200 hr (15 hours) during the study periods. ECG tapes were analyzed using a Delmar Avionics model Strata Scan 563 (Irvine, CA). A complete 5-min segment of the normal-to-normal (NN) interval was taken for HRV analysis, including the standard deviation of NN intervals (SDNN) and the square root of the mean of the sum of the squares of differences between adjacent NN intervals (r-MSSD). Each participant obtained approximately 720 successful 5-min HRV measurements during the four visits (12 measurements for each hour, 180 measurements for each visit) for data analysis. We obtained approximately 32,432 5-min measurements of HRV indices for 50 participants in our data analysis (missing data rate = 9.9%).

Statistical analysis

Mixed-effects models were used to examine the association between household $PM_{2.5}$ and log_{10}-transformed HRV indices by running R statistical software version 2.15.1. The independent variables were the 1-, 2-, 3- and 4-h mean household $PM_{2.5}$, whereas the dependent variables were SDNN and r-MSSD. We treated participant age, BMI, hour of day, temperature, relative humidity, noise, household $PM_{2.5}$ and household activity periods (yes vs. no) including indoor tobacco smoke exposure, indoor chemical dispersion, burning incense, stir-frying and cleaning with detergent as fixed effects and fitted the participant identity number as a random intercept term in our mixed-effects models.

Effect modification by indoor tobacco smoke exposure, burning incense, stir-frying and cleaning with detergent were assessed in separate mixed-effects models by including interaction terms between household $PM_{2.5}$ effects and each potential effect modifier. Household $PM_{2.5}$ effects are expressed as percent changes by interquartile range (IQR) changes as $[10^{(\beta \times IQR)}-1] \times 100\%$ for log_{10}-transformed HRV indices, where β is the estimated regression coefficient. Power analysis and sample sizes calculation were performed with power analysis and sample size (PASS) (NSCC, Kaysville, UT, USA). A significance level of 0.05 was used to determine statistical significance in our models.

Results

Thirty-six thousand 5-min measurements of indoor environmental variables and 32,423 5-min measurements of HRV indices were included in the data analyses. As shown in Table 1, the age range of the 50 housewives was 25–64 years. The mean BMI was 23.2 kg/m^2. Of the 50 participants, 44 with 157 home visits cooked by stir-frying, 38 participants with 129 home visits cleaned with detergent, and 24 participants with 96 home visits burned incense. Only 2 participants with 6 home visits went out shopping, and 50 participants with 194 home visits stayed home during the study periods. All participants used gas for cooking during the study period.

The participants' household $PM_{2.5}$ exposure, meteorological conditions, noise exposure and HRV indices are summarized in Table 2. When the participants' HRV indices were measured during the study periods of the years 2010 to 2012, they demonstrated relatively normal PM levels (WHO air quality guidelines for 24-h mean $PM_{2.5}$: 25 μg/m^3) [11], with a household $PM_{2.5}$ of 23.5 μg/m^3 (SD = 19.4 μg/m^3). The mean noise level was under 50 dBA, which may not enhance sympathetic activity [12]. The mean values (SD) of the log_{10}-transformed HRV indices were 1.62 msec (0.32) for SDNN and 1.11 msec (0.28) for r-MSSD.

The associations between household $PM_{2.5}$ and log_{10}-transformed HRV indices estimated by the mixed-effects models are

Table 1. Basic characteristics of the 50 participants (Mean ± SD).

Variables	
Age, years	
Mean	38.0±10.5
Range	25–64
Body mass index, kg/m²	
Mean	23.2±2.5
Range	19.0–31.0
Household activities among the 50 participants, no (%)	
Indoor tobacco smoke exposure	7 (14)
Incense burning	24 (48)
Cooking with stir-fry	44 (88)
Cleaning with detergent	38 (76)
Indoor chemical dispersion	5 (10)
Shopping	2 (4)
Household activities during 200 home visits, no (%)	
Indoor tobacco smoke exposure	14 (7)
Incense burning	96 (48)
Cooking with stir-fry	157 (78.5)
Cleaning with detergent	129 (64.5)
Chemical indoor dispersion	8 (4)
Shopping	6 (3)

Table 2. Summary statistics of 5-min household $PM_{2.5}$, meteorological conditions, noise and HRV indices for the 50 participants experienced during 200 visits.

Variables	No.	Mean ± SD	Range
Household $PM_{2.5}$, $\mu g/m^3$	36,000	23.5±19.4	12.0–121.0
Noise, dBA	36,000	47.5±22.5	26.0–78.0
Temperature, °C	36,000	24.7±3.6	14.0–33.2
Relative humidity, %	36,000	69.5±3.0	65.1–78.2
Log_{10} SDNN, msec	32,423	1.62±0.32	1.01–2.00
Log_{10} r-MSSD, msec	32,423	1.11±0.28	0.55–1.65

interaction was found between indoor tobacco smoke exposure and household $PM_{2.5}$ for HRV indices.

Discussion

To our knowledge, this is the first study to evaluate the impact of personal exposure to household $PM_{2.5}$ and the impact of household activities on acute changes in HRV indices among housewives. In general, our findings suggest that personal exposure to household $PM_{2.5}$ may impair autonomic function and result in decreased HRV indices. Few studies have investigated the relationship between personal $PM_{2.5}$ exposure and HRV indices [5,13,14], and the majority of those have examined effects of ambient $PM_{2.5}$ exposure on autonomic function [15–17] in human subjects. Our $PM_{2.5}$-induced HRV reductions are in agreement with previous findings [13–17]. The findings support the statement of the American Heart Association's expert panel regarding the biological mechanisms of the effects of $PM_{2.5}$ on cardiovascular events, which are thought to occur through a neural mechanism, altering central nervous system functions [3].

We found that stir-frying and burning incense increased indoor $PM_{2.5}$ levels and that the increase may modify the association between household $PM_{2.5}$ and HRV indices. Epidemiological studies have reported that individuals exposed to indoor cooking oil fumes have a high risk of respiratory diseases [18] and lung cancer [19]. Few panel studies have reported the association between air pollution due to cooking and cardiopulmonary endpoints. In a panel of 387 nonsmoking Chinese restaurant workers, exposure to cooking oil fumes in kitchens was associated with increased urinary 8-OHdG levels. Female workers had a greater oxidative stress response to cooking oil fumes than male

shown in Table 3. After adjusting the models for age, BMI, hour of day, temperature, relative humidity, noise and household activity periods including indoor tobacco smoke exposure, indoor chemical dispersion, burning incense, stir-frying and cleaning with detergent, household $PM_{2.5}$ exposures significantly decreased SDNN and r-MSSD. Interquartile increases in the 1-, 2-, 3- and 4-h mean household $PM_{2.5}$ (19.8, 17.4, 16.5, and 16.2 $\mu g/m^3$, respectively) were associated with 1.25–4.31% decreases in SDNN. The 2-, 3- and 4- means were associated with 1.96–3.71% decreases in the r-MSSD. The greatest decreases in log_{10}-transformed HRV indices occurred at the 4-h mean. Age, BMI, temperature and household activity periods including indoor tobacco smoke exposure, burning incense, stir-frying and cleaning with detergent were significantly associated with decreased SDNN and r-MSSD. No association was observed between relative humidity, noise, indoor chemical dispersion and decreased HRV indices.

We found a consistent effect modification for household $PM_{2.5}$ by different household activity periods, including stir-frying, cleaning with detergent and burning incense (Table 4). Participants showed changes of −4.52% and −2.94% in SDNN and r-MSSD, respectively, associated with increased household $PM_{2.5}$ during stir-frying, whereas participants showed no significant change in HRV indices during study periods without cooking. We also found relatively stronger effects of household $PM_{2.5}$ on participants during cleaning with detergent compared to study periods without cleaning. A similar result was observed in a model including an interaction term between burning incense and household $PM_{2.5}$, although the statistical significance of the interaction was weaker than those in models evaluating effect modifications by cooking and cleaning. However, no significant

Table 3. Percentage changes (95% CI)[a] in HRV indices for interquartile range changes in household $PM_{2.5}$.

Outcome	1-hr mean	2-hr mean	3-hr mean	4-hr mean
Log_{10} SDNN	−1.25	−2.15	−3.02	−4.31
	(−2.00, −0.50)	(−2.77, −1.53)	(−3.87, −2.17)	(−6.50, −2.12)
Log_{10} r-MSSD	−0.12	−1.96	−2.64	−3.71
	(−2.78, 2.54)	(−3.01, −0.91)	(−4.80, −0.48)	(−5.11, −2.30)

[a]Coefficients are expressed as % changes for interquartile range changes in household $PM_{2.5}$ exposure in models adjusting for age, BMI, hour of day, temperature, relative humidity, noise and household activity periods including indoor tobacco smoke exposure, indoor chemical dispersion, burning incense, stir-frying and cleaning with detergent.

Table 4. Effect modification[a] of the association of HRV indices with interquartile range increases in 4-h mean household $PM_{2.5}$ by household activities.

	Log$_{10}$ SDNN (95% CI)	Log10 r-MSSD (95% CI)
Indoor tobacco smoke exposure		
Yes	−0.99 (−2.22, 0.24)	−1.49 (−2.84, −0.14)
No	0.31 (−1.50, 2.12)	1.52 (−3.07, 6.11)
P-value, interaction[b]	0.458	0.227
Burning incense		
Yes	−2.25 (−4.02, −0.48)	−1.99 (−3.42, −0.56)
No	1.89 (−0.45, 4.23)	0.65 (−1.00, 2.30)
P-value interaction	0.025	0.087
Stir-frying		
Yes	−4.52 (−5.37, −3.67)	−2.94 (−3.86, −2.02)
No	−1.15 (−3.08, 0.78)	0.42 (−1.58, 2.42)
P-value interaction	<0.001	<0.001
Cleaning with detergent		
Yes	−3.38 (−4.78, −1.98)	−2.44 (−4.31, −0.57)
No	−0.68 (−1.25, −0.11)	2.68 (1.57, 3.79)
P-value interaction	0.012	<0.001

[a]Coefficients are expressed as % changes for interquartile range changes in household $PM_{2.5}$ exposure in models adjusting for age, BMI, hour of day, temperature, relative humidity, noise and interaction terms.
[b]P-value is for effect modification.

workers [20]. Another panel study observed the association between household wood smoke exposure and ST-segment depression in a panel of 70 women using open wood fires for cooking [21]. The present study showed the effect modification of stir-frying on the association between household PM and HRV reduction. Burning incense is a long-standing Asian tradition used to give respect to ancestors. It has been reported that Taiwan households worship twice per day and are exposed to high levels of particulate air pollution [22]. A recent epidemiological study has reported that exposure to incense smoke in the home may increase the risk of lung cancer among smokers [23]. An *in vitro* study showed that exposure of human coronary artery endothelial cells to burning incense particles induced cytokine production and reduced nitric oxide formation [24]. Studies evaluating incense PM-induced autonomic dysfunction are lacking. Our study suggested some caution in the use of incense for housewives due to incense PM-induced decreases in HRV indices. Overall, our findings add to the growing evidence that air pollutants from stir-frying and incense burning can induce autonomic dysfunction in human subjects similar to those from vehicle and industrial emissions [13–17]. The public health implication is grave because high levels of exposure to indoor air pollution from stir-frying and burning incense are common in Asian countries.

Another interesting finding in our study was that use of detergents when cleaning appeared to modify the effects of household $PM_{2.5}$ on HRV indices; greater household $PM_{2.5}$ effects on HRV indices were observed when participants cleaned with detergent. A recent cross-sectional study used the indoor air quality checklist published by the Department of Occupational Health and Safety to evaluate the health risk of 102 building occupants in a nonindustrial workplace setting. The results showed that the main factors influencing the high number of complaints regarding indoor air quality included indoor detergent and chemical dispersion. Cleanliness led to high pollutant levels and complaints from occupants due to health risks when working inside [10]. Although a limited understanding of the indoor dispersion of detergents and chemicals can make them the primary source of indoor air pollution, odor-related complaints are an example of the human sense of the existence of indoor chemical pollutants. The present study indicated that cleaning with detergent increased the levels of household $PM_{2.5}$ and modified the association between household $PM_{2.5}$ and HRV reduction. These findings have important implications for the feasibility of reliably investigating the associations between cleanliness, indoor air quality and health effects in large-scale epidemiological and intervention studies of household PM. We recommend further studies to investigate the clinical significance of the association between household particle control and cardiovascular health improvement.

Some possible limitations may confound our findings of HRV reduction by household $PM_{2.5}$, including unavailable data on associations with outcomes and some key physiologic and environmental information. First, we could not adjust for respiration-modulated autonomic activity in our study because we were unable to measure key respiration parameters, such as nasal and mouth airflow, chest wall movement and abdominal movement, during the study periods. Second, medication and comorbidity among older housewives could still confound our findings for household $PM_{2.5}$ effects on HRV reduction even though we used strict criteria to exclude cases with chronic cardiopulmonary diseases and specific medication from our study. Third, other unmeasured indoor air pollutants, such as ozone, carbon monoxide and total volatile organic compounds, may have confounded our findings. Fourth, non-randomized recruitment may result in selection bias and confound the association of HRV with household $PM_{2.5}$. Last, the effects of noise and household activities on HRV require further clarification because the sample size of our study may not be large enough to adjust for their effects completely.

Conclusions

We believe our findings generally indicate that household $PM_{2.5}$ was associated with autonomic function in housewives. Household activities including stir-frying, burning incense and cleaning with detergent were associated with increased levels of household $PM_{2.5}$ and modify its effects on HRV reduction.

Author Contributions

Conceived and designed the experiments: LL KC. Performed the experiments: YH HC BH CL. Analyzed the data: HC KC. Contributed reagents/materials/analysis tools: LL HC KC. Wrote the paper: LL KC.

References

1. Pope CA 3rd, Burnett RT, Thurston GD, Thun MJ, Calle EE, et al. (2004) Cardiovascular mortality and long-term exposure to particulate air pollution: epidemiological evidence of general pathophysiological pathways of disease. Circulation 109:71–77.
2. Pope CA 3rd, Dockery DW (2006) Health effects of fine particulate air pollution: lines that connect. J Air Waste Manag Assoc 56:709–742.
3. Brook RD, Rajagopalan S, American Heart Association Council on Epidemiology, Prevention, Council on the Kidney in Cardiovascular Disease, Council on Nutrition, Physical Activity, Metabolism, et al. (2010) Particulate matter air pollution and cardiovascular disease: An update to the scientific statement from the American Heart Association. Circulation 121:2331–2378.
4. Chan CC, Chuang KJ, Shiao GM, Lin LY (2004) Personal exposure to submicrometer particles and heart rate variability in human subjects. Environ Health Perspect 112:1063–1067.
5. Chuang KJ, Chan CC, Chen NT, Su TC, Lin LY (2005) Effects of particle size fractions on reducing heart rate variability in cardiac and hypertensive patients. Environ Health Perspect 113:1693–1697.
6. Lanki T, Ahokas A, Alm S, Janssen NA, Hoek G, et al. (2007) Determinants of personal and indoor PM2.5 and absorbance among elderly subjects with coronary heart disease. J Expo Sci Environ Epidemiol 17:124–133.
7. Klepeis NE, Nelson WC, Ott WR, Robinson JP, Tsang AM, et al. (2001) The National Human Activity Pattern Survey (NHAPS): a resource for assessing exposure to environmental pollutants. J Expo Sci Environ Epidemiol 11:231–252.
8. Lim SS, Vos T, Flaxman AD, Danaei G, Shibuya K, et al. (2012) A comparative risk assessment of burden of disease and injury attributable to 67 risk factors and risk factor clusters in 21 regions, 1990–2010: a systematic analysis for the Global Burden of Disease Study 2010. Lancet 380:2224–2260.
9. Polidori A, Turpin B, Meng QY, Lee JH, Weisel C, et al. (2006) Fine organic particulate matter dominates indoor-generated PM2.5 in RIOPA homes. J Expo Sci Environ Epidemiol 16:321–331.
10. Syazwan A, Rafee BM, Juahir H, Azman A, Nizar A, et al. (2012) Analysis of indoor air pollutants checklist using environmetric technique for health risk assessment of sick building complaint in nonindustrial workplace. Drug Healthc Patient Saf 4:107–126.
11. WHO-Europe (2006) Air Quality Guidelines, Global Update 2005: Particulate Matter, Ozone, Nitrogen Dioxide, and Sulfur Dioxide: World Health Organization Europe.

12. Lee GS, Chen ML, Wang GY (2010) Evoked response of heart rate variability using short-duration white noise. Auton Neurosc 155:94–97.
13. Magari SR, Hauser R, Schwartz J, Williams PL, Hauser R, et al. (2002) Association between personal measurements of environmental exposure to particulates and heart rate variability. Epidemiology 13:305–310.
14. Magari SR, Hauser R, Schwartz J, Williams PL, Smith TJ, et al. (2001) Association of heart rate variability with occupational and environmental exposure to particulate air pollution. Circulation 104:986–991.
15. Gold DR, Litonjua A, Schwartz J, Lovett EG, Larson AC, et al. (2000) Ambient pollution and heart rate variability. Circulation 101:1267–1273.
16. Chuang KJ, Chan CC, Su TC, Lee CT, Tang CS (2007) The effect of urban air pollution on inflammation, oxidative stress, coagulation, and autonomic dysfunction in young adults. Am J Respir Crit Care Med 176:370–376.
17. Chuang KJ, Chan CC, Su TC, Lin LY, Lee CT. (2007) Associations between particulate sulfate and organic carbon exposures and heart rate variability in patients with or at risk for cardiovascular diseases. J Occup Environ Med 49:610–617.
18. Svendsen K, Sjaastad AK, Siverstsen I (2003) Respiratory symptoms in kitchen workers. Am J Ind Med 43:436–439.
19. Behera D, Balamugesh T (2006) Dose-response relationship between cooking fumes exposures and lung cancer among Chinese nonsmoking women. Cancer Res 66:4961–4967.
20. Pan CH, Chan CC, Wu KY (2008) Effects on Chinese restaurant workers of exposure to cooking oil fumes: a cautionary note on urinary 8-hydroxy-2'-deoxyguanosine. Cancer Epidemiol Biomarkers Prev 17:3351–3357.
21. McCracken J, Smith KR, Stone P, Díaz A, Arana B, et al. (2011) Intervention to lower household wood smoke exposure in Guatemala reduces ST-segment depression on electrocardiograms. Environ Health Perspect 119:1562–1568.
22. Lung SC, Mao IF, Liu IJ (1999) Community air quality monitoring and resident's personal exposure assessment in large metropolitan areas in Taiwan. Taiwan Environmental Protection Administration.
23. Tse LA, Yu IT, Qiu H, Au JS, Wang XR (2011) A case-referent study of lung cancer and incense smoke, smoking, and residential radon in Chinese men. Environ Health Perspect 119:1641–1646.
24. Lin LY, Lin HY, Chen HW, Su TL, Huang LC, et al. (2012) Effects of temple particles on inflammation and endothelial cell response. Sci Total Environ 414:68–72.

Permissions

The contributors of this book come from diverse backgrounds, making this book a truly international effort. This book will bring forth new frontiers with its revolutionizing research information and detailed analysis of the nascent developments around the world.

We would like to thank all the contributing authors for lending their expertise to make the book truly unique. They have played a crucial role in the development of this book. Without their invaluable contributions this book wouldn't have been possible. They have made vital efforts to compile up to date information on the varied aspects of this subject to make this book a valuable addition to the collection of many professionals and students.

This book was conceptualized with the vision of imparting up-to-date information and advanced data in this field. To ensure the same, a matchless editorial board was set up. Every individual on the board went through rigorous rounds of assessment to prove their worth. After which they invested a large part of their time researching and compiling the most relevant data for our readers.

The editorial board has been involved in producing this book since its inception. They have spent rigorous hours researching and exploring the diverse topics which have resulted in the successful publishing of this book. They have passed on their knowledge of decades through this book. To expedite this challenging task, the publisher supported the team at every step. A small team of assistant editors was also appointed to further simplify the editing procedure and attain best results for the readers.

Apart from the editorial board, the designing team has also invested a significant amount of their time in understanding the subject and creating the most relevant covers. They scrutinized every image to scout for the most suitable representation of the subject and create an appropriate cover for the book.

The publishing team has been an ardent support to the editorial, designing and production team. Their endless efforts to recruit the best for this project, has resulted in the accomplishment of this book. They are a veteran in the field of academics and their pool of knowledge is as vast as their experience in printing. Their expertise and guidance has proved useful at every step. Their uncompromising quality standards have made this book an exceptional effort. Their encouragement from time to time has been an inspiration for everyone.

The publisher and the editorial board hope that this book will prove to be a valuable piece of knowledge for researchers, students, practitioners and scholars across the globe.

List of Contributors

Denis Bard, Wahida Kihal, Christophe Fermanian and Sophie Glorion
Department of Epidemiology and Biostatistics, École des Hautes Étudesen Santé Publique, Rennes and Sorbonne Paris Cité, Paris, France

Charles Schillinger
Association pour la Surveillance de la Qualitéde l'Air en Alsace-ASPA, Schiltigheim, France

Claire Ségala
SEPIA-Santé, Baud, France

Dominique Arveiler
Department of Epidemiology and Public Health (EA3430), University
of Strasbourg, Strasbourg, France

Christiane Weber
Laboratoire Image, Ville, Environnement (LIVE UMR7362 CNRS), Faculté de géographie et damenagement, University of Strasbourg, Strasbourg, France

Kirsten C. Verhein
Department of Physiology and Pharmacology, Oregon Health and Science University, Portland, Oregon, United States of America

Allison D. Fryer and David B. Jacoby
Division of Pulmonary and Critical Care Medicine, Oregon Health and Science University, Portland, Oregon, United States of America

Francesco G. Salituro and Mark W. Ledeboer
Vertex Pharmaceuticals, Inc., Cambridge, Massachusetts, United States
of America

Jennifer D. Roberts and Brandon Knight
Department of Preventive Medicine and Biometrics, F. Edward Hebert School of Medicine, Uniformed Services University, Bethesda, Maryland, United States of America

Jameson D. Voss
Department of Preventive Medicine and Biometrics, F. Edward Hebert School of Medicine, Uniformed Services University, Bethesda, Maryland, United States of America
Epidemiology Consult Service, United States Air Force School of Aerospace Medicine, Wright-Patterson Air Force Base, Ohio, United States of America

Yang Xiang, Xiao-Qun Qin, Hui-Jun Liu, Yu-Rong Tan, Chi Liu and Cai-Xia Liu
Xiangya School of Medicine, Central South University, Changsha, Hunan, China

Raja Jurdak
Commonwealth Scientific Industrial and Research Organisation, Brisbane, QLD, Australia
University of Queensland, St. Lucia, Brisbane, QLD, Australia

Madeleine Chalfant and Karen K. Bernd
Department of Biology, Davidson College, Davidson, North Carolina, United States of America

Jan Sundell
Institute of Built Environment, Department of Building Science, Tsinghua University, Beijing, China

Lihui Huang
Institute of Built Environment, Department of Building Science, Tsinghua University, Beijing, China
Key Laboratory of Eco Planning & Green Building, Ministry of Education (Tsinghua University), Beijing, China

Jinhan Mo and Yinping Zhang
Institute of Built Environment, Department of Building Science, Tsinghua University, Beijing, China
Key Laboratory of Eco Planning & Green Building, Ministry of Education (Tsinghua University), Beijing, China
Built Environmental Test Center, Tsinghua University, Beijing, China

Zhihua Fan
Department of Environmental and Occupational Medicine, Robert Wood Johnson Medical School, Environmental and Occupational Health Sciences Institute, Rutgers University, Piscataway, New Jersey, United States of America

Julien Tolaini and Marion Guillaume
IRMES (bioMedical Research Institute of Sports Epidemiology), INSEP, Paris, France

Geoffroy Berthelot and Andy Marc
IRMES (bioMedical Research Institute of Sports Epidemiology), INSEP, Paris, France
Université Paris Descartes, Sorbonne Paris Cité, Paris, France

Nour El Helou
IRMES (bioMedical Research Institute of Sports Epidemiology), INSEP, Paris, France
Université Paris Descartes, Sorbonne Paris Cité, Paris, France
Faculté de Pharmacie, Département de Nutrition, Université Saint Joseph, Beirut, Lebanon

Muriel Tafflet
IRMES (bioMedical Research Institute of Sports Epidemiology), INSEP, Paris, France
INSERM, U970, Paris Cardiovascular Research Center – PARCC, Paris, France

Jean-François Toussaint
IRMES (bioMedical Research Institute of Sports Epidemiology), INSEP, Paris, France
Université Paris Descartes, Sorbonne Paris Cité Paris, France
Hôtel-Dieu Hospital, CIMS, AP-HP, Paris, France

Christophe Hausswirth
Research Department, INSEP, Paris, France

Candice C. Clay, Kinjal Maniar-Hew, Joan E. Gerriets, Theodore T. Wang and Justin H. Fontaine
California National Primate Research Center, University of California Davis, Davis, California, United States of America

Michael J. Evans and Lisa A. Miller
California National Primate Research Center, University of California Davis, Davis, California, United States of America
Department of Anatomy, Physiology, and Cell Biology, School of Veterinary Medicine, University of California Davis, Davis, California, United States of America

Edward M. Postlethwait
Department of Environmental Health Sciences, School of Public Health, University of Alabama, Birmingham, Alabama, United States of America

Nagisa Ishinabe
Department of Agricultural Economics, Purdue University, West Lafayette, Indiana, United States of America

Hidemichi Fujii
Graduate School of Environmental Studies, Tohoku University, Sendai, Japan

Shunsuke Managi
Graduate School of Environmental Studies, Tohoku University, Sendai, Japan
Institute for Global Environmental Strategies, Hayama, Kanagawa, Japan

Martin Christner, Holger Rohde, Manuel Wolters, Martin Aepfelbacher and Moritz Hentschke
Department of Medical Microbiology, Virology and Hygiene, University Medical Center Hamburg-Eppendorf, Hamburg, Germany

Maria Trusch
Institute of Organic Chemistry, Mass Spectrometry Facility, University of Hamburg, Hamburg, Germany

Marcel Kwiatkowski and Hartmut Schlüter
Department of Clinical Chemistry, Mass Spectrometric Proteomics, University Medical Center Hamburg-Eppendorf, Hamburg, Germany

Krisztián Papp and Zoltán Szittner
Department of Immunology, Eötvös Loránd University, Budapest, Hungary
Diagnosticum Ltd., Budapest, Hungary

Péter Végh
Diagnosticum Ltd., Budapest, Hungary

Renáta Hóbor and László Czirják
Department of Rheumatology and Immunology, Clinic Center, University of Pécs, Pécs, Hungary

Zoltán Vokó
Department of Health Policy and Health Economics, Eötvös Loránd University, Budapest, Hungary
Syreon Research Institute, Budapest, Hungary

János Podani
Department of Plant Systematics, Ecology and Theoretical Biology, Eötvös Loránd University, Budapest, Hungary

József Prechl
Diagnosticum Ltd., Budapest, Hungary
Immunology Research Group of the Hungarian
Academy of Sciences, Eötvös Loránd University,
Budapest, Hungary

Elena Belousova and Brett G. Toelle
Woolcock Institute of Medical Research, Sydney,
Australia

**Christine T. Cowie, Wafaa Ezz, Wei Xuan,
Adriana Cortes-Waterman and Guy B. Marks**
Woolcock Institute of Medical Research, Sydney,
Australia
Cooperative Research Centre for Asthma and
Airways, Sydney, Australia

Nectarios Rose
Woolcock Institute of Medical Research, Sydney,
Australia
NSW Health Department, Sydney, Australia

Vicky Sheppeard
Western Sydney and Nepean Blue Mountains
Public Health Unit, Sydney, Australia

George Filippidis
Institute of Electronic Structure and Laser,
Foundation for Research and Technology, Crete,
Greece

**George J. Tserevelakis, Barbara Petanidou and
Costas Fotakis**
Institute of Electronic Structure and Laser,
Foundation for Research and Technology, Crete,
Greece
Physics Department, University of Crete, Crete,
Greece

Evgenia V. Megalou
Institute of Molecular Biology and Biotechnology,
Foundation for Research and Technology, Crete,
Greece

Nektarios Tavernarakis
Instituteof Molecular Biology and Biotechnology,
Foundation for Research and Technology, Crete,
Greece
Medical School, University of Crete, Crete, Greece

Yasutomo Hoshika and Kenji Omasa
Graduate School of Agricultural and Life Sciences,
The University of Tokyo, Tokyo, Japan

Elena Paoletti
Institute of Plant Protection, National Research
Council, Sesto Fiorentino,
Florence, Italy

**Isabelle Beck, Jeroen T. M. Buters and Carsten
Schmidt-Weber**
ZAUM – Center of Allergy & Environment,
Member of the German Center for Lung Research
(DZL), Technische Universität München/Helmholtz
Center, Munich, Germany

Stefanie Gilles and Heidrun Behrendt
ZAUM – Center of Allergy & Environment,
Member of the German Center for Lung Research
(DZL), Technische Universität München/Helmholtz
Center, Munich, Germany
Christine-Kühne-Center for Allergy Research and
Education (CK Care), Davos, Switzerland

Susanne Jochner and Annette Menzel
Department of Ecology and Ecosystem Management,
Ecoclimatology, Technische Universität München,
Freising, Germany

Mareike McIntyre and Johannes Ring
Department of Dermatology and Allergy, Technische
Universität München, Munich, Germany

Claudia Traidl-Hoffmann
ZAUM – Center of Allergy & Environment,
Member of the German Center for Lung Research
(DZL), Technische Universität München/Helmholtz
Center, Munich, Germany
Department of Dermatology and Allergy, Technische
Universität München, Munich, Germany
Christine-Kühne-Center for Allergy Research and
Education (CK Care), Davos, Switzerland

**Eman S. Zarie, Viktor Kaidas, Dawit Gedamu,
Yogendra K. Mishra and Rainer Adelung**
Functional Nanomaterials, Institute for Materials
Science, Christian Albrechts University Kiel,
Kaiserstrasse, Germany

**Franz H. Furkert, Regina Scherließ and Hartwig
Steckel**
Department of Pharmaceutics and Biopharmaceutics,
Christian Albrechts University Kiel, Grasweg, Kiel,
Germany

Birte Groessner-Schreiber
Department of Operative Dentistry and Periodon-
tology, Christian Albrechts University Kiel, Arnold-
Heller- Kiel, Germany

Rafaela T. P. Sant'Anna, Emily V. Monteiro, Juliano R. T. Pereira and Roberto B. Faria
Instituto de Química, Universidade Federal do Rio de Janeiro, Rio de Janeiro, RJ, Brazil

Jillian P. Fry
Center for a Livable Future, Johns Hopkins University, Baltimore, Maryland, United States of America
Department of Environmental Health Sciences, Johns Hopkins
Bloomberg School of Public Health, Baltimore, Maryland, United States of America

Linnea I. Laestadius
Center for a Livable Future, Johns Hopkins University, Baltimore, Maryland, United States of America
Joseph J. Zilber School of Public Health, University of Wisconsin-Milwaukee,
Milwaukee, Wisconsin, United States of America

Clare Grechis
Center for a Livable Future, Johns Hopkins University, Baltimore, Maryland, United States of America
Department of Health Policy and Management, Johns Hopkins Bloomberg School of Public Health, Baltimore, Maryland, United States of America

Keeve E. Nachman and Roni A. Neff
Center for a Livable Future, Johns Hopkins University, Baltimore, Maryland, United States of America
Department of Environmental Health Sciences, Johns Hopkins Bloomberg School of Public Health, Baltimore, Maryland, United States of America
Department of Health Policy and Management, Johns Hopkins Bloomberg School of Public Health, Baltimore, Maryland, United States of America

Shun-an Zheng, Xiangqun Zheng and Chun Chen
Agro-Environmental Protection Institute, Ministry of Agriculture, Tianjin, People's Republic of China
Key Laboratory of Production Environment and Agro-Product Safety, Ministry of Agriculture, Tianjin, People's Republic of China
Tianjin Key Laboratory of Agro-Environment and Agro-Product Safety, Tianjin, People's Republic of China

Juan A. Chen, Martin Simon, Feihua Wu, Wenjun Hu and Juan B. Chen
Key Laboratory of the Ministry of Education for

Coastal and Wetland Ecosystems, College of the Environment and Ecology, Xiamen University, Xiamen, Fujian, P. R. China

Tingwu Liu
Key Laboratory of the Ministry of Education for Coastal and Wetland Ecosystems, College of the Environment and Ecology, Xiamen University, Xiamen, Fujian, P. R. Chin
Department of Biology, Huaiyin Normal University, Huaian, Jiangsu, P. R. China

Wenhua Wang
Key Laboratory of the Ministry of Education for Coastal and Wetland Ecosystems, College of the Environment and Ecology, Xiamen University, Xiamen, Fujian, P. R. China
Department of Biology, Duke University, Durham, North Carolina, United States of America

Hailei Zheng
Key Laboratory of the Ministry of Education for Coastal and Wetland Ecosystems, College of the Environment and Ecology, Xiamen University, Xiamen, Fujian, P. R. China
State Key Laboratory of Marine Environmental Science, Xiamen University, Xiamen, Fujian, P. R. China

Kristina J. Anderson-Teixeirab and Sarah C. Davisc
Energy Biosciences Institute, University of Illinois at Urbana-Champaign, Urbana, Illinois, United States of America

Benjamin D. Duvala
Energy Biosciences Institute, University of Illinois at Urbana-Champaign, Urbana, Illinois, United States of America
Global Change Solutions, Urbana, Illinois, United States of America

Stephen P. Long and Evan H. DeLucia
Energy Biosciences Institute, University of Illinois at Urbana-Champaign, Urbana, Illinois, United States of America
Global Change Solutions, Urbana, Illinois, United States of America
Department of Plant Biology, University of Illinois at Urbana- Champaign, Urbana, Illinois, United States of America

Cindy Keogh and William J. Parton
Natural Resource Ecology Laboratory, Fort Collins, Colorado, United States of America

Lirong Song
Key Laboratory of Algal Biology, Institute of Hydrobiology, Chinese Academy of Sciences, Wuhan, China

Yi Ding
Key Laboratory of Algal Biology, Institute of Hydrobiology, Chinese Academy of Sciences, Wuhan, China
University of the Chinese Academy of Sciences, Beijing, China

Bojan Sedmak
Department of Genetic Toxicology and Cancer Biology, National Institute of Biology, Ljubljana, Slovenia

Matthew W. Dray, Stephen M. Thomas, A. Donald ÁBear, Steve J. Ormerod and T. Hefin Jones
Cardiff School of Biosciences, Cardiff University, Cardiff, United Kingdom

Thomas W. Crowther
Cardiff School of Biosciences, Cardiff University, Cardiff, United Kingdom
School of Forestry and Environmental Studies, Yale University, New Haven, Connecticut, United States of America

Douglas L. Godbold
Institute of Forest Ecology, University of Natural Resources and Life Sciences (BOKU), Vienna, Austria

Susan E. Hartley
York Environmental Sustainability Institute, University of York, York, United Kingdom

Ya-Li Huang and Kai-Jen Chuang
Department of Public Health, School of Medicine, College of Medicine, Taipei Medical University, Taipei, Taiwan
School of Public Health, College of Public Health and Nutrition, Taipei Medical University, Taipei, Taiwan

Bor-Cheng Han
School of Public Health, College of Public Health and Nutrition, Taipei Medical University, Taipei, Taiwan

Hua -Wei Chen
Department of Cosmetic Application and Management, St. Mary's Junior College of Medicine, Nursing and Management, Yilan, Taiwan

Chien-Wei Liu
Department of Information Management, St. Mary's Junior College of Medicine, Nursing and Management, Yilan, Taiwan

Hsiao-Chi Chuang
Division of Pulmonary Medicine, Department of Internal Medicine, Shuang Ho Hospital, Taipei Medical University, Taipei, Taiwan
School of Respiratory Therapy, College of Medicine, Taipei Medical University, Taipei, Taiwan

Lian-Yu Lin
Department of Internal Medicine, Division of Cardiology, National Taiwan University Hospital, Taipei, Taiwan

Index

www.ingramcontent.com/pod-product-compliance
Lightning Source LLC
Chambersburg PA
CBHW082058190326
41458CB00010B/3523